Population Biology of Vector-Borne Diseases

ECOLOGY AND EVOLUTION OF
INFECTIOUS DISEASES SERIES

Series Editor: *John M. Drake*

Infectious diseases are embedded in a complex of social, environmental, and biological forces. This series provides an up-to-date synthesis of contemporary issues in the ecology and evolution of infectious diseases. Topics covered include aspects of population ecology, vector biology, animal behavior and movement, methods for inference and disease forecasting, ecoimmunology, evolution, biogeography, and others. Each book makes topical connections to pathogens of current interest. The series is intended for students and researchers working in ecology, evolution, and epidemiology with applications to the disciplines of one health (animal health, public health, and environmental health), environmental management, health policy, therapeutics, and wildlife conservation.

Pertussis: Epidemiology, Immunology, & Evolution
Pejman Rohani and Samuel V. Scarpino

Population Biology of Vector-Borne Diseases
Edited by John M. Drake, Michael B. Bonsall, and Michael R. Strand

Population Biology of Vector-Borne Diseases

Ecology and Evolution of Infectious Diseases Series

John M. Drake
Center for the Ecology of Infectious Diseases, University of Georgia, USA

Michael B. Bonsall
Department of Zoology, University of Oxford, UK

Michael R. Strand
Department of Entomology, University of Georgia, USA

OXFORD
UNIVERSITY PRESS

OXFORD

UNIVERSITY PRESS

Great Clarendon Street, Oxford, OX2 6DP,
United Kingdom

Oxford University Press is a department of the University of Oxford.
It furthers the University's objective of excellence in research, scholarship,
and education by publishing worldwide. Oxford is a registered trade mark of
Oxford University Press in the UK and in certain other countries

© Oxford University Press 2021

The moral rights of the authors have been asserted

First Edition published in 2021

Impression: 1

Published in the United States of America by Oxford University Press
198 Madison Avenue, New York, NY 10016, United States of America

British Library Cataloguing in Publication Data
Data available

Library of Congress Control Number: 2020932785

ISBN 978–0–19–885324–4 (hbk.)
ISBN 978–0–19–885325–1 (pbk.)

DOI: 10.1093/oso/9780198853244.001.0001

Printed and bound by
CPI Group (UK) Ltd, Croydon, CR0 4YY

Contents

List of Contributors   vii

1   Introduction: Current Topics in the Population Biology of Infectious Diseases   **1**
John M. Drake, Michael B. Bonsall, and Michael R. Strand

Section I   Theory of Population Biology   11

2   Heterogeneity, Stochasticity and Complexity in the Dynamics and Control of Mosquito-Borne Pathogens   **13**
Robert C. Reiner Jr. and David L. Smith

3   Seven Challenges for Spatial Analyses of Vector-Borne Diseases   **29**
T. Alex Perkins, Guido España, Sean M. Moore, Rachel J. Oidtman,
Swarnali Sharma, Brajendra Singh, Amir S. Siraj, K. James Soda,
Morgan Smith, Magdalene K. Walters, and Edwin Michael

4   Infectious Disease Forecasting for Public Health   **45**
Stephen A. Lauer, Alexandria C. Brown, and Nicholas G. Reich

5   Force of Infection and Variation in Outbreak Size in a Multi-Species Host-Pathogen System: West Nile Virus in New York City   **69**
John M. Drake, Krisztian Magori, Kevin Knoblich, Sarah E. Bowden,
and Waheed I. Bajwa

Section II   Empirical Ecology   83

6   Environmental Drivers of Vector-Borne Diseases   **85**
Marta S. Shocket, Christopher B. Anderson, Jamie M. Caldwell,
Marissa L. Childs, Lisa I. Couper, Songhee Han, Mallory J. Harris,
Meghan E. Howard, Morgan P. Kain, Andrew J. MacDonald, Nicole Nova,
and Erin A. Mordecai

7   Population Biology of *Culicoides*-Borne Viruses of Livestock In Europe   **119**
Simon Gubbins

8   **Ecological Interactions Influencing the Emergence, Abundance, and Human Exposure to Tick-Borne Pathogens**   135
Maria A. Diuk-Wasser, Maria del Pilar Fernandez, and Stephen Davis

9   **Carry-Over Effects of the Larval Environment in Mosquito-Borne Disease Systems**   155
Michelle V. Evans, Philip M. Newberry, and Courtney C. Murdock

10 **Incorporating Vector Ecology and Life History Into Disease Transmission Models: Insights from Tsetse (*Glossina* spp.)**   175
Sinead English, Antoine M. G. Barreaux, Michael B. Bonsall, John W. Hargrove, Matt J. Keeling, Kat S. Rock, and Glyn A. Vale

Section III Ecological Interactions   189

11 **Mosquito–Virus Interactions**   191
Christine M. Reitmayer, Michelle V. Evans, Kerri L. Miazgowicz, Philip M. Newberry, Nicole Solano, Blanka Tesla, and Courtney C. Murdock

12 **Kindling, Logs, and Coals: The Dynamics of *Trypanosoma cruzi*, the Etiological Agent of Chagas Disease in Arequipa, Peru**   215
Michael Z. Levy

13 **Gut Microbiome Assembly and Function in Mosquitoes**   227
Kerri L. Coon and Michael R. Strand

Section IV Applications   245

14 **Direct and Indirect Social Drivers and Impacts of Vector-Borne Diseases**   247
Sadie J. Ryan, Catherine A. Lippi, Kevin L. Bardosh, Erika F. Frydenlund, Holly D. Gaff, Naveed Heydari, Anthony J. Wilson, and Anna M. Stewart-Ibarra

15 **Vector Control, Optimal Control and Vector-Borne Disease Dynamics**   267
Michael B. Bonsall

Index   289

List of Contributors

Christopher B. Anderson Stanford University, USA

Waheed I. Bajwa New York City Department of Health and Mental Hygiene, USA

Kevin L. Bardosh University of Washington, USA

Antoine M. G. Barreaux University of Bristol, UK

Michael B. Bonsall University of Oxford, UK

Sarah E. Bowden University of Georgia, USA

Alexandria C. Brown University of Massachusetts Amherst, USA

Jamie M. Caldwell Stanford University, USA

Marissa L. Childs Stanford University, USA

Lisa I. Couper Stanford University, USA

Kerri L. Coon University of Wisconsin-Madison, USA

Stephen Davis RMIT University, Australia

Maria A. Diuk-Wasser Columbia University, USA

John M. Drake University of Georgia, USA

Sinead English University of Bristol, UK

Guido España University of Notre Dame, USA

Michelle V. Evans University of Georgia, USA

Maria del Pilar Fernandez Columbia University, USA

Erika F. Frydenlund Old Dominion University, USA

Holly D. Gaff Old Dominion University, USA

Simon Gubbins The Pirbright Institute, UK

Songhee Han Stanford University, USA

John W. Hargrove Stellenbosch University, South Africa

Mallory J. Harris University of Georgia and Stanford University, USA

Naveed Heydari SUNY Upstate Medical University, USA

Meghan E. Howard Stanford University, USA

Michael Z. Levy University of Pennsylvania, USA

Stephen A. Lauer Johns Hopkins University, USA

Catherine A. Lippi University of Florida, USA

Morgan P. Kain Stanford University, USA

Matt J. Keeling University of Warwick, UK

Kevin Knoblich University of Georgia, USA

Andrew J. MacDonald Stanford University and University of California Santa Barbara, USA

Krisztian Magori Eastern Washington University, USA

Kerri L Miazgowicz University of Georgia, USA

Edwin Michael University of Notre Dame, USA

Sean M. Moore University of Notre Dame, USA

Erin A. Mordecai Stanford University, USA

Courtney C. Murdock University of Georgia, USA

Philip M. Newberry University of Georgia, USA

Nicole Nova Stanford University, USA

Rachel J. Oidtman University of Notre Dame, USA

T. Alex Perkins University of Notre Dame, USA

Nicholas G. Reich University of Massachusetts Amherst, USA

Robert C. Reiner Jr. University of Washington, USA

Christine M. Reitmayer University of Georgia, USA and The Pirbright Institute, UK

Kat S. Rock University of Warwick, UK

Sadie J. Ryan University of Florida, USA

Swarnali Sharma University of Notre Dame, USA

Marta S. Shocket Stanford University and University of California Los Angeles, USA

Brajendra Singh University of Notre Dame, USA

Amir S. Siraj University of Notre Dame, USA

David L. Smith University of Washington, USA

K. James Soda University of Notre Dame, USA

Nicole Solano University of Georgia, USA

Morgan Smith University of Notre Dame, USA

Anna M. Stewart-Ibarra InterAmerican Institute for Global Change Research, Uruguay

Michael R. Strand University of Georgia, USA

Blanka Tesla University of Georgia, USA

Glyn A. Vale University of Greenwich, UK

Magdalene K. Walters University of Notre Dame, USA

Anthony J. Wilson The Pirbright Institute, UK

Current Topics in the Population Biology of Infectious Diseases

John M. Drake, Michael B. Bonsall, and Michael R. Strand

1.1 Introduction

This book concerns the population biology of vector-borne diseases. Vector-borne diseases of people are a perennial challenge for public health. Although recent decades have enjoyed major declines in the incidence of diseases like malaria and onchocerciasis (river blindness), vector-borne diseases continue to claim the lives of more than 700,000 people per year and exact costs of tens of billions of dollars in expenses for control and through lost productivity (WHO 2017). Not just people are infected. Vector-borne illnesses of livestock like Rift Valley Fever, African Swine Fever, and Bluetongue Virus impose costs of similar magnitude and the effects of novel pathogens on wildlife may be devastating (Warner 1968; George et al. 2015).

More so, perhaps, than directly transmitted diseases, the dynamics of vector-borne diseases are intrinsically ecological. Environmental dependence of ectotherm vectors like mosquitoes, ticks, and black flies means that vector-borne pathogens are acutely sensitive to changing environmental conditions. Host alternation (i.e. a transmission cycle that alternates between arthropod vectors and vertebrate hosts) means that evolution and interspecific interactions are interwoven. Promiscuity of association between viruses and vector hosts means that pathogen evolution and transmission cannot be isolated from the context of the larger ecological community.

Perennially important vector-borne diseases such as malaria and dengue have deeply informed our scientific understanding of the maintenance of vector-borne pathogens, for instance giving rise to the standard mathematical theory for vector-borne disease transmission (the Ross–Macdonald equations). New emerging viruses such as West Nile virus, Chikungunya virus, and Zika virus have presented new scientific questions and practical problems. Thus, the challenges presented by vector-borne diseases have been a rich source of ecological questions, while ecological theory has provided the conceptual tools for thinking about the evolution, transmission, and spatial extent of vector-borne diseases.

1.2 Origin and Organization of this Book

The chapters collected here review much of this ecological knowledge about vector-borne diseases while reporting on the state of the science from the viewpoints of many researchers working on its frontier. Several of these chapters were originally presented at two special symposia held at the Royal Entomological Society, UK (17 May 2017) and the University of Georgia, USA (24 February 2018). Authors were asked to provide an up-to-date presentation of classical concepts, review emerging trends, synthesize existing knowledge, and share their perspectives about the most important topics for future research. Here, the chapters are organized

John M. Drake, Michael B. Bonsall, and Michael R. Strand, *Introduction: Current Topics in the Population biology of Infectious Diseases* In: *Population Biology of Vector-Borne Diseases*. Edited by: John M. Drake, Michael B. Bonsall, and Michael R. Strand: Oxford University Press (2021). © John M. Drake, Michael B. Bonsall, and Michael R. Strand.
DOI: 10.1093/oso/9780198853244.003.0001

according to four primary sections: 1. Theory, 2. Environmental drivers, 3. Species interactions, and 4. Management.

1.2.1 Theory

This section starts, in Chapter 2, with an introduction by Reiner & Smith (2020) to the standard quantitative theory of vector-borne disease transmission, known as the Ross–Macdonald equations, named for Ronald Ross and George Macdonald, who developed and applied these equations first in the early twentieth century (Ross 1911) and then in the 1950s (Macdonald 1956, 1957). Reiner and Smith derive and define many of the key ideas of transmission theory in vector-host interactions such as 'vectorial capacity' (the number of infectious bites that would arise from all the mosquitoes biting a single infectious human on a single day), 'entomological inoculation rate' (the number of infectious bites, per person, per day), and the basic reproduction number, R_0 (the number of secondary infections arising from each primary infection, a threshold for persistence of the pathogen in a population of hosts). They conclude by identifying some key problems for further research, such as the role for fluctuations in the vector population size (and the population ecology of the vector more generally), a need for better links between environmental conditions and transmission via their impacts on mosquito ecology and behavior, heterogeneous biting (which leads to aggregation of bites on particular hosts), and the special challenges associated with translating this knowledge to control.

In Chapter 3, Perkins et al. (2020) take up this challenge by focusing on the fact that systems of vectors and hosts are spatially extended and that this spatial extension typically covers a wide range of environmental conditions. Nature is 'patchy' from the points of view of both vectors and hosts and this spatial heterogeneity matters to the understanding of disease dynamics at meaningful spatial scales. Particularly, Perkins et al. identify seven key challenges. The first three challenges have more to do with the interpretation of incidence data, particularly when it is derived from multiple (spatially distant) sources:

- Age can confound estimates based on multiple, uncoordinated studies.
- Population immunity complicates the relationship between transmission potential and disease.
- Herd immunity limits the population at risk of immunizing pathogens.

The next four challenges focus on the spread of a pathogen from one place to another, challenges that are particularly acute when the properties of those places differ (e.g. 'sources' and 'sinks') with respect to pathogen amplification:

- Surveillance sensitivity is spatially heterogeneous and can distort spatial patterns of disease.
- Spatial coupling means that control in one area can have impacts across multiple areas.
- Spatial heterogeneity means that spatially uniform goals for control may not be successful.
- Hot spots of transmission can be spatially hierarchical and temporally dynamic.

The main ideas of this chapter are that (i) one needs to understand not just the pattern of incidence over space, but also the dynamics of incidence (how incidence changes over *time* and why), and (ii) that the exchange of infectious agents among locations (spatial coupling), may ultimately be key to understanding spatial pattern. These ideas are neatly illustrated with some thought experiments that illuminate the depth of the intellectual problem as seen by these authors.

Chapter 4 provides an interesting contrast to Chapter 3. Whereas the goal of Chapter 3 is to understand patterns in space, the goal of Chapter 4 is to understand patterns through time. Here, Reich et al. (2020) take up the modern challenge of providing infectious disease transmission *forecasts*, i.e. predictions about future transmission that might be useful for making decisions, either on an individual basis (e.g. protective measures) or collectively (i.e. policy). Infectious disease forecasting is a relatively new, but highly active, area of research. Reich et al. (2020) review the history of forecasting in various scientific domains and identify some special challenges for forecasting infectious diseases, including the complexity of the problem (involving multiple actors—pathogens, vectors, hosts—at multiple scales, with interactions and nonlinear feedbacks among all of them), the sparsity of data with respect to this complexity, and the 'forecasting feedback loop', which is the idea that forecasts can result in changes to individual behavior, which then changes the rules governing the underlying system being forecasted. Also in contrast to the first two chapters, which emphasize models reflecting known biological realities, Reich et al. observe that a fundamental principle of forecasting is to use as simple a model as possible. There is, therefore, a potential tension between statistical models made to provide accurate predictions and biological models that provide understanding because they 'mirror the ecological complexity of the system'.

Finally, Chapter 5 concludes the Theory section with a case study on West Nile virus in New York City during the first decade of the twenty-first century. The purpose of this study was to investigate possible explanations for the considerable inter-annual variation in human cases. The number of human cases per year is relatively small (ranging from around five cases to thirty), but highly variable. Why? The authors argue that it cannot just be due to sampling error. Naturally, one expects that it may have to do with environmental variation, but there was no evidence for a correlation between either temperature or precipitation with cases, i.e. evidence for conditions that might make for particularly 'good' or 'bad' West Nile seasons. To try and get at the root of the problem, Drake et al. (2020) attempt to measure the force of infection as directly as possible, by translating measured variation in virus prevalence in mosquitoes into a quantity that might directly relate to human disease risk. To do this, they present a model that reflects the contributions of the multiple species of mosquitoes that transmit West Nile virus in New York city, partitioning the relative contributions of each to the force of infection. This is accomplished with the aid of a meta-analysis and leads to the conclusion that only an additional 15 percent of the variation in infection can be attributed to variation in the force of infection. A large part, it seems, must be due to something else. In line with the argument of Chapter 3, and an earlier paper by the same group (Magori 2011), Drake et al. suggest that it is the patchy nature of the urban environment that ampli-

fies random spatial variation as annual outbreaks of the pathogen spread throughout the city.

1.2.2 Environment

The section on environment begins with Chapter 6 by Shocket et al. (2020) on how environmental factors mediate the transmission of pathogens from vectors to hosts. They focus on three particular features: temperature, humidity, and precipitation. Interestingly, temperature and precipitation are found typically to have a unimodal relationship to important features of vector biology or transmission, i.e. transmission is maximized at an intermediate level not too cold (where activity of the ectothermic vector is impossible or compromised) or too hot (where the vector is at risk of desiccation), not too dry (where there is no larval habitat) or too wet (where habitats may be washed out). In contrast, humidity (the least studied of the three) seems to have a consistently positive relationship with transmission. Second, Shocket et al. review the importance of habitat, especially edge habitats, which they note to be an important part of transmission for vivax malaria in South America, Chagas disease (which is promoted by palm trees), and the black flies that carry the filarial worms causing river blindness. The major challenges identified by Shocket et al. align with those raised in Chapter 1 and Chapter 2: nonlinear relationships among important causal variables, interacting and correlated environmental drivers, and the different relationships that may be observed at different scales of measurement. These challenges lead to the key gaps these authors propose for future research, particularly the challenge of integrating information about multiple drivers along with a diverse array of information collected at multiple scales.

Chapter 7 continues this section by reviewing the ecology of viruses spread by biting midges in the genus *Culicoides*. As Gubbins (2020) reports, *Culicoides* primarily transmit viruses to animals (Oropouche virus being the primary exception) and the impact on livestock animals, particularly, may be significant. This chapter primarily focuses on two such viruses, Bluetongue virus (BTV) and Schmallenberg virus (SBV). Both viruses are transmitted by multiple species. BTV is complicated by the fact that it exists as many genetically diverse strains. Despite the many practical difficulties in working out the ecology of these viruses, much has been learned in recent years. Arguably, the biggest question at the moment is understanding how the virus is maintained from one transmission season to the next, for instance via overwintering vectors or transplacental transmission in hosts.

This tour of host-vector systems continues in Chapter 8, where Diuk-Wasser et al. (2020) review the ecology of tick-borne diseases, focusing on Lyme disease in the United States. Lyme disease has been growing in importance in the U.S. over the last half century as it follows the expansion of its primary vector *Ixodes scapularis*. Development of this vector through three life stages—larva, nymph, and adult over two years—together with shifts in host feeding preferences (adults primarily feed on deer, which are poor hosts, whereas larvae and nymphs preferentially feed on rodents), presents an interesting and complex system in the sense of many interacting parts. The fact that multiple host species are competent, but not always present, complicates the matter. A key ecological focus of research, then, has been the study of forest fragmentation, which has the effect of increasing edge habitat (ideal for amplifying hosts like rodents) but also increasing the proximity to open habitat (lethal for ticks, which are at risk of desiccation) and propensity for local extinction. Unsurprisingly, the findings of empirical studies are mixed. Diuk-Wasser et al. conclude that the explanation, not yet in hand, will invariably come down to the complexity of this system, explaining that 'the apparently contradictory findings of different studies evaluating the fragmentation effects on the enzootic hazard and human disease risk may be partly due to differences in the spatial and temporal scales, the components of the system considered, and the measured outcome.'

Chapter 9, by Evans et al. (2020) concludes the section on environment with a review of how the local, early-life environment experienced by mosquitoes as larvae in aquatic habitats may have long-lasting impacts on their contribution to infectious disease transmission as adults in the terrestrial environment. Such 'carryover effects' are now widely recognized to be important, but difficult to characterize in the field because the microhabitats

occupied by larval mosquitoes are so variable. For instance, Evans et al. note that so-called 'silver spoon effects' may occur wherein individuals from high quality larval environments exhibit high lifetime reproductive success (contributing to the potential to amplify disease transmission) even when the environment, as experienced by adult mosquitoes, is of poor quality. Alternatively, low quality larval environments (for instance, due to larval crowding) can have the opposite effect. Evans et al. review the literature on carryover effects, focusing their findings on those measurements that impact vectorial capacity. Key factors include temperature, nutrition, competition, and microbiome (also reviewed in Chapter 12). Although relatively little theoretical work has been done on this topic, Evans et al. present a simple model, parameterized with laboratory data, that suggests carryover effects may have very substantial impact on R_0. Placed in the context of the other chapters in this book (especially Chapters 2, 3, 5, and 6), it is an open question whether such fine-grained variation, as is experienced by mosquitoes in nature, may account for a substantial portion of the spatial heterogeneity in transmission that is one of the hallmarks of vector-borne diseases.

Sleeping sickness is an often neglected infectious disease of livestock and humans in Africa. It is thought that the economic development of the African continent has been restricted because of this vector-borne disease (Alsan 2015) and, in Chapter 10, English et al. explain how a richer understanding of Tsetse fly life history is critical to understanding the ecological and epidemiological consequences of sleeping sickness. Following on from earlier chapters (Chapter 2), English et al. (2020) emphasize the need to incorporate more relevant information into expressions for R_0—rather than focusing on temporal or spatial aspects—especially variability in vector life histories. Variation in vector life expectancies, age structure and the time since the last blood meal alter trypanosome infection prevalence in both host and vectors. This variation in life history coupled with the peculiarities of Tsetse reproductive biology has important implications for developing predictive epidemiology frameworks for this vector-borne disease.

1.2.3 Species interactions

Interaction among species is a foundational topic of ecology and the issue of species interactions is central to understanding the population biology of vector-borne diseases. Of course, the host-parasite relationship *is* a species interaction. In the context of vector-borne diseases, the interactions among species are (at least) three-fold: pathogen-vector, pathogen-host, and vector-host.

Chapter 11 introduces the subject of species interactions by reviewing the state of the science concerning virus-mosquito interactions. Murdock et al. (2020) first note that all arboviruses acquired by blood feeding on a vertebrate host initially infect the mosquito midgut but must disseminate to the salivary glands in order to be transmitted. A diversity of immunological and non-immunological processes within mosquitoes may impede this multi-step process, which defines whether or not a given species is a competent vector. Reciprocally, arboviruses have evolved a diversity of strategies for promoting infection and transmission by vector species that range from high mutation rates to circumventing or manipulating mosquito antiviral defenses). Interestingly, Murdock et al. (2020) report that there are also several known instances where infection of *Aedes* and *Culex* cell lines in the laboratory by viruses that do not infect vertebrates results in blocking infection or reducing replication of arboviruses. Thus, certain viruses that specifically infect mosquitoes could potentially be used as biological control agents to block or reduce transmission of viruses pathogenic to humans or other animals. Such approaches could also potentially have less ecological impact than, say, sterile male release, while potentially offering a complementary strategy to other factors, such as the bacterium *Wolbachia*, that can also disrupt the acquisition and transmission of certain arboviruses by mosquitoes.

Chapter 12 takes a different approach to understanding the role of species interactions in vector-borne diseases. Here, Levy (2020) looks at Chagas disease in primarily urban but also rural environments. Chagas disease is caused by the protozoan parasite *Trypanosoma cruzi*, which is vectored by insects in the genus *Triatoma* (kissing bugs). Infection is actually due to contamination of the bite

site with fecal material from the insect that contains the parasite. This transmission mechanism is very inefficient, but many vertebrate species are permissive to *T. cruzi*—including domesticated guinea pigs, dogs, and chickens—leading to a situation where persistence of the pathogen in a (human) population may be ensured by the rich ecological community of vertebrates living in close proximity to one another. Standard models like the Ross–Macdonald model do not account for the various time and spatial scales of infection and transmission as well as multiple host interactions which include guinea pigs reared at high densities in pens and human domiciles that may be occupied by people and dogs. Since the Ross–Macdonald model does not reflect these various scales, Levy asks us to consider the analogy of a campfire, where several permissive vertebrate host species (humans, dogs, chickens, and guinea pigs) represent fuels (logs, kindling, and smoldering coals) that play roles in the continuation of the epidemic. Developing a proper, multi-scale model of this process is an open problem in the epidemiology of Chagas disease.

The last chapter in this section takes up a relatively recent development in ecological understanding—the role of the microbiome in the functioning of organisms in their environments. In Chapter 13, Coon & Strand (2020) review the state of knowledge specifically concerning the gut microbiome of mosquitoes. It turns out that this gut community (comprised mainly of bacteria in the phyla Actinobacteria, Bacteroidetes, Firmicutes, and Proteobacteria) is both essential to complete development of the mosquito and is acquired environmentally (rather than from a direct endowment from the parent). Although the broad outlines of this community are quite predictable (i.e. composition at the phylum level), there is much more variation at lower taxonomic levels, presumably because the different environments in which larval mosquitoes acquire their associated symbionts provide very different, albeit functionally complementary, bacterial species pools from which to select. The ways in which variation in this community impact disease transmission are various. The effect on proper development is obvious. But, given the functional redundancy among various bacterial species, it ensures that virtually all natural populations of mosquitoes are able to obtain the required species. However, there may be more subtle effects as the specific composition of the microbiota available to larvae from different larval sources may affect adult traits such as body size, teneral reserves (lipid, glucose, glycogen, and protein retained during the transition from pupae to adult), fecundity, longevity and even vector competence—all of which are components of vectorial capacity. It follows that there will be increasing interest in determining whether vectorial capacity might be altered by manipulating the environmental conditions under which mosquitoes and other vector species acquire their bacterial symbionts.

1.2.4 Management

The final section of the book examines the translation of an ecological understanding of vector-borne diseases into applications of knowledge for management. In the first of these chapters (Chapter 14), Ryan et al. (2020) look at the social context in which vector-borne diseases occur. Their starting point is the widely accepted idea that pervasive infectious disease in a population is maintained in a cycle whereby the costs of disease (e.g. lost income) and conditions conducive to transmission (e.g. low quality housing) are mutually reinforcing, creating a 'poverty trap'. But, as Ryan et al. explain, the real situation is considerably more complicated and involves factors at multiple scales, including households (e.g. water storage), cities (e.g. vector control, emergence of insecticide resistance), and nation states (e.g. health policy). As a result some interesting spatial issues arise, wherein poverty may be clustered in neighborhoods that are adjacent to and share disease vectors with wealthier neighborhoods. For instance, they write 'It is not hard to imagine a situation in which a hotel guest sun-bathing poolside is bitten by an Ae. *aegypti* mosquito that has emerged from the cistern of a neighboring household.' It follows that 'improving equity across local sites—that is, bringing whole neighborhoods and cities into an improved standard of living, with improved access to public health and vector control interventions—will be needed when considering economic and public health improvements to fully mitigate [vector-borne disease] transmission, given

the flight radius of the Ae. *aegypti* is around 200 m.' It is exactly this kind of scenario that is envisioned in Chapter 3, where Perkins et al. (2020) write about spatial coupling among heterogeneous sites of transmission. Another way in which the concept of the poverty trap fails to address key barriers to disease management in poor communities is the key role of trust, a non-monetary currency that Ryan et al. argue is crucial to effective management of vector-borne diseases in many settings. This role of trust is most usefully explored in a case study wherein the authors partnered with local communities in Haiti to operate vector control programs aimed at reducing lymphatic filariasis, Chikungunya, Zika, and other vector-borne diseases. What this case study makes clear is how operational success depends as much on social factors as it does ecological ones. In the end, the authors seem to adopt a view that retains important insights of the poverty trap idea, but contextualized within a complex social-environmental system. Successful management, it seems, will depend on conceptual innovations. The success of these innovations, they explain 'will rest on economic latitude' (i.e. elements of the poverty trap idea) and 'social acceptance.' (i.e. trust).

The final chapter (Chapter 15) in this section explores recent advances in the use of genetics-based approaches for vector control. Building on classic sterile insect technique (SIT), modern genetic and biotechnological approaches are providing a suite of novel tools for managing vector populations. With the advent of modern molecular methods (such as gene editing) this is rapidly developing field of research for vector, particularly, mosquito control. Reviewing some of these different genetics-based methods, Bonsall (2020) highlights the importance of understanding different ecological principles (e.g. spatial heterogeneity, species interactions) as critical to the success of vector management. However, Bonsall emphasizes that vector population management is not divorced from economics, and integrating genetic and ecological thinking into economic models for the management of vector borne diseases is essential and can highlight how to develop appropriate management actions in time and/or space. With these novel genetic approaches, as with all genetic modification, regulation and legislation predicated on the precautionary principle might be thought as stifling or restricting biotechnological innovation. Developments in the policy on the release of genetically-modified vectors to manage unintentional risks to biodiversity and/or human health is rapidly developing—risk assessments on these novel genetic technologies could make more use of relevant statistical frameworks to incorporate ecology (and potentially social contexts) and is argued as a necessity to provide a logically consistent and proportionate approach to the genetic modification of vectors for managing vector-borne diseases.

1.3 Key Themes

Taken together, these contributions identify a number of key themes that are current topics in the ecology of vector-borne diseases. The first of these themes is the importance of *complexity*, understood as the combination of nonlinear feedbacks, interactions among multiple processes, and spatial heterogeneity. The predominance of nonlinear phenomena in infectious disease transmission is exceedingly well known, for instance through the familiar family of compartmental models for directly transmitted, immunizing infections (i.e. SIR/SEIR models). The vector-borne disease counterpart to this theory is the Ross–Macdonald model reviewed by Reiner and Smith (2020) in Chapter 1. But, as Reiner and Smith point out, as useful as the Ross–Macdonald model has been for generating a conceptual repertoire to guide control efforts—concepts like vectorial capacity and entomological inoculation rate—it has not provided us with a quantitatively reliable theory of disease transmission, almost certainly because of the many environmental factors (Chapters 6, 8, 9, and 13) and social factors (Chapter 14) that impinge on transmission and are heterogeneously distributed in space (Chapter 3).

The second overarching theme is the *multi-scale nature of the conditions driving transmission* of vector-borne diseases. Such hierarchical scales occur with respect to both spatial (Chapters 3, 5, and 12) and temporal (Chapters 12) phenomena. Even a single environmental variable (temperature) may have different manifestations at different scales, ranging from the microclimate variations of particular larval habitats at the scale of centimeters (Chapter 9) to the gross climate features that determine the distributions of vector species (Chapter 7).

An interesting commonality among many of the data-oriented chapters is a focus on R_0 (Chapters 5, 6, 7, and 8). It is understandable that, as a threshold condition for disease elimination, parameterizing models for R_0 for specific systems has been a key goal of empirical analysis. The history of application of the Ross–Macdonald model in malaria elimination retold by Reiner and Smith (Chapter 2) shows how useful such quantitative criteria may be in applications. What is absent, however, is a reliable theory of *dynamics*. As Reich et al. point out (Chapter 4), in many cases our 'mechanistic understanding' of disease transmission may be no more useful for prediction than a biology-agnostic statistical model trained on the history of incidence in a population. We submit that developing an improved dynamical theory should be one of the key goals for future research. This is a point also made directly by Chapter 3.

Finally, it is increasingly apparent that *ecological details* are often crucial to understanding the transmission and persistence of vector-borne diseases. Edge effects and fragmentation, topics of long-standing importance in ecology, are now seen to be central to understanding the distribution of several vector-borne diseases (Chapter 6) and the intensity of Lyme disease at a landscape level in particular (Chapter 8). From Chapters 9 and 13 we learn that ontogenetic niche shift—the occupancy of different habitats by some animals during different stages of the life cycle—has important implications for the ability of a vector population to continue the infection cycle. The obscure details of overwintering by the insect vector drives the spatial distribution of Bluetongue and Schmallenberg viruses (Chapter 7). Integrating *ecological details* is then essential to both the successful application of methods (Chapter 15) and the social contexts (Chapter 14) for vector control. At the beginning of this introductory chapter, we noted that the management challenges presented by vector-borne diseases have been a rich source of ecological ideas. It would seem that ongoing innovation in ecological theory and empirical studies has much to contribute to the control and management of vector-borne diseases in humans, livestock, and even wildlife in the future.

References

Alsan, M. (2015). The effect of the TseTse -y on African development. American Economic Review, 105, 382–410.

Christine M. Reitmayer, Michelle V. Evans, Kerri L. Miazgowicz, Philip M. Newberry, Nicole Solano, Blanka Tesla, and Courtney C. Murdock, *Mosquito – Virus Interactions* In: *Population Biology of Vector-Borne Diseases*. Edited by John M. Drake, Michael B. Bonsall, and Michael R. Strand: Oxford University Press (2020). © Christine M. Reitmayer, Michelle V. Evans, Kerri L. Miazgowicz, Philip M. Newberry, Nicole Solano, Blanka Tesla, and Courtney C. Murdock.

George, T.L. et al. (2015). Persistent impacts of West Nile virus on North American bird populations. *Proceedings of the National Academy of Sciences of the United States of America*, 112, 14290–4.

John M. Drake, Krisztian Magori, Kevin Knoblich, Sarah E. Bowden, and Waheed I. Bajwa, *Force of Infection and Variation in Outbreak Size in a Multi-Species Host-Pathogen System: West Nile Virus in New York City* In: *Population Biology of Vector-Borne Diseases*. Edited by: John M. Drake, Michael B. Bonsall, and Michael R. Strand: Oxford University Press (2020). © John M. Drake, Krisztian Magori, Kevin Knoblich, Sarah E. Bowden, and Waheed I. Bajwa.

John M. Drake, Michael B. Bonsall, and Michael R. Strand, *Introduction: Current topics in the population biology of infectious diseases* In: *Population Biology of Vector-Borne Diseases*. Edited by: John M. Drake, Michael B. Bonsall, and Michael R. Strand: Oxford University Press (2020). © John M. Drake, Michael B. Bonsall, and_Michael R. Strand.

Kerri_L._Coon and Michael_R._Strand, *Gut microbiome assembly and function in mosquitoes* In: *Population Biology of Vector-Borne Diseases*. Edited by: John M. Drake, Michael B. Bonsall, and Michael R. Strand: Oxford University Press (2020). © Kerri L. Coon and Michael R. Strand.

Macdonald, G. (1956). Theory of the eradication of malaria. *Bulletin of the World Health Organization*, 15(3–5), 369–87.

Macdonald, G. (1957). *The Epidemiology and Control of Malaria*. London: Oxford University Press.

Magori, K., Bajwa, W.I., Bowden, S., & Drake, J.M. (2011). Decelerating spread of West Nile virus by percolation in a heterogeneous urban landscape. *PLoS Computational Biology*, 7, e1002104.

Maria A. Diuk-Wasser, Maria del Pilar Fernandez, and Stephen Davis, *Ecological interactions in_uencing the emergence, abundance and human exposure to tick-borne pathogens* In: *Population Biology of Vector-Borne Diseases*. Edited by: John M. Drake, Michael B. Bonsall, and Michael R. Strand: Oxford University Press (2020). © Maria A. Diuk-Wasser, Maria del Pilar Fernandez, and Stephen Davis.

Marta S. Shocket, Christopher B. Anderson, Jamie M. Caldwell, Marissa L. Childs, Lisa I. Couper, Songhee Han, Mallory J. Harris, Meghan E. Howard, Morgan P._Kain, Andrew J. MacDonald, Nicole Nova, and Erin A. Mordecai, *Environmental Drivers of Vector-Borne Diseases*. In: *Population Biology of Vector-Borne Diseases*. Edited by: John M. Drake, Michael B. Bonsall, and Michael R. Strand: Oxford University Press (2020). © Marta S. Shocket, Christopher B. Anderson, Jamie M. Caldwell, Marissa L. Childs, Lisa I. Couper, Songhee Han, Mallory J. Harris, Meghan E. Howard, Morgan P. Kain, Andrew J. MacDonald, Nicole Nova, and Erin A. Mordecai.

Michael B. Bonsall, *Vector control, optimal control and vector-borne disease dynamics* In: *Population Biology of Vector-Borne Diseases*. Edited by: John M. Drake, Michael B. Bonsall, and Michael R. Strand: Oxford University Press (2020). © Michael B. Bonsall.

Michael Z. Levy, *Kindling, Logs and Coals: The Dynamics of Trypanosoma cruzi, the etiological agent of Chagas Disease, in Arequipa, Peru* In: *Population Biology of Vector-Borne Diseases*. Edited by: John M. Drake, Michael B. Bonsall, and Michael R. Strand: Oxford University Press (2020). © Michael Z. Levy.

Michelle V. Evans, Philip M. Newberry, and Courtney C. Murdock, *Carry-over effects of the larval environment in mosquito-borne disease systems* In: *Population Biology of Vector-Borne Diseases*. Edited by: John M. Drake, Michael B. Bonsall, and Michael R. Strand: Oxford University Press (2020). © Michelle V. Evans, Philip_M. Newberry, and Courtney C. Murdock.

Robert C. Reiner Jr. and David L. Smith, *Heterogeneity, Stochasticity and Complexity in the Dynamics and Control of Mosquito-Borne Pathogens* In: *Population Biology of Vector-Borne Diseases*. Edited by: John M. Drake, Michael B. Bonsall, and Michael R. Strand: Oxford University Press (2020). © Robert C. Reiner Jr. and David L. Smith.

Ross, R. (1911). *The Prevention of Malaria*. 2nd ed., London: John Murray.

Sadie_J._Ryan, Catherine_A._Lippi, Kevin_L._Bardosh, Erika_F._Frydenlund, Holly_D._Gaff, Naveed Heydari, Anthony_J._Wilson, Anna_M._Stewart-Ibarra, *Direct and indirect social drivers and impacts of vector-borne diseases* In: *Population Biology of Vector-Borne Diseases*. Edited by: John M. Drake, Michael B. Bonsall, and Michael R. Strand: Oxford University Press (2020). © Sadie_J._Ryan, Catherine_A._Lippi, Kevin_L._Bardosh,

Erika_F._Frydenlund, Holly_D._Gaff, Naveed Heydari, Anthony_J._Wilson, Anna_M._Stewart-Ibarra.

Simon Gubbins, *Population biology of Culicoides-borne viruses of livestock in Europe*. In: *Population Biology of Vector-Borne Diseases* In: *Population Biology of Vector-Borne Diseases*. Edited by: John M. Drake, Michael B. Bonsall, and Michael R. Strand: Oxford University Press (2020). © Simon Gubbins.

Sinead English, Antoine M.G. Barreaux, Michael B. Bonsall, John W. Hargrove, Matt J. Keeling, Kat S. Rock, and Glyn A. Vale, *Incorporating vector ecology and life_history into disease transmission models: insights from tsetse (Glossina spp.)* In: *Population Biology of Vector-Borne Diseases*. Edited by: John M. Drake, Michael B. Bonsall, and Michael R. Strand: Oxford University Press (2020). © Sinead English, Antoine M.G. Barreaux, Michael B. Bonsall, John W. Hargrove, Matt_J. Keeling, Kat S. Rock, and Glyn A. Vale.

Stephen A. Lauer, Alexandria C. Brown, and Nicholas G. Reich, *Infectious Disease Forecasting for Public Health* In: *Population Biology of Vector-Borne Diseases*. Edited by: John M. Drake, Michael B. Bonsall, and Michael R. Strand: Oxford University Press (2020). © Stephen A. Lauer, Alexandria C. Brown, and Nicholas G. Reich.

T. Alex Perkins, Guido España, Sean M. Moore, Rachel J. Oidtman, Swarnali Sharma, Brajendra Singh, Amir S. Siraj, K. James Soda, Morgan Smith, Magdalene K._Walters, and Edwin Michael, *Seven Challenges for Spatial Analyses of Vector-Borne Diseases* In: *Population Biology of Vector-Borne Diseases*. Edited by: John_M._Drake, Michael B. Bonsall, and Michael R._Strand: Oxford University Press (2020). © T. Alex Perkins, Guido España, Sean M. Moore, Rachel_J._Oidtman, Swarnali Sharma, Brajendra Singh, Amir S. Siraj, K. James Soda, Morgan Smith, Magdalene K. Walters, and Edwin Michael.

Warner, R.E. (1968). *The role of introduced diseases in the extinction of the endemic* Hawaiian Avifauna. *Condor,* 70, 101–20.

WHO, 2017. *Global Vector Control Response 2017–2030.* Geneva: World Health Organization.

Theory of Population Biology

Heterogeneity, Stochasticity and Complexity in the Dynamics and Control of Mosquito-Borne Pathogens

Robert C. Reiner Jr. and David L. Smith

2.1 Introduction

A very old hypothesis, proven finally by Ronald Ross, is that mosquitoes transmit malaria (Service 1978; Ross 1897). While mosquito populations are logically necessary for mosquito-borne pathogen transmission, the study of transmission since then shows it is noisy, heterogeneous, and complex (Smith et al. 2014). Humans travel, mosquitoes fly, pathogens incubate, and human care seeking is unpredictable, so human malaria cases often go uncounted, or they get counted far away from the point where infection occurred (Tatem & Smith 2010; Tatem et al. 2017). These same problems apply to most mosquito-borne diseases. Countries with no mosquito vectors record travel-related cases of malaria, dengue fever, Zika, chikungunya, and other diseases. If there are no potential vectors nearby (and if there is no alternative mode of transmission, such as the sexual transmission mode for Zika virus), it is obvious that the case must have been acquired somewhere else, but where vector population distributions are heterogeneous, attributing a case detected here and now to transmission occurring there and then is far more difficult. Quantitative studies of mosquitoes have been neglected, in part, because the costs of counting mosquitoes and clinical cases are comparable. Studies with the spatial granularity and temporal coverage required to understand epidemics are consequently rare, so comparatively few studies have examined even the most basic associations between mosquitoes and disease incidence among populations over space and time. While it seems intuitively obvious that there must be some quantitative relationship between mosquito populations and the intensity of mosquito-borne pathogen transmission, measuring that relationship has been a challenge.

There are, in fact, at least four distinct factors driving patterns in mosquito-borne pathogen dynamics that present a challenge: 1. mosquito populations are heterogeneous over time and space; 2. mosquito population densities and the resulting pathogen transmission dynamics are highly stochastic and somewhat unpredictable within a population; 3. interactions between mosquito populations, pathogen exposure, and the immune status of the pathogen's host population are non-linear; and 4. the accuracy of various metrics, including mosquito and case counts, is poorly characterized. As transmission and its relation to mosquito populations is so complicated, making progress through observational studies alone is limited. Modeling the association between mosquito population density over time and space and resulting disease dynamics using a combination of statistical and mechanistic models may be the most promising way to understand this complicated problem.

In the following section, we examine the history of using models and data to drive some aspects of

Robert C. Reiner Jr. and David L. Smith, *Heterogeneity, Stochasticity and Complexity in the Dynamics and Control of Mosquito-Borne Pathogens* In: *Population Biology of Vector-Borne Diseases*. Edited by: John M. Drake, Michael B. Bonsall, and Michael R. Strand: Oxford University Press (2021). © Robert C. Reiner Jr. and David L. Smith. DOI: 10.1093/oso/9780198853244.003.0002 research and policy agendas for mosquito-borne pathogens. We begin with a historical narrative covering development of mechanistic models of mosquito-borne pathogen transmission (Smith et al. 2012). Models of mosquito-borne pathogens can be traced to Ronald Ross; a model in 1905 described mosquito movement (Ross 1905), and the first malaria transmission models had been published by 1911 (Ross 1908; 1911a,b). George Macdonald played a pivotal role in development of the model in publications over a twenty-year period, starting in 1950 (Macdonald 1950a,b; Macdonald 1951; 1952a,b; 1953; 1955a,b; 1956a,b; 1957; WHO 1957; Macdonald & Göeckel 1964). We present one Ross–Macdonald model and give simple formulas for most of the metrics and quantities that are now part of the theory. In 1970, the scope of topics addressed by theory started to expand, driven partly by the end of global disease control programs and expansion of infectious disease dynamics and control and mathematical ecology in academic departments (Reiner et al. 2013). High-speed desktop computing has now made it possible to develop ever more complicated simulation models to explore how some of these factors interact, and to design and implement sophisticated statistical methods for analyzing data. Despite these advances, some challenges remain in tackling the complexity of modeling mosquito-borne pathogen dynamics and control (Smith et al. 2012; Smith et al. 2014). Our third section is an eclectic discussion of the expansion of theory from 1970 through the present, covering some topics relating to heterogeneity, stochasticity, complexity, and some of the unmet challenges for using models and data to explore the dynamics and control of mosquito-borne pathogens.

2.2 Developing the Theory, 1905–1980

Early models for mosquito-borne pathogen transmission were motivated by problems in malaria control. Ronald Ross, who won the 1904 Nobel Prize in medicine for his discovery in 1897 that mosquitoes transmit malaria (Ross 1897), played a prominent role in advancing the science and policy. Ross followed his discovery by advocating for innovative efforts to control malaria through larval

source management (Ross 1899). When outcomes of these control efforts ranged from surprisingly successful to ineffective, Ross struggled with the question of why (Ross 1907). Because of his prominence as a scientist, Ross was asked to advise Mauritius on malaria control, but he did so with little theory and few metrics (Ross 1908). In light of the mixed success of larval control and the availability (even then) of bed nets and quinine prophylaxis, Ross turned to mathematical models to understand the complex world of malaria.

In fact, Ross published two distinct malaria transmission models (Ross 1908; Ross 1911b). Both models were motivated by understanding how malaria transmission would vary in response to control and covered the same themes, but they used different species of math (Lotka 1923a). The models sought to describe the relationship between exposure and infection, and they posited that there must exist a threshold on the mosquito densities required to support transmission. A clear explanation and thorough analysis of both models was done later, mostly by Alfred Lotka (Waite 1910; Lotka 1923a–d). With Sharpe, Lotka also extended the models to consider the effect of delays caused by parasite incubation periods for humans and mosquitoes (Sharpe & Lotka 1923). Though Ross's models were simple and the quantities in his models were poorly defined, the models served an important purpose. Most critically, the models pointed to quantities it would be important to try and measure, which were followed by studies that developed well-defined metrics to measure them.

The next generation of mathematical innovation came from George Macdonald, whose lasting impact on modeling transmission by vectors was mainly to reformulate Ross's model using meaningful parameters based on decades of malaria research. Macdonald synthesized data describing epidemiology and infection in humans (Macdonald 1950a,b), as well as survival, infection, the extrinsic incubation period (EIP, the time required for malaria parasites to complete sporogony to become infectious), and the proportion of mosquitoes that have sporozoites in their salivary glands, called the sporozoite rate (Macdonald 1952b). In papers that followed over the next three years,

Macdonald developed the basic reproductive number, discussed how to measure it, and explained the relevance of the basic reproductive rate for control (Macdonald 1952a; 1953; 1955a,b; 1956a,b; 1957).

A likely motivation for Macdonald's research was a burgeoning effort to control adult mosquito populations with indoor residual spraying (IRS), the practice of spraying insecticides on the walls of houses to kill adult mosquitoes when they landed to rest. Though entomologists had long been spraying pyrethrum-based insecticides on walls to control insects (Potter, 1938), the method was more potent when combined with dichlorodiphenyltrichloroethane (DDT). In the decade after World War II ended, IRS trials with DDT had demonstrated dramatic effects on pathogen transmission (Brown 1998), and enthusiasm rapidly built to organize the first global campaign to eradicate malaria (Dobson et al. 2000). Macdonald's research and Macdonald himself played important roles in the formation of the World Health Organization (WHO) strategy for malaria eradication (WHO 1957).

An important idea, articulated mathematically through Macdonald's model for the sporozoite rate, which was a *post hoc* explanation for the early success of DDT-based IRS spray programs, was that malaria transmission was highly sensitive to the mortality rate of adult mosquitoes (Macdonald 1952a). The ideas contributed to a new wave of methods to measure mosquito longevity (Draper & Davidson 1953; Davidson 1954; Davidson & Draper 1953; Gillies 1954; Gillies & Wilkes 1965).

Ironically, though the basic reproductive numbers are one of Macdonald's most influential contributions to theory, these played a minimal role in the design of the Global Malaria Eradication Programme (GMEP), which was based more on the ideas of Fred Soper (Gladwell 2002; Dobson et al. 2000); the GMEP aimed for the highest coverage possible with the goal of interrupting transmission completely (WHO 1957). There was, nevertheless, an active research program to measure the effect sizes of vector control. Adding to generations of entomological data measuring mosquito blood feeding preferences, Garrett-Jones codified a parameter called the human blood index, the proportion of blood fed mosquitoes who had human blood (Garrett-Jones 1964b). He discussed a method for estimating it and used it to improve an understanding of Macdonald's formula. Garrett-Jones called it vectorial capacity or the 'daily reproductive rate' (Garrett-Jones & Shidrawi 1969; Garrett-Jones 1964a), because it was a measure of the daily output from mosquitoes. The formula he developed was extracted from Macdonald's formula for the reproductive numbers. Here, finally, was a formula that defined measurable quantities that could be used to establish threshold criteria for malaria elimination. Garrett-Jones then set out to test the theory by measuring changes in vectorial capacity in response to control (Garrett-Jones & Shidrawi 1969).

2.2.1 The Ross–Macdonald Model

Since the Ross–Macdonald model played such a prominent role in the formation of theory for mosquito-borne pathogens, we have presented it here (Smith et al. 2012). Several versions of the model exist. This particular formulation borrows heavily from one first presented by Aron and May (Aron & May 1982). Table 2.1 lists all the parameters, terms, and variables used in the following section.

The Ross–Macdonald model assumes there is a constant ratio of mosquitoes to humans m, and the parasite requires n days to complete development in the mosquito, called the extrinsic incubation period (EIP). Mosquitoes are assumed to die at a constant per capita rate g. (Macdonald used a slightly different notation to describe mosquito survival; p denotes the probability a mosquito survives one day, so $p^n = e^{-gn}$ and $g = -\ln p$.) Mosquitoes blood feed at a daily per capita rate f. A proportion, Q, of blood meals are taken on humans. (Macdonald used a single parameter for the human blood feeding rate, $a = fQ$.) Macdonald's models assume a fraction, b, of bites by infectious mosquitoes actually cause an infection. Herein, we also assume that, on average, a fraction, c, of bites on infectious humans infects a mosquito.

Macdonald also introduced a model for superinfection: what happens when people are infected and re-infected faster than parasites clear? Classical versions of the Ross–Macdonald model neverthe-

Table 2.1 Name and description of parameters, variables, and terms in the Ross–Macdonald Model

---

m The ratio of mosquitoes to humans

λ The daily emergence rate of adult mosquitoes per human

f The daily blood feeding rate of mosquitoes

Q The fraction of bites taken on humans

a The daily human blood feeding rate, $a = fQ$

n The extrinsic incubation period (# days).

g The adult mosquito death rate

p The probability a mosquito survives one day, $p = e^{-g}$

S The stability index, $S = fQ/g$

P The probability of surviving the EIP, $P = e^{-gn}$

v The number of female eggs laid per feeding cycle

G The number of female eggs laid per female, per lifespan, $G = vf/g$

r The per-capita rate that malaria infections clear

h The force of infection, $h = bE$

b The proportion of bites by infectious mosquitoes that infect humans

c The proportion of bites on infectious humans that infect mosquitoes

m The ratio of mosquitoes to humans

x Proportion of humans who are infected

y Proportion of mosquitoes who are infected

z Proportion of mosquitoes who are infectious

E The entomological inoculation rate (EIR), $E = maz$

---

less follow Ross's simple model for human malaria infections, in which humans recover at a constant rate, r. After clearing an infection, humans recover and are susceptible to infection again.

Given these parameters, it is possible to write down a model tracking changes in the proportion of infected mosquitoes (y) and humans (x). Mosquitoes become infected after biting an infectious host and they remain infected until they die. The ratio of mosquitoes to humans is assumed to remain constant (m). To get infected a mosquito must bite (f) an infectious human (Qx) and get infected (c). Mosquitoes would remain infected until they die (g). Changes in the fraction of infected mosquitoes thus follow an equation:

$$\bar{y} = fQcx(1-y) - gy. \qquad [2.1]$$

Exposure to malaria is related to the fraction of mosquitoes that are infectious. Let x_{t-n} and y_{t-n}

denote the values of the variables at time $t - n$. After n days, only a fraction $P = e^{-gn}$ would remain alive. The fraction of infectious mosquitoes follows a delay differential equation for infectious mosquitoes:

$$\bar{z} = fQe^{-gn}cx_{t-n}(1-y_{t-n}) - gz. \qquad [2.2]$$

From these equations, it is possible to derive a few basic relationships that have played an important historical role. First, the expected number of human blood meals, per mosquito, over its whole life is $S = fQ/g$, which Macdonald called 'a convenient measure of stability of insect-borne disease' (Macdonald, 1953). Note also that the fraction of infectious mosquitoes is related to the number of human bites per mosquito at the steady state (denoted with a super bar):

$$\bar{y} = \frac{cS\bar{x}}{1 + cS\bar{x}}. \qquad [2.3]$$

Note that at the steady state, $\bar{x} = \bar{x}_{t-n}$ and $\bar{y} = \bar{y}_{t-n}$ so $\bar{z} = \bar{y}e^{-gn}$. Among the most important features of this formula is that under the assumptions of the model, the distribution of blood meals per mosquito would have a geometric distribution, so it is possible to compute S from the fraction of mosquitoes that have never blood fed, a fact used by Davidson and others to estimate it (Draper & Davidson 1953).

Finally, the following equation describes changes in the fraction of infected humans. The rate of infection is the product of the product of the number of bites received by a person in a day (fQm) the proportion of those that are infectious (z) and the proportion of infectious bites that would cause an infection (b). Infections are assumed to clear at a per capita rate, r. Changes in the prevalence of malaria infection would be described by the following:

$$\bar{x} = fQbmz(1-x) - rx. \qquad [2.4]$$

The daily entomological inoculation rate (EIR), the number of infectious bites, per person, per day, in this model is $E = fQmz$. The force of infection (FOI), the number of infections, per person, per day, is $h = bE$. In this equation, the prevalence of infection at equilibrium is:

$$\bar{x} = \frac{h}{h + r}. \qquad [2.5]$$

These simple functional relationships are the starting point for understanding the more complicated relationships between models, data, and malaria epidemiology.

2.2.2 The Basic Reproductive Number and Vectorial Capacity

One point of Macdonald's analysis was to define a threshold criterion for endemic malaria and relate it to control measures. In the paragraphs that follow, we will describe Macdonald's analysis, but we will introduce a new term that Macdonald did not consider. Rather than treat m as a parameter, we convert it to a variable, and let λ denote the number of adult female mosquitoes emerging per human per day (Smith & McKenzie 2004). To keep m constant at its steady state, we assume human population density remains constant, and we model mosquito population densities using the equation:

$$\bar{m} = \lambda - gm \qquad [2.6]$$

At the steady state of Eq. 6, $\bar{m} = \lambda/g$. This makes it possible to point out some important oversights of Macdonald's analysis.

The basic reproductive number can be derived in a number of different ways, but to be consistent with Macdonald's formulation, we define it as the number of human malaria cases arising from each human malaria case (or mosquito cases per mosquito case). To be perfectly clear, by 'case' we mean an incident infection with malaria parasites, whether or not accompanied by clinical symptoms:

$$R_0 = \frac{bc\,ma^2}{r(-\ln p)}p^n = \lambda\frac{bcf^2Q^2}{rg^2}e^{-gn} = \frac{bc}{r}\lambda S^2 P. \quad [2.7]$$

(The first formula for R_0 is closest to Macdonald's, who omitted the parameter c.)Note that Macdonald ignored imperfect transmission, and his formula used the probability of surviving one day, $p = e^{-g}$, or equivalently $g = -\ln p$.

R_0 provides a threshold condition. If $R_0 < 1$, each case would generate less than one case under ideal conditions and malaria could not persist. If, on the other hand, each case generated more than one case, $R_0 > 1$, then malaria prevalence would tend to increase until the system comes to a malaria-endemic equilibrium. In this model, the steady state can be expressed in terms of just a few terms. At the steady state, the fraction of infected humans is:

$$\bar{x} = \frac{R_0 - 1}{R_0 + cS} \qquad [2.8]$$

The fraction of infectious mosquitoes, called the sporozoite rate, is:

$$\bar{z} = \frac{cS\bar{x}}{1 + cS\bar{x}}e^{-gn} = \frac{cS(R_0 - 1)}{cS + R_0 + cS(R_0 - 1)}e^{-gn}. \quad [2.9]$$

These formulas, which are only valid if $R_0 > 1$, provide a useful way of expressing all the relationships implied by the variables in terms of mainly R_0. The dynamics of the model were simple, but they have played an important role in theory for malaria epidemiology and transmission.

From Macdonald's formula from R_0, Garrett-Jones named vectorial capacity or the 'daily reproductive rate', the number of infectious bites that would arise from all the mosquitoes biting a single infectious human on a single day. Garrett-Jones, who had also notably devised the human blood index and added the parameter describing human blood feeding to the formula, Q, extracted a formula for vectorial capacity.

$$V = \frac{ma^2}{-\ln p}p^n = \lambda\frac{f^2Q^2}{g^2}e^{-gn} = \lambda S^2 P \qquad [2.10]$$

By assuming humans were 'perfectly infectious', the formula isolated mosquito traits from the problem of understanding variability in the infectiousness of humans, which is better treated in other parts of the model. The formula, written in this last way, has a very simple interpretation. Each mosquito emerging (λ) must blood feed on a human to become infected (the first S), and then after surviving to become infectious (with probability P), a mosquito must blood feed again to transmit (the second S). It reflects the simple fact that transmission requires two blood meals and survival through the EIP.

2.2.3 Sensitivity to Parameters

Among the most important conclusions of Macdonald's work was analysis of his formula for the sporozoite rate, or in a formula only slightly modified from the way he wrote it:

$$\bar{z} = \frac{a\bar{x}p^n}{a\bar{x} - \ln p} \qquad [2.11]$$

Note that this is one part of a model-based formula for a term (defined later) that is now called the entomological inoculation rate (EIR, or E), defined as the number of infectious bites per person per day. When the EIR is estimated under field conditions, it is computed by taking the product of the human biting rate (ma) times the sporozoite rate:

$$E = ma\bar{z} = \frac{ma^2\bar{x}p^n}{a\bar{x} - \ln p} \qquad [2.12]$$

In his formulas, Macdonald noted that killing adult mosquitoes (reducing p) would have an extraordinarily large effect on malaria transmission intensity because survival was so critical to the process of transmission. The effect is more apparent in the formula for vectorial capacity (see Eq. 2.10). A mosquito must survive for n days after becoming infected, with probability $P = p^n$, and then it would bite a number of humans:

$$S = \frac{a}{-\ln p} = \frac{fQ}{g}. \qquad [2.13]$$

Macdonald produced a table to illustrate the principle that reductions in transmission arising from strategies that kill adult mosquitoes will tend to be more efficient than other strategies, all else equal. This was a clear case where a mathematical formula influenced policy, though perhaps not in the way most people suppose. The formula and table provided an explanation for something that had already been demonstrated: DDT-based spraying with IRS had already been shown to be highly effective. Macdonald's formula provided an explanation for why.

The formulas for vectorial capacity (Eq. 2.10) make it clear that Macdonald ignored effects on the human biting rate (Smith & McKenzie 2004), which in this notation is $ma = \lambda fQ/g = \lambda S$. Macdonald's formula would underestimate the total effects of changing mortality rate: $V \propto g^2 e^{-gn}$. Mosquito survival affects vectorial capacity in three ways.

If anything, it has been argued that even this formula under-estimates the effects of mortality because the number of adults emerging is related, in some way, to the number of eggs laid per mosquito life, which is related to blood feeding rates, f, mosquito mortality g, and female eggs laid per

blood meal, v: $G = vf/g$. If λ is a function of G, then vectorial capacity is:

$$V = \lambda \left(\frac{vf}{g}\right)\frac{f^2Q^2}{g^2}e^{-gn} = \lambda(G)S^2P. \qquad [2.14]$$

Survival affects transmission intensity in four ways, blood feeding in three, human blood preferences in two, and mosquito density in one way (Brady et al. 2016; Brady et al. 2015). Factors affecting the EIP have one effect, though it is compounded through the entire length of EIP. A problem that we will address below is that there is great heterogeneity and uncertainty about the functional form for $\lambda(G)$.

2.2.4 Measuring Transmission

An important function of Ross's model was to guide development of the metrics used to measure transmission. In his earliest work, Ross focused on the relationship between exposure and infection, so Ross worked to improve the accuracy of estimates of the prevalence of infection. Since Laveran, malaria had been detected by light microscopy, and the prevalence of infection was called the malaria rate and later the parasite rate. An alternative measure of malaria prevalence was the proportion of the population with a palpably enlarged spleen, called the spleen rate. The malaria parasite rate was, at least by comparison to the spleen rate, a fairly specific and accurate measure of malaria in populations (Ross 1903). Ross continued to work on the mathematics of epidemics, eventually publishing three papers, the last two with the mathematician Hilda Hudson (Ross 1916; Ross & Hudson 1917a-b).

Ross had established a basic set of equations outlining how it would be possible to conduct a quantitative study of the relationship between exposure and infection. He considered the hazard rate for infections, or the force of infection, which he called the 'happenings' rate, h, and wrote down an equation describing the relationship between exposure, recovery, and infection. Letting age stand in for time, it is possible to write down an equation that describes infection in a cohort of children as they age (A, compare to Eq. 2.4):

$$dx/dA = h(1-x) - rx.$$ [2.15]

Under the initial condition, $x(0) = 0$, the solution to Eq. 2.15 is:

$$x(A) = \left(1 - e^{-(h+r)A}\right)\frac{h}{h+r}$$ [2.16]

This formalizes the notion that the force of infection rises in children to a steady state. It would take decades to formalize statistical methods to estimate the hazard rate in relation to malaria infection (Pull & Grab 1974). Similar methods have been used to examine serological data: the patterns of seropositivity by age are highly informative of the history of transmission (Drakeley et al. 2005).

Ross had also suggested, however, that a more accurate measure of exposure would come from counting infectious mosquitoes, which motivated some early studies that measured exposure to the bites of infectious mosquitoes (Barber et al. 1931; Davey & Gordon 1933). One early study referred to the 'inoculation rate', and measured 'household biting density', and 'household infective density' and their relations to measures of individual risk (Davey & Gordon 1933). A feature of these malaria data, documented in 1933 for the first time and reanalysed by Macdonald in 1950, was that the estimates of malaria exposure made entomologically dramatically overestimated the force of infection (Macdonald 1950).

A challenge for Macdonald's formula for R_0 and Garrett-Jones's formula for vectorial capacity, which had motivated research in malaria transmission through the duration of the Global Malaria Eradication Programme, was that it required several independent measures. While the approach was mathematically sound, it was difficult and expensive to measure all the parameters accurately, and there were concerns about compounding errors (Dye 1986). After promoting development of vectorial capacity during the GMEP, the WHO effectively abandoned it and adopted the old metric, 'infective biting density' rebranded under a new name, the 'entomological inoculation rate' (Onori & Grab 1980a; Onori & Grab 1980b), a change and endorsement that made entomological measures of transmission far more common (Hay et al. 2000).

The WHO endorsement of the EIR as a complete metric of transmission intensity was pragmatic. The

EIR could be measured as a single, comprehensive measure of transmission from a single sampling method (as already introduced). First, vectors are caught and counted as they attempt to feed, which is the basis of the metric now called the human biting rate (HBR, aka 'man biting rate'). Next, those same mosquitoes are examined to see what fraction had sporozoites in their salivary glands, which is the metric called the sporozoite rate. The product of these two is defined the EIR. Both metrics appear in the models, and it is conceptually straightforward to relate the models and the data, though questions remain about the accuracy of various field methods used to catch mosquitoes (Silver 2008).

It was much easier to measure the EIR, which can be calculated as the product of two other quantities estimated using a variety of field methods: ma, called the human biting rate; and the sporozoite rate. The formula for the EIR is:

$$E = mfQ\bar{z} = mfQ \; e^{-gn}\left(\frac{c\bar{x}S}{1+c\bar{x}S}\right) =$$ [2.17]

$$\lambda S^2 e^{-gn}\frac{c\bar{x}}{1+c\bar{x}S} = V\frac{c\bar{x}}{1+c\bar{x}S}$$

The last formula exposes a close conceptual relationship between the EIR and vectorial capacity: one measures the number of infectious bites arising if hosts were perfectly infectious, and the other measures the number of infectious bites received in a population, given the ambient levels of net infectiousness (and superinfection of mosquitoes). Under the assumptions of this model, after correcting for net infectiousness, infectious bites arising, and infectious bites received must approximately balance out.

2.3 Heterogeneous Transmission, 1970 to the Present

By 1970, the Ross–Macdonald model existed as a set of models, concepts, and basic metrics for measuring malaria transmission (Smith et al. 2012). Ross acknowledged the models were missing some basic features that were necessary for applying the ideas in context: topics such as spatial dynamics, seasonality, and heterogeneous transmission. Macdonald acknowledged the shortcomings of the simplifying assumption of constant mosquito populations, in part, through an extensive discussion of the classification of transmission (Macdonald 1953; Macdonald 1957). The models lacked an explicit connection to mosquito population dynamics, which could be important for vector control. Limitations of the statistical methodology and metrology have since been revealed as common practices have been slow to evolve. The following paragraphs are an eclectic overview of the history of the development of these concepts.

2.3.1 Mosquitoes, Seasonality, and Noise

Of all the models published describing mosquito-borne transmission (Reiner et al. 2013), a distinction can be made about the way disease transmission models consider mosquitoes: some include only a term or parameter describing exogenous forcing; some include variables that describe pathogen infection dynamics in mature mosquito populations (with either a stable population size or an emergence rate); and finally, some also include mosquito population dynamics with feedbacks between egg laying and immature population densities. In the latter, mosquito population density itself has a threshold on reproduction required for mosquito population persistence, and mosquito density declines in response to egg laying and feedbacks through aquatic population dynamics.

Many models ignore mosquito population dynamic feedbacks, in part, because the intensity of transmission by mosquito populations on disease dynamics can be summarized effectively by changes in vectorial capacity. Among the first to recognize this fact was a model developed as part of the Garki Project, which modeled transmission intensity by mosquitoes using a term that described the EIP-lagged seasonally forced vectorial capacity (Dietz et al. 1974). Supporting the Garki project was a set of trigonometric functions for simulating seasonal forcing on malaria transmission dynamics (Dietz 1976).

Macdonald's model for the sporozoite rate fell into the second class of models—those that consider infection dynamics in mosquitoes. While Macdonald did not do so, it is possible to consider the emergence rate of adult mosquitoes as a parameter (as we have done above), which can vary over time (which we did not do previously) to describe or simulate average mosquito density and its seasonal pattern to environmental covariates (e.g. temperature and rainfall), especially for modeling seasonality in transmission. Note that these models must be formulated in terms of the density of mosquitoes, rather than proportions (Smith et al. 2004). If mosquito population density were not constant, Macdonald's formula for the sporozoite rate would require another term that describes how the sporozoite rate changes as mosquito population density changes, depending on whether mosquito population size is getting larger and younger or smaller and older (Smith et al. 2004; Reiner, Guerra, et al. 2015).

Models with mosquito population dynamics can also consider population dynamics as part of the total effect size of vector control. Models with structured aquatic habitats and habitat heterogeneity can give a very different picture of the dynamic feedbacks. While there may be a response to crowding for some mosquito species in some habitats, it remains unclear what is the balance between density independent mortality factors and density dependent factors overall. Habitats can be ephemerally available, appearing and disappearing with some complex patterns on a landscape, driven by hydrology (Hardy et al. 2013). What effect, if any, does this sort of habitat heterogeneity have on mosquito populations and transmission?

Two broad problems lie at the frontier for modeling. One challenge is calibrating models to local environmental conditions. Attempts to find simple, general relationships between mosquito abundance, weather, and climate have been largely unsatisfying (Reiner, Geary, et al. 2015). The problem is not that temperature or humidity do not matter, but that there are often other factors driving mosquito ecology and pathogen transmission that have even larger effects (Gething et al. 2010). One notable example is hydrology (Bomblies et al. 2008; Bomblies et al. 2009). There has been some attempt to identify the sorts of landscapes that respond to weather in different ways (Hardy et al. 2013), but detailed studies in situ often reveal site-specific features that matter for transmission (Bidlingmayer 1985). All vector mosquito populations require blood hosts and aquatic habitat, but sugar sources are required for all male mosquito populations and they affect female mosquito populations to varying degrees. These resources can be available seasonally or ephemerally, driven by hydrology, topography, and the way humans use landscapes. While there are dozens of individual studies showing how small details matter in context, and while some mathematical models have been developed to understand the local dynamics of mosquitoes in relation to disease transmission, there is no general theory covering the heterogeneity in these studies. It is this local contextualization of mosquito ecology and vector control that represents a challenge for mathematical modeling in the future; what are reasonable ways of dealing with the contextual heterogeneity?

The second problem is understanding all the factors contributing to noise in systems. While most mechanistic models for mosquitoes would predict a canonical seasonal pattern in mosquito density with even exposure, mosquito counts in the vast majority of mosquito studies follow a negative binomial distribution (Taylor 1984). Moreover, counts follow a power law: the variances of these distributions follow a power law with respect to the means. How do such distributions arise? One possibility is that factors such as wind, the joint distribution of resources, and habitat heterogeneity affect how mosquitoes move around (Midega et al. 2012; Service 1997). Weak winds can exacerbate other factors affecting mosquito dispersion and could affect how mosquitoes aggregate around resources. Strong winds can decimate local mosquito populations. How important is environmental noise, and how much does it affect transmission?

2.3.2 Mosquitoes and Spatial Dynamics

Transmission models were not Ross's first attempt to use mathematics to understand malaria. One practical question arising early in the history of larval control was how large an area would need to be kept clear of mosquito habitat to eliminate malaria (or yellow fever) transmission in a region. Answering the question required knowing something about the movement and spatial distribution of adult mosquitoes around aquatic habitat. Ross's first model considered diffusive movement of adult

mosquitoes into an area that had been cleared of habitat (Ross 1905). The figures painted an intuitive picture of adult mosquito density declining sigmoidally from outside a control zone, through the edge and into its center. At least part of the spatial dynamics of mosquito-borne diseases is related to mosquito movement, but perhaps not always in the most obvious ways.

Ross's model, and the vast majority of models since then, model mosquito movement as a diffusion-based process, and they largely consider dispersion on homogeneous landscapes. If mosquitoes searched only for blood, this might be a useful approximating model, but female mosquitoes also lay eggs requiring aquatic habitats (Le Menach et al. 2005). In fact, mosquitoes search for at least four distinct resources: vertebrate blood, aquatic habitats, sugar sources, and (at least early in life) mates. A fifth resource would be a set of haunts, where mosquitoes will land and rest. What follows immediately from this principle is that mosquitoes will often begin their search for one resource at a site where they have just recently found another: for example, a search for blood hosts that begins at an aquatic habitat where they have laid eggs (Bidlingmayer 1985). Mosquitoes use the wind to bring them cues about the distribution of blood resources on landscapes, and there is some evidence that when risk is distributed around aquatic habitats, that it is affected by the wind (Midega et al. 2012). Some habitats tend to be permanent, but others have an ephemeral or seasonal quality, such that a landscape could be a complex spatio-temporal oscillator: e.g. habitat is available in the Northeast in the spring, the South in the summer, and the West in the fall.

The problem of modeling mosquito search in relation to wind has been tackled at an individual level by plume models; mosquitoes tack across wind until they find a signal, then follow that signal towards its source (Cummins et al. 2012). The flight patterns are of some interest, per se, and represent an interesting branch of science. The problem for mosquitoes in natural situations is where they end up in relation to where they started in a context where wind direction and speed both affect dispersion. In a population with multiple sources, which one does a mosquito find to follow, and where does it end up (Hassanali et al. 2008)? The problem is probably subject to even more heterogeneity, as mosquitoes are not willing to move through just any habitat (Reisen & Lothrop 1995). There is, moreover, the fact that mosquitoes are often carried up away from near the ground (i.e. the boundary layer) and carried up into the air, where they can travel very long distances (Service 1997). Even if they travel short distances, mosquitoes fall out of these massive air flows at locations they cannot control. Over short distances, this means that winds will tend to generate patterns in the distribution of mosquitoes, as mosquito movement through heterogeneous landscapes combined with wind tend to generate heterogeneity. Even the best models of wind and other factors fail to take landscape heterogeneity and all its effects into account, and mosquito density and dispersal are usually modeled as mean field models with diffusive movement. A few exceptions to this rule have shown (at least conceptually) that even aquatic habitats that fail to produce any emergent adults can still affect adult behavior, so long as the adult females go there to lay eggs (Le Menach et al. 2005). What is even more likely is that heterogeneity in landscape features gives rise to idiosyncratic patterns in mosquito blood feeding behavior. At least some of this is amenable to study, if only to identify useful approximating models that relate the statistical properties of mosquitoes on landscapes to dynamics of disease.

2.3.3 Heterogeneous Biting

An important feature of malaria is heterogeneity in exposure. Since exposure to mosquito-borne pathogens requires the presence of a competent mosquito host, the distribution of mosquitoes has always been recognized as a predominant factor explaining the geographical distribution patterns of mosquito-borne pathogens. It has also been recognized, at least implicitly, that the distribution of mosquitoes can be highly heterogeneous at a very fine grain (Bidlingmayer 1985).

The burdens of mosquito-transmitted filarial worms are highly aggregated on humans, and their distributions are well-described by the negative binomial family of probability distribution functions (Hairston & de Meillon 1968). These distribution patterns could be explained, in part, by a model in which the propensity to be bitten has a Gamma distribution; the negative binomial distributions describing mosquito counts would arise as a mixture process of this Gamma distributed propensity with a Poisson. The question is how to quantify this Gamma distribution.

Heterogeneous biting probably plays a role in determining thresholds on transmission. Heterogeneous biting among individuals within a population should affect the transmission dynamics by amplifying transmission during an invasion (Dye & Hasibeder 1986; Smith et al. 2014; Lloyd et al. 2007). The number of mosquito bites per human per day is a critical measure of exposure overall, so it is not surprising that thresholds for invasion should be lowered if a segment of the population is consistently bitten at a higher rate. In a series of papers, this among-human variation in transmission was shown to scale with the coefficient of variation of biting rates, α. Heterogeneous biting would lower the threshold for invasion by a factor that was equal to $1+\alpha$. A challenge for understanding such invasions is the difference between heterogeneous biting among individuals in a well-mixed population, invasion of a spatially structured population, and the spatial scales that characterize transmission.

Mosquito counts data are highly aggregated, suggesting it might be possible to identify those who are bitten most and target them for control (Woolhouse et al. 1997). If biting follows something like a 20–80 rule, in which 20 percent of the population gets 80 percent of the bites, then finding the 20 percent would dramatically increase the efficiency of malaria control. What has not been explored sufficiently is the degree to which heterogeneous biting patterns are stable enough to enhance transmission overall, and the extent to which these effects are nullified by poor mixing. As mentioned before, environmental noise *can* be a source of heterogeneous biting and that it can strongly affect the degree to which heterogeneous biting on individuals matters for transmission.

2.3.4 Heterogeneity in Responses to Vector Control

Macdonald's analysis cast a long shadow on the study of mosquito population dynamics in medical entomology. Macdonald's argument about sensitivity to parameters gave support to the DDT-based IRS spray programs of the GMEP, but mosquito population densities in his models are assumed to remain constant. The effect sizes in Macdonald's analysis reflect changes in the sporozoite rate, or in other words, changes to survival and human blood feeding after mosquitoes become infected (i.e. SP in Eq. 2.14). Macdonald's analysis ignored the fact that the effects of vector control come from population dynamic effects through changes to mosquito population density and their indirect consequences, which are also sensitive to mosquito longevity (terms that affect the human biting rate, $\lambda(G) \cdot S$ in Eq. 2.14).

On the one hand, the focus on sensitivity to parameters drew attention to mosquito longevity, and later engendered a focus on the question of whether mosquitoes senesce and a quest for methods to age-grade mosquitoes. On the other hand, this emphasis on sensitivity over variability has, perhaps, had some unintended consequences. Sensitivity to parameters draws attention to old mosquitoes and overshadows the role of mosquito ecology: most of the variability in the potential intensity of malaria transmission by a species over its geographical range is driven by differences in mosquito population densities. At least some part of the effect sizes of control must be due to changes in population density of mosquitoes, as well (Kiware et al. 2012; Killeen & Chitnis 2014).

Like the earliest studies of larval control, randomized control trials and large-scale implementation of vector control have had highly variable outcomes. While there are some potentially obvious reasons for variability in the outcome of these trials (e.g. differences in baseline transmission, poor operational coverage or insecticide resistance), one of the ideas that has not been adequately tested is whether Macdonald was right; were the effects of DDT due mostly to changes in the sporozoite rate? Macdonald's formula dramatically under-estimated some of the effects of mosquito longevity on control, but a belief in that analysis would tend to attribute all of the effects of IRS to changes in mosquito longevity after infection and the sporozoite rate, while there may have been equally large changes in mosquito population density and the human biting rate. Where IRS and other forms of vector control have worked, it is apparent that a fraction of the response to IRS and other forms of vector control has come from massive changes in mosquito density, not just to longevity.

The role of ecology in control has been systematically neglected, and so has its role as a potential source of heterogeneity in the responses. An informal justification often given is that declines in mosquito population density would have only a linear effect, but as pointed out previously, changes in mosquito longevity would affect the number of eggs laid with additional reductions coming through a non-linearity introduced by mosquito population dynamics. While it is comparatively easy to study sensitivity to parameters in mathematical equations, it is far more challenging to understand and model the ecological factors giving rise to heterogeneity in the outcomes of studies (Brady et al. 2016; Brady et al. 2015). More to the point, an uncritical look at Macdonald's theory means that the question of whether heterogeneity in the outcome of studies was attributable to large changes in the ecology.

Perhaps the most important critique of Macdonald's hypothesis is that it was highly persuasive and has remained largely untested. What if Macdonald was, in fact, wrong in attributing the large effect sizes to changes in the proportion of very old mosquitoes? What if, instead, very large differences in transmission arising from IRS had more to do with changes in population size due to the local extinction of mosquito populations? In at least some examples, this has been the case (e.g. Meyers et al. 2016). A problem for medical entomology is that these unrecognized and therefore unmeasured factors could have been at least part of the explanation for most of the observed heterogeneity in observed responses to vector control.

2.4 Summary and Conclusions

A recent systematic review of mark-release-recapture studies for mosquitoes highlighted some endemic problems with the study of mosquito-borne pathogens (Guerra et al. 2014). The studies, which were done for their own reasons, highlighted interesting features of mosquito populations, but taken all together, they failed to paint a coherent picture. The studies differed so much in their designs, it was not possible to use the studies to try and forge a synthesis. What seems obvious now, a fact that was highlighted by some studies, is that mosquito dispersal is related to their behavioral state and available resources. Similar arguments can be made for virtually every aspect of mosquito biology affecting transmission. The context dependence of mosquito dispersal and other studies stands in sharp contrast to the lack of reports from these studies about the context.

There are some notable exceptions to this critique. Foremost among these efforts is the evidence-based approach to variability and uncertainty in the prevalence of malaria from the databases maintained by Malaria Atlas Project, along with their Bayesian geostatistical analysis (Sinka et al. 2012; Wiebe et al. 2017; Bhatt et al. 2015; Gething et al. 2011; Gething et al. 2012; Hay et al. 2009). This, at least, is an attempt to pull together all the data to characterize the epidemiological and entomological patterns. A key observation is the difference in the quality and quantity of the data. It is common to take blood and test for parasites. On the entomological side, the bulk of the data come as vector occurrence data. Quantitative studies of mosquito abundance are comparatively uncommon. They are highly heterogeneous in their methods, and most of the studies fail to report even the most basic data describing sampling effort.

There is also very little theory suggesting how heterogeneity in mosquito populations would affect transmission of mosquito-borne, acute, immunizing infections. For SIR diseases, such as dengue transmission by *Aedes aegypti* in the tropics, does mosquito abundance or the availability of susceptible hosts limit disease dynamics, or is it a combination of the two? There are many open questions remaining in the study of mosquitoes and transmission. While there are some important challenges to advancing these studies, the topics of greatest interest now are various sorts of heterogeneities and their causes and consequences.

References

Aron, J.L. & May, R.M. (1982). The population dynamics of malaria. In *Population Dynamics and Infectious Disease*, 139–79. London, UK: Chapman and Hall.

Barber, M.A., Olinger, M.T., & Putnam, P. (1931). Studies on malaria in southern Nigeria. *Annals of Tropical Medicine and Parasitology*, 25, 461–508.

Bhatt, S., Weiss, D.J., Cameron, E. et al. (2015). The effect of malaria control on *Plasmodium falciparum* in Africa between 2000 and 2015. *Nature*, 526, 207–11.

Bidlingmayer, W.L. (1985). The measurement of adult mosquito population changes—some considerations. *Journal of the American Mosquito Control Association*, 1, 328–48.

Bomblies, A., Duchemin, J.B., & Eltahir, E.A.B. (2008). Hydrology of malaria: Model development and application to a Sahelian village. *Water Resources Research*, 44, W12445.

Bomblies, A., Duchemin, J.B., & Eltahir, E.A.B. (2009). A mechanistic approach for accurate simulation of village scale malaria transmission. *Malaria Journal*, 8, 223.

Brady, O.J., Godfray, H.C.J., Tatem, A.J. et al. (2015). Adult vector control, mosquito ecology and malaria transmission. *International Health*, 7, 121–9.

Brady, O.J., Godfray, H.C.J., Tatem, A.J. et al. (2016). Vectorial capacity and vector control: Reconsidering sensitivity to parameters for malaria elimination. *Transactions of the Royal Society of Tropical Medicine and Hygiene*, 110, 107–17.

Brown, P.J. (1998). Failure-as-success: Multiple meanings of eradication in the Rockefeller Foundation Sardinia project, 1946–1951. *Parassitologia*, 40, 117–30.

Cummins, B., Cortez, R., Foppa, I.M., Walbeck, J., & Hyman, J. (2012). A spatial model of mosquito host-seeking behavior. *PLoS Computational Biology*, 8, e1002500.

Davey, T.H. & Gordon, R.M. (1933). The estimation of the density of infective anophelines as a method of calculating the relative risk of inoculation with malaria from different species or in different localities. *Annals of Tropical Medicine and Parasitology*, 27, 27–52.

Davidson, G. (1954). Estimation of the survival-rate of anopheline mosquitoes in nature. *Nature*, 174, 792–3.

Davidson, G. & Draper, C.C. (1953). Field studies of some of the basic factors concerned in the transmission of malaria. *Transactions of the Royal Society of Tropical Medicine and Hygiene*, 47, 522–35.

Dietz, K. (1976). The incidence of infectious disease under the influence of seasonal fluctuations. In *Mathematical Models in Medicine: Workshop, Mainz, March 1976*, 1–15. Berlin; New York: Springer Verlag.

Dietz, K., Molineaux, L., & Thomas, A. (1974). A malaria model tested in the African savannah. *Bulletin of the World Health Organization*, 50, 347–57.

Dobson, M.J., Malowany, M., & Snow, R.W. (2000). Malaria control in East Africa: The Kampala Conference and the Pare-Taveta Scheme: A meeting of common and high ground. *Parassitologia*, 42, 149–66.

Drakeley, C.J., Corran, P.H., Coleman, P.G. et al. (2005). Estimating medium- and long-term trends in malaria transmission by using serological markers of malaria exposure. *Proceedings of the National Academy of Sciences of the United States of America*, 102, 5108–13.

Draper, C.C. & Davidson, G. (1953). A new method of estimating the survival-rate of anopheline mosquitoes in nature. *Nature*, 172, 503.

Dye, C. (1986). Vectorial capacity: Must we measure all its components? *Parasitology Today*, 2, 203–9.

Dye, C. & Hasibeder, G. (1986). Population dynamics of mosquito-borne disease: Effects of flies which bite some people more frequently than others. *Transactions of the Royal Society of Tropical Medicine and Hygiene*, 80, 69–77.

Garrett-Jones, C. (1964a). Prognosis for interruption of malaria transmission through assessment of the mosquito's vectorial capacity. *Nature*, 204, 1173–5.

Garrett-Jones, C. (1964b). The human blood index of malaria vectors in relation to epidemiological assessment. *Bulletin of the World Health Organization*, 30, 241–61.

Garrett-Jones, C. & Shidrawi, G.R., 1969. Malaria vectorial capacity of a population of *Anopheles gambiae*: An exercise in epidemiological entomology. *Bulletin of the World Health Organization*, 40, 531–45.

Gething, P.W., Elyazar, I.R.F., Moyes, C.L. et al. (2012). A long-neglected world malaria map: *Plasmodium vivax* endemicity in 2010, J.M. Carlton, ed. *PLoS Neglected Tropical Diseases*, 6, e1814.

Gething, P.W., Patil, A.P., Smith, D.L. et al. (2011). A new world malaria map: *Plasmodium falciparum* endemicity in 2010. *Malaria Journal*, 10, 378.

Gething, P.W., Smith, D.L., Patil, A.P., Tatem, A.J., Snow, R.W., & Hay, S.I. (2010). Climate change and the global malaria recession. *Nature*, 465, 342–5.

Gillies, M.T. (1954). The recognition of age-groups within populations of *Anopheles gambiae* by the pre-gravid rate and the sporozoite rate. *Annals of Tropical Medicine and Parasitology*, 48, 58–74.

Gillies, M.T. & Wilkes, T.J. (1965). A study of the age-composition of populations of *Anopheles gambiae* Giles and *A. funestus* Giles in North-Eastern Tanzania. *Bulletin of Entomological Research*, 56, 237–63.

Gladwell, M. (2002). Fred Soper and the global malaria eradication programme. *Journal of Public Health Policy*, 23, 479–97.

Guerra, C.A., Reiner, R.C.J., Perkins, T.A. et al. (2014). A global assembly of adult female mosquito mark-release-recapture data to inform the control of mosquito-borne pathogens. *Parasites & Vectors*, 7, 276.

Hairston, N.G. & de Meillon, B. (1968). On the inefficiency of transmission of *Wuchereria bancrofti* from mosquito to human host. *Bulletin of the World Health Organization*, 38, 935–41.

Hardy, A.J., Gamarra, J.G.P., Cross, D.E., et al. (2013). Habitat hydrology and geomorphology control the distribution of malaria vector larvae in rural Africa. *PLoS One*, 8, e81931.

Hassanali, A., Nedorezov, L.V., & Sadykou, A.M. (2008). Zooprophylactic diversion of mosquitoes from human to alternative hosts: A static simulation model. *Ecological Modelling*, 212, 155–61.

Hay, S.I., Guerra, C.A., Gething, P.W., et al. (2009). A world malaria map: *Plasmodium falciparum* endemicity in 2007. I. Müeller, ed. *PLoS Medicine*, 6, e1000048.

Hay, S.I., Rogers, D.J., Toomer, J.F., & Snow, R.W. (2000). Annual *Plasmodium falciparum* entomological inoculation rates (EIR) across Africa: Literature survey, Internet access and review. *Transactions of the Royal Society of Tropical Medicine and Hygiene*, 94, 113–27.

Killeen, G.F. & Chitnis, N. (2014). Potential causes and consequences of behavioural resilience and resistance in malaria vector populations: A mathematical modelling analysis. *Malaria Journal*, 13, 97.

Kiware, S.S., Chitnis, N., Moore, S.J. et al. (2012). Simplified models of vector control impact upon malaria transmission by zoophagic mosquitoes. *PLoS One*, 7, e37661.

Le Menach, A., McKenzie, F.E., Flahault, A., & Smith, D.L. (2005). The unexpected importance of mosquito oviposition behaviour for malaria: Non-productive larval habitats can be sources for malaria transmission. *Malaria Journal*, 4, 23.

Lloyd, A.L., Zhang, J., & Root, A.M. (2007). Stochasticity and heterogeneity in host-vector models. *Journal of The Royal Society Interface*, 4, 851–63.

Lotka, A.J. (1923a). Contribution to the analysis of malaria epidemiology I General part. *American Journal of Hygiene*, 3, 1–37.

Lotka, A.J. (1923b). Contribution to the analysis of malaria epidemiology II General part (continued) comparison of two formulae given. *American Journal of Hygiene*, 3, 38–54.

Lotka, A.J. (1923c). Contribution to the analysis of malaria epidemiology III Numerical part. *American Journal of Hygiene*, 3, 55–95.

Lotka, A.J., (1923d). Contribution to the analysis of malaria epidemiology V Summary. *American Journal of Hygiene*, 3, 113–21.

Macdonald, G. (1950a). The analysis of infection rates in diseases in which superinfection occurs. *Tropical diseases Bulletin*, 47, 907–15.

Macdonald, G. (1950b). The analysis of malaria parasite rates in infants. *Tropical Diseases Bulletin*, 47, 915–38.

Macdonald, G. (1951). Community aspects of immunity to malaria. *British Medical Bulletin*, 8, 33–6.

Macdonald, G. (1952a). The analysis of equilibrium in malaria. *Tropical Diseases Bulletin*, 49, 813–1129.

Macdonald, G. (1952b). The analysis of the sporozoite rate. *Tropical Diseases Bulletin*, 49, 569–86.

Macdonald, G. (1953). The analysis of malaria epidemics. *Tropical Diseases Bulletin*, 50, 871–89.

Macdonald, G. (1955a). A new approach to the epidemiology of malaria. *Indian Journal of Malariology*, 9, 261–70.

Macdonald, G. (1955b). The measurement of malaria transmission. *Proceedings of the Royal Society of Medicine*, 48, 295–301.

Macdonald, G. (1956a). Epidemiological basis of malaria control. *Bulletin of the World Health Organization*, 15, 613–26.

Macdonald, G. (1956b). Theory of the eradication of malaria. *Bulletin of the World Health Organization*, 15, 369–87.

Macdonald, G. (1957). *The Epidemiology and Control of Malaria*. Oxford University Press, London.

Macdonald, G. & Göeckel, G. (1964). The malaria parasite rate and interruption of transmission. *Bulletin of the World Health Organization*, 31, 365–77.

Meyers, J.I. Pathikonda, S., Popkin-Hall, Z.R. et al. (2016). Increasing outdoor host-seeking in *Anopheles gambiae* over 6 years of vector control on Bioko Island. *Malaria Journal*, 15, 239.

Midega, J.T. Smith, D.L., Olotu, A. et al. (2012). Wind direction and proximity to larval sites determines malaria risk in Kilifi District in Kenya. *Nature Communications*, 3, 674.

Onori, E. & Grab, B. (1980a.) Indicators for the forecasting of malaria epidemics. *Bulletin of the World Health Organization*, 58, 91–8.

Onori, E. & Grab, B. (1980b). Quantitative estimates of the evolution of a malaria epidemic in Turkey if remedial measures had not been applied. *Bulletin of the World Health Organization*, 58, 321–6.

Potter, C. (1938). The use of protective films of insecticide in the control of indoor insects, with special reference to *Plodia interpunctella* Hb. and *Ephestia elutella* Hb. *Annals of Applied Biology*, 25, 836–54.

Pull, J.H. & Grab, B. (1974). A simple epidemiological model for evaluating the malaria inoculation rate and the risk of infection in infants. *Bulletin of the World Health Organization*, 51, 507–16.

Reiner, R.C.J., Perkins, T.A., Barker, C.M. et al. (2013). A systematic review of mathematical models of mosquito-borne pathogen transmission: 1970–2010. *Journal of The Royal Society Interface*, 10, 20120921–1.

Reiner, R.C.J., Geary, M., Atkinson, P.M., Smith, D.L., & Gething, P.W. (2015). Seasonality of *Plasmodium falciparum* transmission: A systematic review. *Malaria Journal*, 14, 343.

Reiner, R.C.J., Guerra, C.A., Donnelly, M.J., Bousema, T., Drakeley, C., & Smith, D.L. (2015). Estimating malaria transmission from humans to mosquitoes in a noisy landscape. *Journal of The Royal Society Interface*, 12, p.20150478.

Reisen, W.K. & Lothrop, H.D. (1995). Population ecology and dispersal of *Culex tarsalis* (Diptera: Culicidae) in the Coachella Valley of California. *Journal of Medical Entomology*, 32, 490–502.

Ross, R. (1897). On some peculiar pigmented cells found in two mosquitos fed on malarial blood. *British Medical Journal*, 2, 1786–8.

Ross, R. (1899). Inaugural lecture on the possibility of extirpating malaria from certain localities by a new method. *The British Medical Journal* 2, 1–4.

Ross, R. (1903). *The Thick Film Process for The Detection of Organisms in the Blood*. University Press of Liverpool, Liverpool, UK.

Ross, R. (1905). The logical basis of the sanitary policy of mosquito reduction. *Science*, 22, 689–99.

Ross, R. (1908). *Report on the Prevention of Malaria in Mauritius*. Waterlow & Sons Limited, London.

Ross, R. (1907). The prevention of malaria in British possessions, Egypt, and parts of America. *Lancet*, 879–87.

Ross, R. (1911a). Some quantitative studies in epidemiology. *Nature*, 87, 466–7.

Ross, R. (1911b). *The Prevention of Malaria*. 2nd John Murray, London.

Ross, R. (1916). An application of the theory of probabilities to the study of a priori pathometry. Part I. *Proceedings of the Royal Society of London Series a-Mathematical Physical and Engineering Sciences*, 92, 204–30.

Ross, R. & Hudson, H.P. (1917a). An application of the theory of probabilities to the study of a priori pathometry. Part II. *Proceedings of the Royal Society of London Series a-Mathematical Physical and Engineering Sciences*, 93, 212–25.

Ross, R. & Hudson, H. (1917b). An application of the theory of probabilities to the study of a priori pathometry. Part III. *Proceedings of the Royal Society of London Series a-Mathematical Physical and Engineering Sciences*, 93, 225–40.

Service, M. W. (1978). A short history of early medical entomology. *Journal of Medical Entomology*, 14, 603–26.

Service, M. W. (1997). Mosquito (Diptera: Culicidae) dispersal—the long and short of it. *Journal of Medical Entomology*, 34, 579–88.

Sharpe, F.R. & Lotka, A.J. (1923). Contribution to the analysis of malaria epidemiology IV Incubation lag. *American Journal of Hygiene*, 3, 96–112.

Silver, J.B. (2008). *Mosquito Ecology: Field Sampling Methods*. 3rd ed., Springer, New York.

Sinka, M.E., Bangs, M.J., Manguin, S. et al., (2012). A global map of dominant malaria vectors. *Parasites & Vectors*, 5, 69.

Smith, D.L. & McKenzie, F.E., (2004). Statics and dynamics of malaria infection in *Anopheles* mosquitoes. *Malaria Journal*, 3, 13.

Smith, D.L., Perkins, T.A., Reiner, R.C.J. et al. (2014). Recasting the theory of mosquito-borne pathogen transmission dynamics and control. *Transactions of the Royal Society of Tropical Medicine and Hygiene*, 108, 185–97.

Smith, D.L. Battle, K.E., Hay, S.I., Barker, C.M., Scott, T.W., & McKenzie, F.E. (2012). Ross, Macdonald, and a theory for the dynamics and control of mosquito-transmitted pathogens. *PLoS Pathogens*, 8, e1002588.

Smith, D.L., Dushoff, J., & McKenzie, F.E. (2004). The risk of a mosquito-borne infection in a heterogeneous environment. *PLoS Biology*, 2, e368.

Tatem, A.J. & Smith, D.L. (2010). International population movements and regional *Plasmodium falciparum* malaria elimination strategies. *Proceedings of the National Academy of Sciences of the United States of America*, 107, 12222–7.

Tatem, A.J. Jia, P., Ordanovich, D. et al. (2017). The geography of imported malaria to non-endemic countries: A meta-analysis of nationally reported statistics. *Lancet Infectious Diseases*, 17, 98–107.

Taylor, L.R. (1984). Assessing and interpreting the spatial distributions of insect populations. *Annual Reviews of Entomology*, 29, 321–57.

Waite, H., (1910). Mosquitoes and malaria. A study of the relation between the number of mosquitoes in a locality and the malaria rate. *Biometrika*, 7, 421–36.

WHO (1957). *Expert Committee on Malaria*. World Health Organization, Geneva, Switzerland.

Wiebe, A., Longbottom, J., Gleave, K. et al. (2017). Geographical distributions of African malaria vector sibling species and evidence for insecticide resistance. *Malaria Journal*, 16, 85.

Woolhouse, M.E.J. Dye, C., Etard, J.F. et al., (1997). Heterogeneities in the transmission of infectious agents: Implications for the design of control programs. *Proceedings of the National Academy of Sciences of the United States of America*, 94, 338–42.

CHAPTER 3

Seven Challenges for Spatial Analyses of Vector-Borne Diseases

T. Alex Perkins, Guido España, Sean M. Moore, Rachel J. Oidtman, Swarnali Sharma, Brajendra Singh, Amir S. Siraj, K. James Soda, Morgan Smith, Magdalene K. Walters, and Edwin Michael

3.1 Introduction

Spatial analyses have played a role in epidemiology since at least the mid nineteenth century. Although historical details of the famous example of John Snow's use of a map to discover the pump handle that was the source of a cholera outbreak in London have been debated (Brody et al. 2000), visualizing the locations of cholera cases on maps during outbreaks in the nineteenth century was done on numerous occasions and proved useful for public health planning (Gilbert 1958). Since then, these same basic ideas have been expanded upon to inform descriptions of epidemiologic patterns and to provide crucial insights for disease control.

The most basic form of spatial analysis involves placing data directly on maps, perhaps as individual points or aggregated across administrative units (Lawson 2006). This type of analysis is straightforward, and it may be sufficient for some situations, such as when a high proportion of cases are observed and the spatial signal of exposure is clear. In situations where it is suspected that there might be spatial gaps in data coverage, a common approach is to make predictions for locations where observations have not been made by performing spatial regressions across locations where observations have been made (Elliott et al. 2000; Lawson et al. 2016). This basic idea can be implemented with any number of statistical or machine learning approaches, including those that explicitly account

for correlations among nearby locations, those that consider relationships between environmental factors and the epidemiologic quantity of interest, or both (Diggle et al. 2003; Ostfeld et al. 2005; Waller & Carlin 2010).

Some of the most common quantities that are mapped in spatial analyses in epidemiology include incidence of disease over some time frame, prevalence of infection by age or some other group, and occurrence of the disease (Lawson et al. 2016). Often, the goal of these analyses is to produce maps that can guide surveillance or control efforts (Eisen & Lozano-Fuentes 2009; Eisen & Eisen 2011). They can also be overlaid with maps of population density to generate aggregated estimates of disease burden or population at risk (Hay et al. 2004). For approaches that include spatial descriptions of uncertainty, they can also be used to guide investments in future data collection that would have the greatest impact on reducing uncertainty (Kraemer et al. 2016).

Spatial approaches have been particularly popular for studying the epidemiology of vector-borne diseases, due in large part to their notable sensitivity to spatially variable environmental factors (Reisen 2010; Shocket et al. 2020). Temperature is one of the most widely appreciated environmental factors, as it affects numerous aspects of vector life history in addition to parasite development rates in the vector (Thomas & Blanford 2003; Shocket et al. 2020). Precipitation and satellite-derived wetness or vegetation indices are also commonly

T. Alex Perkins, Guido España, Sean M. Moore, Rachel J. Oidtman, Swarnali Sharma, Brajendra Singh, Amir S. Siraj, K. James Soda, Morgan Smith, Magdalene K. Walters, and Edwin Michael, *Seven Challenges for Spatial Analyses of Vector-Borne Diseases* In: *Population Biology of Vector-Borne Diseases*. Edited by: John M. Drake, Michael B. Bonsall, and Michael R. Strand: Oxford University Press (2021). © T. Alex Perkins, Guido España, Sean M. Moore, Rachel J. Oidtman, Swarnali Sharma, Brajendra Singh, Amir S. Siraj, K. James Soda, Morgan Smith, Magdalene K. Walters, and Edwin Michael.
DOI: 10.1093/oso/9780198853244.003.0003

incorporated, due to their association with breeding habitats for many vectors (Kalluri et al. 2007). In some cases, spatial analyses go so far as to incorporate hydrological data or models, depending on the ecology of the vector (Bomblies et al. 2008). For similar reasons, various classifications of land use can also be informative in many cases (Norris 2004; Kilpatrick 2011). Although sometimes more challenging to work with, human socioeconomic data can also be insightful, due to their potential association with vector contact (Béguin et al. 2011). In situations where it has been mapped, data on intervention usage can also be informative (Garske et al. 2014; Bhatt et al. 2015).

Although spatial analyses of vector-borne diseases often take statistical approaches similar to analyses of many other diseases, there is an altogether different approach that is mostly unique to vector-borne diseases. This approach, which we refer to as 'mechanistic', combines results from laboratory studies of environmental factors, usually temperature, with mathematical models to make spatial predictions of vector-borne disease transmission (Parham et al. 2015). Many studies in this category frame their predictions with a derivation of the basic reproduction number, R_0, from the Ross-Macdonald model of malaria transmission (Parham & Michael 2010; Moore et al. 2012; Liu-Helmersson et al. 2014; Mayo et al. 2016; Reiner & Smith 2020). A number of predictions can be derived from the value of R_0, but the one that receives the most attention in mapping relies on the use of R_0 as a threshold to predict whether transmission can be sustained in an area ($R_0 > 1$) (Smith & McKenzie 2004). Component parameters of R_0 that are commonly thought to depend on environmental factors include the ratio of vectors to humans, vector biting rate, vector mortality, the development rate of the parasite inside the vector, and the probability of transmission when an infectious vector bites a susceptible host, or vice versa (Mordecai et al. 2019; Reiner & Smith 2020).

Statistical and mechanistic approaches each have their distinct strengths and weaknesses (Parham et al. 2015). The former come with the assurance that they are based directly on empirical data, but they lack process-based explanations for the relationships between environmental factors and epidemiologic quantities that they seek to predict. The latter are well-grounded in process-based explanations and tend to be informed by laboratory data, but their predictions about epidemiologic quantities tend not to be rigorously validated on data from field studies or surveillance. For informing control, both make predictions about areas that might be good targets. Unless copious data on intervention usage are available, however, only the mechanistic approaches are capable of predicting the impact that a given control strategy might have.

In this chapter, we draw attention to two sets of challenges for these approaches to spatial analysis of vector-borne diseases. The first section presents four challenges that demonstrate why spatial patterns of disease may not always provide an accurate depiction of spatial patterns of transmission. In systems where these challenges manifest, failing to account for them could lead to biased, or simply incorrect, descriptions of spatial patterns of transmission potential. The second section presents three challenges that underscore the importance of appropriate data and modeling to inform spatial approaches to disease control. Failing to address these challenges could reduce the impact of control and put disease elimination targets out of reach. Together, these seven challenges are presented through a series of case studies drawn from a variety of vector-borne disease systems.

3.2 Challenges for Spatial Analyses of Epidemiologic Patterns

3.2.1 Age can Confound Estimates Based on Multiple, Uncoordinated Studies

At a global scale, malaria has the largest burden of any vector-borne disease (Breman 2001). To make international control of this disease effective, it is critical to understand the global distribution of *P. falciparum* and other malarial parasites (Hay & Snow 2006). High-resolution spatial estimates of parasite prevalence can be used to help allocate resources and track the progress of control efforts over time. The Malaria Atlas Project has been a model for how spatial epidemiological approaches may inform real-world policy and decision making. As a foundational piece of that effort, Hay et al. (2009) used data from 7953 parasite prevalence surveys to create an initial map of the fine-scale global distribution of *P. falciparum*. These prevalence surveys measured the *P. falciparum* parasite rate (PfPR); i.e. the proportion of the population carrying asexual blood-stage parasites. Identification of fine-scale spatial heterogeneity in malaria prevalence was only possible because such a large number of surveys were available; however, PfPR varies with age, and the majority of surveys did not report age-stratified parasite rates. Therefore, a critical step in synthesizing these thousands of studies into a single, unified output was to generate age-standardized PfPR estimates across all surveys, even those that reported an unstratified PfPR (Smith et al. 2007).

The PfPR is known to be a function of both age and local transmission intensity (Smith et al. 2012). As a function of age, PfPR increases during infancy and early childhood as individuals acquire their first infection, reaches a plateau in older children, and then declines in older adolescents and adults as immunity to *P. falciparum* develops (Baird et al. 1991; Trape et al. 1994; Gupta et al. 1999). The shape and timing of age-specific PfPR curves vary spatially, because this relationship depends on local transmission intensity as measured by the local entomological inoculation rate (EIR); i.e. the per capita number of bites by sporozoite-positive vectors (Smith et al. 2005; Reiner & Smith 2020). Smith et al. (2007) fitted several different statistical and mechanistic models to twenty-one age-stratified PfPR survey datasets and then tested the fitted models on two age ranges from an additional 121 studies to determine the best function to generate an age-standardized PfPR estimate from any survey where the underlying age distribution of the study population could be estimated. This resulted in a model of true prevalence, as well as the sensitivity of the *P. falciparum* microscopy test, which is assumed to decline with age as the build-up of blood-stage immunity over the course of repeated infections reduces parasite densities below the test's detection threshold (McKenzie et al. 2003). Based on these results, Hay et al. (2009) modeled the probability that an individual of a given age in a given study population was detected as *P. falciparum* positive as a function of the local epidemiological parameters, local detection-probability parameters, and the age distribution of study participants. This allowed for estimation of the PfPR for individuals between ages two and ten ($PfPR_{2-10}$) for each survey, which was then incorporated into a Bayesian geostatistical framework to generate a global map of $PfPR_{2-10}$ using a Gaussian process model (Hay et al. 2009).

These initial maps of $PfPR_{2-10}$ were used to describe the level of malaria endemicity across all regions at risk of malaria transmission, to identify the number of people living in areas falling within different endemicity categories, and to target areas where malaria control needed to be improved and where elimination was most feasible (Hay et al. 2009). Later, updated annual estimates of $PfPR_{2-10}$ from 2000 to 2015 showed that control efforts had reduced malaria prevalence by 50 percent and incidence of clinical disease by 40 percent in Africa over the last 15 years (Bhatt et al. 2015). Bhatt et al. (2015) coupled these annual $PfPR_{2-10}$ estimates with detailed reconstructions of changing intervention coverage to estimate the relative impacts of different interventions. The most widespread intervention, insecticide-treated nets, was estimated to be responsible for 68 percent (95 percent interval: 62–72 percent) of the reduction in clinical incidence, while the use of artemisinin-based combination therapy and indoor residual spraying were estimated to account for 19 percent (15–24 percent) and 13 percent (11–16 percent) of the reduction, respectively. Standardizing PfPR estimates across studies conducted in populations with a wide variety of age distributions was a critical innovation for synthesizing these disparate studies into unified outputs that have been highly influential to malaria elimination policy.

3.2.2 Population Immunity Complicates the Relationship Between Transmission Potential and Disease

Much like how age standardization of parasite prevalence data has been essential for obtaining spatially unified estimates of malaria transmission, accounting for population immunity is an important consideration for accurate estimation of spatial variation in the epidemiology of viral diseases vectored by *Aedes* mosquitoes, which confer strong, long-lasting immunity (Simmons et al. 2012; Leta et al. 2018). Although it is tempting to measure transmission with patterns of disease incidence, the force of infection (FoI)—i.e., the rate at which susceptible people become infected—is typically viewed as a more direct estimate of transmission, given that it reflects the per capita rate at which susceptible individuals become infected (Muench 1934; Grenfell & Anderson 1985). This makes FoI a more robust measure of transmission intensity among regions with significant population immunity. For instance, low-transmission areas could display higher incidence than high-transmission areas in certain situations, such as when conditions for transmission happen to be favorable in both areas at a given time but immunity is much higher in the historically high-transmission area. Scenarios like this are increasingly common as weather anomalies, such as El Niño (Lowe et al. 2017), and shifting climates alter the landscape of transmission over time. To estimate FoI, 'catalytic' models are often applied to serological data or age-specific incidence data (Hens et al. 2010). In the case of dengue, estimating FoI involves several challenges, such as variability of FoI over time and the existence of four dengue virus serotypes with complex interactions mediated by human immunity (Balmaseda et al. 2010; Morrison et al. 2010; Simmons et al. 2012). These immunological interactions can obscure the identification of infecting serotypes from serological studies, contributing bias to estimates of FoI from serological data (Rodriguez-Barraquer et al. 2011). Also, a large proportion of undetected infections can impact FoI estimates based on clinical records (King et al. 2008). To reconcile these issues, FoI can be estimated from longitudinal studies to validate serological results while deriving serotype-specific information (Reiner et al. 2014).

One recent study nicely illustrates the extent to which predictions of spatial variation in transmission intensity can vary when based on FoI versus incidence. As Zika, dengue, and chikungunya viruses are all vectored by the same mosquito (primarily *Aedes aegypti*), it is reasonable to expect that spatial patterns of the transmission of these three viruses should be largely similar. To assess the potential for prior knowledge of the epidemiology of dengue and chikungunya to inform projections of Zika's epidemiology in the Americas, Rodriguez-Barraquer et al. (2016) used historic information about dengue and chikungunya epidemics to identify regions at risk of Zika. The authors fitted catalytic models to age-specific dengue incidence data to estimate the FoI of previous dengue epidemics sub-nationally in Brazil and Colombia. For each region, the incidence of Zika was compared to incidence during the 2014–15 chikungunya epidemic and the average incidence of dengue over the previous five years, as well as to the FoI of dengue. From this analysis, the FoI of dengue showed a relatively high correlation with Zika incidence (R^2 ranged 0.36–0.41 across regions), whereas Zika incidence had a poor correlation with dengue incidence ($R^2 = 0.01$ or less across all regions) and chikungunya incidence ($R^2 = 0.01$ in Brazil and 0.09 in Colombia). The high correlation between Zika incidence and dengue FoI supports the hypothesis that FoI is a more direct measure of transmission intensity than incidence due to the fact that it is more robust to feedbacks from population immunity. One of the most important implications of this study for the Zika epidemic was that its methodology allowed for rapid identification of populations at risk of microcephaly in portions of Latin America where Zika had not yet spread, serving as a possible early warning of impending Zika risk (Asher et al. 2017).

Whereas the analysis by Rodriguez-Barraquer et al. (2016) focused on using FoI to predict the spatial distribution of an emerging disease, estimates of FoI can also be useful for understanding long-term changes in endemic diseases due to shifts in the demographic composition of populations. Cummings et al. (2009) performed pioneering work on this topic, showing that an increase in the average age of dengue incidence in Thailand could be explained by changes in the demographic structure of the population. As immune people live longer, they indirectly protect younger age groups by increasing herd immunity due to their increased numerical prominence compared to those same age groups in past years. Similarly, as birth rates decline,

---

> **Box 3.1 How can incidence be lower in an area with higher suitability for transmission?**
>
> Consider two patches i and j, wherein a pathogen is transmitted homogeneously in each according to a susceptible-infected-recovered (SIR) model of transmission. The suitability for transmission in these patches can be described by their basic reproduction numbers, $R_{0,i}$ and $R_{0,j}$. Given some initial fraction of a population susceptible to infection, $S(0)$, theory predicts that the fraction of the population infected over the course of an epidemic, $S(0) - S(\infty)$, can be found through an implicit solution of $S(\infty) = exp(ln(S(0)) + R(S(0) - S(\infty)))$ (Keeling & Rohani 2011). Let patch i have the higher value of $R_{0,i} = 3$, as compared with $R_{0,j} = 1.5$. The SIR dynamic of this disease means that epidemics will be transient and likely to fade out once herd immunity develops. One consequence of this is that the extent of population immunity in patch i or j may depend not only on $R_{0,i}$ and $R_{0,j}$ but also on how long ago a prior epidemic occurred. Over time, births and deaths will increase $S(0)$ for the next epidemic. For patches i and j, consider a scenario where past epidemics occurred at different times, such that $S_i(0) = 0.6$ and $S_j(0) = 0.2$. Combined with their values of R_0 and the aforementioned solution for epidemic size, this suggests that a new epidemic would infect 12.6 percent of the population in patch i and 25.1 percent of the population in patch j. This hypothetical example illustrates that a population with double the basic reproduction number of another can experience an epidemic half as large. While this may seem surprising, such a scenario may explain why Asia experienced very little Zika incidence over the same timeframe when hundreds of thousands of Zika cases were reported across the Americas.
>
> Source: Siraj & Perkins 2017

the susceptible population replenishes more slowly. With respect to Zika, Ferguson et al. (2016) proposed that similar effects are likely to dictate the timing and characteristics of Zika re-emergence in the future when births replenish the pool of susceptible hosts to the point that large-scale Zika virus transmission can once again be supported.

3.2.3 Herd Immunity Limits the Population at Risk of Immunizing Pathogens

Species distribution modeling approaches (Elith & Leathwick 2009) are being applied increasingly to map risk for a wide range of infectious diseases, including many endemic and emerging vector-borne diseases. These approaches have some advantages in the context of emerging diseases, such as their reliance on historical data about occurrence and their ability to assimilate spatially rich information about covariates of transmission. At the same time, there are limitations of these approaches in this context, such as their assumption that data about a species' range reflect equilibrium conditions, when in fact disease emergence events are associated with species whose ranges are definitively not at equilibrium (Kearney & Porter 2009).

Soon after the 2015–2016 Zika epidemic was declared a public health emergency of international concern by the World Health Organization, numerous models were published that projected the future spread of Zika and quantified the population at risk continentally and even globally. In more than one example (Messina et al. 2016; Bogoch et al. 2016; Carlson et al. 2016), species distribution modeling approaches were used to map either environmental suitability for Zika virus or some variant thereof. This quantity ranges from 0–1, with high suitability indicating that values of spatial covariates at a given location are similar to values of those covariates at locations of known Zika occurrence. Authors of those studies aggregated this spatially rich information into compact summaries of the population at risk by multiplying suitability by local population size and summing the resulting product across continents or the globe. One estimate (Messina et al. 2016) arrived at a total population at risk of 2.17 billion. Although there is a clear basis for this estimate, it is unclear how such a number should be interpreted or used to inform public health decision making.

Perkins et al. (2016) developed an approach that also leverages spatially rich covariates of transmission but puts spatial estimates of populations at risk in a clearer context. The first step in their application of this approach to Zika involved relating spatial covariates to the basic reproduction number, R_0, based on empirically estimated relationships between temperature and constituent parameters that appear in the R_0 formula (Chan & Johansson 2012; Brady et al. 2014). The second step in this process involved calibrating an unknown but important relationship between socioeconomic conditions (captured by the G-Econ index, (Nordhaus 2006)) and mosquito-human contact (captured in part by mosquito occurrence probabilities, (Kraemer et al. 2015)). This was done by completing a third step in which a classic relationship between R_0 and final epidemic size defined by the susceptible-infectious-recovered (SIR) model (Kermack & McKendrick 1927) was used to translate spatially granular estimates of R_0 into estimates of how many people would become infected during the Zika epidemic before it inevitably burned out due to the build-up of herd immunity.

The outputs of the approach by Perkins et al. (2016) resulted in a continent-wide projection that as many as 93 million people could become infected with Zika virus across the Americas during the first wave of the epidemic. Both this total and the spatial distribution of projected infections have since been found to be consistent with independent estimates that made use of data collected during the epidemic (Zhang et al. 2017; Moore et al. 2019). An advantage of the approach by Perkins et al. is that projections could be made prior to any data whatsoever being collected during the Zika epidemic. While this strength presents potential weaknesses by omitting crucial information collected during the epidemic, it possesses similar attributes as species distribution models but produces more specific and readily interpretable outputs. Along with two other models (Zhang et al. 2017; Rodriguez-Barraquer et al. 2019), these outputs were used to guide the selection of sites for Zika vaccine trials (Asher et al. 2017) and have the potential to play an important role in assessing future areas at risk of Zika virus epidemics as the epidemiology of this disease transitions into a new era (Perkins 2017; Siraj & Perkins 2017).

3.2.4 Surveillance Sensitivity is Spatially Heterogeneous and Can Distort Spatial Patterns of Disease

Spatial heterogeneity is apparent not only in patterns of disease incidence, but also in disease reporting rates, which can vary dramatically within and between regions. Following the 2015-2017 Zika epidemic in the Americas, Moore et al. (2019) estimated national-level reporting rates for suspected Zika cases that ranged from 1 percent in Peru to 93 percent in Puerto Rico. They also identified considerable heterogeneity in reporting rates within countries. Sub-nationally, Colombia had the greatest heterogeneity in reporting rates, with 70 percent of heterogeneity in reporting rates attributed to differences in reporting among Colombia's thirty-two departments (Moore et al. 2019). Not accounting for differences in reporting rates such as these could lead to biased estimates of spatial patterns of disease incidence.

Guarding against bias due to spatial heterogeneity in reporting requires understanding the underlying drivers of that heterogeneity. One explanation is that surveillance systems in some areas may be more sensitive. For example, consistently high dengue reporting rates in Singapore are thought to be attributable to its highly developed dengue surveillance system (Imai et al. 2016). In part, differences in the sensitivity of surveillance systems to a given disease may reflect whether there is a history of—and, relatedly, an awareness of—that disease in that area. Another explanation for spatial heterogeneity in reporting is spatial heterogeneity in geographic and socioeconomic factors, which are known to influence care seeking patterns of individuals affected by disease (Alegana et al. 2018). For example, in a malaria-endemic area in Ethiopia, children from poorer families were more likely to receive care at a later time than their wealthier counterparts and were also more likely to seek care from community health workers or try home treatments rather than utilizing public facilities or private clinics (Deressa et al. 2007).

To the extent that drivers of spatial heterogeneity in reporting rates can be understood, models have the potential to help resolve these observational challenges. In an analysis of leishmaniasis incidence from the Indian state of Bihar, Mubayi et al. (2010) found that a classification of districts as high- or low-risk was incorrect when based on reported data taken at face value. By estimating district-specific reporting rates in a dynamic transmission model, along with other parameters related to transmission, this study produced estimates of the true incidence of leishmaniasis, which differed both in magnitude and relative spatial pattern from reported data. Interestingly, estimates of district-specific reporting rates were found to be negatively associated with literacy and population density, with the latter hypothesized to result from the

increasing availability of private health facilities (which do not report to the public health surveillance system) in urban areas (Mubayi et al. 2010).

3.3 Challenges for Spatial Analyses to Inform Control

3.3.1 Spatial Coupling Means that Control in One Area Can Have Impacts Across Multiple Areas

With increasing urbanization and growth in transportation networks, it has become increasingly necessary to consider the role of human mobility in vector-borne disease epidemiology (Stoddard et al. 2009). As the spatial scales of vector movements are generally very limited compared to those of humans, acknowledgement of human mobility is essential for capturing spatially coupled dynamics (Perkins et al. 2013; Smith et al. 2014). Sub-nationally across Kenya, Wesolowski et al. (2012) demonstrated how human mobility leads to a source-sink dynamic for malaria, whereby 'source' regions (i.e. capable of sustained local transmission) enable malaria persistence in 'sink' regions (i.e. incapable of sustained local transmission). These source-sink dynamics arise from infectious individuals taking overnight trips for any number of purposes (Marshall et al. 2016; Marshall et al. 2018). An important consequence of this source-sink dynamic is that elimination efforts need to be coordinated across interconnected regions to be successful (Tatem & Smith 2010).

As human mobility can enable persistence in sink regions, monitoring parasite importation into sink areas is important for malaria elimination in these settings. In one example of how this can be accomplished, Ruktanonchai et al. (2016) used call data records from mobile phones in combination with malaria prevalence maps to characterize the malaria transmission landscape before and after malaria control campaigns in Namibia. This analysis showed that the transmission landscape shifted from being dominated by high incidence in two large source regions to a patchier source-sink transmission landscape with low-level incidence in many regions and high-level incidence in

---

Box 3.2 How can there be fewer reported cases in an area with higher transmission?

Consider two patches i and j, wherein a pathogen is transmitted homogeneously in each according to a susceptible-infected-susceptible (SIS) model of transmission. Let the rate of recovery from infection, γ, be the same in i and j, but the transmission coefficients, β_i and β_j, be different. For a vector-borne disease following SIS dynamics, transmission coefficients are equivalent to vectoral capacity, meaning that differences in any of several aspects of vector ecology between i and j could account for a difference between β_i and β_j. Provided that $\beta > \gamma$ infection prevalence in both populations would exhibit stable, non-zero equilibria of $\bar{I}_i = 1 - \beta_i / \gamma$ and $\bar{I}_j = 1 - \beta_j / \gamma$ respectively. Combined with the differential equations describing the model's dynamics, this implies that the rates of infection of susceptible individuals are $\beta_i - \gamma$ and $\beta_j - \gamma$. Hence, if $\beta_i > \beta_j$, then the rate at which individuals in patch i are infected is clearly greater than the rate in patch j. Even so, the probabilities that infections are reported to passive surveillance systems, ρ_i and ρ_j, play a key role in determining the rates at which cases are reported, $\rho_i (\beta_i - \gamma)$ and $\rho_j (\beta_j - \gamma)$, over time. In the event that $\rho_i / \rho_j > (\beta_j - \gamma) / (\beta_i - \gamma)$, more cases will be reported in patch j, despite a higher incidence of infection in patch i. Positive associations between vector-borne disease transmission and underreporting of disease—mediated by associations of each with socioeconomic factors—make this scenario all too plausible.

more isolated hotspots (Ruktanonchai et al. 2016). These inferences provide information about impacts of control that go beyond what can be gleaned from more simplistic analyses of epidemiological data. In a similar example, Le Menach et al. (2011) found that elimination of malaria from the island of Zanzibar should be achievable by implementing border screening and other measures that would reduce malaria importation from portions of mainland Africa where malaria is endemic.

While source-sink dynamics are clearest for malaria, due to its behavior of locally stable transmission around an endemic equilibrium, there is similar importance of host mobility for vector-borne diseases with locally unstable epidemic behavior, such as dengue. For example, Kraemer et al. (2018) found that differing assumptions about patterns of human

---

Box 3.3 From the perspective of a sink population, is it more important to control the source or the sink?

Consider the same scenario as in Box 3.2–i.e., SIS model dynamics in two patches i and j–but with the patches coupled by human mobility. This can be achieved by assuming that individuals from a given patch spend a fraction c of their time visiting the other patch. Because $S=1-I$, this yields a single equation governing infection dynamics in patch i,

$$\frac{dI_i}{dt} = ((1-c)\Lambda_i + c\Lambda_j)(1-I_i) - \gamma I_i,$$

where $\Lambda_i = \beta_i((1-c)I_i + cI_j)$ and $\Lambda_j = \beta_j(cI_i + (1-c)I_j)$. An analogous equation for patch j can be obtained by switching i and j subscripts in the equation for patch i. For purposes of illustration, consider a single set of parameters: $\beta_i = 0.1$, $\beta_j = 0.001$, $\gamma = 0.025$, $c = 0.05$. In the absence of coupling between i and j, it can be shown that $R_0 = \beta/\gamma$ in each patch, meaning that i is indeed a source ($R_{0,i} = 4$) and j is a sink ($R_{0,j} = 0.04$). Results of numerical solutions of this model with and without control after day 500 are shown in the figure.

Source: code for generating this figure is available at https://github.com/TAlexPerkins/PopBioVBD_Spatial.

$$\frac{dI_i}{dt} = ((1-c)\Lambda_i + c\Lambda_j)(1-I_i) - \gamma I_i,$$

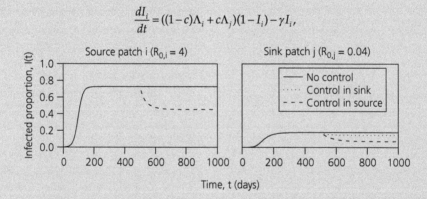

In the presence of spatial coupling and the absence of control, I_i and I_j approach equilibria of 0.73 and 0.18, respectively. When β is halved in the sink population (e.g., due to insecticide-treated nets), the equilibrium value of I_j is reduced somewhat (0.15) and I_i is nearly unaffected. When β is halved in the source population, the equilibrium value of I_i is reduced considerably more (0.07) and so is I_j (0.45). While surveillance and control in sink patches can be important for maintaining their status as sinks (Gerardin et al. 2017), these results illustrate the considerable benefits that can be obtained by reducing transmission in a source patch.

connectivity within the large urban area of Lahore, Pakistan led to vastly different inferences about spatial heterogeneity in transmission potential in different districts of the city. This is problematic given that those inferences could lead to different approaches for targeting resources for surveillance and control, underscoring the importance of high-quality data on human connectivity. Expanding out into the region surrounding Lahore, another analysis by Kraemer et al. (2015) found evidence of different degrees of mixing—which encompasses the frequency and fine-scale structure of vector-host contacts—within urban and rural areas. Again, these differences led to different inferences about transmission potential in these areas and had different implications for control. At a national scale in Pakistan, Wesolowski et al. (2015) found that call data records from mobile phones played a crucial role in enabling predictions of epidemic risk in different cities within the country.

3.3.2 Spatial Heterogeneity Means that Spatially Uniform Goals for Control May Not be Successful

There is an increasing appreciation for the need to consider spatial heterogeneity in decisions about vector-borne disease control. In part, this sentiment has developed in response to growing experience from the field, where health officials are beginning to confront significant spatial heterogeneity in the impacts of uniformly applied interventions (Liang et al. 2007; Gurarie & Seto 2009; Bousema et al. 2012). Particularly for macro-parasitic vector-borne diseases (e.g. lymphatic filariasis, onchocerciasis, and schistosomiasis), one of the most important aspects of control programs for which there may be need for a spatially heterogeneous strategy is the specification of infection thresholds that serve as benchmarks for decisions about investments in control (Michael et al. 2006; Duerr et al. 2005). Whether driving infection below a specified threshold results in parasite extinction or persistence depends on nonlinear dynamics of the system (Michael & Singh 2016; Michael et al. 2006; Gambhir & Michael 2008; Singh & Michael 2015). As such, mathematical models that can reliably capture local

transmission conditions have much to offer to the development of spatially heterogeneous control strategies in these systems (Michael & Singh 2016; Michael et al. 2017; Singh & Michael 2015; Gambhir et al. 2010).

One major difficulty in reliably capturing local transmission conditions in models is deciding how to incorporate spatiotemporal variables into a model's structure (Michael et al. 2017; Cushman & Huettmann 2010b; Spear et al. 2002) such that the system's behavior recapitulates epidemiological patterns across a range of control strategies (Cushman 2010; Cushman & Huettmann 2010a). Some of the specific challenges that these modeling efforts face include accurately portraying multidimensional initial conditions (e.g. descriptions of infection prevalence by age) and accounting for spatial structure and inhomogeneous mixing in transmission dynamics (e.g. some people live in closer proximity to breeding habitats) (Cushman 2010; Cushman & Huettmann 2010a; Beven & Alcock 2012). Confronting these challenges requires specifying a model's structure in a sufficient level of detail to capture these behaviors and fitting the model to location-specific data that are informative of these relationships (Beven & Alcock 2012; Beven 2002). Bayesian frameworks can be particularly useful for these problems, given their emphasis on using information contained within data to constrain uncertainty about model parameters and to flexibly describe covariation in posterior estimates of model parameters. Whenever possible, it is helpful for models to assimilate information from multiple data types, including not only epidemiological data but also data on vector infection rates and vector ecology (Michael & Singh 2016; Singh & Michael 2015; Gambhir et al. 2010; Singh et al. 2013). For macro-parasitic vector-borne diseases, such model fitting approaches have enabled estimation of infection thresholds applicable to disease eradication.

Infection thresholds can be thought about either in terms of thresholds of infection prevalence in humans or in relation to vector abundances required to sustain transmission (Michael et al. 2006; Duerr et al. 2005). Some critical findings about these thresholds for lymphatic filariasis elimination are that they can be as low as 0.0002 to 0.08 percent for the human infection indicator, micro-filaraemia (mf) prevalence (Michael & Singh 2016; Michael et al. 2017; Singh & Michael 2015), and that these values may decline in a power-law fashion with increasing vector abundance. As these values are considerably lower than the World Health Organization's recommendation of 1 percent mf prevalence (World Health Organization 2011), a control strategy seeking to drive mf prevalence below this 1 percent threshold would be insufficient to eliminate the parasite in many areas. Another important finding from models fitted to location-specific data was that thresholds can vary significantly across different vector species and, consequently, across different sites with different abundances of those vectors (Michael & Singh 2016; Michael et al. 2017; Singh & Michael 2015). In addition to quantifying locally appropriate thresholds, models have been used to show that timelines for elimination are expected to vary significantly across sites in a complex manner depending on baseline prevalence, vector abundance, and mf threshold (Michael & Singh 2016; Michael et al. 2017; Singh & Michael 2015).

These observations point to important consequences of using non-spatial approaches in a top-down paradigm for controlling complex spatial phenomena (Gambhir et al. 2010; Holling & Meffe 1996; Levin 1999). Instead, they indicate that new control paradigms using spatial heterogeneity as the basis for management are needed. One approach is to use models fitted to data from a location of interest as the basis for developing locally appropriate management strategies for that location (Beven & Alcock 2012). Another approach is to identify strategies that reduce between-site variability in the system's response to interventions. For example, supplementing mass drug administration programs with vector control has been shown in models to homogenize the spatial outcomes of intervention, resulting in a significant reduction in between-site variability in predicted elimination timelines (Michael & Singh 2016). Field testing of these approaches is now needed to validate them and to determine which approach might be a more cost-effective management strategy for accomplishing the control or elimination of macro-parasitic vector-borne diseases, and indeed other parasitic diseases, across spatially heterogeneous landscapes.

3.3.3 Hotspots of Transmission can be Spatially Hierarchical and Temporally Dynamic.

As is typical for vector-borne diseases, the risk of malaria is spatially heterogeneous. Bousema et al. (2012) proposed a conceptual outline of how heterogeneity influences transmission within foci—i.e. regions capable of sustained local transmission—and how this could influence control efforts. Within a focus, there can often be one or more locations where environmental, social, or behavioral conditions are more favorable for transmission and, as a result, where transmission is more intense than the average for the focus as a whole. Bousema et al. (2012) refer to these locations as 'hotspots'. Not surprisingly, malaria control may be more effective when control efforts target hotspots, a strategy that can have an impact not only in the hotspot but also in surrounding areas where transmission is fueled by infections originating in the hotspot. Targeted campaigns are, therefore, only feasible if hotspots can be identified in the first place, and if those hotspots persist long enough for interventions to have an impact on them.

To address these questions about the scale and permanence of hotspots, Bejon et al. (2014) performed an analysis of fine-scale patterns of multi-annual malaria incidence in Kilifi County, Kenya. In this study, the malaria positive fraction (MPF) of children 15 years old and younger that presented with fever was documented, and hotspots of MPF were identified using spatial scan statistics. Unexpectedly, hotspots were identified not only at one spatial scale, but in a hierarchy across multiple spatial scales. In other words, there was clear evidence of hotspots within hotspots, all the way down to the scale of an individual homestead. Further, because the region was monitored for a long period (nine years), this study was able to assess the temporal stability of these hotspots, finding that these hotspots exhibited turnover during the study. From a control perspective, these results show that the spatial and temporal scales at which hotspots should be sought are not obvious. Simulation results from this study suggested that targeting hotspots 4–8 km in diameter and aggregating data over a one-month period may have been the most effective scale for identifying hotspots in this setting

(Bousema et al. 2012). While that particular scale is not translatable outside that setting, the general conclusion that the scale at which hotspots occur is unique to a given setting is a result with clear implications for other settings.

While Bejon et al. (2014) showed that it may be possible to identify hotspots, campaigns that have targeted hotspots have yielded mixed results. Bousema et al. (2016) conducted a trial involving a campaign targeting five hotspots in Kenya. Eight weeks into this campaign, a greater reduction in parasite prevalence among residents of the targeted hotspots was evident. In the end though, the parasite was never completely eliminated in any targeted hotspot, and the initially encouraging difference between targeted and non-targeted hotspots dissipated with time. The reason that the targeted campaigns did not have a more lasting impact is unclear. One possibility is that control efforts did not impact the vector population or parasite reservoir to the degree required to push transmission beyond a threshold required to achieve elimination. Another possibility is that factors similar to those that drove the temporal instability of hotspots observed by Bejon et al. (2014) may have meant that only the hotspots at the time of the campaign were impacted, whereas those that had not yet occurred were more or less unaffected by the campaign. Along similar lines, it is possible that hotspots too small to detect provided a refuge for the parasite during the campaign, allowing it to be quickly reintroduced upon cessation of the campaign. Regardless of the cause, the spatially hierarchical and temporally dynamic nature of hotspots present a challenge to control efforts.

3.4 Conclusion

Performing an analysis that is naive about nuanced relationships between an epidemiologic quantity for which data are available (e.g. disease incidence) and a different epidemiologic quantity that is of primary interest (e.g. transmission potential) can result in inaccurate or incorrect inferences. Negative feedbacks from immunity–wherein a given amount of transmission results in less disease than would be expected in the absence of immunity–are one major source of these challenges. In this chapter, we have

highlighted how negative feedbacks from immunity can impact epidemiologic patterns across age groups (Smith et al. 2007), alter predictions of relative risk across different spatial areas (Rodriguez-Barraquer et al. 2019), and affect estimates of the magnitude of populations at risk across different areas (Perkins et al. 2016). Another important source of these challenges relates to differences in the sensitivities of surveillance systems across different areas. Surveillance sensitivity and vector-borne disease share some drivers in common, which offers hope for confronting these challenges, provided that those shared relationships with common drivers can be understood well enough (Mubayi et al. 2010).

This first set of challenges affects both statistical and mechanistic approaches to spatial analysis (Parham et al. 2015). Applying statistical approaches directly to available epidemiologic data is not inherently problematic. Rather, problems arise in the interpretation of the statistical model. One such problem relates to predictions based on these models. If these models fail to account for factors that gave rise to the patterns in the data to which they were fitted, then they are likely to perform poorly at making predictions outside that context. Another such problem relates to inferences from these models. For example, if understanding the relationship between temperature and transmission is of interest, a statistical model's estimated relationship between temperature and disease may bear little resemblance to the relationship of interest. In the case of mechanistic approaches, problems arise when validation of the model against spatial data occurs. The primary issue is that comparing the model's predictions to patterns based on available data may be a poor substitute for comparisons to the true patterns of interest.

In addition to flawed assessments of spatial epidemiologic patterns, the challenges we described may pose a risk of providing unhelpful guidance to control programs. Many spatial analyses, both statistical and mechanistic, do not account for spatial coupling, contributing further to problems with identifying spatial patterns of transmission potential based on spatial patterns of disease (Wesolowski et al. 2012). Even if spatial patterns of local transmission potential can be accurately quantified though, there must be political will for those predictions to be acted on; if not, spatially uniform policies will likely result in failure to meet public health objectives (Michael et al. 2017). In some cases though, acting on spatial risk assessments may be very difficult if the best spatial scale for targeting interventions is unknown or if the best spatial targets for control change over time (Bejon et al. 2014).

While addressing the challenges we have described is not easy, it is possible. To this end, we offer three recommendations. First, there is no substitute for high-quality data. Aspects of data that are important for addressing these challenges include the availability of multiple data types (e.g. one or more types of epidemiologic data, vector data, relevant social data) and sufficiently high spatial and temporal resolution to inform control at spatial and temporal scales that will maximize its impact. Second, spatial patterns must be analyzed with models that are structurally appropriate, meaning that they account for all factors that play a meaningful role in generating spatial patterns and that they account for interactions among those factors in a logical way. The case studies we have described highlight several examples of factors that may be appropriate for inclusion in spatial analyses, including negative feedbacks from immunity, surveillance sensitivity, and spatial coupling. Third, integrating causal inference methods with infectious disease mapping would represent an important development (Kraemer et al. 2019). In particular, such a development could enhance confidence in predicted outcomes of alternative control strategies, empowering decision makers with better information to guide control efforts.

References

Alegana, V.A., Maina, J., Ouma, P.O. et al. (2018). National and sub-national variation in patterns of febrile case management in sub-Saharan Africa. *Nature Communications*, 9, 4994.

Asher, J., Barker, C., Chen, G. et al. (2017). Preliminary results of models to predict areas in the Americas with increased likelihood of Zika virus transmission in 2017. *bioRxiv*, doi:10.1101/187591.

Baird, J.K., Jones, T.R., Danudirgo, E.W. et al. (1991). Age-dependent acquired protection against Plasmodium falciparum in people having two years exposure to hyperendemic malaria. *American Journal of Tropical Medicine and Hygiene*, 45, 65–76.

Balmaseda, A., Standish, K., Mercado, J.C. et al. (2010). Trends in patterns of dengue transmission over 4 years in a pediatric cohort study in Nicaragua. *Journal of Infectious Diseases*, 201, 5–14.

Béguin, A. Hales, S., Rocklov, J., Astrom, C., Louis, V.R., & Sauerborn, R. (2011). The opposing effects of climate change and socio-economic development on the global distribution of malaria. *Global Environmental Change: Human and Policy Dimensions*, 21, 1209–14.

Bejon, P., Williams, T.N., Nyundo, C. et al. (2014). A micro-epidemiological analysis of febrile malaria in Coastal Kenya showing hotspots within hotspots. *eLife*, 3, e02130.

Beven, K. (2002). Towards a coherent philosophy for modelling the environment. *Proceedings of the Royal Society of London A: Mathematical, Physical and Engineering Sciences*, 458, 2465–84.

Beven, K.J. & Alcock, R.E. (2012). Modelling everything everywhere: A new approach to decision-making for water management under uncertainty. *Freshwater Biology*, 57, 124–32.

Bhatt, S. Weiss, D.J., Cameron, E. et al. (2015). The effect of malaria control on Plasmodium falciparum in Africa between 2000 and 2015. *Nature*, 526, 207–11.

Bogoch, I.I. Brady, O.J., Kraemer, M.U.G. et al. (2016). Potential for Zika virus introduction and transmission in resource-limited countries in Africa and the Asia-Pacific region: A modelling study. *Lancet Infectious Diseases*, 16, 1237–45.

Bomblies, A., Duchemin, J.-B., & Eltahir, E.A.B. (2008). Hydrology of malaria: Model development and application to a Sahelian village. *Water Resources Research*, 44, 1157.

Bousema, T. Griffin, J.T., Sauerwein, R.W. et al. (2012). Hitting hotspots: Spatial targeting of malaria for control and elimination. *PLoS Medicine*, 9, e1001165.

Bousema, T. Stresman, G., Baidjoe, A.Y. et al. (2016). The impact of hotspot-targeted interventions on malaria transmission in Rachuonyo South District in the Western Kenyan Highlands: A cluster-randomized controlled trial. *PLoS Medicine*, 13, e1001993.

Brady, O.J. Golding, N., Pigott, D.M. et al. (2014). Global temperature constraints on *Aedes aegypti* and *Ae. Albopictus* persistence and competence for dengue virus transmission. *Parasites & Vectors*, 7, 338.

Breman, J.G. (2001). The ears of the hippopotamus: Manifestations, determinants, and estimates of the malaria burden. *American Journal of Tropical Medicine and Hygiene*, 64, 1–11.

Brody, H., Rip, M.R., Vinten-Johansen, P., Paneth, N., & Rachman, S. (2000). Map-making and myth-making in

Broad Street: The London cholera epidemic, 1854. *Lancet*, 356, 64–8.

Carlson, C.J., Dougherty, E.R., & Getz, W. (2016). An ecological assessment of the pandemic threat of Zika virus. *PLoS Neglected Tropical Diseases*, 10, e0004968.

Chan, M. & Johansson, M.A. (2012). The incubation periods of dengue viruses. *PLoS One*, 7, e50972.

Cummings, D.A.T., Iamsirithaworn, S., Lessler, J.T. et al. (2009). The impact of the demographic transition on dengue in Thailand: Insights from a statistical analysis and mathematical modeling. *PLoS Medicine*, e1000139.

Cushman, S.A. (2010). Space and time in ecology: Noise or fundamental driver? In S.A. Cushman & F. Huettmann, eds. *Spatial Complexity, Informatics, and Wildlife Conservation*, pp. 19–41. Springer Japan, Tokyo.

Cushman, S.A. & Huettmann, F. (2010a). Introduction: Ecological knowledge, theory and information in space and time [Chapter 1]. In: Cushman, Samuel A.; Huettmann, Falk, eds. *Spatial Complexity, Informatics, and Wildlife Conservation*.

Cushman, S.A. & Huettmann, F. eds. (2010b). *Spatial Complexity, Informatics, and Wildlife Conservation*. Springer, New York.

Deressa, W., Ali, A., & Berhane, Y. (2007). Household and socioeconomic factors associated with childhood febrile illnesses and treatment seeking behaviour in an area of epidemic malaria in rural Ethiopia. *Transactions of the Royal Society of Tropical Medicine and Hygiene*, 101, 939–47.

Diggle, P.J., Ribeiro, P.J., & Christensen, O.F. (2003). An introduction to model-based geostatistics. In J. Møller, ed. *Spatial Statistics and Computational Methods*. pp. 43–86. Springer New York, New York, NY.

Duerr, H.-P., Dietz, K., & Eichner, M. (2005). Determinants of the eradicability of filarial infections: A conceptual approach. *Trends in Parasitology*, 21, 88–96.

Eisen, L. & Eisen, R.J. (2011). Using geographic information systems and decision support systems for the prediction, prevention, and control of vector-borne diseases. *Annual Review of Entomology*, 56, 41–61.

Eisen, L. & Lozano-Fuentes, S. (2009). Use of mapping and spatial and space-time modeling approaches in operational control of Aedes aegypti and dengue. *PLoS Neglected Tropical Diseases*, 3, e411.

Elith, J. & Leathwick, J.R. (2009). Species distribution models: Ecological explanation and prediction across space and time. *Annual Review of Ecology, Evolution, and Systematics*, 40, 677–97.

Elliott, P. et al. (2000). *Spatial Epidemiology: Methods and Applications*, Oxford: Oxford University Press.

Ferguson, N.M., Cucunuba, Z.M., Dorigatti, I. et al. (2016). Countering the Zika epidemic in Latin America. *Science*, 353, 353–4.

Gambhir, M., Bockarie, M., Tisch, D. et al. (2010). Geographic and ecologic heterogeneity in elimination thresholds for the major vector-borne helminthic disease, lymphatic filariasis. *BMC Biology*, 8, 22.

Gambhir, M. & Michael, E. (2008). Complex ecological dynamics and eradicability of the vector borne macroparasitic disease, lymphatic filariasis. *PLoS One*, 3, e2874.

Garske, T., Van Kerkhove, M.D., Yactayo, S. et al. (2014). Yellow fever in Africa: Estimating the burden of disease and impact of mass vaccination from outbreak and serological data. *PLoS Medicine*, 11, e1001638.

Gerardin, J. Bever, C.A., Bridenbecker, D. et al. (2017). Effectiveness of reactive case detection for malaria elimination in three archetypical transmission settings: A modelling study. *Malaria Journal*, 16, 248.

Gilbert, E.W. (1958). Pioneer maps of health and disease in England. *The Geographical Journal*, 124, 172–83.

Grenfell, B.T. & Anderson, R.M. (1985). The estimation of age-related rates of infection from case notifications and serological data. *Journal of Hygiene*, 95, 419–36.

Gupta, S., Snow, R.W., Donnelly, C.A., Marsh, K., & Newbold, C. (1999). Immunity to non-cerebral severe malaria is acquired after one or two infections. *Nature Medicine*, 5, 340–3.

Gurarie, D. & Seto, E.Y.W. (2009). Connectivity sustains disease transmission in environments with low potential for endemicity: Modelling schistosomiasis with hydrologic and social connectivities. *Journal of the Royal Society Interface*, 6, 495–508.

Hay, S.I., Guerra, C.A., Gething, P.W. et al. (2009). A world malaria map: Plasmodium falciparum endemicity in 2007. *PLoS Medicine*, 6(3), e1000048.

Hay, S.I. et al. (2004). The global distribution and population at risk of malaria: Past, present, and future. *Lancet Infectious Diseases*, 4, 327–36.

Hay, S.I. & Snow, R.W. (2006). The Malaria Atlas Project: Developing global maps of malaria risk. *PLoS Medicine*, 3, e473.

Hens, N., Aerts, M., Faes, C. et al. (2010). Seventy-five years of estimating the force of infection from current status data. *Epidemiology and Infection*, 138, 802–12.

Holling, C.S. & Meffe, G.K. (1996). Command and control and the pathology of natural resource management. *Conservation Biology*, 10, 328–37.

Imai, N., Dorigatti, I., Cauchemez, S., & Ferguson, N.M. (2016). Estimating dengue transmission intensity from case-notification data from multiple countries. *PLoS Neglected Tropical Diseases*, 10, e0004833.

Kalluri, S., Gilruth, P., Rogers, D., & Szczur, M. (2007). Surveillance of arthropod vector-borne infectious diseases using remote sensing techniques: A review. *PLoS Pathogens*, 3, 1361–71.

Kearney, M. & Porter, W. (2009). Mechanistic niche modelling: Combining physiological and spatial data to predict species ranges. *Ecology Letters*, 12, 334–50.

Keeling, M.J. & Rohani, P. (2011). *Modeling Infectious Diseases in Humans and Animals*, Princeton University Press, Princeton.

Kermack, W.O. & McKendrick, A.G. (1927). A contribution to the mathematical theory of epidemics. *Proceedings of the Royal Society A, Mathematical and Physical Sciences*, 115, 700–21.

Kilpatrick, A.M. (2011). Globalization, land use, and the invasion of West Nile virus. *Science*, 334, 323–7.

King, A.A. Ionides, E.L., Pascual, M., & Bouma, M.J. (2008). Inapparent infections and cholera dynamics. *Nature*, 454, 877–80.

Kraemer, M.U.G., Perkins, T.A., Cummings, D.A.T. et al. (2015). Big city, small world: Density, contact rates, and transmission of dengue across Pakistan. *Journal of the Royal Society Interface*, 12, 20150468.

Kraemer, M.U.G., Bisanzio, D., Reiner, R.C. et al. (2018). Inferences about spatiotemporal variation in dengue virus transmission are sensitive to assumptions about human mobility: A case study using geolocated tweets from Lahore, Pakistan. *EPJ Data Science*, 7, 16.

Kraemer, M.U.G., Hay, S.I., Pigott, D.M., Smith, D.L., Wint, G.R.W., & Golding, N. (2016). Progress and challenges in infectious disease cartography. *Trends in Parasitology*, 32, 19–29.

Kraemer, M.U.G., Reiner, R.C., & Bhatt, S. (2019). Causal inference in spatial mapping. *Trends in Parasitology* 35, 743–746.

Lawson, A.B. et al. (2016). *Handbook of Spatial Epidemiology*, CRC Press/Taylor & Francis, New York.

Lawson, A.B. (2006). *Statistical Methods in Spatial Epidemiology*, Wiley, Chichester, UK.

Le Menach, A., Tatem, A.J., Cohen, J.M. et al. (2011). Travel risk, malaria importation and malaria transmission in Zanzibar. *Scientific Reports*, 1, 93.

Leta, S., Beyene, T.J., De Clercq, E.M., Amenu, K., Kraemer, M.U.G., & Revie, C.W. (2018). Global risk mapping for major diseases transmitted by *Aedes aegypti* and *Aedes albopictus*. *International Journal of Infectious Diseases*, 67, 25–35.

Levin, S. (1999). Towards a science of ecological management. *Conservation Ecology*, 3, 6.

Liang, S., Seto, E.Y.W., Remais, J.V. et al. (2007). Environmental effects on parasitic disease transmission exemplified by schistosomiasis in western China. *Proceedings of the National Academy of Sciences of the United States of America*, 104, 7110–15.

Liu-Helmersson, J., Stenlund, H., Wilder-Smith, A., & Rocklov, J. (2014). Vectorial capacity of *Aedes aegypti*: Effects of temperature and implications for global dengue epidemic potential. *PLoS One*, 9, e89783.

Lowe, R., Stewart-Ibarra, A.M., Petrova, D. et al. (2017). Climate services for health: Predicting the evolution of the 2016 dengue season in Machala, Ecuador. *Lancet Planetary Health*, 1, e142–51.

Marshall, J.M. Toure, M., Ouedraogo, A.L. et al. (2016). Key traveller groups of relevance to spatial malaria transmission: A survey of movement patterns in four sub-Saharan African countries. *Malaria Journal*, 15, 200.

Marshall, J.M. Wu, S.L., Sanchez, H.M. et al. (2018). Mathematical models of human mobility of relevance to malaria transmission in Africa. *Scientific Reports*, 8, 7713.

Mayo, C., Shelley, C., MacLachlan, N.J., Gardner, I., Hartley, D., & Barker, C. (2016). A deterministic model to quantify risk and guide mitigation strategies to reduce bluetongue virus transmission in California dairy cattle. *PLoS One*, 11, e0165806.

McKenzie, F.E., Sirichaisinthop, J., Miller, R.S., Gasser, R.A., & Wongsrichanalai, C. (2003). Dependence of malaria detection and species diagnosis by microscopy on parasite density. *American Journal of Tropical Medicine and Hygiene*, 69, 372–6.

Messina, J.P., Kraemer, M.U.G., Brady, O.J. et al. (2016). Mapping global environmental suitability for Zika virus. *eLife*, 5, e15272.

Michael, E., Singh, B.K., Mayala, B.K., Smith, M.E., Hampton, S., & Nabrzyski, J. (2017). Continental-scale, data-driven predictive assessment of eliminating the vector-borne disease, lymphatic filariasis, in sub-Saharan Africa by 2020. *BMC Medicine*, 15, 176.

Michael, E., Malecela-Lazaro, M.N., Kabali, C., Snow, L.C., & Kazura, J.W. (2006). Mathematical models and lymphatic filariasis control: Endpoints and optimal interventions. *Trends in Parasitology*, 22, 226–33.

Michael, E. & Singh, B.K. (2016). Heterogeneous dynamics, robustness/fragility trade-offs, and the eradication of the macroparasitic disease, lymphatic filariasis. *BMC Medicine*, 14, 14.

Moore, S.M., Shrestha, S., Tomlinson, K.W., & Vuong, H. (2012). Predicting the effect of climate change on African trypanosomiasis: Integrating epidemiology with parasite and vector biology. *Journal of the Royal Society Interface*, 9, 817–30.

Moore, S.M., Oidtman, R.J., Soda, J. et al. (2019). Leveraging multiple data types to estimate the true size of the Zika epidemic in the Americas. *medRxiv*, 19002865.

Mordecai, E.A., Caldwell, J.M., Grossman, M.K. et al. (2019). Thermal biology of mosquito-borne disease. *Ecology Letters* 22, 1690–1708.

Morrison, A.C., Minnick, S.L., Rocha, C. et al. (2010). Epidemiology of dengue virus in Iquitos, Peru 1999 to 2005: Interepidemic and epidemic patterns of transmission. *PLoS Neglected Tropical Diseases*, 4, e670.

Mubayi, A., Castillo-Chavez, C., Chowell, G. et al. (2010). Transmission dynamics and underreporting of Kala-azar in the Indian state of Bihar. *Journal of Theoretical Biology*, 262, 177–85.

Muench, H. (1934). Derivation of rates from summation data by the catalytic curve. *Journal of the American Statistical Association*, 29, 25–38.

Nordhaus, W.D. (2006). Geography and macroeconomics: New data and new findings. *Proceedings of the National Academy of Sciences of the United States of America*, 103, 3510–17.

Norris, D.E. (2004). Mosquito-borne diseases as a consequence of land use change. *EcoHealth*, 1, 19–24.

World Health Organization. (2011). Monitoring and epidemiological assessment of mass drug administration in the global programme to eliminate lymphatic filariasis: A manual for national elimination programmes.

Ostfeld, R.S., Glass, G.E., & Keesing, F. (2005). Spatial epidemiology: An emerging (or re-emerging) discipline. *Trends in Ecology & Evolution*, 20, 328–36.

Parham, P.E., Waldock, J., Christophedes, G.K. et al. (2015). Climate, environmental and socio-economic change: Weighing up the balance in vector-borne disease transmission. *Philosophical Transactions of the Royal Society of London B*, 370, 20130551.

Parham, P.E. & Michael, E. (2010). Modeling the effects of weather and climate change on malaria transmission. *Environmental Health Perspectives*, 118, 620–6.

Perkins, T.A., Scott, T.W., Le Menach, A., & Smith, D.L. (2013). Heterogeneity, mixing, and the spatial scales of mosquito-borne pathogen transmission. *PLoS Computational Biology*, 9, e1003327.

Perkins, T.A., Siraj, A.S., Ruktanonchai, C.W., Kraemer, M.U.G., & Tatem, A.J. (2016). Model-based projections of Zika virus infections in childbearing women in the Americas. *Nature Microbiology*, 1, 16126.

Perkins, T.A. (2017). Retracing Zika's footsteps across the Americas with computational modeling. *Proceedings of the National Academy of Sciences of the United States of America*, 114, 5558–60.

Reiner, R.C., Stoddard, S.T., Forshey, B.M. et al. (2014). Time-varying, serotype-specific force of infection of dengue virus. *Proceedings of the National Academy of Sciences of the United States of America*, 111, E2694–702.

Reiner, R.C. & D.L. Smith. (2020). Heterogeneity, stochasticity and complexity in the dynamics and control of mosquito-borne pathogens. In J.M. Drake, M.R. Strand, and M. Bonsall, eds. *Population Biology of Vector-borne Diseases*. Oxford University Press, Oxford.

Reisen, W.K. (2010). Landscape epidemiology of vector-borne diseases. *Annual Review of Entomology*, 55, 461–83.

Rodriguez-Barraquer, I., Cordeiro, M.T., Braga, C., de Souza, W.V., Marques, E.T., & Cummings, D.A.T. (2011). From re-emergence to hyperendemicity: The natural history of the dengue epidemic in Brazil. *PLoS Neglected Tropical Diseases*, 5, e935.

Rodriguez-Barraquer, I., Salje, H., Lessler, J., & Cummings, D.A.T. (2016). Predicting intensities of Zika infection and microcephaly using transmission intensities of other arboviruses. *bioRxiv*, 041095.

Rodriguez-Barraquer, I., Salje, H., & Cummings, D.A.T. (2019). Opportunities for improved surveillance and control of dengue from age-specific case data. *eLife*, 8, e45474.

Ruktanonchai, N.W., DeLeenheer, P., Tatem, A.J. et al. (2016). Identifying malaria transmission foci for elimination using human mobility data. *PLoS Computational Biology*, 12, e1004846.

Shocket, M. et al. (2020). Environmental drivers of vector-borne diseases. In *Population Biology of Vector-borne Diseases*. J.M. Drake, M.R. Strand, and M. Bonsall, eds. Oxford University Press, Oxford.

Simmons, C.P. Farrar, J.J., van Vinh Chau, N., & Wills, B. (2012). Dengue. *New England Journal of Medicine*, 366, 1423–32.

Singh, B.K. Bockarie, M.J., Gambhir, M. et al. (2013). Sequential modelling of the effects of mass drug treatments on anopheline-mediated lymphatic filariasis infection in Papua New Guinea. *PLoS One*, 8, e67004.

Singh, B.K. & Michael, E. (2015). Bayesian calibration of simulation models for supporting management of the elimination of the macroparasitic disease, lymphatic filariasis. *Parasites & Vectors*, 8, 522.

Siraj, A.S. & Perkins, T.A. (2017). Assessing the population at risk of Zika virus in Asia: Is the emergency really over? *BMJ Global Health*, 2, e000309.

Smith, D.L., Perkins, T.A., Reiner, R.C.J. et al. (2014). Recasting the theory of mosquito-borne pathogen transmission dynamics and control. *Transactions of the Royal Society of Tropical Medicine and Hygiene*, 108, 185–97.

Smith, D.L., Battle, K.E., Hay, S.I., Barker, C.M., Scott, T.W., & McKenzie, F.E. (2012). Ross, Macdonald, and a theory for the dynamics and control of mosquito-transmitted pathogens. *PLoS Pathogens*, 8, e1002588.

Smith, D.L., Guerra, C.A., Snow, R.W., & Hay, S.I. (2007). Standardizing estimates of the *Plasmodium falciparum* parasite rate. *Malaria Journal*, 6, 131.

Smith, D.L., Dushoff, J., Snow, R.W., & Hay, S.I. (2005). The entomological inoculation rate and *Plasmodium falciparum* infection in African children. *Nature*, 438, 492–5.

Smith, D.L. & McKenzie, F.E. (2004). Statics and dynamics of malaria infection in Anopheles mosquitoes. *Malaria Journal*, 3, 13.

Spear, R.C., Hubbard, A., Liang, S., & Seto, E. (2002). Disease transmission models for public health decision making: toward an approach for designing intervention strategies for Schistosomiasis japonica. *Environmental Health Perspectives*, 110, 907–15.

Stoddard, S.T., Morrison, A.C., Vazquez-Prokopec, G.M. et al. (2009). The role of human movement in the transmission of vector-borne pathogens. *PLoS Neglected Tropical Diseases*, 3, e481.

Tatem, A.J. & Smith, D.L. (2010). International population movements and regional *Plasmodium falciparum* malaria elimination strategies. *Proceedings of the National Academy of Sciences of the United States of America*, 107, 12222–7.

Thomas, M.B. & Blanford, S. (2003). Thermal biology in insect-parasite interactions. *Trends in Ecology & Evolution*, 18, 344–50.

Trape, J.F., Rogier, C., Konate, L. et al. (1994). The Dielmo project: a longitudinal study of natural malaria infection and the mechanisms of protective immunity in a community living in a holoendemic area of Senegal. *American Journal of Tropical Medicine and Hygiene*, 51, 123–37.

Waller, L.A. & Carlin, B.P. (2010). *Disease Mapping*. London: Chapman & Hall/CRC, 2010, pp. 217–43.

Wesolowski, A., Qureshi, T., Boni, M.F. et al. (2015). Impact of human mobility on the emergence of dengue epidemics in Pakistan. *Proceedings of the National Academy of Sciences of the United States of America*, 112, 11887–92.

Wesolowski, A., Eagle, N., Tatem, A.J. et al. (2012). Quantifying the impact of human mobility on malaria. *Science*, 338, 267–70.

Zhang, Q., Sun, K., Chinazzi, M. et al. (2017). Spread of Zika virus in the Americas. *Proceedings of the National Academy of Sciences of the United States of America*, 114, E4334–43.

CHAPTER 4

Infectious Disease Forecasting for Public Health

Stephen A. Lauer, Alexandria C. Brown, and Nicholas G. Reich

… diviners employ art, who, having learned the known by observation, seek the unknown by deduction.

Cicero (44 BCE, as quoted in (McCloskey 1992))

We may regard the present state of the universe as the effect of its past and the cause of its future. An intellect which at a certain moment would know all forces that set nature in motion, and all positions of all items of which nature is composed, if this intellect were also vast enough to submit these data to analysis, it would embrace in a single formula the movements of the greatest bodies of the universe and those of the tiniest atom; for such an intellect nothing would be uncertain and the future just like the past would be present before its eyes.

LaPlace (1825, as quoted in (Silver 2012))

4.1 Background

4.1.1 A Brief History of Forecasting

The ability to foretell or divine future events for millennia has been seen as a valued skill. While there are records of Babylonians attempting to predict weather patterns as early as 4000 BCE based on climatological observations (Milham 1918), early attempts at divination were just as likely to be driven by unscientific observation. However, in the last 150 years, rapid technological advancements have made data-driven forecasting a reality across a number of scientific and mathematical fields.

The science of forecasting was pushed forward especially in the second half of the 20th century by the fields of meteorology and economics, but more recently other fields have started to build on this research. Examples include world population projections (Gerland et al. 2014; Raftery et al., 2012), political elections, (Campbell 1996; Graefe 2015; Lewis-Beck & Stegmaier 2014), seismology (Field et al. 2009; Bray & Schoenberg 2013; Chambers et al. 2012), as well as infectious disease epidemiology (Biggerstaff et al. 2016; Lowe et al. 2014; Viboud et al. 2017; Held et al. 2017; Reich et al. 2016).

Forecasting has been an active and growing area of research for over a century, (Fig. 4.1), with particular acceleration observed since 1980. While research focused on forecasting infectious diseases started in earnest in the 1990s, since 2005 the number of articles on infectious disease forecasting has increased seven-fold, at a faster pace than research on general forecasting during that time, which increased by a factor of three. In 1991, forecasting was the topic of one of every thousand published academic papers, based on counts from the Science Citation Index and the Social Science Citation Index, obtained from the Web of Science. In 2017, over four of every thousand indexed publications were about forecasting.

Stephen A. Lauer, Alexandria C. Brown, and Nicholas G. Reich, *Infectious Disease Forecasting for Public Health* In: *Population Biology of Vector-Borne Diseases.*
Edited by: John M. Drake, Michael B. Bonsall, and Michael R. Strand: Oxford University Press (2021). © Stephen A. Lauer, Alexandria C. Brown, and Nicholas G. Reich.
DOI: 10.1093/oso/9780198853244.003.003.0004

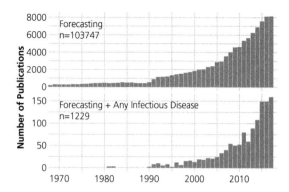

Figure 4.1 Publication trends from 1970 through 2016
The y-axis in each panel shows the number of publications, according to the Web of Science, for papers with the topic of (A) 'forecast*', and (B) 'forecast*' + any of a list of infectious diseases taken from WHO (World Health Organization, n.d.). There are 1,989 and 0 publications, respectively, that were published prior to 1970.

Source: All counts are taken from the Science Citation Index and the Social Science Citation Index, obtained via the Web of Science database.

4.1.2 What is a Forecast?

In common parlance, there is not a strong distinction between the terms 'prediction' and 'forecast'. Nor does there exist a strong consensus in the biomedical, ecological, or public health literature on the distinction. Nate Silver has suggested that etymologically, the term forecast 'implied planning under conditions of uncertainty' in contrast to prediction, which was a more ancient idea associated with superstition and divination (Silver 2012). In the modern scientific world, some fields, such as seismology, use the term forecast to refer to a probabilistic statement in contrast to a prediction which is a 'definitive and specific' statement about a future event. In other fields, the difference in meaning is even less clearly defined, with forecasting often connoting the prediction of a future value or feature of a time-series (Diebold 2001). We note that conventionally in biomedical research the term 'prediction' refers to an individual-level clinical outcome (e.g. risk-scores give individualized predictions of heart attack or stroke risk), but both 'prediction' and 'forecast' are used interchangeably to refer to events or outcomes that may impact more than one person. The term 'forecast' often refers broadly to a quantitative estimate about a future trend or event observable or experienced by many individuals as opposed to than a local, individualized outcome. Our use of the term forecasting takes aspects from several of these definitions. Specifically, we define a forecast as *a quantitative statement about an event, outcome, or trend that has not yet been observed, conditional on data that has been observed.*

Probabilistic forecasts represent an important class of forecasts. They are forecasts that have a statement about the uncertainty surrounding an event, outcome, or trend, and are not just providing a single 'best guess' estimate about the event. For example, a 'point' forecast for the number of dengue hemorrhagic fever cases in Singpore in the fourteenth calendar week of 2024 could be simply the number 25. However, a probabilistic forecast for the same target might say that while 25 cases is the most likely specific outcome, the probability of seeing exactly 25 cases is only 0.10 or 10 percent.

Note that a forecasted event or outcome need not necessarily be in the future, as events in the past that have not yet been quantitatively measured may also be forecasted. For example, on 1 May we may have data for a particular time-series available through 1 April. We could make a 'forecast' of the time-series for 15 April even though this event is in the past. This type of forecast has been referred to as a 'nowcast', (Brooks et al. 2018; Reich et al. 2019) although more generally, this is a special case of a forecast.

4.1.3 Forecasting Challenges that are Specific to Infectious Disease

There are operational and statistical challenges in forecasting that are specific to the setting of infectious disease. These challenges in and of themselves may not be unique to the field, but taken together, they describe obstacles that forecasters face when taking on a problem in infectious disease.

4.1.3.1 Challenge 1: System Complexity

When attempting to forecast the transmission of an infectious disease, and in particular a vector-borne disease, researchers need to account for processes on micro and macro scales. Behaviors of and interactions between viruses, vectors, hosts, and the environment each play a part in determining the transmission of a disease. For example, it has been hypothesized that rapid changes in climate may lead to unforeseen shifts in vector population dynamics, highlighting the fragility of models that attempt to generate forecasts based on existing knowledge about these complex systems (Mandal et al. 2011; Jewell & Brown 2015; Ostfeld & Brunner 2015).

Researchers have developed mechanistic models based on biological and behavioral principles that encode the processes by which diseases spread (Section 4.2.1.1). For vector-borne diseases, these models often include complex dynamics (see, e.g. (Mandal et al. 2011 and Reiner et al. 2013). That said, vector dynamics are often omitted from models of vector-borne disease for simplicity and tractability (Reiner et al. 2013). While numerous mechanistic models have been fit to data to provide inference about disease transmission parameters for vector-borne diseases (Reiner et al. 2013; Lourenço & Recker 2014), far fewer studies have examined prospective forecasting performance of such models (Jewell & Brown 2015).

The need for models to mirror the ecological complexity of the system stands in conflict with a fundamental principle of forecasting which is to use as simple a model as necessary. Providing a good 'fit' to in-sample data will not guarantee that such a model will generate accurate or even reasonable forecasts. Due to the dearth of high-resolution data on vector populations and host infections, understanding whether and how such detailed biological data can improve the accuracy of forecasts of population-level transmission largely remains to be seen (Section 4.2.3).

4.1.3.2 Challenge 2: Data Sparcity

A central challenge in forecasting vector-borne disease, and infectious diseases more generally, is how to balance the complexity of the biological and social models used with the coarseness of available data (Moran et al. 2016). For forecasting weather, scientists rely on tens of thousands of sensors across the world collecting continuous real-time data. There are no analogues to these rich and highly accurate data streams for infectious disease researchers. The gold-standard data in epidemiological surveillance arise from systems that typically capture only a fraction of all cases and often are reported with substantial delays and/or revisions to existing data. Technology shows some promise to provide more richly detailed data in a timely fashion about humans, climate, and vectors alike (George et al. 2019). However, new methods and data streams will be required to develop and implement increasingly refined forecasting models for vector-borne diseases.

4.1.3.3 Challenge 3: The Forcasting Feedback Loop

Forecasts of disease incidence can encourage governments and public health organizations to intervene to slow transmission. If forecasts of infectious disease are used to inform targeted interventions or risk communication strategies and the interventions change the course of the epidemic, then the forecast itself becomes enmeshed in the causal pathway of an outbreak. This feedback loop has been identified as the single most important challenge separating infectious disease forecasting from forecasting natural phenomena such as weather (Moran et al. 2016).

In settings where forecasts will be used to inform interventions, this feedback loop of infectious disease forecasting should be taken into account in the forecasts. Without such accounting, if a forecast predicts an outbreak and triggers an intervention that prevents the epidemic from occurring, then the forecast itself would be seen as wrong, despite this being a public health victory. This implies that forecasting models should, when in these settings, create multiple forecasts under different intervention scenarios. Mechanistic forecasting models, that use explicit disease transmission parameters, may be best suited for these types of forecasts, since intervention effects could be incorporated directly as impacting these parameters. However, any forecasts from such a model should be subjected to intense scrutiny, since it will necessarily be based on very strong assumptions about the intervention and transmission patterns.

Methodological development is needed in this area to address open scientific questions. What model frameworks can best balance forecast accuracy with the ability to incorporate multiple potential future scenarios? Can forecast models be used to assess intervention effectiveness?

4.1.4 Definitions and Basic Notation

Here, we introduce some basic mathematical notation for time-series forecasting that we use throughout this chapter. In many forecasting applications, the available data are often a time series of observed values for a particular location or setting. For infectious disease applications, these data are often a measure of incidence, such as case counts or the percentage of all doctor visits with primary complaint about a particular disease. In the text that follows, we use language specific to that of spatio-temporal disease incidence data, although much of what we describe can be applied more generally as well.

4.1.4.1 Data

We start with a simple example and later extend the notation to more realistic scenarios. In our example, we have a complete (i.e. no missing data) time series of infectious disease case counts from a single location, such as a school or hospital. We define y_t as an observed value of this incidence in time interval t from our time series $\{y_1, y_2, y_3, \ldots y_t, \ldots, y_T\}$. We assume that these observations are draws from random variables $Y_1, Y_2, Y_3, \ldots, Y_t, \ldots, Y_T$, where the probability distribution of Y_{t+1} may be dependent on t, prior values of y represented as $y_{1:t}$, and a matrix of other covariates $\mathbf{x}_t$. (Often, the analyst may wish to include multiple different lagged values of a single covariate. In this notation, for simplicity, these would all be considered to be part of $\mathbf{x}$.) We use T throughout to refer to the total number of time points in the time-series and t to refer to a specific time point relative to which a forecast is generated.

Two important features of our observed data are frequency and scale. In our example, incidence is recorded at regular time intervals. Furthermore, many infectious disease time series have a cyclical element. We define the frequency of a time series as the number of observations within a single cycle. For example, if we have monthly incidence data and know that there are annual weather patterns that influence incidence in our observed data, the frequency of our time series would be 12 months or observations.

4.1.4.2 Targets

Targets are the as-yet-unknown features of the data that are the subject of forecasting. In our example, we may want to forecast incidence at a certain future time—but targets can be a variety of endpoints extrapolated from the observed data. For forecasts of the time-series values itself, i.e. when a target is defined to be a past or future value of the time-series Y_{t+k}, we use a special nomenclature, referring to them as 'k-step-ahead' forecasts. In general, we define $Z_{i|t}$ as a random variable for target i positioned relative to time t. For example, in the infectious disease context, $Z_{i|t}$ could be:

- incidence at time t, or Y_t,
- incidence at time $t + k$ either in the future or past relative to time t, or Y_{t+k}, where k is a positive or negative integer,
- peak incidence within some period of time or season, or $\max_{t \in \mathcal{S}}(Y_t)$ where t are defined to be within season $\mathcal{S}$,
- the time(s) at which a peak occurs within some season, or $\{t' \in \mathcal{S} : Y_{t'} = \max_{t \in \mathcal{S}}(Y_t)\}$
- a binary indicator of whether incidence at time $t + k$ is above a specified threshold C, or $\mathbb{1}\{Y_{t+k} > C\}$.

4.1.4.3 Forecasts

A forecast, as defined in Section 4.1.2, must provide *quantitative* information about an outcome. In the context of this notation, a probabilistic forecast can be represented as a predictive density function for a target, or $f_{z_{i|t}}(z \mid y_{1:t}, t, \mathbf{x}_t)$. The form of this density function depends on the type of variable that Z is, and it could be derived from a known parametric distribution or specified directly. For example, if the target is a binary outcome (e.g. whether in week four, the observed incidence will be above ten cases) the density could be specified as a Bernoulli distribution with a parameter associated with the probability of the outcome occurring. It could also be specified directly as a probability that the incidence is >10 and the probability that the incidence is ≤10. For an integer-valued target (e.g. the number of new cases occurring in February), the predictive density could be represented by a Poisson distribution with a given mean or as a vector of probabilities associated with all possible integer values of cases.

To enable clear definitions for forecasting in real-time, forecasts must be associated with a specific time t. This time t represents the point relative to which targets are defined. For example, if a forecast is associated with week 45 in 2013, then a '-1-step-ahead' (read 'minus-one-step-ahead') forecast would be associated with incidence in week 44 of 2013 and a '3-step-ahead' forecast would be associated with week 48.

4.1.4.4 Forecast for Time Scale

Another consideration for infectious disease forecasting is the forecast horizon, the temporal range that the forecast predicts (Myers et al. 2000; Soyiri & Reidpath 2013). Regardless of the model type, many recent infectious disease forecasting efforts have focused on short time scales (weeks or months) (Reich et al. 2016; Birrell et al. 2011; Buczak et al. 2012; Gerardi & Monteiro 2011; Hii et al. 2012; Lowe et al. 2011; Lu et al. 2010; Nishiura 2011; Shaman et al. 2013; Shaman et al. 2014; Sumi & Kamo 2012; Yan et al. 2010). These studies demonstrated the importance of recent case counts and seasonality on the immediate trajectory of infectious disease incidence. In 2015, the National Oceanic and Atmospheric Administration (NOAA) and the Centers for Disease Control and Prevention (CDC) hosted a competition to make within-season forecasts for longer forecast horizons, such as annual dengue incidence, epidemic peak, and peak height, for San Juan, Puerto Rico and Iquitos, Peru (National Oceanographic and Atmospheric Administration 2015). Prior to these competitions, long-term forecasts were more commonly used for chronic disease prevalence than for non-chronic infectious disease incidence (Soyiri & Reidpath 2013).

4.2 Models Used for Forecasting Infectious Diseases

4.2.1 Mechanistic vs. Statistical: A Taxonomy of Forecasting Models

According to Myers et al. (2000) forecasting models for infectious diseases take either a 'biological approach' or a 'statistical approach'. Others have phrased this distinction as one of mechanistic (i.e. biological) and phenomenological (i.e. statistical) models. A model based on disease biology can account for previously unforeseen scenarios that are possible due to transmission dynamics, however these models often require specification of a large number of parameters and covariates in order to make forecasts. On the other hand, statistical forecasting models are restricted by the assumption that future incidence will follow the patterns of incidence observed in the past, but can be specified without full knowledge of the disease process or interactions between members of the population. In this section, we discuss the major modeling methods across the biological-statistical spectrum (Table 4.1).

4.2.1.1 Mechanistic Models

Compartmental models are the standard biological, or mechanistic, approach for modeling infectious disease (Keeling & Rohani 2007; Siettos & Russo 2013; Lessler & Cummings 2016). Kermack and McKendrick proposed one such model, now known as the susceptible-infectious-recovered (SIR) model, in which members of a population transition through each compartment (susceptible to infectious to recovered) over the course of an epidemic (Kermack & McKendrick 1927). While this process mimics the behavior of an outbreak, the simplest model assumes that the population is 'well mixed', such that each individual is equally likely to encounter any other individual. Since this is unlikely, researchers can add more compartments (e.g. adults and children), along with contact rates between compartments, or individually model each member of the population (i.e. agent-based modeling) (Eubank et al. 2004). Though greater complexity requires more modeling assumptions, compartmental models have been effective at estimating underlying infectious disease processes and the potential impact of interventions (Keeling & Rohani 2007; Lessler & Cummings 2016; Ferguson et al. 2006; Reich et al. 2013).

Compartmental models have been extended for use in forecasting infectious disease incidence. Variations on the SIR model have been used to develop forecast models for influenza (Birrell et al. 2011; Nishiura 2011; Osthus et al. 2017). Some of

Table 4.1 A summary of selected forecasting papers organized by vector-borne diseases (VBD) and other. Additionally, for each paper, we indicate which method is used, whether the method was mechanistic or statistical, whether the vector was specifically modeled, and the disease of interest.

	Method	SIR model	Vector modeled?	Disease
VBD	EAKF	SI	Yes	West Nile (Defelice et al. 2017)
	Ordinary diff. eqs.	SEIR	Yes	dengue (Lourenço & Recker 2014) malaria (Tompkins et al. 2019)
	Ross-MacDonald NN	SIR	Yes	dengue (Dinh et al. 2016)
	Ross-MacDonald ODE	SIR	Yes	dengue (Amaku et al. 2016; Zhu et al. 2018)
	Stochastic diff. eqs.	SI	Yes	*Theileria orientalis* (Jewell & Brown 2015)
	ARIMA	-	No	dengue (Johansson et al, 2016), malaria (Anwar et al. 2016; Zinszer et al. 2015)
	Cellular automaton	SIR	No	dengue (Gerardi & Monteiro 2011)
	EAKF	SIR	No	dengue (Yamana et al. 2016)
	GLM/Regression	-	No	dengue (Lowe et al. 2014; Reich et al. 2016; Hii et al. 2012; Lauer et al, 2018) malaria (Sewe et al. 2017), other (Paul et al. 2011; Liu et al. 2005–2010; Moore et al. 2012)
	Holt-Winters	-	No	dengue (Buczak et al. 2018)
	KCDE	-	No	dengue (Ray et al. 2017)
	Neural networks	-	No	malaria (Thakur & Dharavath 2019), Zika (Akhtar et al, 2018)
Other	EAKF	SEIRX	-	Ebola (Shaman et al. 2014)
	Markov chain	SEIR	-	Ebola (Gaffey & Viboud 2017)
	Stochastic diff. eqs.	SEIR	-	Ebola (Funk et al. 2016; Asher 2018)
	Agent-based	SEIR	-	influenza (Hyder et al. 2013; Nsoesie et al. 2013)
	Chain binomial	SEIR	-	influenza (Nishiura 2011)
	Compartmental	SEIR	-	influenza (Birrell et al. 2011)
	Dynamic Bayesian	SIR	-	influenza (Osthus et al. 2017; Osthus et al. 2019)
	EAKF	SIRS	-	influenza (Shaman & Karspeck 2012; Pei & Shaman 2017)
	Expert opinion	-	-	influenza (Farrow et al. 2017)
	GLM/Regression	-	-	influenza (Held & Paul 2012; Goldstein et al. 2011)
	Random forest	-	-	influenza (Kane et al. 2014)
	ARIMA	-	-	influenza (Dugas et al, 2013), other (Lu et al. 2010; Yan et al, 2010)
	Neural network	-	-	influenza (Xu et al. 2017), hepatitis A (Guan et al. 2004)
	Holt-Winters	-	-	leprosy (Deiner et al. 2017)

these approaches incorporate humidity into a SIRS compartmental model (Shaman et al. 2013; Shaman & Karspeck 2012). The Ross–MacDonald model, a compartmental model originally developed for malaria that accounts for interactions between mosquitoes and humans, (Smith et al. 2012) has been used to make forecasts of dengue fever (Dinh et al. 2016; Amaku et al. 2016; Zhu et al. 2018). However, other mechanistic approaches to forecast

dengue outbreaks have not modeled mosquitoes specifically (Yamana et al. 2016).

4.2.1.2 Classical Statistical Models

On the statistical side of the modeling spectrum, many regression-style methods have been used for forecasting. Perhaps the most well-known statistical method for time series is the auto-regressive integrated moving average, or ARIMA (Box &

Jenkins 1962). ARIMA models use a linear, regression-type equation in which the predictors are lags of the dependent variable and/or lags of the forecast errors. ARIMA and seasonal ARIMA (SARIMA) models are frequently applied to infectious disease time series (Soyiri & Reidpath 2013; Johansson et al. 2016; Ray et al. 2017; Siettos & Russo 2013; Steffen Unkel et al. 2012). Lu et al. combined a SARIMA model with a Markov switching model (a type of compartmental model) to account for anomalies in the surveillance process (Lu et al. 2010).

Also under the subheading of trend and seasonal estimation are simple exponential smoothing strategies, known as Holt–Winters models (Holt 2004; Winters 1960). Exponential smoothing techniques involve taking weighted averages of past observations with exponentially decreasing weights further from the present. Holt–Winters in particular is known for its efficient and accurate predictive ability (Gelper et al. 2010; Goodwin et al. 2010). These approaches have been used successfully in forecasting dengue fever (Buczak et al. 2018) and leprosy (Deiner et al. 2017).

Some researchers have used generalized linear regression models to develop infectious disease forecasts. In some cases, researchers used lagged covariates (e.g. temperature, rainfall, or prior incidence) to predict future incidence (Reich et al. 2016; Hii et al. 2012; Lowe et al. 2011; Paul et al. 2011; Moore et al. 2012). Held and Paul also combined statistical and biological theory by building a regression model that consisted of three components of disease incidence: endemic, epidemic, and spatio-temporal epidemic (to account for spread of disease across locations) (Held et al. 2005). This has become a well-established framework for forecasting infectious disease surveillance data (Held et al. 2017; Ray et al. 2017; Höhle & Matthias an der 2014), and is accompanied by open-source software implementing the methods (Meyer et al. 2017).

4.2.1.3 Modern Statistical Methods

Modern statistical methods, i.e. not the classical time-series and regression-based approaches, are an increasingly popular way to forecast infectious disease incidence. These methods include non-parametric approaches as well as more black-box machine-learning style algorithms. We focus in this

section on stand-alone forecasting methods, for a discussion on ensemble methods, see Section 4.2.4.

Statistical or machine-learning approaches have been in existence for decades. While machine-learning and statistical methods are sometimes classified separately (Siettos & Russo 2013), we group them together as 'statistical', as both terms encapsulate approaches that use patterns from past incidence in order to forecast future incidence (Myers et al. 2000). These approaches can be used for 'data mining', by which large amounts of data are extracted from various online sources for pattern-recognition tasks, or for modeling, using empirical methods such as random forests, neural networks, or or support vector machines that do not make any parametric model assumptions. These techniques came about in the computer science and artificial intelligence communities (see, e.g. (Ho 1995)), but can also be expressed statistically (Hastie et al. 2009).

Several papers have found that machine-learning modeling methods can outperform standard statistical models for infectious disease forecasting: random forests outperformed ARIMA forecasting avian influenza (Kane et al. 2014), a maximum entropy model outperformed logistic regression forecasting hemorrhagic fever with renal syndrome (Liu et al. 2005–2010), and fuzzy association rule mining outperformed logistic regression forecasting dengue outbreaks (Buczak et al. 2012). Additionally, kernel conditional density estimation, a semi-parametric method, was shown to have more well-calibrated probabilistic forecasts than SARIMA and other regression-based approaches for forecasting highly variable dengue outbreaks in San Juan, Puerto Rico (Ray et al. 2017). Heteroskedastic Gaussian processes showed better forecast accuracy than generalized linear models in forecasting dengue fever (Johnson et al. 2018). Neural networks have also been used for forecasting influenza (Xu et al. 2017; Wu et al. 2018), Zika (Akhtar et al, 2018), and Hepatitis A (Guan et al. 2004).

4.2.1.4 Comparisons between Mechanistic and Statistical Models

From an epidemiological perspective, mechanistic models have several clear advantages over statistical models. They are biologically motivated, and therefore have parameters that relate to well-established theory and can be interpreted by experts in the field. They have been adapted in previous work to include vector dynamics, which could be important for forecasting outbreaks of vector-borne diseases (Jewell & Brown 2015; Reiner et al. 2013; Lourenço & Recker 2014). Mechanistic models can flexibly incorporate features such as interventions or behavioral changes, which can be critical, especially if forecasts are desired for different intervention scenarios (see Section 4.1.3). While mechanistic models can be built to rely heavily on previously observed data, they also can be instantiated with very little prior data, such as in emerging outbreaks (see Section 4.2). Additionally, while forecasts from statistical models are typically bounded by trends that have been previously observed, mechanistic models can forecast outside of previously observed trends if the underlying states of the models call for such dynamics.

Despite these advantages, in forecasting settings where substantial historical data is available, statistical models may prove more effective at using past observed trends to forecast the future. Many statistical models were designed to be either more flexibly or parsimoniously parameterized, meaning that they may be able to more easily capture dynamics common to infectious disease time-series such as auto-regressive correlation and seasonality. Additionally, they can be built to rely less heavily on specific assumptions about a particular biological disease transmission model, giving them flexibility to adapt when the data does not follow a particular pattern. In other words, since any specified mechanistic model is necessarily a simplification of the true underlying disease process, the question is how much will forecast accuracy suffer as a result of the inevitable model misspecification. In many cases, heavily parameterized mechanistic models may be more sensitive to model misspecification than a more flexible statistical model.

In practice, however, the distinction between statistical and mechanistic models is not always sharply defined. Due to complexities of real-time surveillance systems, forecasting targets of public health interest often represent a mixture of different signals that would be challenging in practice to be forecasted accurately by a single mechanistic model.

For example, surveillance data of case counts may include distorted signals due to figments of partially automated reporting systems. This might be manifested in decreased case counts during holidays, or an increase in case counts at the end of the season when final reports are being prepared (Reich et al. 2016). In other settings, the actual target of interest may be a composite measure, such as with influenza-like illness (ILI) in the US (Reich et al. 2019). In both of these settings, the signal which is being predicted may be driven by factors that are not directly relevant to trends in disease transmission (e.g. clinical visitation or reporting patterns, changes in the processes used for diagnosis or case reporting). In these settings, statistical models that can have a more flexible understanding of the trends they are using to fit data may be at an advantage over mechanistic models. Research has demonstrated the value of coupling flexible statistical formulations with mechanistic curve-fitting to produce accurate forecasts (Funk et al. 2016; Asher 2018; Osthus et al. 2019; Pei & Shaman 2017).

Despite many unanswered questions about when and in what settings one type of model will generally do better than the other, studies that make explicit, data-driven comparisons are fairly uncommon. Multi-team infectious disease forecasting challenges provide some of the best data available on this important question. For forecasting both seasonal influenza and dengue outbreaks, what limited data there are suggest that mechanistic and statistical approaches show fairly similar performance, with statistical models showing a slight advantage (Reich et al. 2019; McGowan et al. 2019; Johansson et al. 2019). A collaborative effort during the 2014 West Africa Ebola outbreak to forecast synthetic data showed fairly comparable results from mechanistic and statistical models and did not make an explicit comparison between the two (Viboud et al. 2017). Summary analyses from other challenges have not been published, but we note that a very simple quasi-mechanistic model was the best performing model in forecasting the pattern of emergence of Chikunguya in the Americas (Lega & Brown 2016; DARPA 2015). In summary, more research is needed to improve our understanding about whether different types of forecasting methods can be shown to be more reliable than others, especially as data availability and resolution improves over time.

4.2.2 Forecasting in Emergent Settings

In emerging outbreak scenarios, where limited data is available, mechanistic models may be able to take advantage of assumptions about the underlying transmission process, enabling rudimentary forecasts even with minimal data. On the other hand, many statistical models without assuming a mechanistic structure rely on past data to be able to make forecasts. That said, any forecasts made in settings with limited data must be subjected to rigorous sensitivity analyses, as such forecasts will necessarily be heavily reliant on model assumptions.

A wide range of different mechanistic models have been used in settings where infectious disease forecasts are desired for an emerging threat. A simple non-linear growth model performed the best in a prospective challenge for forecasting Chikungunya in the Americas (Lega & Brown 2016). This growth model made the assumption that the rate of change in the cumulative number of cases through the course of an outbreak follows a parabolic rise and fall, by which it can quickly approximate the parameters of an Susceptible-Infected-Recovered (SIR) model and estimate the peak incidence, duration, and total number of cases during an epidemic. Unlike many other mechanistic forecasting approaches, this model has a small number of parameters, is easy to fit, and makes only a few assumptions about the underlying disease process. A deterministic Susceptible-Exposed-Infected-Recovered (SEIR) model was used to forecast synthetic Ebola epidemic data, showing comparable performance to other methods on the same data (Viboud et al. 2017; Gaffey & Viboud 2017). A stochastic SEIR model also forecasted synthetic Ebola data, and showed somewhat less reliable performance compared to other methods (Funk et al. 2016). Data-driven agent-based models have also been shown to be a viable forecasting tool for emerging infectious diseases (Venkatramanan et al. 2017). An SIR model, similar to one used to forecast seasonal dengue fever and influenza, was used with a more complex compartmental structure to forecast the spread of Ebola during the outbreak in West Africa in 2014 (Shaman et al. 2014). A set of quasi-mechanistic models were used to forecast Zika virus transmission during 2017 to help plan for vaccine trials (Asher et al. 2017).

4.2.3 Using External Data Sources to Inform Forecasts

Traditional approaches to infectious disease forecasting often have relied on a single time-series, or multiple similar time-series (e.g. incidence from multiple locations). However, other types of epidemiological data may provide important information about current transmission patterns.

4.2.3.1 Vector Data

The use of data on prevalence and abundance of disease vectors in forecasting models has not been extensively explored in the literature. This is likely due in part to climate data being more widely available and the belief that such data could serve as a good proxy for actual data on vectors. That said, perhaps the most well-developed area for incorporating vector data into forecasts is the use of prevalence data in forecasting mosquito-borne diseases such as West Nile Virus (Defelice et al. 2017; Kilpatrick & Pape 2013; Davis et al. 2017), and dengue fever (Shi et al. 2015). However, none of these studies have explicitly quantified the added value of vector data on forecast accuracy. While the hypothetical benefits of good vector surveillance data have been clearly quantified (Yamana & Shaman 2019), the benefit of these data in practice (especially when other reasonable proxy data may be available; see 'Climate Data' Section 4.2.3.5) is still unclear.

4.2.3.2 Laboratory Data

Leveraging laboratory data, collected either through passive or active surveillance strategies, may provide crucial data about what specific pathogens are currently being transmitted and could inform forecasting efforts. This is an area that warrants more research, as few efforts have tackled the challenge of having laboratory test data inform forecasts at the population level. One model uses an aggregate measure of genetic distance of circulating influenza strains from the strains in the vaccine as a variable to help forecast peak timing and intensity of seasonal outbreaks in the U.S. (Xiangjun et al. 2017; Xiangjun & Pascual 2018). Some efforts have also been made to make strain-specific forecasts for influenza (Kandula et al. 2017). Other efforts have focused on longer-term forecasts of what strains will predominate in a given season, with an eye towards providing information to influenza vaccine manufacturers (Morris et al. 2017). These efforts have moved beyond influenza, and forecasting pathogen evolution is being worked on for a variety of different pathogens (Hadfield et al. 2017).

4.2.3.3 Expert Opinion

Another, and very different, kind of epidemiological data for forecasting is expert opinion. Long seen as a useful indicator in business applications (Surowiecki 2004), expert opinion has recently begun to be used in infectious disease applications (Farrow et al. 2017; Deiner et al. 2017). While not traditional clinical data, expert opinion surveys leverage powerful computers, i.e. human brains, that can synthesize historical experience with real-time data (Budescu & Chen 2014). Intuitive interfaces can facilitate the specification of quantitative and digitally entered forecasts from experts who need not be technically savvy, lowering the barriers to participation and subsequent analysis (Farrow et al. 2017). In the 2016/2017 influenza season in the US, a forecast model based on expert opinion was a top-performer in a CDC-led forecasting competition (Reich et al. 2019; McGowan et al. 2019). Human judgment and expert opinion surveys are a promising area for further forecasting research, especially in contexts with limited data availability.

4.2.3.4 Digital Epidemiology

Digital epidemiology has been defined as the use of digital data for epidemiology when the data were 'not generated with the primary purpose of doing epidemiology' (Salathé 2018). Broadly speaking, this might refer to online search query data, social media data, satellite imagery, or climate data, to name a few. These resources may hold promise for forecasters who want to incorporate 'Big Data' streams into their models. In the past ten years, much research has explored the potential for leveraging multiple data streams to improve forecasting

efforts, but this practice is still in its nascent stages. So far, the utility of digital epidemiological data for forecasting has been somewhat limited, perhaps due to challenges in our understanding of how digital data generated by human behavior and interactions with the digital world relate to epidemiological targets (Moran et al. 2016; Salathé 2018; Priedhorsky et al. 2017).

Perhaps the most famous and controversial example of using digital data streams to support infectious disease prediction surround the early promising performance (Ginsberg et al. 2009; Dugas et al. 2012) and later dismal failure (Lazer et al. 2014) of Google Flu trends to predict influenza-like-illness in the US. Google Flu trends was based on tracking influenza-related search terms entered into the search engine. Although Google eventually discontinued the public face of the project due to poor performance, criticism of the Google Flu trends approach centered around how data was included or excluded, interpreted, and handled rather than the algorithm that produced the actual forecasts (Santillana et al. 2014; Olson et al. 2013). Ongoing research on using search engine data in forecasting has continued despite the failure of Google Flu trends, producing incremental but consistent improvements to forecast accuracy (Yang et al. 2017; Yang et al. 2015; McGough et al. 2017; Lu et al. 2018; Osthus et al. 2019). More focused search query data, such as data from clinician queries, has also been shown promise for assisting real-time forecasting efforts (Santillana et al. 2014; Thorner et al. 2016).

4.2.3.5 Climate Data

The use of climate data for epidemic forecasting serves as another clear example of re-purposing data for epidemiology. While climate factors are known biological drivers of infection risk (e.g. the impact of absolute humidity on influenza virus fitness (Shaman & Kohn 2009), or temperature and humidity providing optimal conditions for mosquito breeding), the evidence supporting the use of climate data in forecasting models is mixed. Climatological factors such as temperature, rainfall, and relative humidity were used to forecast annual counts of dengue hemorrhagic fever in provinces of Thailand (Lauer et al. 2018). However, only temperature and rainfall were included after a rigorous covariate selection process and neither were included in the final model, although subanalyses showed variation in these associations across different geographic regions of Thailand. Climate factors were shown to improve forecasts of dengue outbreak timing in Brazil (Lowe et al. 2017), but played a less influential role in dengue forecasts in Mexico (Johansson et al. 2016). Aggregated measures of absolute humidity have been incorporated into influenza forecasts in the U.S. (Shaman et al. 2013; Yang et al. 2017). However, without clear standardization across these studies, these mixed results may reflect heterogeneity in the spatial and temporal scales at which forecasts are made, climate factors are measured and aggregated, and disease transmission actually occurs.

4.2.4 Forecasting with Ensembles

Ensemble forecasting models, or models that combine multiple forecasts into a single forecast, have been the industry standard in weather forecasting and a wide array of other prediction-focused fields for decades. By fusing together different modeling frameworks, ensembles that have a diverse library of models to choose from end up incorporating information from multiple different perspectives and sources (Hastie et al. 2009). When appropriate methods are used to combine either point or probabilistic forecasts, the resulting ensemble should in theory always have better long-run performance than any single model (Bates & Granger 1969; Makridakis & Winkler 1983; Clemen & Winkler 2007). However, researchers have suggested that adjustments are necessary to correct for the introduction of bias (Granger & Ramanathan 1984) and miscalibration (Gneiting et al. 2013) in the process of building ensemble models.

Ensembles have been increasingly used in infectious disease applications and have shown promising results. For forecasting influenza, several model averaging approaches have shown improved performance over individual models (Yamana et al. 2017; Ray & Reich 2018; Reich et al. 2019). Similar approaches have yielded similar results for dengue fever (Yamana et al. 2016), lymphatic filariasis (Smith et al. 2017), and Ebola (Viboud et al. 2017).

In many of these examples, however, the number and diversity of distinct modeling approaches was fairly small. To unlock the full potential value of ensemble forecasting, as well as understanding the added value of contributions from new and different data sources or modeling strategies, more scalable frameworks for building forecast models are required. There is a need to develop infrastructure and frameworks that can facilitate the building of ensemble forecast models (George et al. n.d.). This will require clear technical definitions of modeling and forecasting standards. Given the history of improved forecast performance due to improvements in ensemble methodology in other fields that focus on non-temporal data, the prospects for continued ensemble development in the area of temporal and time-series data, and especially infectious disease settings, are bright.

4.3 Components of a Forecasting System

Due to the complex biological, social, and environmental mechanisms underlying infectious disease transmission, the true underlying processes that give rise to the observed data is unknown. This makes choosing one model, or even multiple, from the selection of possible choices quite challenging. How can one decide which model is the best for forecasting future targets?

Deciding on the structure of the forecasting exercise, including the targets and the evaluation metrics, provides critical information to help decide upon appropriate methods. Because models perform differently depending on the forecast target, type of forecast, model training technique, and evaluation metric, it is important to specify the forecasting system prior to fitting the models (Armstrong 1990). This ensures that the system is tailored to the particular design of the exercise.

4.3.1 Forecast Type

When building a forecasting system, the first steps are to choose the forecast target (as described in (Section 4.1.4) and type. Forecast targets are often dictated by the goals of a public health initiative. Researchers and public health officials collaborate to find a forecast target that is most useful for allocating resources and implementing interventions to reduce the severity of an infectious disease outbreak. The forecast target helps inform the selection of the forecast type. Forecasts can typically be classified as either point forecasts or probabilistic forecasts. Some authors classify 'interval' forecasts as a separate entity (Diebold 2001), however, these can be seen as a simplified version of a fully probabilistic forecast.

A point forecast is a forecast of a single value that attempts to minimize a function of the error between that value and the eventually observed value. The mean, median, or mode of a predictive distribution is often used as the point forecast for a specified target. This choice may depend on the specific metric used for evaluation, as the mean is the theoretically optimal choice if point forecasts are being evaluated with mean squared error (MSE) and the median is optimal if being evaluated by mean absolute error (MAE) (Reich et al. 2016). While point forecasts are simpler to produce and interpret, they may make simplifying assumptions about the underlying probability distribution, leading to low-quality forecasts. For example, a point forecast based on the mean may represent a value for which there is actually a small likelihood of occurring if, for example, it lies between the peaks of a multi-modal distribution. This could mislead officials and researchers into forecasting a medium-sized outbreak when the full distribution actually shows that the most likely future scenarios are for either low incidence or an epidemic outbreak.

Interval forecasts supplement point forecasts with a 'prediction interval', or a range of likely values. The nominal level of a prediction interval indicates the percentage of eventually-observed outcomes that should fall within that interval. If a model makes 100 forecasts, about 95 should fall within the 95 percent prediction interval. More generally, a $(1 - \alpha * 100)$ percent prediction interval can be thought of as the interval that has a significance level of α. Interval forecasts are typically derived from some form of a probabilistic model or assessment of in-sample forecast error or uncertainty.

A fully probabilistic forecast must specify a probability distribution function The goal of a probabilistic forecast is to assign the maximum probability to the true future value. Probabilistic forecasts can specify a closed-form parametric density function

(e.g. a Gaussian distribution with a mean and variance) or an empirical distribution, either with an empirical cumulative density function, a set of samples from the predictive density, or a binned density function, with probabilities assigned to a discrete set of possible outcomes. Density estimation often requires simulation-generating methodology, which can be more time-consuming and computationally-intensive than other techniques. Ongoing advances in computing continue to make density forecasting methods more feasible for researchers. Density forecasts contain the most nuanced information of all of the forecasting methods, but are often the most difficult to interpret and communicate to non-expert collaborators.

4.3.2 Evaluation and Scoring

There is a rich literature on scoring and evaluating all types of forecasts. Of course appropriate metrics will depend on the forecasting setting and the scoring criteria for a particular exercise. In general, models should be fit with a loss function or 'goodness of fit' criteria that that is similar or identical to the method that will be used to evaluate forecasts.

4.3.2.1 General Principles for Scoring Forecasts

Research suggests that metrics should be *scale-independent* (Armstrong 1990; Hyndman & Koehler 2006). For example, within a single infectious disease time series, larger incidence values are both more difficult to forecast and often have larger errors on an absolute scale than smaller incidence values simply because they are larger numbers. Thus, incidence values near the seasonal peak are both larger and more variable than incidence near the seasonal nadir and, consequently, forecasting model error will depend on the size of the value it is forecasting. In these situations, something closer to scale-independence can be achieved either by using logged metrics can weight errors more equally across different scales or by using relative measures of accuracy (see, e.g. Equation 4.1 and Reich et al. 2016).

Metrics should be defined and finite in reasonable scenarios. This principle ensures that scores from single forecasts may be combined together, for example with an average. If single forecasts could be infinite in reasonable scenarios, one individual forecast could eclipse all other scores in a summary measure such as an average. However, even non-experts can agree that a model that forecasts negative values of disease incidence should be considered invalid and it could be appropriate to have an infinite scoring value for such a forecast.

Forecasts should be evaluated using *proper scoring rules* (Gneiting & Raftery 2007). A proper scoring rule incentivizes the forecaster to report exactly what the model predicts. An improper scoring rule could incentivize adjusting the forecast reported by a model to 'game' the system only to get a better score. While using an improper scoring rule is unlikely to change the relative performance of a set of models dramatically, it can lead to settings where the best scoring forecast is one that has been modified in undesirable ways. For example, when probabilistic forecasts for season peak week are scored by evaluating the probability assigned to the true peak week plus or minus one week (an improper score used by the US CDC in influenza forecasting competitions (Reich et al. 2019; McGowan et al. 2019)), even high-scoring forecast models can be adjusted to have better scores by adjusting the probabilities assigned to different weeks in a systematic way that is 'dishonest' to the original forecast.

Forecast accuracy should always be *evaluated with out-of-sample observations and with as large of a sample of observations as is feasible.* Since forecasts are by definition predictions of as-yet-unobserved data, it is critical to evaluate forecast accuracy on out-of-sample data. This means that forecasts should only be evaluated on observations that were not used to fit or train the model. Ideally, the out-of-sample data would be 'prospective' in the sense that all of evaluated observations would be from a point in time after the training data. However, in cases with limited data availability, this may not be feasible. Data quantity can be a limiting factor for many real-world datasets, especially for infectious disease surveillance. The analyst must balance competing needs of having sufficient data for training a realistic model with holding out data (ideally prospectively) for cross-validation and testing.

4.3.2.1 Evaluating Point Forecasts

Point forecasts are typically evaluated on their own using metrics such as mean squared error (MSE) or the mean absolute error (MAE). For comparative evaluation of point forecasts in practice, many researchers recommend using a metric that scales the forecasting error against that of a reference model (Reich et al. 2016; Hyndman & Koehler 2006; Gneiting & Raftery 2007).

One example is the relative mean absolute error (rMAE), which divides the mean absolute error of one forecasting model (model A) by the mean absolute error of a second model (model B):

$$\text{rMAE} = \frac{\sum_{t=1}^{n} |z_t - \hat{z}_t^{\text{A}}|}{\sum_{t=1}^{n} |z_t - \hat{z}_t^{\text{B}}|}. \tag{4.1}$$

Recall that z_t is the target of interest forecasted at time t and $\hat{z}_t^A$ represents the forecast from model A at time t. In principle, rMAE may be calculated between any two models, however, it is common for rMAE to be calculated for a set of models against a common 'reference' or basline model in the denominator.

An additional desirable feature of rMAE is that it is interpretable for public health officials. When $rMAE < 1$ this means that the forecasting model has less error than the reference model on the scale of the original data and $rMAE > 1$ means that the forecasting model has more error than the reference model. For example, if model A has an rMAE of 0.9 compared to a reference seasonal model of case counts for a particular disease that means that the predictions from model A were 10 percent closer to the observed value than predictions from the reference model were.

4.3.2.2 Evaluating Interval Forecasts

Interval forecasts can be evaluated by their coverage rate and their width. Prediction intervals should be as narrow as possible while covering a proportion of forecasts approximately equal to that expected by its level.

Perhaps the most commonly used interval metric is the coverage rate (CR) which is simply the fraction of all $(1 - \alpha) * 100$ percent prediction intervals that cover the true value. Therefore,

$$CR_\alpha = \frac{1}{T} \sum_{t=1}^{T} \mathbb{I}(l_t^\alpha \leq z_t \leq u_t^\alpha)$$

where $\mathbb{I}$ is the indicator function (equalling 0 if the expression inside is FALSE and 1 if TRUE), and l_t^α and u_t^α are the lower and upper bounds of a $(1-\alpha)*100$ percent prediction interval for observation z_t. The observed CR_α can be evaluated for its proximity to $(1-a)$.

An interval evaluation metric recommended for its being a 'proper' scoring metric is (Gneiting & Raftery 2007)

$$S_\alpha^{\text{int}}(z_t, u_t^\alpha, l_t^\alpha) = \frac{1}{T}\sum_{t=1}^{T}(u_t^\alpha - l_t^\alpha) + \frac{2}{\alpha}(l_t^\alpha - z_t)\mathbb{I}(z_t < l_t^\alpha)$$
$$+ \frac{2}{\alpha}(z_t - u_t^\alpha)\mathbb{I}(z_t > u_t^\alpha)$$

where it is desirable to minimize the score S_α^{int}. Forecasting models are penalized for having wider intervals and for having observed values that fall far outside of the intervals. Observations that fall outside of large prediction intervals (small α) are penalized more than those that fall outside of small prediction intervals (large α).

4.3.2.3 Evaluating Probabilistic Forecasts

In the long run, probabilistic forecasts should have distributions that are consistent with the distribution of the observed values. Models that assign more weight to the eventually observed values should be scored better than those that do not (Gneiting & Raftery 2007). A commonly used proper scoring rule for probabilistic forecasts is the log score. Aggregated across many predictions, this metric is defined as:

$$\text{LogS} = \frac{1}{T}\sum_{t=1}^{T}\log\hat{p}(z_t)$$

where $\hat{p}(\cdot)$ is the estimated probability of observing the target z_t. However, this metric is sensitive to outliers, as any observation with a forecasted probability of zero causes the metric to go to negative infinity (though adjustments can be made to avoid this). As an alternative, Funk et al. recommend using multiple metrics to evaluate the unbiasedness, calibration, and sharpness of infectious disease forecasts (Funk et al. 2017).

The continuous ranked probability score (CRPS) is a proper scoring rule that measures the difference between the forecasted and observed cumulative distributions (Hersbach 2000). This metric measures both the bias and the uncertainty of the forecasted density and thus rewards forecast that assign weight closer to the observed value, even if it does not assign much weight exactly on the observed value. A point forecast with no uncertainty will have a CRPS equal to the absolute error of the forecast. Unbiased forecasts with more uncertainty will have a higher CRPS than for unbiased forecasts with less uncertainty, however biased forecasts with more uncertainty can have a smaller CRPS than biased forecasts with less uncertainty. While CRPS is scale-dependent, dividing the CRPS of a forecasting model by the MAE of a benchmark model (as in the relative mean absolute error) yields a scale-independent continuous ranked probability skill score (Bradley & Schwartz 2011; Bogner et al. 2018).

4.3.2.4 Tests for Comparing Models

Comparing the relative performance of different models can yield important insights about the benefits of a unique data source or one approach over another. However, in a time-series forecasting context (the most common for infectious disease forecasts) several clear statistical challenges are present when making forecast comparisons. Most importantly, outcomes and forecasts at multiple nearby timepoints will be correlated with each other, reducing and complicating the understanding of the power of these tests to detect 'significant' differences. The Diebold–Mariano test is the most well-known test to compare the errors from two forecasting models (Diebold & Mariano 2002). This method is implemented in software packages, for example, the forecast package in R (Hyndman et al. 2019; Hyndman & Khandakar 2008). Other permutation based approaches have also been used to compare the significance between the forecast errors for two models (Ray & Reich 2018).

However, in the infectious disease forecasting literature it has not yet become common practice to run such tests. Instead, authors have tended to rely on simple numeric comparisons of forecast errors between two or more models. Not running a formal test allows for the possibility that the observed differences are due to chance. However, from a practical perspective, as long as the forecasts are truly prospective in nature and the comparisons presented were the only ones made, such a comparison can provide tangible information about which model to choose for decision-making. In situations where a definitive statement about the predominance of one model over another is desired, a formal test will likely be the best evidence available.

4.3.3 Model Training and Testing

In order for a forecasting model to be useful for researchers or officials it needs to be generalizable to data beyond the observations that were used for fitting. For instance, a model that perfectly forecasts monthly dengue incidence over the past ten years, but performs worse than a reasonable guess—e.g. the average monthly incidence—over the next five years is not very useful. We would be better off using the reasonable guess instead of the forecasting model. Though we can never be certain that our best model will perform well outside of our dataset, we can get a better idea of its *out-of-sample* performance using cross-validation with a training and testing set. We illustrate this concept with an example from Lauer et al. (Lauer et al. 2018), in which we forecasted annual dengue hemorrhagic fever (DHF) incidence in Thailand for seventy-six provinces.

A central challenge for forecasting in general is to train a model in such a way that we minimize the error on the unobserved data, i.e. the test data. For real-time forecasts, the test data will be unobserved at the time a model is specified. When forecasting is performed retrospectively for data that has already been collected, the test data will already have been observed. Strictly speaking, such an experiment is not forecasting at all, as it does not involve making predictions about data that have not yet been observed. Nonetheless, it can be an important part of forecasting research to understand model performance in these settings. To ensure the validity and generalizability of such findings, it is critical to only use the test data once the models have been specified from the training phase.

Typically, forecasters use cross-validation methods to evaluate and estimate error on not-yet-seen observations (Hastie et al. 2009). There is a rich literature on cross-validation methods, including

some techniques specific to time-series applications (Bergmeir et al. 2018). These methods tend to reward slightly more complex models that may have more error on the testing data than a smaller or simpler model would (Shao 1993). Thus, in addition to selecting the model that performs best in the training phase by a pre-specified information criterion or cross-validation metric, forecasters should also choose a more parsimonious model that has more error in the training phase as a check against overfitting (Ng 1997).

Prior to fitting any model, we split our data into a 'training' sample (for initial model selection) and a 'testing' sample (for final model evaluation) (Hastie et al. 2009; Stone 1974). These steps are standard practice in the field, and similar to formal recommendations for modeling disease surveillance data (Althouse et al. 2015). In this example, data from years 2000 through 2009 (760 observations) served as the training phase data and years 2010 through 2014 (380 observations) served as the test phase data (Fig. 4.2a). The training sample is used for model experimentation and parameter tuning.

There is no one right answer for how to split data into training and testing sets, however the choice may be informed by prior knowledge about the modeling setting. We chose to model the training phase data using leave-one-year-out cross-validation, so that each year's training forecast would be conditional on the remaining nine years of data. While this does not preserve strict ordering of data (e.g. the data from 2000 is predicted based on a model fit to data from 2001 through 2009), it ensures that each of the training period forecasts is based on the same amount of data. The alternative would have been to implement a training regimen that would have predicted 2001 based only on 2000 data, 2002 based only on 2000–2001 data. Due to the limited length of this dataset, this would mean that early forecasts would be based on substantially less data. Using leave-one-year-out cross-validation ensures that each of the 10 years of training forecasts will have the same amount of data, and a roughly similar amount of data that we expect to have in the test phase. However, if substantially more data were available prior to 2000 (say, more than five years of data) then it might have been desirable to implement prospective cross-validation in the training phase as well.

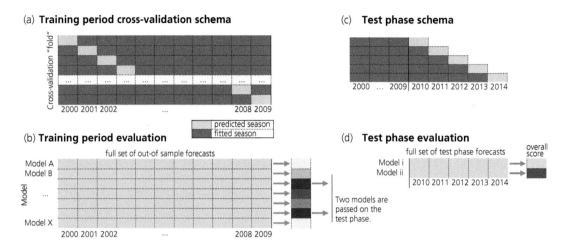

Figure 4.2 A schematic showing the process of model training, cross-validation, and testing for making predictions of annual incidence of dengue hemorrhagic fever using province-level data from Thailand. (a) A diagram of the cross-validation experiment for predicting annual dengue incidence. One year is left out at a time, models are fit to data from the other seasons, and out-of-sample predictions are obtained for the left-out season. (b) A suite of models are fit in the training period, with a full set of out-of-sample forecasts obtained for each year. These scores are then summarized into a single summary out-of-sample, cross-validated score. Two models are chosen and passed into the testing phase. (c) A diagram of the way in which forecasts are made for each of the five testing years. The data from all previous years are fit and the forecasts for the current year are made in a prospective fashion. (d) Only the two models selected in the training phase are implemented in the test phase. The prospective forecasts across all five test-phase seasons are aggregated and summarized into a final overall score.

The training period is complete once all candidate models have generated out-of-sample forecasts for each of the training years. Typically, a small number of models are selected to pass into the test phase. In our example, we ran leave-one-year-out cross-validation on the training phase data to select our model. In this procedure, we fit a model on nine of the ten years to predict the final year—e.g. fitting on 2001–2009 to predict 2000. We repeated this to predict the province-level DHF incidence in each of the ten years, recorded the error for each prediction, and then took the mean absolute error across all predictions and called it the 'cross-validation (CV) error' for a given model (Fig. 4.2b). We performed cross-validation for 202 models with different specifications and covariate combinations. The model that minimized the CV error had five covariates, while the model that minimized the in-sample residual error across the entire training phase had fourteen covariates (Fig. 4.3). In addition to the five-covariate model, we also selected the smallest model within one standard deviation of the smallest CV error—in this case it was a univariate model—to forecast the test phase.

In the test phase, we implemented a prospective testing schema to more realistically simulate real-time

forecasts (Fig. 4.2c). This *rolling-origin-recalibration* window, as it has been called (Bergmeir & Benítez 2012), is implemented by first fitting the model to the training data to forecast the first test phase observation. Then the first observation from the test phase is moved into the training data, the model is re-fit and the second test phase observation is forecasted. We used this method in the testing phase of our example as it is good for evaluating how a model might perform in real-time, as more data is collected and assimilated into the model fitting process. In our example, we evaluated the testing phase forecasts using relative mean absolute error (rMAE) of our model over baseline forecasts based on the ten-year median incidence rate for each province (Fig. 4.2d). The univariate model had less testing phase error than the best cross-validation model (Fig. 4.3). Additionally, the univariate model had about 20 percent less error than a baseline forecast, a rolling ten-year-median for each province (data not shown).

When a forecaster is interpreting the results, he/she should recommend for future use the model that performed best in the test phase. The goal at the outset of this exercise was to find the model that

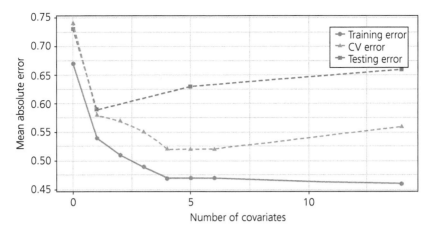

Figure 4.3 Error in forecasting annual incidence of dengue hemorrhagic fever in Thailand, across 76 provinces. Mean absolute errors are measured as the absolute difference between the log observed incidence and the log predicted incidence, where lower values indicate better forecast accuracy. The training phase in-sample error (orange, solid line), out-of-sample cross-validation (CV) error (blue, dotted), and testing phase error (green, dashed) plotted as a function of the number of model covariates (x-axis). The training error monotonically decreases as the number of covariates increases. The CV error is minimized at 5 covariates and better approximates the testing error than the training error, especially for fewer covariates. The univariate model (1 covariate) had the least error in the testing phase.

makes the best forecasts that are generalizable outside of our data, which is defined by the performance on the testing phase. Specific times or places where a different model showed good results could be areas for future forecasting activities, however analysts should be cautious about over-interpreting small-sample size results in training or test phase results.

Prior to splitting the data into training and testing, it is critical to think about how many observations are needed for each sample. There need to be enough training observations to properly fit the model and there need to be enough testing observations to properly evaluate the model. With a short time-series, there may be too few data to split and thus only cross validation can be conducted on all of the data; in this scenario, interpretations about the model performance will be weaker than those with a separate testing phase.

4.4 Operationalizing Forecasts for Public Health

Making forecasts on infectious disease data is difficult due to the culmination of a variety of factors, from the microscopic to the population level, which are difficult to do in any circumstances, but are compounded by logistical issues when trying to make forecasts in real time. Logistical challenges include assimilating newly-collected data into the forecasting framework, accounting for delays in case reporting, and effectively communicating the model results to public health officials.

4.4.1 Reporting Delays

Making forecasts in real time introduces the dimension of reporting delays into our forecasting models. From a disease surveillance perspective, reporting delays are a timeliness issue which varies by features of the disease (ease of diagnosis, incubation time), surveillance entity (local, state, or national government), transmission type (electronic or not), and case load, as well as variability in reporting between surveillance systems (Jajosky & Groseclose 2004; Marinovic et al. 2015). From a data perspective, this means that observed values in the recent past are subject to change.

Since most forecasting models make the assumption that values used for fitting are fixed, we need to adapt our forecasting process by 'nowcasting' observed values in the recent past. One method of nowcasting is to only include 'sufficiently complete' data up such that a forecasting model can make stable forecasts. For instance, 75 percent of dengue hemorrhagic fever cases in Thailand were reported to the Thai Ministry of Public Health within ten weeks of infection (Reich et al. 2016). To account for this, we ignored the last 12 weeks (actually six biweeks, to be exact) before forecasting forward. In our notation, we fit our model to data $y_{1:(t+k-1)}$ to make a k-step forecast, y_{t+k}, where $k = -6$. Another method of nowcasting is to use past reporting delays to model recent incomplete counts. Several frameworks have been proposed to nowcast infectious disease incidence based on past reporting rates (Höhle & Matthias an der 2014; Bastos et al. 2019). Other approaches for nowcasting have incorporated digital surveillance data (Brooks et al. 2018; Osthus et al. 2019).

When case counts for prior time periods are subject to change, it is important for researchers to have a collection of data 'snapshots', so that past situations can be investigated retrospectively with the information that was available at the time. Thus, the database should contain records of cases as they are reported, containing the date of illness and incidence that are timestamped upon deposit into the database.

4.4.2 Communication of Results

Public health authorities have shown increasing interest in working with infectious disease forecasters in the light of recent important public health crises. Starting in 2009 with the pandemic influenza A outbreak, public health officials turned to forecasters for estimates of burden and burden averted due to vaccines and antivirals. During the Ebola outbreak in 2014, public health officials again turned to prediction for specific information regarding the potential outbreak size and intervention impacts. These efforts highlight how infectious disease forecasting can support public health practice now and in the future.

4.4.2.1 What Makes a Good Forecast?

Previous work in meteorology has outlined three distinct forecast attributes of a forecast that contribute to its usefulness, or 'goodness' (Murphy 1993). If we apply these guidelines to infectious disease forecasting, we can surmise that a forecast is good if it is (a) *consistent*: reflecting the forecaster's best judgment, (b) *quality*: forecasts conditions that are actually observed during the time being forecasted, and (c) *valuable*: informs policy or other decision-making that results in increased benefits to individuals or society.

For a forecast to reflect the forecaster's 'best judgment' means that the forecast is reasonable based on the forecaster's expert knowledge base, prior experience, and best and current methodology. The forecaster's internal judgments are not usually available for evaluation or quantification, but could say that a forecast is not a reflection of best judgment if we discover that a forecasting model contains an error or under some conditions produces values outside the range of possible values.

To meet the conditions for high quality, forecasted values must correspond closely to observed values. The field of forecast verification is so vast and specialized that we could not possibly give it a comprehensive treatment here. Suffice it to say that reducing error is the central goal of the field of forecasting. Examples of quality measurement approaches include the mean absolute error and the mean-squared error, which reflect forecast accuracy. Other examples include measures of bias, skill (often a comparison to reference models), and uncertainty (Jolliffe & Stephenson 2003).

Infectious disease forecasts are valuable if they are used to influence decisions. Sometimes value can be accessed in quantitative units (e.g. lives or money saved or lost). Forecast quality influences value to a large extent, but so do other more qualitative features of how the forecast is communicated. For example, a forecast will have a larger impact on decision-making if it is timely, presented clearly, and uses meaningful units in addition to being accurate or improving on a previous system.

4.5 Conclusion and Future Directions

There has been a great deal of progress made in infectious disease forecasting, however the field is very much still in its infancy. Forecasts of epidemics can inform public health response and decision-making, including risk communication to the general public, and timing and spatial targeting of interventions (e.g. vaccination campaigns or vector control measures). However, to maximize the impact that forecasts can have on the practice of public health, interdisciplinary teams must come together to tackle a variety of challenges, from the technological and statistical, to the biological and behavioral. To this end, the field of infectious disease forecasting should emphasize the development and integration of new theoretical frameworks that can be directly linked to tangible public health strategies.

To facilitate the development of scalable forecasting infrastructure and continued research on improving forecasting, the field should focus on developing data standards for both surveillance data and forecasts themselves. This will foster continued methodological development and facilitate scientific inquiry by enabling standard comparisons across forecasting efforts. One key barrier to entry to this field is that the problems are operationally complex: a model may be asked to forecast multiple targets at multiple different times, using only available data at a given time. Converging on standard language and terminology to describe these challenges is key to growing the field and will accelerate discovery and innovation for years to come.

References

Akhtar, M., Kraemer, M.U.G., & Gardner, L. (2018). A dynamic neural network model for real-time prediction of the Zika epidemic in the Americas. *BioRxiv*, 466581.

Althouse, B.M., Scarpino, S.V., Ancel, M., et al. (2015). Enhancing disease surveillance with novel data streams: Challenges and opportunities. *EPJ Data Science*, 4, 17.

Amaku, M., Azevedo, F., Burattini, M.N., et al. (2016). Magnitude and frequency variations of vector-borne infection outbreaks using the Ross–MacDonald model: Explaining and predicting outbreaks of dengue fever. *Epidemiology and Infection*, 144, 3435–50.

Anwar, M.Y., Lewnard, J.A., Parikh, S., & Pitzer, V.E. (2016). Time series analysis of malaria in Afghanistan: Using ARIMA models to predict future trends in incidence. *Malaria Journal*, 15, 566.

Asher, J., Barker, C., Chen, G., et al. (2017). Preliminary results of models to predict areas in the Americas with increased likelihood of Zika virus transmission in 2017. *BioRxiv*, 1875–91.

Asher, J. (2018). Forecasting Ebola with a regression transmission model. *Epidemics*, 22, 50–5.

Ayako, S., & Ken-ichi, K. (2012). Mem spectral analysis for predicting influenza epidemics in Japan. *Environmental Health and Preventive Medicine*, 17, 98–108.

Bastos, L.S., Economou, T., Gomes, M.F.C., et al. (2019). A modelling approach for correcting reporting delays in disease surveillance data. *Statistics in Medicine*, 38, 4363–77.

Bates, J.M. & Granger, C.W.J. (1969). The combination of forecasts. *Journal of the Operational Research Society*, 20, 451–68.

Bergmeir, C. & Benítez, J.M. (2012). On the use of cross-validation for time series predictor evaluation. *Information Sciences*, 191, 192–213.

Bergmeir, C., Hyndman, R.J., & Koo, B. (2018). A note on the validity of cross-validation for evaluating autoregressive time series prediction. *Computational Statistics & Data Analysis*, 120, 70–83.

Biggerstaff, M., Alper, D., Dredze, M., et al. (2016). Results from the centers for disease control and preventions predict the 2013–2014 influenza season challenge. *BMC Infectious Diseases*, 16, 357.

Birrell, P.J., Ketsetzis, G., Gay, N.J., et al. (2011). Bayesian modeling to unmask and predict influenza A/H1N1PDM dynamics in London. *Proceedings of the National Academy of Sciences of the United States of America*, 108, 18238–43.

Bogner, K., Liechti, K., Bernhard, L., Monhart, S., & Zappa, M. (2018). Skill of hydrological extended range forecasts for water resources management in Switzerland. *Water Resources Management*, 32, 969–84.

Box, G.E.P. & Jenkins, G.M. (1962). Some statistical aspects of adaptive optimization and control. *Journal of the Royal Statistical Society. Series B*, 24, 297–343.

Bradley, A.A. & Schwartz, S. (2011). Summary verification measures and their interpretation for ensemble forecasts. *Monthly Weather Review*, 139, 3075–89.

Bray, A. & Schoenberg, F.P. (2013). Assessment of point process models for earthquake forecasting. *Statistical Science*, 510–20.

Brooks, L.C., Farrow, D.C., Hyun, S., Tibshirani, R.J., & Rosenfeld, R. (2018). Nonmechanistic forecasts of seasonal influenza with iterative one-week-ahead distributions. *PLoS Computational Biology*, 14, e1006134.

Buczak, A.L., Baugher, B., Moniz, L.J., Bagley, T., Babin, S.M., & Guven, E. (2018). Ensemble method for dengue prediction. *PLoS one*, 13, e0189988.

Buczak, A.L., Koshute, P.T., Babin, S.M., Feighner, B.H., & Lewis, S.H. (2012). A data-driven epidemiological prediction method for dengue outbreaks using local and remote sensing data. *BMC Medical Informatics and Decision Making*, 12, 124.

Budescu, D.V. & Chen, E. (2014). Identifying expertise to extract the wisdom of crowds. *Management Science*, 61, 267–80.

Campbell, J.E. (1996). Polls and votes: The trial-heat presidential election forecasting model, certainty, and political campaigns. *American Politics Quarterly*, 24, 408–33.

Chambers, D.W., Baglivo, J.A., Ebel, J.E., & Kafka, A.L. (2012). Earthquake forecasting using hidden Markov models. *Pure and Applied Geophysics*, 169, 625–39.

Clemen, R.T. & Winkler, R.L. (2007). Aggregating probability distributions. *Advances in decision analysis: From foundations to applications*, 154.

DARPA. (2015). CHIKV Challenge Announces Winners, Progress toward Forecasting the Spread of Infectious Diseases, 2015.

Davis, J.K., Vincent, G., Hildreth, M.B., et al. (2017). Integrating environmental monitoring and mosquito surveillance to predict vector-borne disease: Prospective forecasts of a West Nile virus outbreak. *PLoS Currents*, 9.

Defelice, N.B., Little, E., Campbell, S.R., & Shaman, J. (2017). Ensemble forecast of human West Nile virus cases and mosquito infection rates. *Nature Communications*, 8, 14592.

Deiner, M.S., Worden, L., Rittel A., et al. (2017). Short-term leprosy forecasting from an expert opinion survey. *PloS ONE*, 12, e0182245.

Diebold, F.X. & Mariano, R.S. (2002). Comparing predictive accuracy. *Journal of Business & Economic Statistics*, 20, 134–44.

Diebold, F.X. (2001). *Elements of Forecasting*. Pennsylvania: Department of Economics, University of Pennsylvania, 4th ed.

Dinh, T.Q., Le, H.V., Cao, T.H., Luong, W.Q.C., & Diep, H.T. (2016). Forecasting the magnitude of dengue in Southern Vietnam. In *Lecture Notes in Computer Science (including subseries Lecture Notes in Artificial Intelligence and Lecture Notes in Bioinformatics)*, 9621, 554–63.

Du, X. & Pascual, M. (2018). Incidence prediction for the 2017–2018 influenza season in the United States with an evolution-informed model. *PLoS Currents*, ecurrents.out breaks.6f03b36587ae74b11353c1127cbe7d0e.

Du, X., King, A.A., Woods, R.J., & Pascual, M. (2017). Evolution-informed forecasting of seasonal influenza A (H3N2). *Science Translational Medicine*, 9, eaan5325.

Dugas, A.F., Hsieh, Y.-H., Levin, S.R., et al. (2012). Google flu trends: Correlation with emergency department influenza rates and crowding metrics. *Clinical Infectious Diseases*, 54, 463–9.

Dugas, A.F., Jalalpour, M., Gel, Y., Levin, S., Torcaso, F., Igusa, T., & Rothman, R.E. (2013). Influenza forecasting with Google flu trends. *PLoS One*, 8, e56176.

Farrow, D.C., Brooks, L.C., Hyun, S., Tibshirani, R.J., Burke, D.S., & Rosenfeld, R. (2017). A human judgment approach to epidemiological forecasting. *PLoS Computational Biology*, 13, e1005248.

Field, E.H., Dawson, T.E., Felzer, K.R., et al. (2009). Uniform California earthquake rupture forecast, version 2 (UCERF 2). *Bulletin of the Seismological Society of America*, 99, 2053–107.

Funk, S., Camacho, A., Kucharski, A.J., Eggo, R.M., Edmunds, W.J. (2018). Real-time forecasting of infectious disease dynamics with a stochastic semimechanistic model. *Epidemics*, 22, 21–62.

Funk, S., Camacho, A., Kucharski, A.J., et al. (2017). Assessing the performance of real-time epidemic forecasts. *BioRxiv*, 17, 7451.

Gaffey, R.H. & Viboud, C. (2017). Application of the CDC ebolaresponse modeling tool to disease predictions. *Epidemics*, 22, 22–8.

Gelper, S., Fried, R., & Croux, C. (2010). Robust forecasting with exponential and holt–winters smoothing. *Journal of Forecasting*, 29, 285–300.

George, D.B., Taylor, W., Shaman, J., et al. (2012). Technology to advance infectious disease forecasting for outbreak management. *Nature Communications*, 10, 3932.

Gerardi, D.O. & Monteiro, L.H.A. (2011). System identification and prediction of dengue fever incidence in Rio De Janeiro. *Mathematical Problems in Engineering*, 20, 11.

Gerland, P., Raftery, A.E., Ševčíková, H., et al. (2014). World population stabilization unlikely this century. *Science*, 346, 234–7.

Ginsberg, J., Mohebbi, M.H., Patel, R.S., et al. (2009). Detecting influenza epidemics using search engine query data. *Nature*, 457, 1012–14.

Gneiting, T. & Raftery, A.E. (2007). Strictly proper scoring rules, prediction, and estimation. *Journal of the American Statistical Association*, 102, 359–78.

Gneiting, T. & Ranjan, R. (2013). Combining predictive distributions. *Electronic Journal of Statistics*, 7, 1747–82.

Goldstein, E., Cobey, S., Takahashi, S., Miller, J.C., & Lipsitch, M. (2011). Predicting the epidemic sizes of influenza A/H1N1, A/H3N2, and B: A statistical method. *PLoS Medicine*, 8, e1001051.

Goodwin, P. (2010). The holt-winters approach to exponential smoothing: 50 years old and going strong. *Foresight*, 19, 30–3.

Graefe, A. (2015). German election forecasting: Comparing and combining methods for 2013. *German Politics*, 24, 195–204.

Granger, C.W.J. & Ramanathan, R. (1984). Improved methods of combining forecasts. *Journal of Forecasting*, 3, 197–204.

Guan, P., Huang, D.-S., & Zhou, B.-S. (2004). Forecasting model for the incidence of hepatitis a based on artificial neural network. *World Journal of Gastroenterology: WJG*, 10, 3579.

Guclu, S., Eubank, H., Kumar, V.S., et al. (2004). Modelling disease outbreaks in realistic urban social networks. *Nature*, 429, 180–4.

Hadfield, J., Megill, C., Bell, S.M., et al. (2017). Nextstrain: Real-time tracking of pathogen evolution. *bioRxiv*, 2240–48.

Haemig, P.D., de Luna, S., Sjöstedt, G.A., et al. (2011). Forecasting risk of tick-borne encephalitis (TBE), using data from wildlife and climate to predict next year's number of human victims. *Scandinavian Journal of Infectious Diseases*, 43, 366–72.

Hastie, T., Tibshirani, R., & Friedman, J. (2009). *The Elements of Statistical Learning: Data Mining, Inference, and Prediction*. Number 2. New York: Springer.

Held, L. & Paul, M. (2012). Modeling seasonality in space-time infectious disease surveillance data. *Biometrical Journal*, 54, 824–43.

Held, L., Höhle, M., & Hofmann, M. (2005). A statistical framework for the analysis of multivariate infectious disease surveillance counts. *Statistical Modelling*, 5, 187–99.

Held, L., Meyer, S., & Bracher, J. (2017). Probabilistic forecasting in infectious disease epidemiology: The 13th Armitage Lecture. *Statistics in Medicine*, 36, 3443–60.

Hersbach, H. (2000). Decomposition of the continuous ranked probability score for ensemble prediction systems. *Weather and Forecasting*, 15, 559–70.

Hii, Y.L., Zhu, H., Ng, N., Ng, L.C., & Rocklöv, J. (2012). Forecast of dengue incidence using temperature and rainfall. *PLoS Neglected Tropical Diseases*, 6: e1908.

Ho, T.K. (1995). Random decision forests. In *Proceedings of 3rd International Conference On Document Analysis and Recognition*, volume 1, pp. 278–82. New York: IEEE.

Höhle, M. & an der Heiden, M. (2014). Bayesian nowcasting during the STEC O104:H4 outbreak in Germany, 2011. *Biometrics*, 70, 993–1002.

Holt, C.C. (2004). Forecasting seasonals and trends by exponentially weighted moving averages. *International Journal of Forecasting*, 20, 5–10.

Hyder, A., Buckeridge, D.L., & Leung, B. (2013). Predictive validation of an influenza spread model. *PLoS One*, 8, e65459.

Hyndman, R.J. & Khandakar, Y. (2008). Automatic time series forecasting: The forecast package for R. *Journal of Statistical Software*, 26, 1–22.

Hyndman, R.J. & Koehler, A.B. (2006). Another look at measures of forecast accuracy. *International Journal of Forecasting*, 22, 679–88.

Hyndman, R.A., George, B., et al. (2019). *Forecast: Forecasting Functions for Time Series and Linear Models*, (2019). R package version 8.7.

Jajosky, R.A. & Groseclose, S.L. (2004). Evaluation of reporting timeliness of public health surveillance systems for infectious diseases. *BMC Public Health*, 4, 29.

Jewell, C.P. & Brown, R.G. (2015). Bayesian data assimilation provides rapid decision support for vector-borne diseases. *Journal of the Royal Society Interface*, 12, 20150367.

Johansson, M.A. (2020). Advancing probabilistic epidemic forecasting through an open challenge: The dengue forecasting project. *Under review*.

Johansson, M.A., Reich, N.G., Hota, A., Brownstein, J.S., & Santillana, M. (2016). Evaluating the performance of infectious disease forecasts: A comparison of climate-driven and seasonal dengue forecasts for Mexico. *Scientific Reports*, 6, 33707.

Johnson, L.R., Gramacy, R.B., Cohen, J., et al. (2018). Phenomenological forecasting of disease incidence using heteroskedastic gaussian processes: A dengue case study. *The Annals of Applied Statistics*, 12, 27–66.

Jolliffe, I.T. & Stephenson, D.B. (2003). Introduction. In Ian T. Jolliffe &, David B. Stephenson, eds, *Forecast Verification: A Practitioner's Guide In Atmospheric Science*. Chichester, West Sussex, England: John Wiley & Sons Ltd, pp. 1–12.

Kandula, S., Yang, W., & Shaman, J. (2017). Type- and sub-type-specific influenza forecast. *American Journal of Epidemiology*, 185, 395–402.

Kane, M.J., Price, N., Scotch, M., & Rabinowitz, P. (2014). Comparison of ARIMA and random forest time series models for prediction of avian influenza H5N1 outbreaks. *BMC Bioinformatics*, 15, 276.

Keeling, M.J. & Pejman, R. (2007). *Modeling Infectious Diseases in Humans and Animals*, Princeton University Press, Princeton.

Kermack, W.O. & McKendrick, A.G. (1927). A contribution to the mathematical theory of epidemics. *Proceedings of the Royal Society of London A: Mathematical, Physical and Engineering Sciences*, 115, 700–21.

Kilpatrick, M.A. & Pape, J.W. (2013). Predicting human West Nile virus infections with mosquito surveillance data. *American Journal of Epidemiology*, 178, 829–35.

Lauer, S.A., Sakrejda, K., Ray, E. L., et al. (2018). Prospective forecasts of annual dengue hemorrhagic fever incidence in Thailand, 2010–2014. *Proceedings of the National Academy of Sciences of the United States of America*, 115, E2175–82.

Lazer, D., Kennedy, R., King, G., & Vespignani, A. (2014). The parable of google flu, traps in big data analysis. *Science*, 343, 1203–5.

Lega, J. & Brown, H.E. (2016). Data-driven outbreak forecasting with a simple nonlinear growth model. *Epidemics*, 17, 19–26.

Lessler, J. & Cummings, D.A.T. (2016). Mechanistic models of infectious disease and their impact on public health. *American Journal of Epidemiology*, 183, 415–22.

Lewis-Beck, M.S. & Stegmaier, M. (2014). Us presidential election forecasting. *PS: Political Science & Politics*, 47, 284–8.

Liu, H.-N., Gao, L.-D., Gerardo, S.-X., et al. (2014). Time-specific ecologic niche models forecast the risk of hemorrhagic fever with renal syndrome in Dongting lake district, China, 2005–2010. *PLoS One*, 9, e106839.

Lourenço, J. & Recker, M. (2014). The 2012 madeira dengue outbreak: Epidemiological determinants and future epidemic potential. *PLoS Neglected Tropical Diseases*, 8, e3083.

Lowe, R., Bailey, T.C., Stephenson, D.B., et al. (2011). Spatio-temporal modelling of climate-sensitive disease risk: Towards an early warning system for dengue in Brazil. *Computers and Geosciences*, 37, 371–81.

Lowe, R., Barcellos, C.. Coelho, C.A.S., et al. (2014). Dengue outlook for the world cup in Brazil: An early warning model framework driven by real-time seasonal climate forecasts. *The Lancet Infectious Diseases*, 14, 619–26.

Lowe, R., Stewart-Ibarra, D., Petrova, A.M., et al. (2017). Climate services for health: Predicting the evolution of the 2016 dengue season in Machala, Ecuador. *The Lancet Planetary Health*, 1, e142–51.

Lu, F.S., Hou, S., Baltrusaitis, K., et al. (2018). Accurate influenza monitoring and forecasting using novel internet data streams: A case study in the Boston metropolis. *JMIR Public Health and Surveillance*, 4, e4.

Lu, H.-M., Zeng, D., & Chen, H. (2010). Prospective infectious disease outbreak detection using Markov switching models. *IEEE Transactions on Knowledge and Data Engineering*, 22, 565–77.

Makridakis, S. & Winkler, R.L. (1983). Averages of forecasts: Some empirical results. *Management Science*, 29, 987–96.

Mandal, S., Sarkar, R.R., & Sinha, S. (2011). Mathematical models of malaria-a review. *Malaria Journal*, 10, 202.

Marinovic, A.B., Swaan, C., van Steenbergen, J., & Kretzschmar, M. (2015). Quantifying reporting timeliness to improve outbreak control. *Emerging Infectious Diseases*, 21, 209–16.

McCloskey, D.N. (1992). The art of forecasting: From ancient to modern times. *Cato J.*, 12, 23.

McGough, S.F., Brownstein, J.S., Hawkins, J.B., & Santillana, M. (2017). Forecasting Zika Incidence in the 2016 Latin America outbreak combining traditional disease surveillance with search, social media, and news report data. *PLoS Neglected Tropical Diseases*, 11, e0005295.

McGowan, C.J., Biggerstaff, M., Johansson, M., et al. (2019). Collaborative efforts to forecast seasonal influenza in the United States, 2015–2016. *Scientific Reports*, 9, 683.

Meyer, S., Held, L., & Höhle, M. (2017). Spatio-temporal analysis of epidemic phenomena using the R package surveillance. *Journal of Statistical Software*, 77, 1–55.

Milham, W.I. (1918). *Meteorology: A Textbook on the Weather, the Causes of its Changes, and Weather Forecasting for the Student and General Reader*. Norwood, New York: Norwood The Macmillan Company.

Moore, S.M., Monaghan, A., Griffith, K.S., et al. (2012). Improvement of disease prediction and modeling through the use of meteorological ensembles: Human plague in Uganda. *PLoS*, 7, e44431.

Moran, K.R., Fairchild, G., Generous, N., et al. (2016). Epidemic forecasting is messier than weather forecasting: The role of human behavior and internet data streams in epidemic forecast. *The Journal of Infectious Diseases*, 214, S404–8.

Morris, D.H., Gostic, K.M., Pompei, S., et al. (2017). Predictive modeling of influenza shows the promise of applied evolutionary biology. *Trends in Microbiology*, 26, 102–18.

Murphy, A.H. (1993). What is a good forecast? An essay on the nature of goodness in weather forecasting. *Weather and Forecasting*, 8, 281–93.

Myers, M.F., Rogers, D.J., Cox, J., Flahault, A., & Hay, S.I. (2000). Forecasting disease risk for increased epidemic preparedness in public health. *Advances in Parasitology*, 47, 309–30.

National Oceanographic and Atmospheric Administration. (June 2015). Dengue forecasting project website, June 2015.

Neil, M., Ferguson, C., Derek, A.T., et al. (2006). Strategies for mitigating an influenza pandemic. *Nature*, 442, 448–52.

Ng, A.Y. (1997). Preventing 'overfitting' of cross-validation data. In *Proceedings of the Fourteenth International Conference on Machine Learning*. New York: Morgan Kaufmann, pp. 245–53.

Nishiura, H. (2011). Real-time forecasting of an epidemic using a discrete time stochastic model: A case study of pandemic influenza (h1n1-2009). *BioMedical Engineering OnLine*, 10, 15.

Nsoesie, E., Mararthe, M., & Brownstein, J. (2013). Forecasting peaks of seasonal influenza epidemics. *PLoS Currents*, 21, 5.

Olson, D.R., Konty, K.J., Paladini, M., Viboud, C., & Simonsen, L. (2013). Reassessing google flu trends data for detection of seasonal and pandemic influenza: A comparative epidemiological study at three geographic scales. *PLoS Computational Biology*, 9, e1003256.

Ostfeld, R.S. & Brunner, J.L. (2015). Climate change and ixodes tick-borne diseases of humans. *Philosophical*

Transactions of the Royal Society B: Biological Sciences, 370, 20140051.

Osthus, D., Daughton, A.R., & Priedhorsky, R. (2019). Even a good influenza forecasting model can benefit from internet-based nowcasts, but those benefits are limited. *PLoS Computational Biology*, 15, e1006599.

Osthus, D., Gattiker, J., Priedhorsky, R., & Del Valle, S.Y. (2019). Dynamic bayesian influenza forecasting in the United States with hierarchical discrepancy (with Discussion). *Bayesian Analysis*, 14, 261–312.

Osthus, D., Hickmann, K.S., Caragea, P.C., Higdon, D., & Del Valle, S.Y. (2017). Forecasting seasonal influenza with a state-space sir model. *The Annals of Applied Statistics*, 11, 202.

Pei, S. & Shaman, J. (2017). Counteracting structural errors in ensemble forecast of influenza outbreaks. *Nature Communications*, 8, 925.

Raftery, A.E., Li, N., Ševčíková, H., Gerland, P., & Heilig, G.K. (2012). Bayesian probabilistic population projections for all countries. *Proceedings of the National Academy of Sciences of the United States of America*, 109, 13915–21.

Ray, E.L. & Reich, N.G. (2018). Prediction of infectious disease epidemics via weighted density ensembles. *PLoS Computational Biology*, 14, e1005910.

Ray, E.L., Krzysztof, S. Lauer, S.A., Johansson, M.A., & Reich, N.G. (2017). Infectious disease prediction with kernel conditional density estimation. *Statistics in Medicine*, 36, 4908–29.

Reich, N.G., Brooks, L.C., Fox, S.J., et al. (2019). A collaborative multiyear, multimodel assessment of seasonal influenza forecasting in the United States. *Proceedings of the National Academy of Sciences of the United States of America*, 116, 3146–54.

Reich, N.G., Lauer, S.A., Sakrejda, K., et al. (2016). Challenges in real-time prediction of infectious disease: A case study of dengue in Thailand. *PLoS Neglected Tropical Diseases*, 10, e0004761.

Reich, N.G., Lessler, J., Sakrejda, K., Lauer, S.A., Iamsirithaworn, S., & Cummings, D.A.T. (2016). Case study in evaluating time series prediction models using the relative mean absolute error. *The American Statistician*, 70, 285–92.

Reich, N.G., McGowan, C.J., Yamana, T.K., et al. (2019). A collaborative multi-model ensemble for real-time influenza season forecasting in the US. *bioRxiv*, 566604.

Reich, N.G., Shrestha, S., King, A.A., et al. (2013). Interactions between serotypes of dengue highlight epidemiological impact of cross-immunity. *Journal of the Royal Society, Interface*, 10, 20130414.

Reiner, R.C. Jr, Perkins, T.A., Barker, C.M., et al. (2013). A systematic review of mathematical models of mosquito-borne pathogen transmission: 1970–2010. *Journal of The Royal Society Interface*, 10, 20120921.

Salathé, M. (2018). Digital epidemiology: What is it, and where is it going? *Life Sciences, Society and Policy*, 14, 1.

Santillana, M.D., Zhang, W., Benjamin, A.M., & Ayers, J.W. (2014). What can digital disease detection learn from (an external revision to) google flu trends? *American Journal of Preventive Medicine*, 47, 341–7.

Santillana, M., Nsoesie, E.O., Mekaru, S.R., Scales, D., & Brownstein, J.S. (2014). Using clinicians' search query data to monitor influenza epidemics. *Clinical Infectious Diseases*, 59, 1446.

Scott, A.J. (2001). Evaluating forecasting methods. pp 343–372 in J.S. Armstrong, Principles of Forecasting, Springer, Boston.

Sewe, M.O., Tozan, Y. Ahlm, C., & Rocklöv, J. (2017). Using remote sensing environmental data to forecast malaria incidence at a rural district hospital in Western Kenya. *Scientific Reports*, 7, 25–89.

Shaman, J. & Karspeck, A. (2012). Forecasting seasonal outbreaks of influenza. *Proceedings of the National Academy of Sciences of the United States of America*, 109, 20425–30.

Shaman, J. & Kohn, M. (2009). Absolute humidity modulates influenza survival, transmission, and seasonality. *Proceedings of the National Academy of Sciences of the United States of America*, 106, 3243–8.

Shaman, J., Karspeck, A., Yang, W. ,Tamerius, J., & Lipsitch, M. (2013). Real-time influenza forecasts during the 2012–2013 season. *Nature Communications*, 4, 28–37.

Shaman, J., Yang, W., & Kandula, S. (2014). Inference and forecast of the current West African Ebola outbreak in Guinea, Sierra Leone and Liberia. *PLoS Currents*, 6:e877c8b8f.

Shao, J. (1993). Linear model selection by cross-validation. *Journal of the American Statistical Association*, 88, 486–94.

Shi, Yuan, L., Xu, K., Suet-Yheng, T., et al. (2015). Three-month real-time dengue forecast models: An early warning system for outbreak alerts and policy decision support in Singapore. *Environmental Health Perspectives*, 124, 1369–75.

Siettos, C.I. & Russo, L. (2013). Mathematical modeling of infectious disease dynamics. *Virulence*, 4, 295–306.

Silver, N. (2012). *The Signal and the Noise: Why So Many Predictions Fail but Some Don't*. The Penguin Group, New York, New York, USA: The Penguin Group.

Smith, D.L., Battle, K.E., Hay, S.I., Barker, C.M., Scott, T.W., & McKenzie, F.E. (2012). Ross, Macdonald, and a theory for the dynamics and control of mosquito-transmitted pathogens. *PLoS Pathogens*, 8.

Smith, M.E., Singh, B.K., Irvine, M.A., et al. (2017). Predicting lymphatic filariasis transmission and elimination dynamics using a multi-model ensemble framework. *Epidemics*, 18, 16–28.

Soyiri, I.N. & Reidpath, D.D. (2010). An overview of health forecasting. *Environmental Health and Preventive Medicine*, 18, 1–9.

Stone, M. (1974). Cross-validatory choice and assessment of statistical predictions. *Journal of the Royal Statistical Society Series B*, 36, 111–47.

Surowiecki, J. (2004). *The Wisdom of Crowds: Why the Many are Smarter than the Few and how Collective Wisdom Shapes Business, Economies, Societies, and Nations*. Doubleday, New York.

Thakur, S. & Dharavath, R. (2019). Artificial neural network based prediction of malaria abundances using big data: A knowledge capturing approach. *Clinical Epidemiology and Global Health*, 7, 121–6.

Thorner, A.R., Cao B., Jiang, T., Warner, A.J., & Bonis, P.A. (2016). Correlation between up-to-date searches and reported cases of Middle East respiratory syndrome during outbreaks in Saudi Arabia. *HMS Scholarly Articles*, 3, ofw043.

Tompkins, A.M., Colón, F.J., Francesca, Di G.G., & Namanya, D.B. (2019). Dynamical malaria forecasts are skillful at regional and local scales in Uganda up to 4 months ahead. *GeoHealth*, 3, 58–66.

Ukel, S., Farrington, C.P., Garthwaite, P.H., Robertson, C., & Andrews, N. (2012). Statistical methods for the prospective detection of infectious disease outbreaks: A review. *Journal of the Royal Statistical Society: Series A (Statistics in Society)*, 175, 49–82.

Venkatramanan, S., Lewis, B., Chen, J., et al. (2017). Using data-driven agent-based models for forecasting emerging infectious diseases. *Epidemics*, 12, 37–49.

Viboud, C., Sun, K., Gaffey, R., et al. (2017). The rapid ebola forecasting challenge: Synthesis and lessons learnt. *Epidemics* 34, 12.

Winters, P.R. (1960). Forecasting sales by exponentially weighted moving averages. *Management Science*, 6, 324–42.

World Health Organization. (2010). Fact sheets: *Infectious Diseases*. Geneva: WHO.

Wu, Y., Yang, Y., Nishiura, H., & Saitoh, M. (1999). Deep learning for epidemiological predictions. In *The 41st International ACM SIGIR Conference on Research & Development in Information Retrieval*, pp. 1085–8. ACM, 2018.

Xu, Q., Yulia, G.R., Ramirez, L.L., et al. (2017). Forecasting influenza in Hong Kong with google search queries and statistical model fusion. *PLoS One*, 12, e0176690.

Yamana T.K. Kandula, S., & Shaman, J. (2017). Individual versus superensemble forecasts of seasonal influenza outbreaks in the United States. *PLoS Computational Biology*, 13, e1005801.

Yamana, T.K., Kandula, S., & Shaman, J. (2016). Superensemble forecasts of dengue outbreaks. *Journal of The Royal Society Interface*, 13, 20160410.

Yamana, T.K. & Shaman, J. (2019). A framework for evaluating the effects of observational type and quality on vector-borne disease forecast. *Epidemics*, 100359.

Yan, W., Xu, Y., Yang, X., & Zhou, Y. (2010). A hybrid model for short-term bacillary dysentery prediction in Yichang city, China. *Japanese Journal of Infectious Diseases*, 63, 264–70.

Yang, S., Kou, S.C., Lu, F., et al. (2017). Advances in using Internet searches to track dengue. *PLoS Computational Biology*, 13, e1005607.

Yang, S., Santillana, M., & Kou, S.C. (2015). Accurate estimation of influenza epidemics using Google search data via ARGO. *Proceedings of the National Academy of Sciences of the United States of America*, 112, 14473–8.

Zhu, G., Xiao, J., Zhang, B., et al. (2018). The spatiotemporal transmission of dengue and its driving mechanism: A case study on the 2014 dengue outbreak in Guangdong, China. *Science of the Total Environment*, 622–3, 252–9.

Zinszer, K., Kigozi, R., Charland, K., et al. (2015). Forecasting malaria in a highly endemic country using environmental and clinical predictors. *Malaria Journal*, 14, 245.

CHAPTER 5

Force of Infection and Variation in Outbreak Size in a Multi-Species Host-Pathogen System: West Nile Virus in New York City

John M. Drake, Krisztian Magori, Kevin Knoblich, Sarah E. Bowden, and Waheed I. Bajwa

5.1 Introduction

Understanding of the dynamics of emerging infectious diseases during the transition to endemicity is an important public health concern and a challenging problem for epidemic theory (Dobson & Foufopoulos 2001; Woolhouse & Gaunt 2007; Lloyd-Smith et al. 2009). Key quantities of interest include the size of primary and secondary outbreaks and prevalence at the endemic equilibrium. These quantities differ according to the epidemiological characteristics of the host-pathogen system. Due to the inherent complexities of transmission in vector-borne disease systems (i.e. often multi-host and/or multi-vector), such quantities are difficult to fully characterize (Hartemink et al. 2008).

West Nile virus (WNV) is a globally emerging infectious disease. Although first isolated in 1937 in Uganda (Smithburn et al. 1940), the modern global pandemic originated in New York City in 1999 (Lanciotti et al. 1999; Nash 2001). Due to subsequent spread throughout the Americas, WNV is now considered to be the most widespread arbovirus worldwide (Kramer, Styer & Ebel 2008). A member of the Japanese Encephalitis antigenic complex, WNV is transmitted to humans through the bite of an infectious mosquito, particularly *Culex* spp. and (in North America) is maintained in a sylvatic cycle involving multiple species of passerine birds (LaDeau et al. 2007). Humans are terminal hosts and do not transmit WNV to susceptible mosquitoes (Hayes et al. 2005). Since the beginning of the WNV pandemic in 1999, considerable effort has been invested in understanding the ecology, immunology, and genetics of WNV (Kramer, Styer & Ebel 2008). As an emerging pathogen in a major urban center, the epidemic of WNV in New York City provides an incomparable model for understanding the ecological and evolutionary processes by which vector-borne pathogens emerge. Here we report a study that combines theory and data to elucidate the key contributors to variation in the size of seasonal outbreaks.

WNV has recurred annually in outbreaks of different sizes since 1999. During the period from 2000–2008, annual differences of up to twenty-six cases were observed (Fig. 5.1). What explains this difference in outbreak sizes? WNV is highly explosive: numerical estimates for R_0 range from $R_0 \approx 25$ to $R_0 > 1200$ (Wonham et al. 2006). Comparison with

John M. Drake, Krisztian Magori, Kevin Knoblich, Sarah E. Bowden, and Waheed I. Bajwa, *Force of Infection and Variation in Outbreak Size in a Multi-Species Host-Pathogen System: West Nile Virus in New York City* In: *Population Biology of Vector-Borne Diseases*. Edited by: John M. Drake, Michael B. Bonsall, and Michael R. Strand: Oxford University Press (2021). © John M. Drake, Krisztian Magori, Kevin Knoblich, Sarah E. Bowden, and Waheed I. Bajwa.
DOI: 10.1093/oso/9780198853244.003.0005

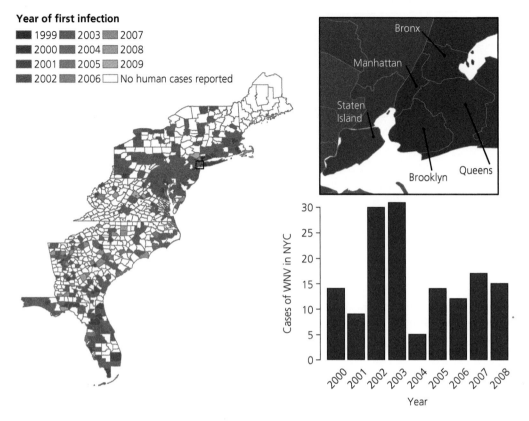

Figure 5.1 Spread of West Nile virus in North American began in the five boroughs of New York City in 1999 (inset) before spreading south and west (map). Epidemic intensity of West Nile virus in New York City (human case reports) varied six-fold from 2000 to 2008 (bar graph). This study aims to determine if this variation is due to inter-annual or spatial variation in the force of infection.

other infectious diseases suggests two explanations. First, it is known for other host-pathogen systems that seasonal periodicity in transmission confers a propensity to multi-annual cycles, especially in short-lived hosts with persistent immunity (Aron & Schwartz 1984, Altizer et al. 2006). Second, randomly fluctuating environmental conditions such as temperature (Kilpatrick et al. 2008; Hashizume, Chaves & Minakawa 2012), humidity (Soverow et al. 2009; van Noort et al. 2012), and precipitation (Wang et al. 2010; Wandiga et al. 2010) can cause variation in transmission intensity from year to year, especially for arthropod-borne pathogens where the vector is dependent on locally fluctuating environmental conditions for habitat (Altizer et al. 2006). WNV is a candidate for inter-annual variation in transmission intensity on both counts. Importantly, both of these hypotheses imply that variation in WNV outbreak sizes will correlate with variation in the annual exposure of susceptible

hosts to potentially infectious contacts. In theoretical epidemiology, this quantity is known as the force of infection (Keeling & Rohani, 2008).

An omnibus test of these two hypotheses can be constructed from an analysis of the force of infection if estimated independently of the epidemic trajectory. Since WNV is a multi-vector pathogen, its force of infection is the sum of the forces of infection of the system's constituent vector species (Dobson 2004, Kilpatrick et al. 2005; Kilpatrick 2011). Here, we show that the total force of infection can be decomposed into individually measurable components (up to a coefficient of proportionality) via a mechanistic model for the infection process (Dobson & Foufopoulos 2001; Kilpatrick & Altizer 2010; Magori et al. 2011). Using surveillance data on the prevalence of WNV in mosquitoes and biting rates on different hosts together with meta-analyses of published experiments on virus dissemination and infectivity, we obtained monthly estimates of the force of infection from 2000–2008 at two spatial scales in New York City: the city entire (786 km²) and separately for each of its five boroughs (mean: 157 km², sd: 84.4 km²). Comparison of these estimates with the observed number of human cases provides little evidence that variation in outbreak size is related to variation in the force of infection. We conclude by proposing that inter-annual variation in WNV cases in New York City is due to the percolation-like spread of the virus in the urban environment and the heterogeneity of human concentration along realized epidemic paths.

5.2 Methods

5.2.1 Magnitude of Variation in West Nile virus Outbreaks

During 2000–2008, the number of persons reported to be infected with West Nile virus ranged from five (in 2004) to 31 (in 2003). What accounts for these differences? It does not appear to be variation in weather (Fig. 5.2) despite the well-known fact that weather variations significantly affect the transmission of mosquito-borne pathogens (Tesla et al. 2018; Shocket et al. 2019). Due to under-reporting, these observations are an underestimate of the number of human infections by a factor of ~140 (Mostashari et al. 2001). Further, in finite populations there is intrinsic variation in the number of infected persons due to the stochastic infection process (demographic stochasticity). Both of these processes invariably contribute to the observed variation. To estimate the expected proportion of variation due to these two factors, we first calculated the estimated per capita annual probability (p) of a resident reporting infection by taking the total number of reported cases over the nine years of this study and dividing by the number of exposed person years (i.e. nine times the approximate population size of New York City, $N = 8{,}175{,}100$). Assuming the reporting rate to be constant, the rule for conditional binomials implies the theoretical distribution

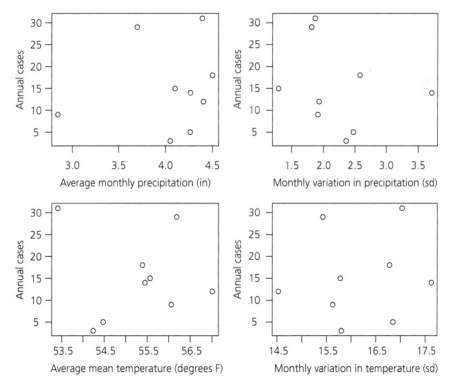

Figure 5.2 There is no evidence that human cases of WNV correlate with variation in four weather variables. Pearson correlation coefficients: average monthly precipitation in inches ($r = 0.14$, $p = 0.71$), standard deviation in precipitation ($r = -0.25$, $p = 0.52$), average monthly temperature ($r = -0.07$, $p = 0.85$), standard deviation in temperature ($r = 0.13$, $p = 0.74$).

of epidemic sizes to be a binomial random variate with parameters p and N. We then calculated the coefficient of variation of nine simulated binomial variates (corresponding to the nine-year period of the study) 10,000 times to obtain the theoretical distribution for the observable coefficient of variation over a nine-year observation window assuming only demographic stochasticity and under-reporting. This result can be compared with the observed coefficient of variation to quantify the magnitude of the discrepancy.

We then derived an expression for the force of infection to humans in a multi-species vector-borne pathogen (Ross 1910; Macdonald 1957; Smith et al. 2012; Magori & Drake 2013). We start with the multi-species Ross–Macdonald model with spillover transmission to humans:

$$\frac{dS_R}{dt} = \mu_R N_R + \delta_R I_R - \alpha_R \beta c_V I_V \frac{(1-\eta)S_R}{(1-\eta)N_R + \eta N_H} - \mu_R S_R, \quad (5.1a)$$

$$\frac{dI_R}{dt} = \alpha_R \beta c_V I_V \frac{(1-\eta)S_R}{(1-\eta)N_R + \eta N_H} - \mu_R I_R - \delta_R I_R - \gamma_R I_R, \quad (5.1b)$$

$$\frac{dR_R}{dt} = \gamma_R I_R - \mu_R R_R, \quad (5.1c)$$

$$\frac{dS_V}{dt} = \mu_V N_V - \mu_V S_V - \alpha_V \beta S_V \frac{(1-\eta)I_R}{(1-\eta)N_R + \eta N_H}, \quad (5.1d)$$

$$\frac{dE_V}{dt} = \alpha_V \beta S_V \frac{(1-\eta)I_R}{(1-\eta)N_R + \eta N_H} - \mu_V E_V - \kappa_V E_V, \quad (5.1e)$$

$$\frac{dI_V}{dt} = \kappa_V E_V - \mu_V I_V, \quad (5.1f)$$

$$\frac{dS_H}{dt} = -\alpha_H \beta c_V I_V \frac{\eta S_H}{(1-\eta)N_R + \eta N_H}, \quad (5.1g)$$

$$\frac{dI_H}{dt} = \alpha_H \beta c_V I_V \frac{\eta S_H}{(1-\eta)N_R + \eta N_H} - \gamma_H I_H - \delta_H I_H, \quad (5.1h)$$

$$\frac{dR_H}{dt} = \gamma_H I_H. \quad (5.1i)$$

where $S_R, I_R, R_R, S_V, E_V, I_V, S_H, I_H$ and R_H are the number of susceptible, infectious and recovered reservoir hosts; susceptible, exposed and infected vectors, and susceptible, infected and recovered humans,

respectively. The parameters μ_R and μ_V are the mortality rates of reservoir hosts and vectors; a_R, a_V, a_H are the probability of transmission from an infectious vector to a reservoir, from an infectious reservoir to a vector, and from an infectious vector to a human host, respectively; β is the biting rate of vectors; $1/\kappa_v$ is the length of the incubation period in the vectors; γ_R and γ_H are the recovery rates of infectious reservoir hosts and infected humans, respectively; δ_R and δ_H are the excess mortality rates of infectious reservoir hosts and humans (i.e. disease-induced mortality), η is the biting preference of vectors for humans, and c_V is competence, defined as the proportion of WNV-positive mosquitoes that become infectious. It is important to note that humans are dead-end hosts of WNV, i.e. they do not contribute to mosquito infections, and they have been treated accordingly in this model. In a more general sense, Eq. (5.1g–5.1i) can represent the spillover transmission of WNV to any non-competent species, in addition to humans, e.g. horses. For simplicity, Eq. (5.1a–5.1i) describe the epidemiological dynamics with a single species of reservoir host and mosquito vector, but can be easily extended, using appropriate indexing, to multiple species of both.

The basic reproductive number of a simplified version of this model, excluding human hosts, was obtained using the 'spectral radius method', and is given by the expression (Magori et al. 2011):

$$R_0 = \sqrt{\alpha_V \alpha_R \beta^2 \frac{N_V}{N_R} \frac{\kappa_V}{\kappa_V + \mu_V} \frac{1}{\mu_V (\delta_R + \mu_R + \gamma_R)}} \quad (5.2)$$

where N_R and N_V are the total number of reservoir hosts and vectors, respectively.

The force of infection to humans is defined as the per capita rate at which susceptible individuals become infected by the pathogen. The force of infection of WNV for human hosts (λ_h) is in this model given by the ratio of the positive terms in Eq. (5.1h) to the susceptible human population:

$$\lambda_h = \sum_{i=1}^{n} \alpha_H \beta_i c_{V,i} I_{V,i} \frac{\eta_i}{(1-\eta_i)N_R + \eta_i N_H} \quad (5.3)$$

This expression decomposes the force of infection into components due to the n different species of biting mosquitoes and may be calculated over arbitrary time intervals to illustrate the dynamics of the evolving epidemic within each year.

This expression for the force of infection may be calculated up to a coefficient of proportionality from observations generated by standard surveillance activities, particularly:

- *Catch Per Unit Effort (CPUE)*—an effort-standardized measure of the density of mosquitoes in a local area;
- *Minimum Infection Rate (MIR)*—a lower bound on the prevalence of infection within a population of mosquitoes;
- *Vector Competence (VC)*—the proportion of infected mosquitoes that become infectious and transmit in a standard assay;
- *Mammal Biting Rate (MBR)*—the proportion of infected mosquitoes that bite mammalian hosts, as a proxy for human biting frequency.

Specifically, assuming $I_v \propto CPUE \times MIR$ and $\frac{\eta N_H}{(1-\eta)N_R + \eta N_H} \propto MBR$, identifying VC with $\beta \times c_v$, and substituting in eqn (5.3), we have the estimator for the partial (species-specific) force of infection:

$$\hat{\lambda}_i \propto \hat{CPUE}_i \times \hat{MIR}_i \times \hat{VC}_i \times \hat{MBR}_i \qquad (5.4)$$

where hats indicate an estimate, assumed to be approximately constant over the time interval of interest, obtained from data on species i. In the case of the substitution for I_v, the proportionality assumption amounts to assuming that prevalence is sufficiently low that the probability of having more than one infected individual in a test pool is negligible and that catchability does not differ between infected and uninfected mosquitoes. In the case of MBR, the proportionality is assumed because the actual frequency of human biting is small yielding a very high variance (low precision) estimate of the human biting rate. It is known, however, that mosquitoes that bite mammals often bite humans, too (Apperson et al. 2004). Thus, the partial force of infection may be calculated up to a coefficient of proportionality from measurable quantities including the catch per unit effort, minimum infection rate of mosquitoes, vector competence, and species-specific mammal biting rate. The coefficient of proportionality is a_H/N_H: the transmission probability

divided by the number of available human hosts. Further, a may be assumed to be relatively constant across species, since the competent species of mosquitoes would all typically take blood meals every two to three days in accordance with their gonotrophic cycles. Thus, the total force of infection may be estimated (up to a coefficient of proportionality) by summing the partial forces of infection.

Our measure of force of infection (λ_i) is conceptually very similar (the estimator is identical) to the measure of risk defined by Kilpatrick et al. (2005) as the product of vector abundance, vector competence, biting preference and WNV infection prevalence. The conceptual value of our derivation is that we demonstrate that this empirically estimable quantity is related to the force of infection to humans in a multi-species vector-borne epidemiological model. This measure is related to, but different from, the force of infection defined by Hamer et al. (2011), which is defined as the number of infectious *Culex pipiens* mosquitoes from vector feeding on a community of bird hosts, and therefore considers force of infection to mosquitoes and not to humans.

5.2.2 Parameter Estimates

The variables *CPUE*, *MIR*, *VC*, and *MBR* were estimated as follows.

CPUE was calculated at the scale of the entire city and individually within the five boroughs from routine surveillance data at one-month intervals by summing the number of mosquitoes collected by species within the geographic unit and dividing by the number of trapping instances (nightly deployments of CDC light traps and Reiter's gravid traps). See Magori et al. (2011) for more details. Number of trapping instances was not recorded in 2000. Therefore, for this year only, the number of trapping instances was estimated by counting the number of unique date-mosquito pool combinations. This number is an underestimate, since traps not collecting any mosquitoes would not be recorded.

MIR was calculated at monthly intervals from the number of WNV+ mosquito pools and the number of individuals in each pool using the bias-corrected maximum likelihood estimate for binomial sampling with unequal pool sizes and skewness-corrected score interval confidence interval (Hepworth 2005) implemented by Biggerstaff (2006).

To estimate *VC*, we performed a statistical meta-analysis of published experimental studies of vector competence (Turell et al. 2000; Turell et al. 2001; Sardelis et al. 2001; Turell et al. 2005; Tiawsirisup et al. 2005). Published studies were surveyed for reports of two quantities: *dissemination* (the proportion of mosquitoes fed on infected blood in which virus is disseminated to the legs in detectable quantities) and *infectivity* (the proportion of mosquitoes with WNV in their salivary glands that successfully infect a susceptible chicken). To be included in the meta-analysis, we required that dissemination trials be based on virus concentrations exceeding the presumptive infective dose of 6.0 pfu ml^{-1} (Turell et al. 2005). Proportion of mosquitoes disseminating and proportion of infected mosquitoes infectious to a susceptible host were calculated across studies using a random effects model and logit transformation of the observed proportions. Since *VC* is defined as the conditional probability of successful transmission given infection and subsequent biting of a susceptible host, species-specific *VC* may then be quantified as the product of these two proportions. Uncertainty in *VC* was propagated by assuming the sampling distributions (after logit-transformation) to be normally distributed, obtaining Monte Carlo samples of the estimated dissemination and infectivity distributions according to Gaussian distributions with the appropriate mean and variance, multiplying these samples and back-transforming the product. This procedure effectively propagates uncertainty in the estimated dissemination and infectivity rates in the calculation of *VC*. *VC* has not been measured in three mosquito species in which WNV has been detected: *Anopheles quadrimaculatus*, *Orthopomodyia signifera*, and *Uranotenia sapphira*. However, it is believed that none of these species transmits WNV (pers. comm. Mike Turrell) so *VC* was set to 0 in these cases. No information could be found on dissemination in *Coquillettidia perturbans*. We therefore conservatively assume that WNV disseminates at the maximum estimated dissemination rate for any species (that of *Ae. albopictus*). Finally, in some collections the identity of *Cx. pipiens* and *Culex restuans* were not distinguished. In

these cases, the collected mosquitoes were included in the *Cx. pipiens* pool, which has the higher *VC*.

To estimate *MBR* we performed a statistical meta-analysis of published studies aimed at identifying the composition of blood meals for mosquitoes collected in the field. Criteria for inclusion in this analysis were that the study had to be published since the introduction of WNV, and it had to analyze blood meals of mosquitoes collected in the northeastern US to at least the taxonomic rank of class (Gingrich & Williams 2005; Apperson et al. 2002; Apperson et al. 2004; Molaei et al. 2006). As for vector competence, multiple studies were fit simultaneously using a random effects model. However, in this case data were pooled across studies after Freeman–Tukey double arcsine transformation. Data on the composition of blood meals were not available for three other mosquito species: *Ochlerotatus trivittatus*, *Orthopomodyia signifera*, and *Uranotenia sapphira*. It is known from the literature, however, that these species are feeding specialists (mammal: *O. trivittatus*; reptile: *O. signifera* and *U. sapphira*; Pinger & Rowley, 1975; Carpenter & La Casse, 1974). *MBR* was therefore set to 1 (all mosquitoes bite mammals) or 0 (no mosquitoes bite mammals) accordingly. Both meta-analyses were performed using the 'meta' package in R (Schwarzer et al. 2015).

5.2.3 Aggregation

To obtain monthly point estimates of the force of infection by species, we multiplied the point estimates of the parameter estimates described in the previous section. To propagate uncertainty, we computed the estimated distribution of this statistic through Monte Carlo simulation from the sampling distributions. Empirical confidence intervals were then obtained as the 2.5 percent and 97.5 percent quantiles from 10,000 samples. Results were summed over species to obtain the total estimated monthly force of infection.

5.2.4 Time Series Analysis

Plots of the estimated force of infection over time were constructed to compare temporal and spatial variation in the force of infection. The city-wide year-to-year consistency in estimated force of infection was quantified by pooling monthly force of infection estimates and performing local linear regression. For comparison with the observed number of human infections, force of infection time series were seasonally decomposed using a loess-based approach (Cleveland et al. 1990). This method reduces the observed series, in our case the monthly estimated force of infection λ_t, into a sum of seasonal (S_t), long-term trend (T_t), and residual components (R_t):

$$\lambda_t = S_t + T_t + R_t \qquad (5.5)$$

We then calculated seasonal anomalies, $z_t = T_t + R_t$ (the sum of trend and residual components) for comparison with the observed number of human cases.

5.3 Results

5.3.1 Magnitude of Variation in West Nile virus Outbreaks

We estimated the coefficient of variation in observed number of human cases of West Nile virus from 2000–2008 to be $CV = 0.537$. The total number of reported cases over the nine years of this study was $n = 147$. Upon dividing by the approximate number of exposed person years $(9 \times 8,175,100)$, we obtained estimated probability of reporting infection per person per year of

$$\hat{p} = \frac{n}{9 \times N} = \frac{147}{9 \times 8,175,100} \approx 0.0000019979. \quad (5.6)$$

Monte Carlo simulations of the coefficient of variation over nine years yielded a median of $CV_{\text{median}} = 0.238$. The 2.5 percent and 97.5 percent quantiles were $CV_{2.5\%} = 0.130$ and $CV_{97.5\%} = 0.368$, respectively. Thus, roughly half of the observed variation $(1 - CV_{\text{median}} / CV = 1 - 0.443 = 0.557)$ remains to be explained. At a maximum, demographic stochasticity and under-reporting explain 68.5 percent of the observed variation in outbreak size.

5.3.2 Force of Infection

Virus dissemination and infectivity were heterogeneous across species (Fig. 5.3a,b). Accordingly,

vector competence also varied across species (Fig. 5.3c). Unsurprisingly, *Cx.* and *Aedes albopictus* were found to be highly competent (Fig. 5.3c). Importantly, *Cx. pipiens* was found to be less competent than *Cx. salinarius* and *Cx. restuans* due to a lower rate of virus dissemination (Fig. 5.3c). Mammal biting rate was highest among species not particularly competent for WNV, e.g. *Aedes vexans* (Fig. 5.3d). *Culex salinarius* is notable for having both relatively high competence and mammal biting rate. These results underscore that the ecological complexity of West Nile virus transmission is important to its epidemiological dynamics.

Substituting these estimates into Eqn. (5.4) provides an estimate of the combined effect of all species on the total force of infection. Visual inspection (Fig. 5.4) leads to the somewhat surprising conclusion that the force of infection, averaged over the City of New York, was quite consistent from year to year, not variable as one would expect if transmission is driven by weather anomalies. How consistent? Pooling all months from all years and fitting a smooth curve of the individual force of infection estimates against time produces a picture of the consistency for the force of infection curve from year to year. The coefficient of determination $(R^2 = 0.41)$, calculated as the squared value of the correlation between predicted and observed values in this regression, provides a quantitative measure of this consistency. A different measure is obtained by comparing the variance of z_t with the variance of λ_t, i.e. $R^2 = 1 - \text{var}(z_t) / \text{var}(\lambda_t) = 0.698$, suggesting that the proportion of variance explained by seasonality is about 70 percent. The finding that force of infection is consistent from year to year is reinforced by inspection of the time series of estimated force of infection for each of the separate boroughs (Fig.5. 5). In these series there is clearly a greater level of variability, but the overall impression of a very consistent seasonal pattern among boroughs remains.

The seasonal consistency demonstrated in these two analyses provides the first indication that interannual variation in the force of infection is at best a contributing factor to the year-to-year variation in outbreak size. Indeed, the strength and timing of the contribution may be quantified by calculating the correlation between anomalies in the force of

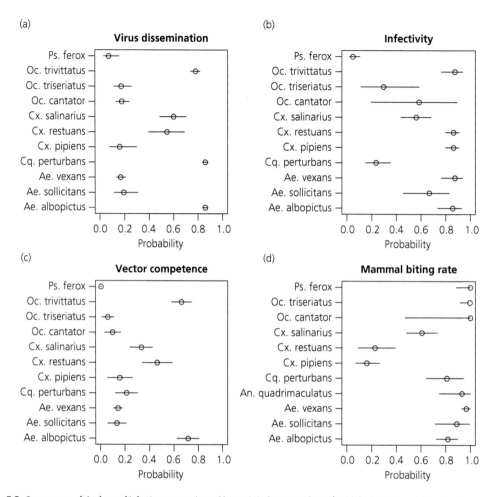

Figure 5.3 Components of the force of infection were estimated by statistical meta-analysis of published experimental studies. Summary results show there to exist substantial variation among species in (a) probability of virus dissemination (b) infectivity, (c) vector competence (the product of virus dissemination and infectivity, estimated through Monte Carlo simulation from results shown in panels (a) and (b)), and (d) mammal biting rate. Error bars are back-transformed one standard error deviations from the mean effect estimated across studies.

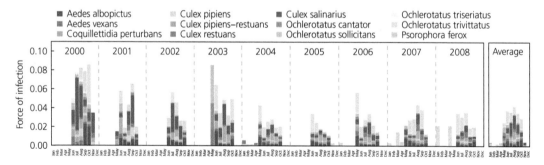

Figure 5.4 Spatially averaged force of infection of West Nile virus consistently varied in a seasonally predictable way during 2000–2008. Colored bars represent the partial forces of infection attributable to twelve different mosquito species. Typically, force of infection first reached substantial levels in June and continued through the fall, although in some years virus was detected throughout the winter and the first month at which force of infection exceeded 0.04 ('high exposure') varied considerably from year to year. Seasonality is highly predictive of the estimated force of infection, accounting for a substantial portion of the total observed variation ($R^2 = 0.41$).

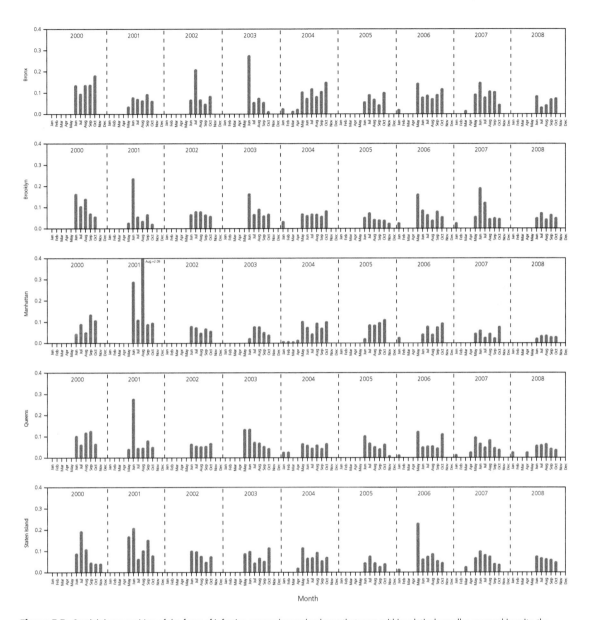

Figure 5.5 Spatial decomposition of the force of infection among boroughs shows that even within relatively smaller geographic units, the course of the WNV epidemic was highly predictable from year to year. There was some variation among boroughs in the average summertime peak in force of infection, but little variation in the duration of detectable transmission. The high estimate in the borough of Manhattan in August 2001 is due to the collection of 1100 *Ochlerotatus sollicitans* in gravid traps at the Trinity Cemetery which were not tested for WNV, and a separate collection of a single *Ochlerotatus sollicitans* which tested positive for WNV.

infection and the number of cases. At the scale of the city, there was no evidence for a correlation in any of the key summer months (June: $\rho = 0.35$, $p = 0.35$; July: $\rho = 0.42$, $p = 0.26$; August: $\rho = 0.29$, $p = 0.46$). Presumably, this is due to the low power of nine observations to reject the null hypothesis of

no effect. The analysis is strengthened at the scale of boroughs, where there was strong evidence for an effect in July ($\rho = 0.39, p = 0.007$), but not June ($\rho = 0.11, p = 0.49$) or August ($\rho = 0.08, p = 0.60$) (Fig. 5.6). Squaring the correlation coefficient for July ($R^2 = 0.39^2 = 0.152$) provides a measure of the

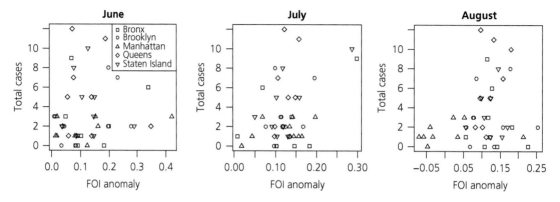

Figure 5.6 Correlation between total number of human cases and the non-seasonal components of the force of infection by borough. Correlation coefficients calculated by pooling across boroughs show there to be a significant correlation in July ($p = 0.39$, $p = 0.007$), but not June ($p = 0.11$, $p = 0.49$), or August ($p = 0.08$, $p = 0.60$).

magnitude of the effect, indicating that seasonal anomalies in force of infection explain about 15 percent of the variation in the number of human cases of West Nile virus. In conclusion, there is strong evidence for only a weak effect of anomalies in the force of infection on the year-to-year variation in the size of West Nile virus outbreaks in New York City.

5.4 Discussion

In summary, New York City experienced a variation of 26 human cases of West Nile virus from 2000–2008. Calculations based on sampling from finite populations indicate that under-reporting and intrinsic noise explain ~50 percent of this variation. Anomalies in the force of infection explain an additional 15 percent. These results are surprising in view of the large size of this system (8 million susceptible human hosts) and the well-established environmental dependence of most vector-borne pathogens (Rohr et al. 2011).

Weaknesses of this analysis include the precision with which we could estimate some parameters and the possibility that our estimation procedures are subject to bias of different kinds. Thus, for instance, we do not have a good estimate of the human biting rate and have instead assumed that the human biting rate is a simple proportion of the propensity of a species to feed on mammals in general. Another possible source of bias is that our cal-

culations are all performed on a data set that pools across different trap types. It is well known, however, that the different kinds of mosquito traps capture species at different rates (Williams & Gingrich 2007). Change in the proportion of traps of each kind therefore inevitably contributes to bias into our estimates of *CPUE*, which directly feeds the calculation of force of infection. We have not attempted to quantify the contribution of this source of bias to our estimates. Given the other very large sources of variation in this system, and especially in view of the seasonal consistency of our estimates, we think it very unlikely that the differences between trap types have detectably contributed to our estimates of the force of infection.

Is it nonetheless possible that there exists considerable inter-annual variation in the force of infection that we have failed to detect? A skeptic of our conclusion that the observed variation in outbreak size is only marginally due to variation in the force of infection might argue that we are, in effect, accepting the null hypothesis (no effect) under circumstances of low power (i.e. $n = 9$ years of observation at the city scale) in the calculation of correlations between anomalies and observed number of cases. The case is strengthened, of course, in the borough-level analysis, which provides stronger evidence and allows calculation of the magnitude of the effect, estimated to be about 15 percent of the total variation. Of course, the skeptic could reply that these results also are subject to limited power, to which we would respond that, if anything, our confidence in this estimate is inflated since we have not accounted for the multiple comparisons in six different correlations calculated in the search for a relationship and the estimated magnitudes of the remaining correlations are relatively small. The conclusion that variation in force of infections is only a minor contributor to variation would appear to be robust.

What, then, remains to explain the observed variation in outbreak size? Here we consider four possibilities, of which three seem implausible. The first is that the number of human cases is responsive to the size of the susceptible population, i.e. that the number of susceptible persons exposed varied substantially from year to year. This would occur if, for instance, there was a decline in the proportion of the population taking defensive measures such as applying insect repellent or avoiding areas with mosquitoes in high density or if herd immunity was built in the population. However, if this were the case then one would expect epidemic size to show a clear trend, for which there is presently no evidence. Alternatively, it is possible that under-reporting and demographic stochasticity explain a greater proportion of the variation than concluded here. For this explanation to be correct requires either that under-reporting itself varies six-fold or that the observed variation is truly anomalous in the sense of being far in the tail of the distribution. A third possible explanation focuses on the years 2002 and 2003 in which the number of cases was greatest. During these years, WN02 viral genotype replaced the NY99 genotype. So, during this time, there was a temporary increase in genetic diversity. Increased genetic diversity, which confers a fitness benefit in mosquitoes (Fitzpatrick et al. 2010), was previously linked to higher prevalence in the state of New York (Ebel et al. 2004). Such a transient increase in diversity might have boosted MIR, leading to a subsequent temporary increase in the force of infection. The WN02 genotype completely replaced NY99 by 2004, however, corresponding to a decrease in genetic diversity, therefore hypothetically causing the system to return to its previous (lower) force of infection. In contradiction to this hypothesis, however, human cases in NYC increased to the greatest on record in 2010 and 2012 without any clear attendant evolution change.

Failing these explanations, we consider one final possibility, namely that the variation in outbreak size is due to spatial heterogeneity in the concentration of susceptible persons across the city. In fact, this explanation is consistent with our previous results (Magori et al. 2011), in which it was shown that the spatial dynamics of West Nile virus are influenced by spatial heterogeneity so that spread within years is percolation-like rather than wave-like. One corollary of this finding is that the spatial path through the city is likely to vary considerably from year to year. If the human population is heterogeneously concentrated at scales smaller than the borough-level aggregations reported here, then it is plausible that variation in outbreak size actually reflects variation in human concentration on realized epidemic paths. Indeed, a weaker version of this hypothesis is that just the age structure of the population is heterogeneous, since persons ≥65 years of age are at substantially greater risk of neuroinvasive disease. We note that this is an imminently testable hypothesis, requiring only more finely resolved spatial information on changes in the force of infection over time.

In conclusion, we have found that variation in the size of annual West Nile virus outbreaks in New York City is not determined by variation in the force of infection at scales of either the entire city or its five constituent boroughs. This finding rules out both endogenous dynamics (e.g. multi-annual cycling) and extrinsic perturbations (e.g. differences in weather across years) as possible explanations of the observed variation. We propose instead, that the observed variation in outbreak size is due to heterogeneity in the susceptible population along stochastic epidemic paths—a phenomenon that could only result from percolation-like spread in heterogeneous environments (Magori et al. 2011).

Acknowledgments

This work was supported by research grant EF-0723601 from the joint NSF-NIH Ecology of Infectious Diseases program.

References

Altizer, S., Dobson, A., Hosseini, P., Hudson, P., Pascual, M., Rohani, P. (2006). Seasonality and the dynamics of infectious diseases. *Ecology Letters*, 9, 467–84.

Apperson, C.S. Hassan, H.K., Harrison, B.A. et al. (2004). Host feeding patterns of established and potential mosquito vectors of West Nile virus in the eastern United States. *Vector-Borne and Zoonotic Diseases*, 4, 71–82.

Apperson, C.S., Harrison, B.A., & Unnasch, T.R. (2002). Host-feeding habits of *Culex* and other mosquitoes (Diptera: Culicidae) in the borough of Queens in New York City, with characters and techniques for identification of *Culex* mosquitoes. *Journal of Medical Entomology*, 39, 777–85.

Aron, J.L. & Schwartz, I.B. (1984). Seasonality and period-doubling bifurcations in an epidemic model. *Journal of Theoretical Biology*, 110, 665–79.

Biggerstaff, B.J. (2006). PooledInfRate, Version 3.0: a Microsoft Excel Add-In to compute prevalence estimates from pooled samples. *Fort Collins, CO: Centers for Disease Control and Prevention*.

Carpenter, S.J. & La Casse, W.J. (1974). *Mosquitoes of North America (north of Mexico)*. California University Press, Berkeley.

Cleveland, R.B. Cleveland, W.S., McRae, J.E. & Terpenning, I. (1990). STL: A seasonal-trend decomposition procedure based on loess. *Journal of Official Statistics*, 6, 3–73.

Dobson, A. (2004). Population dynamics of pathogens with multiple host species. *The American Naturalist*, 164, S64–78.

Dobson, A. & Foufopoulos, J. (2001). Emerging infectious pathogens of wildlife. *Philosophical Transactions of the Royal Society of London, Series B*, 356, 1001–12.

Ebel, G.D., Carricaburu, J., Young, D., Bernard, K.A. & Kramer, L.D., (2004). Genetic and phenotypic variation of West Nile virus in New York, 2000–2003. *American Journal of Tropical Medicine and Hygiene*, 71, 493–500.

Fitzpatrick, K.A., Deardorff, E.R., Pesko, K., et al. (2010). Population variation of West Nile virus confers a host-specific fitness benefit in mosquitoes. *Virology*, 404, 89–95.

Gingrich, J.B. & Williams, G.M. (2005). Host-feeding patterns of suspected West Nile virus mosquito vectors in Delaware, 2001–2002. *Journal of the American Mosquito Control Association*, 21, 194–200.

Hamer, G.L. Chaves, L.F., Anderson, T.K., et al. (2011). Fine-scale variation in vector host use and force of infection drive localized patterns of West Nile virus transmission. *PLoS One*, 6, e23767.

Hartemink, N.A., Randolph, S.E., & Davis, S.A. (2008). The basic reproduction number for complex disease systems: Defining R_0 for tick-borne infections. *American Naturalist*, 171, 743–54.

Hashizume, M., Chaves, L.F. & Minakawa, N. (2012). Indian Ocean Dipole drives malaria resurgence in East African highlands. *Scientific Reports*, 2, 269.

Hayes, E.B., Komar, N., Nasci, R.S., Montgomery, S.P., O'Leary, D.R. & Campbell, G.L. (2005). Epidemiology and transmission dynamics of West Nile virus disease. *Emerging Infectious Diseases*, 11, 1167–73.

Hepworth, G. (2005). Confidence intervals for proportions estimated by group testing with groups of unequal size. *Journal of Agricultural, Biological, and Environmental Statistics*, 10, 478–97.

Keeling, M.J. & Rohani, P. (2008). *Modeling Infectious Diseases in Humans and Animals*. Princeton University Press, Princeton.

Kilpatrick, A.M. (2011). Globalization, land use, and the invasion of West Nile virus. *Science*, 334, 323–7.

Kilpatrick, A.M., Meola, M.A., Moudy, R.M. & Kramer, L.D. (2008). Temperature, viral genetics, and the transmission of West Nile virus by *Culex pipiens* mosquitoes. *PLoS Pathogens*, 4, e1000092.

Kilpatrick, A.M., Kramer, L.D., Campbell, S.R., Alleyne, E.O., Dobson, A.P. & Daszak, P. (2005). West Nile virus risk assessment and the bridge vector paradigm. *Emerging Infectious Diseases*, 11, 425–9.

Kilpatrick, A.M. & Altizer, S. (2010). Disease ecology. *Nature Education Knowledge* 3, 55. Available at: https://www.nature.com/scitable/knowledge/library/disease-ecology-15947677.

Kramer, L.D., Styer, L.M. & Ebel, G.D. (2008). A global perspective on the epidemiology of West Nile virus. *Annual Review of Entomology*, 53, 61–81.

LaDeau, S.L., Kilpatrick, A.M. & Marra, P.P. (2007). West Nile virus emergence and large-scale declines of North American bird populations. *Nature*, 447, 710–13.

Lanciotti, R.S., Roehrig, J.T., Deubel, V. et al. (1999). Origin of the West Nile virus responsible for an outbreak of encephalitis in the northeastern United States. *Science*, 286, 2333–7.

Lloyd-Smith, J.O., George, D., Pepin, K., et al. (2009). Epidemic dynamics at the human-animal interface. *Science*, 326, 1362–7.

Macdonald, G. (1957). *The Epidemiology and Control of Malaria*. Oxford University Press, Oxford.

Magori, K. et al. (2011). Decelerating spread of West Nile virus by percolation in a heterogeneous urban landscape. *PLoS Computational Biology*, 7, e1002104.

Magori, K. & Drake, J.M. (2013). The population dynamics of vector-borne diseases. *Nature Education Knowledge* 4, 14. Available at: https://www.nature.com/scitable/knowledge/library/the-population-dynamics-of-vector-borne-diseases-102042523.

Molaei, G., Andreadis, T.G., Armstrong, P.M., Anderson, J.F. & Vossbrinck, C.R. (2006). Host feeding patterns of *Culex* mosquitoes and West Nile virus transmission, northeastern United States. *Emerging Infectious Diseases*, 12, 468–74.

Mostashari, F., Bunning, M.L., Kitsutani, P.T. et al. (2001). Epidemic West Nile encephalitis, New York, 1999: results of a household-based seroepidemiological survey. *Lancet*, 358, 261–4.

Nash, D., Mostashari, F., Fine, A., et al. (2001). The outbreak of West Nile virus infection in the New York City area in 1999. *The New England Journal of Medicine*, 344, 1807–14.

Pinger, R.R. & Rowley, W.A. (1975). Host preferences of *Aedes trivittatus* (Diptera: Culicidae) in central Iowa. *American Journal of Tropical Medicine and Hygiene*, 24, 889–93.

Rohr, J.R., Dobson, A.P., Johnson, P.T.J., et al. (2011). Frontiers in climate change–disease research. *Trends in Ecology & Evolution*, 26, 270–7.

Ross, R. (1910). *The Prevention of Malaria*. E.P. Dutton, New York.

Sardelis, M.R., Turrell, M.J., Dohn, D.J. & O'Guinn, M.L. (2001). Vector competence of selected North American Culex and Coquillettidia mosquitoes for West Nile virus. *Emerging Infectious Diseases* 7:1018-1022.

Schwarzer, G., Carpenter, J.R. & Rücker, G. (2015). *Meta-analysis with R*. Springer, New York.

Shocket, M.S., Anderson, C.B., Caldwell, J.M. et al. (2020). Environmental drivers of vector-borne diseases. (Chapter 6, this volume).

Smith, D.L., Battle, K.E., Hay, S.I., Barker, C.M., Scott, T.W. & McKenzie, F.E. (2012). Ross, Macdonald, and a theory for the dynamics and control of mosquito-transmitted pathogens. *PLoS Pathogens*, 8, e1002588.

Smithburn, K.C., Hughes, T.P., Burke, A.W. & Paul, J.H. (1940). A Neurotropic virus isolated from the blood of a native of Uganda. *American Journal of Tropical Medicine and Hygiene*, 20, 471–92.

Soverow, J.E., Wellenius, G.A., Fisman, D.N. & Mittleman, M.A. (2009). Infectious disease in a warming world: how weather influenced West Nile virus in the United States (2001–2005). *Environmental Health Perspectives*, 117, 1049–52.

Tesla, B., Demakovsky, L.R., Mordecai, E.A. et al., 2018). Temperature drives Zika virus transmission: evidence from empirical and mathematical models. *Proceedings of the Royal Society B: Biological Sciences*, 285, 20180795.

Tiawsirisup, S., Platt, K.B., Evans, R.B. & Rowley, W.A. (2005). A comparision of West Nile virus transmission by *Ochlerotatus trivittatus* (COQ.), *Culex pipiens* (L.), and *Aedes albopictus* (Skuse). *Vector-borne and Zoonotic Diseases*, 5, 40–7.

Turrell, M.J., Dohm, D.J., Sardelis, M.R., O'Guinn, M.L., Andreadis, T.G., Blow, J.A. (2005). An update on the potential of north American mosquitoes (Diptera: Culicidae) to transmit West Nile virus. *Journal of Medical Entomology*, 42, 57–62.

Turrell, M.J., O'Guinn, M.L., Dohm, D.J. & Jones, J.W. (2001). Vector competence of North American mosquitoes (Diptera: Culicidae) for West Nile virus. *Journal of Medical Entomology*, 38, 130–4.

Turell, M.J., O'Guinn, M., & Oliver, J. (2000). Potential for New York mosquitoes to transmit West Nile virus. *American Journal of Tropical Medicine and Hygiene*, 62, 413–14.

van Noort, S.P., Águas, R., Ballesteros, S. & Gomes, M.G.M. (2012). The role of weather on the relation between influenza and influenza-like illness. *Journal of Theoretical Biology*, 298, 131–7.

Wandiga, S.O., Opondo, M., Olago, D. & Githeko, A. et al. (2010). Vulnerability to epidemic malaria in the highlands of Lake Victoria basin: The role of climate change/variability, hydrology and socio-economic factors. *Climatic Change*, 99, 473–97.

Wang, G., Minnis, R.B., Belant, J.L. & Wax, C.L. (2010). Dry weather induces outbreaks of human West Nile virus infections. *BMC Infectious Diseases*, 10, 38.

Williams, G.M. & Gingrich, J.B. (2007). Comparison of light traps, gravid traps, and resting boxes for West Nile virus surveillance. *Journal of Vector Ecology*, 32, 285–91.

Wonham, M.J., Lewis, M.A., Rencławowicz, J. & van den Driessche, P. (2006). Transmission assumptions generate conflicting predictions in host–vector disease models: A case study in West Nile virus. *Ecology Letters*, 9, 706–25.

Woolhouse, M. & Gaunt, E. (2007). Ecological origins of novel human pathogens. *Critical Reviews in Microbiology*, 33, 231–42.

SECTION II

Empirical Ecology

CHAPTER 6

Environmental Drivers of Vector-Borne Diseases

Marta S. Shocket, Christopher B. Anderson, Jamie M. Caldwell,
Marissa L. Childs, Lisa I. Couper, Songhee Han, Mallory J. Harris,
Meghan E. Howard, Morgan P. Kain, Andrew J. MacDonald, Nicole Nova,
and Erin A. Mordecai

6.1 Introduction

Infection by vector-borne pathogens emerges from the abundances, traits, and interactions of arthropod vectors, hosts, and pathogens. These physiological and ecological processes are mediated by the biophysical environment. Consequently, the transmission of vector-borne disease is sensitive to environmental conditions. Global environmental change has already triggered rapid shifts in the spatial and temporal patterns of vector-borne diseases (Kilpatrick & Randolph 2012), with important consequences for human health and economic wellbeing. Each year vector-borne diseases cause one-sixth of worldwide human illness and disability, resulting in more than one million deaths (World Health Organization 2014), and cost billions of dollars in agriculture and livestock losses (Forum on Microbial Threats 2008). Understanding the environmental drivers of transmission is critical for predicting and responding to vector-borne disease in an age of accelerated human-driven environmental change.

For sustained transmission of vector-borne disease, vectors must be present and abundant, bite infected hosts to acquire the pathogen, become infectious, and bite and infect new susceptible hosts. These processes are mediated by environmental factors, including temperature, humidity, rainfall, and habitat (Fig. 6.1). Temperature and humidity alter vector physiology, influencing traits such as behavior, reproduction, development rate, survival, biting rate, susceptibility, and ability to transmit pathogens (Bayoh 2001; Mordecai et al. 2019; Ogden et al. 2004; Thomas & Blanford 2003; Vail & Smith 2002). Temperature also moderates pathogen proliferation within vectors (Liu-Helmersson et al. 2014; Mordecai et al. 2019; Paull et al. 2017). Meanwhile, the amount of rainfall can drive the reproduction of vectors that use aquatic habitats for breeding (Cheke et al. 2015; Gao et al. 2017; Koenraadt et al. 2004; Reisen et al. 2008), and alter vector-host contact rates (Knap et al. 2009; Vail & Smith 2002). Vector life history traits influence vector population abundance and contact with hosts, while vector, pathogen, and host traits govern the rate at which vectors become infectious. Land use and habitat influence nearly every component of the transmission cycle: both vector and host traits and abundances, community compositions, and contact rates (Allan et al. 2003; Despommier et al. 2006; Ezenwa et al. 2006;

Marta S. Shocket, Christopher B. Anderson, Jamie M. Caldwell, Marissa L. Childs, Lisa I. Couper, Songhee Han, Mallory J. Harris, Meghan E. Howard, Morgan P. Kain, Andrew J. MacDonald, Nicole Nova, and Erin A. Mordecai, *Environmental Drivers of Vector-Borne Diseases*. In: *Population Biology of Vector-Borne Diseases*. Edited by: John M. Drake, Michael B. Bonsall, and Michael R. Strand: Oxford University Press (2021). © Marta S. Shocket, Christopher B. Anderson, Jamie M. Caldwell, Marissa L. Childs, Lisa I. Couper, Songhee Han, Mallory J. Harris, Meghan E. Howard, Morgan P. Kain, Andrew J. MacDonald, Nicole Nova, and Erin A. Mordecai.
DOI: 10.1093/oso/9780198853244.003.0006

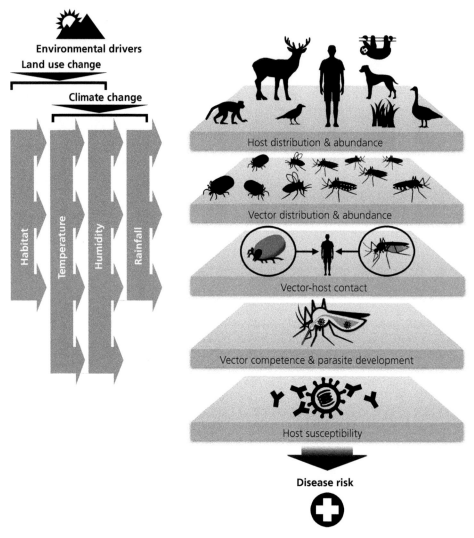

Figure 6.1 Effects of environmental drivers on vector-borne disease transmission. For transmission to occur, hosts and vectors must overlap in space and time, vectors must contact hosts, vectors must be competent to acquire and transmit the pathogen, and hosts must be susceptible. Many of these processes are affected by environmental drivers like habitat type and quality, temperature, humidity, and rainfall. Climate change and land use change in turn impact these drivers. Arrows between the four main environmental drivers and physiological/ecological processes illustrate effects that occur in many vector-borne disease systems. Variation in host susceptibility primarily stems from other factors, such as genetic variation and prior pathogen exposure (although in some cases, including for many plant hosts, environmental variation may additionally affect host susceptibility). Together these processes determine disease risk.

Ferraguti et al. 2016; Lane et al. 2007; Murdock et al. 2017).

As transmission of vector-borne disease is moderated by environmental conditions, shifts in these drivers can alter the spatial and temporal patterns of transmission. Anthropogenic climate change is altering temperature and precipitation patterns

(Wu et al. 2016). These global changes modify local climatic conditions and are predicted to decrease suitability for transmission in some areas while increasing it in others (Altizer et al. 2013; Lafferty 2009; Lafferty & Mordecai 2016). Further, climate change is increasing the frequency of extreme weather events such as droughts, floods, and hurricanes (Cai et al. 2014; Fischer & Knutti 2015), which impact transmission (Caillouët et al. 2008; Pontes et al. 2000; Shultz et al. 2005).

In addition to climate change, people are transforming landscapes at an unprecedented rate in order to support an increasing human population (Foley 2005; Lambin et al. 2010). Economic pressures and the need for resources drive activities such as agriculture, rangeland/grazing, logging, mining, and suburban and urban development (Lambin et al. 2001; Patz et al. 2004). This land use change is a powerful modifier of vector-borne disease transmission (Gottdenker et al. 2014; Patz et al. 2000, 2004) because it fundamentally alters ecosystems and ecological communities, with cascading effects on climate, extreme weather events, hydrology, and soil health (Fu 2003; Houghton 1999; Lambin et al. 2001; Sterling et al. 2013; Tolba & El-Kholy 1992). As a result, land use change can affect disease transmission by impacting the availability and suitability of habitat for vectors, local climatic conditions (Murdock et al. 2017; Vanwambeke et al. 2007), vector-host contact rates (Gottdenker et al. 2014; Larsen et al. 2014; MacDonald et al. 2019), interactions of vectors with competitors and predators (Devictor et al. 2008; Didham 2010), and human behavior (Berry et al. 2018; Larsen et al. 2014).

In this chapter, we synthesize research on environmental drivers of vector-borne disease. We emphasize the horizontal transmission between vectors and hosts that drives most outbreaks, and focus primarily on pathogens that infect humans, referencing pathogens of livestock, wildlife, and plants when they provide unique or contrasting examples of biological mechanisms. We define 'environmental drivers' to include climate factors (including temperature, humidity, and rainfall) and habitat factors (including habitat type, land use, and plant communities). First, we briefly describe the biology of vector life cycles and pathogen transmission to highlight underlying mechanisms for environmental impacts on vector populations and disease transmission (Section 6.2). Next, we synthesize how climate (Section 6.3) and habitat and land use (Section 6.4) impact transmission, highlighting general patterns, contrasts among systems, and connections across biological scales from trait

responses to population dynamics and range limits. We then discuss approaches and major challenges to understanding environmental drivers of vector-borne disease (Section 6.5). Finally, we propose future directions that leverage new data sources and technologies (Section 6.6).

6.2 Biology of Vector Life Cycles and Pathogen Transmission

Vector taxa vary widely in life histories. Here, we briefly describe the biology of dipteran, hemipteran, and ixodid tick vectors, the three taxonomic groups that contain the majority of vectors discussed in this chapter. Then, we highlight the commonalities and differences that influence their sensitivity to environmental drivers.

6.2.1 Vector Life Cycles

Dipteran vectors (including mosquitoes, flies, and midges) follow similar lifecycles but vary substantially in breeding habitat. Mosquitoes lay eggs in or near the margins of standing water, with species-specific preferences for water body characteristics, including size, salinity, and vegetation or algal growth (Day 2016). Blackflies lay their eggs on vegetation above rapidly moving, highly oxygenated streams and rivers (Crump et al. 2012). *Culicoides* midges lay eggs in a variety of terrestrial, semiaquatic, and aquatic habitats (Mellor et al. 2000; Purse et al. 2015). Sand flies lay eggs in soil, in animal burrows, or on trees or other substrates (Feliciangeli 2004; Moncaz et al. 2012). Tsetse flies develop a single larva in their uterus and deposit it into soil (Franco et al. 2014b). Despite the differences in larval habitat, the larvae for all of these dipteran taxa consume organic detritus from their immediate environment (except tsetse flies, which are nourished by maternal secretions), and ultimately emerge as flying adults that can readily disperse. Adults reproduce continuously (as long as environmental conditions are suitable), with overlapping generations in a given season or year. Adults of both sexes feed on sugar from plants (Abbasi et al. 2018; Foster 1995; Kaufmann et al. 2015; Myburgh et al. 2001; Solano et al. 2015;

Stone & Foster 2013). Only female adults bite vertebrate hosts and take blood meals, which are typically required for each cycle of egg production (Abbasi et al. 2018; Crump et al. 2012; Foster 1995; Mellor et al. 2000) (male tsetse flies also blood feed: Franco et al. 2014b).

Hemipterans (including aphids, whiteflies, and leafhoppers) are common vectors for plant diseases (Canto et al. 2009). Like dipterans, they reproduce continuously with overlapping generations. However, both immature and adult hemipterans (of both sexes) live and feed on vascular tissues of the same plant hosts, although only adults can fly and easily move between host individuals (Byrne & Bellows 1991; DeLong 1971). Additionally, many hemipterans can reproduce asexually (Byrne & Bellows 1991; DeLong 1971; Müller et al. 2001). In particular, many aphid species switch between an asexual, wingless morph that reproduces more quickly, and a sexual, winged morph that can disperse much farther (Müller et al. 2001).

In contrast to those of dipterans and hemipterans, ixodid or 'hard' tick life cycles span multiple years. After hatching from eggs, ixodid ticks develop through larval, nymphal, and adult stages that each require a single blood meal from a vertebrate host (Oliver 1989). The time between life stages lasts from several months to multiple years depending on the species and environmental conditions (Padgett & Lane 2001). Although ticks remain physically associated with their vertebrate hosts for longer than their dipteran counterparts—several days for each blood meal—off-host periods (when they are more sensitive to climate factors) comprise over 90 percent of an individual tick's lifespan (Needham & Teel 1991). Rather than flying, all tick life stages 'quest' for passing hosts from vegetation, and vertebrate host movement during the on-host period allows for widespread dispersal.

6.2.2 Pathogen Transmission

Sustained transmission of a vector-borne disease typically requires the presence of vectors and hosts that are both susceptible to infection and competent for transmitting the pathogen. Both vectors and hosts can vary in their susceptibility (i.e. probability of acquiring a disseminated infection) and transmission competence (i.e. probability of transmitting the infection). Much of the variation in transmission competence stems from differing quantities of pathogen circulating within the host or vector (Althouse & Hanley 2015; Nguyen et al. 2013). Further, the duration of the incubation period (time between exposure and becoming infectious) and infectious period (time during which pathogen transmission is possible) in the vector and in the host can also vary.

One notable set of exceptions to this general transmission paradigm is 'nonpersistent' pathogens, which are transmitted without causing a disseminated infection during one stage of transmission, either within the vector or host. Many plant viruses have evolved to be transmitted without infecting the vector, instead becoming transiently associated with vector mouth parts, and being transmitted as vectors move between host plants on the order of minutes or hours (Ng & Falk 2006). Myxoma virus is transmitted to rabbits by a variety of arthropods in a similar way (Kerr et al. 2015), although mosquitoes remain infectious for several days after biting an infected rabbit (Kilham & Woke 1953). Additionally, some tick-borne pathogens are transmitted between ticks that feed simultaneously on the same vertebrate host individual without infecting the host, a process known as 'cofeeding' (Labuda et al. 1997; Voordouw 2015).

Some pathogens are specialized to infect a small number of host or vector species, while others have broader host or vector ranges. Further, different locations support different host and vector communities. These factors produce a spectrum of transmission types, from primary maintenance in cycles between vectors and humans (or other host species of interest) (Sundararaman et al. 2013; Vasilakis et al. 2011) to maintenance in cycles between vectors and reservoir hosts coupled with occasional spillover to humans (Brock et al. 2016; Estrada-Peña & de la Fuente 2014; Petersen et al. 2013). Both types of diseases are expected to respond to variation in climate and habitat; however, spillover infections should depend more strongly on effects through reservoir host communities.

6.3 Climate

6.3.1 Temperature

Vector and pathogen traits typically depend on temperature in predictable, nonlinear ways. In general, ectotherm performance increases with temperature up to an optimum, and then sharply declines (Angilletta 2006; Dell et al. 2011; Deutsch et al. 2008). Many traits related to metabolic rate that influence disease transmission and vector abundance typically follow this pattern, including biting rate, vector and pathogen development rates, and sometimes fecundity (Amarasekare & Savage 2012; Cheke et al. 2015; Mordecai et al. 2019; Ogden et al. 2004; Purse et al. 2015). By contrast, traits related to probabilistic events, such as vector survival in immature and adult stages and transmission competence, usually have a more symmetrical thermal response and peak at lower temperatures than other traits (Alsan 2015; Amarasekare & Savage 2012; Mordecai et al. 2019; Purse et al. 2015; Takaoka 2015). These thermal responses have important effects on disease transmission, especially for diseases transmitted by vectors with continually reproducing populations. In these systems, population abundances can increase rapidly in response to favorable increases in temperature (Ogden & Lindsay 2016). Additionally, as development rates increase, the corresponding developmental periods shorten, including the critically important length of time required for vectors to become infectious. This pathogen extrinsic incubation period, together with vector mortality, determines the probability that a vector survives long enough to become infectious. Thus, predicting the thermal response of disease transmission is a quantitatively complex problem because it depends on combining the nonlinear thermal responses of many traits (Rogers & Randolph 2000).

Trait-based models that use these principles of thermal biology generally predict that disease transmission peaks at intermediate temperatures within the relevant range of temperatures vectors experience (Fig. 6.2). The specific temperature of the optimum depends on both vector and pathogen trait responses, and has been estimated to vary between 19–29°C (Brand & Keeling 2017; Cheke et al. 2015; Moore et al. 2012; Mordecai et al. 2017, 2019; Takaoka 2015). These models are typically based on the assumption that trait values depend on mean temperatures over relatively short time periods that can be approximated by measuring traits in the laboratory at constant temperatures. However, some models also take daily thermal variation into account, because mean temperature can be a poor predictor of performance when the shape of the thermal response is nonlinear, which is usually the case (Bernhardt et al. 2018; Ruel & Ayres 1999). In general, temperature fluctuations increase vector performance and disease transmission when the mean temperature is near the lower thermal margin (because vectors are rescued by spending some of the day in warmer, more favorable temperatures), but decrease them when the mean temperature is near the optimum (because vectors spend relatively little time at the optimal temperature) or near the upper thermal margin (because of harmful effects of heat stress) (Lambrechts et al. 2011; Liu-Helmersson et al. 2014; Paaijmans et al. 2010, 2013). These effects of temperature variation can have important impacts on transmission. For example, in a part of Thailand where mean temperatures are consistent year-round, the high transmission season for dengue is defined by smaller daily fluctuations in temperature (Lambrechts et al. 2011).

Observed patterns of disease support temperature as a nonlinear driver of vector-borne disease transmission. There is a large body of literature showing positive associations between temperature and transmission of various vector-borne diseases (Gao et al. 2017; Lindgren & Gustafson 2001; Naish et al. 2014; Nnko et al. 2017; Paz 2015; Shocket et al. 2018; Subak 2003). This positive relationship between temperature and transmission is particularly clear at high-altitude locations and the edges of range limits where disease transmission is typically limited by cool temperatures. For instance, in the highlands of both Ethiopia and Colombia, human malaria transmission occurs at higher altitudes in years with higher mean temperatures (Siraj et al. 2014). There is also a smaller, but growing set of evidence for reduced transmission of mosquito-borne diseases at extreme, high temperatures (Gatton et al. 2005; Mordecai et al. 2013; Peña-García

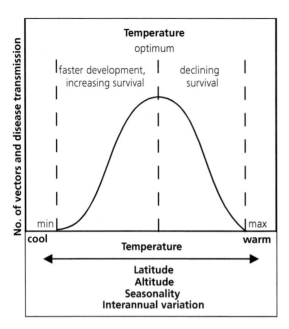

Figure 6.2 For vectors that reproduce continuously, mechanistic models predict that vector population size and disease transmission respond nonlinearly to temperature and are maximized at intermediate temperatures. As temperature increases to the optimum, transmission is promoted by faster development and higher survival of the vector and pathogen; as temperature continues to increase beyond the optimum, transmission is inhibited by lower survival of the vector. Variation in temperature often explains the seasonality of and interannual variation in disease transmission, as well as patterns of disease across latitudinal and altitudinal gradients.

et al. 2017; Perkins et al. 2015; Shah et al. 2019; Shocket et al. 2020 Variation in temperature helps to explain variation in vector density or disease transmission at multiple spatial and temporal scales: latitude (e.g. Geoghegan et al. 2014; Khatchikian et al. 2012), altitude (e.g. Bødker et al. 2003; Zamora-vilchis et al. 2012), seasonality (e.g. Hartley et al. 2012; Shocket et al. 2018), and interannual variation (e.g. Morin et al. 2015; Siraj et al. 2015).

The unimodal response of transmission to temperature suggests that higher temperatures from global climate change will increase transmission of vector-borne disease in cooler areas and seasons and decrease transmission in warmer areas and seasons, shifting geographic range limits and seasonality (Lafferty 2009; Lafferty & Mordecai 2016; Medone et al. 2015; Mordecai et al. 2019). For example, over the next 60 years the hotspot of

Plasmodium falciparum malaria transmission risk in Africa—where a large number of people are exposed to high risk year-round—is predicted to move from the western coastal region (where temperatures are currently optimal) to the eastern highlands (where temperatures are currently cooler than the optimum) (Ryan et al. 2015). However, based on current temperatures and distributions of human populations, climate warming is often predicted to lead to net increases in transmission for many vector-borne pathogens, including malaria (Caminade et al. 2014), dengue (Messina et al. 2019), West Nile virus (Harrigan et al. 2014; Shocket et al. 2020), Ross River virus (Shocket et al. 2018), and trypanosomiasis (Moore et al. 2012). Specifically, transmission is predicted to increase at higher elevations (e.g. Ryan et al. 2015) and extend to more temperate latitudes (e.g. Harrigan et al. 2014). These predictions arise from both increased transmission in areas where vectors already occur and from vectors expanding their geographic ranges into new areas (Brownstein et al. 2005a; Kraemer et al. 2019; Medlock et al. 2013; Moo-Llanes et al. 2013; Ogden et al. 2006). Transmission seasons are also expected to lengthen (e.g. Jones et al. 2019), starting earlier in the spring and extending later into the fall.

Temperature can also impact vector populations (and thus disease transmission) outside of the main transmission season, particularly in highly seasonal environments where overwinter survival is important. Milder winter temperatures increase the overwinter survival of *Culex* mosquitoes (Reisen et al. 2008; Vinogradova 2000), *Aedes* mosquito eggs (Fischer et al. 2011; Thomas et al. 2012), *Culicoides* midges (Wittmann et al. 2001), ixodid ticks (Brunner et al. 2012), aphids (Robert et al. 2000), leafhoppers (Boland et al. 2004), and whiteflies (Canto et al. 2009), and are associated with West Nile virus outbreaks in Russia (Platonov et al. 2008) and increased prevalence of tick-borne diseases in Europe (Lindgren & Gustafson 2001). Additionally, tolerance of temperature extremes often predicts arthropod species range limits better than performance at mean temperatures (e.g. Overgaard et al. 2014). Milder winter temperatures have likely led to expanding northern range limits for ticks in Europe (Lindgren et al. 2000; Medlock et al. 2013) and North America (Brownstein et al. 2005a; Ogden et al. 2006). Thus, rising winter temperatures due to climate change may expand the geographic ranges of vector species and the diseases they transmit. Additionally, other indirect effects are possible: severe winter freezes are associated with outbreaks of St. Louis encephalitis in Florida, likely by killing understory vegetation and increasing reproductive success of bird reservoir hosts (Day & Shaman 2009).

Among continually reproducing vectors, midges and hemipteran vectors of plant diseases provide examples of unique mechanisms for temperature effects on disease transmission. Midge transmission of bluetongue virus has the potential for a 'baton effect': as current vector species expand northward in Europe due to climate change, they may begin to overlap with and 'hand off' the virus to other midge species that are competent but do not currently act as vectors because the virus is not present within their geographic range (Wittmann & Baylis 2000). Accordingly, geographic expansion of the virus may outpace that of current vector species. For aphids, temperature impacts allocation of reproductive output to winged and non-winged forms, both directly and indirectly via aphid density (Müller et al. 2001). In turn, these changes can affect aphid dispersal and the spatial spread of disease (Canto et al. 2009; Newman 2004). In particular, warmer winter temperatures lead to earlier first and last flights of aphid species in a given year (Bell et al. 2015; Sheppard et al. 2016). This timing, and its influence on viral inoculation, is critical, because infection at earlier developmental stages often leads to more severe disease in crop plants (Boland et al. 2004; Thackray et al. 2009). Additionally, immune defenses (which can affect susceptibility to pathogens) and disease severity (i.e. negative consequences for the host given an infection) are more temperature-dependent for ectothermic plant hosts than for endothermic vertebrate hosts (Boland et al. 2004; Garrett et al. 2006; West et al. 2012).

Tick-borne diseases provide the most notable contrasting example for the effects of temperature on transmission of vector-borne disease, stemming from their distinctive life history compared to continually reproducing vectors (Ogden & Lindsay 2016). The relatively fixed reproductive cycle of ticks means their population size cannot increase rapidly via short-term effects of temperature on vital rates. Instead, temperature affects tick abundance through thermal extremes that act as population bottlenecks (Brunner et al. 2012; Ogden & Lindsay 2016; Ogden et al. 2014; Padgett & Lane 2001). Additionally, tick blood meals are so infrequent that pathogen development rate—key for determining the probability of becoming infectious in other systems—is essentially irrelevant (Ogden & Lindsay 2016). Instead, temperature affects the probability that a tick becomes infectious by altering tick phenology and questing behavior. Long-term average temperatures determine when each life stage completes development and becomes active, and thus their temporal overlap (Gatewood et al. 2009; Ogden & Lindsay 2016; Randolph et al. 2000). Pathogens that require co-feeding of larvae and nymphs for transmission (e.g. tick-borne encephalitis and less persistent strains of Lyme pathogens) are only maintained in areas where both life stages are active simultaneously (Gatewood et al. 2009; Kurtenbach et al. 2006; Randolph et al. 2000). Warming from climate change may disrupt current patterns of tick phenology, favoring less persistent pathogens or strains if synchrony increases, or suppressing them if synchrony decreases (Ostfeld & Brunner 2015). Tick questing behavior also responds rapidly to current weather conditions: ticks modify their questing height or cease questing altogether in order to avoid desiccation in hot, dry conditions (Arsnoe et al. 2015; Vail & Smith 2002). These behaviors change the contact rates of ticks with vertebrate hosts, impacting human risk directly via exposure and indirectly via changes in which reservoir host species ticks contact for blood meals (Ginsberg et al. 2017).

6.3.2 Humidity

Humidity can have strong effects on vector performance by maintaining water balance and preventing desiccation. In general, higher humidity increases survival and fecundity, especially in favorable thermal environments where these processes are not limited by temperature (Alsan 2015; Bayoh 2001; Costa et al. 2010; Lyons et al. 2014; Rodgers et al. 2007; Zahler & Gothe 1995). These relationships likely contribute to the positive relationships found between humidity and cases of mosquito-borne diseases like malaria (Zacarias & Andersson 2011), dengue fever (Campbell et al. 2013; Karim et al. 2012; Xu et al. 2014), and Ross River fever (Bi et al. 2009).

Ticks are particularly sensitive to humidity because, unlike other vectors, they cannot hydrate by ingesting plant sugars or standing water (Needham & Teel 1991; Ogden & Lindsay 2016). Instead, ticks regulate water balance by using molecules in their saliva to absorb water from the humid leaf litter (Bowman & Sauer 2004). When humidity is low, ticks must frequently return to the leaf litter to rehydrate, which consumes energy, diverts time away from host-seeking, and reduces the probability of successfully feeding on a host (Ogden & Lindsay 2016). Thus, ixodid tick survival declines sharply in subsaturated air in both laboratory (Rodgers et al. 2007; Stafford 1994) and field conditions (Bertrand & Wilson 1996). Accordingly, low air moisture has been associated with decreased incidence of Lyme disease in the U.S. (McCabe & Bunnell 2004; Subak 2003).

6.3.3 Rainfall

Like temperature, rainfall can have nonlinear impacts on the abundance of vectors with aquatic larval stages and on incidence of the diseases that they transmit (Fig. 6.3). Most simply, rainfall provides more breeding habitat for these vectors as it accumulates in natural hydrological features and pools (e.g. rivers, ponds, and leaf litter) and human-made containers (e.g. bottle caps and jugs). Accordingly, increased precipitation is often positively associated with abundances of mosquitoes (Barton et al. 2004; Reisen et al. 2008; Sang et al. 2017), blackflies (Cheke et al. 2015; Nwoke et al. 1992), and *Culicoides* midges (Gao et al. 2017; Walker & Davies 1971), and with human cases of mosquito-borne diseases like malaria (Alemu et al. 2011; Rozendaal 1992), dengue fever (Li et al. 1985; Morin et al. 2015; Stewart-Ibarra & Lowe 2013), chikungunya fever (Perkins et al. 2015; Riou et al. 2017), Ross River fever (Hu et al. 2006; Kelly-Hope et al. 2004; Whelan et al. 2003), Rift Valley fever (Anyamba et al. 2012; Bicout & Sabatier 2004), and St. Louis encephalitis (Day et al. 1990), and with cases of trypanosomiasis trans-

mitted by blackflies (Adeleke et al. 2010). Flooding in particular is often associated with outbreaks of mosquito-borne disease (Gagnon et al. 2002; Tall & Gatton 2019). This positive rainfall-transmission relationship is important for driving both interannual variation in disease (e.g. Anyamba et al. 2012) and seasonal patterns of disease (e.g. Mabaso et al. 2005). For instance, in locations where temperature is suitable year-round, the high transmission season is often defined by the wet season (Chandran & Azeez 2015; Jacups et al. 2008; Mabaso et al. 2005).

The relationship between rainfall and vector abundance or disease transmission can become negative at very low or high levels of rainfall (Fig. 6.3). High rainfall or flooding events can kill larvae or flush them from aquatic habitats (Benedum et al. 2018; Koenraadt & Harrington 2008; Paaijmans et al. 2007). Accordingly, across portions of the Amazon, most malaria transmission occurs during the dry season and increasing precipitation results in fewer human cases (Barros et al. 2011; Olson et al. 2009). Low rainfall and drought can increase vector abundance and disease transmission by two major mechanisms. First, low rainfall can increase breeding habitat through changes in human behavior. In many regions, it is common to store water in open containers for household use, and water storage often increases during droughts or the dry season (Anyamba et al. 2012; Chandran & Azeez 2015; Pontes et al. 2000; Trewin et al. 2013). This practice increases the reproduction and abundance of container-breeding mosquitoes like the urban specialist vector *Aedes aegypti* (Padmanabha et al. 2010; Stewart Ibarra et al. 2013), leading to associations between drought and the diseases that it transmits, including chikungunya fever (Anyamba et al. 2012) and dengue fever (Chandran & Azeez 2015; Pontes et al. 2000). Second, low rainfall can impact transmission indirectly through reservoir host and vector communities. Drought increases transmission of West Nile virus (Paull et al. 2017; Shaman et al. 2005), likely via physiological stress on the bird reservoir hosts or elevated vector-host contact rates around the few remaining water sources. Additionally, drought can alter mosquito species community composition (Tian et al. 2015), which can impact transmission (Tokarz & Smith 2020). Thus, the relationship between rainfall and disease

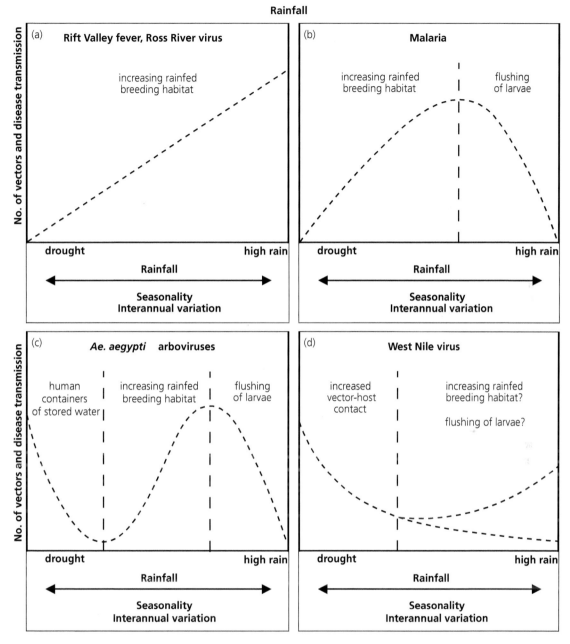

Figure 6.3 Hypothesized responses of vector abundance and disease transmission to rainfall for several mosquito-borne diseases. Responses are often nonlinear, vary across systems and locations, and are less certain than the thermal response of transmission (Fig. 6.2). Variation in rainfall often explains the seasonality of and interannual variation in disease transmission. (A) For pathogens like Rift Valley fever virus and Ross River virus, vector abundance and transmission increase with rainfall over a wide range of rainfall conditions because higher rainfall leads to more rainfed breeding habitat. (B) The relationship between rainfall and *Anopheles* spp. or malaria transmission varies across locations. In settings with lower overall rainfall, increased rain creates breeding habitat, while in settings with higher overall rainfall, increased rain flushes larvae. (C) Drought can increase human water storage behavior, and thus increase breeding habitat for vectors that reproduce in containers like *Ae. aegypti*, often leading to outbreaks of arboviruses. However, rainfall can also increase rainfed breeding habitat, and high rainfall can flush larval habitats. (D) Drought increases transmission of West Nile virus, likely via increased contact between vectors and reservoir hosts. At higher levels of rainfall, the relationship between rainfall and the abundance of vectors for West Nile virus varies across locations.

transmission by vectors with aquatic larval stages is context-dependent based on the rainfall quantity, vector biology, and reservoir host and human behavior.

Rainfall can also interact with other factors or act via multiple mechanisms to influence disease transmission and drive epidemic dynamics in complex ways. In the East African highlands, rainfall in combination with human host immunity and land use change contributes to multi-year cycles in malaria prevalence (Pascual et al. 2008; Zhou et al. 2004). In southern India, dengue cases responded positively to the seasonality of rainfall due to rainfed breeding habitat but responded negatively to interannual variation in rainfall due to water storage practices (Chandran & Azeez 2015). In some systems, specific sequences of weather patterns are the best predictors for disease transmission. For instance, in Texas, transmission of West Nile virus was highest in years when a high rainfall spring was followed by a cool, dry summer (Ukawuba & Shaman 2018). For Ross River virus, these specific weather sequences manifest through reservoir host dynamics: high summer rainfall typically drives more transmission, but this effect can be negated if low rainfall the preceding winter led to low recruitment of juvenile kangaroos (Mackenzie et al. 2000) or enhanced if low rainfall the previous summer left reservoir hosts relatively unexposed and thus more immunologically susceptible to infection (Woodruff et al. 2002). When heavy rainfall occurs in the context of extreme, infrastructure-damaging events like hurricanes, the effects on disease transmission can be even greater due to reduced vector control and damaged housing (Caillouët et al. 2008; Shultz et al. 2005; Sorensen et al. 2017). Thus, rainfall can drive transmission of vector-borne disease through many mechanisms and produce a variety of observed patterns across systems and locations.

In many parts of the globe, rainfall and temperature both vary strongly with the El Niño-Southern Oscillation (ENSO), a multi-year climate oscillation that drives interannual variation in epidemics for many diseases. For example, Rift Valley fever outbreaks in eastern Africa are strongly driven by ENSO-related precipitation that accumulates in low-lying areas and provides breeding habitat for mosquito vectors (Anyamba et al. 2009; Linthicum et al. 1999). Accordingly, these outbreaks can be accurately predicted by remotely-sensed proxies of ENSO: sea surface temperature and vegetation indices (Anyamba et al. 2009; Linthicum et al. 1999). Similarly, remotely-sensed ENSO metrics based on sea surface temperature and atmospheric pressure are associated with outbreaks of dengue fever across many locations in the Americas, Asia, and Oceania (Adde et al. 2016; Cazelles et al. 2005; Colón-González et al. 2011; Hu et al. 2010; van Panhuis et al. 2015), and can increase accuracy of predictions independent of temperature and rainfall (Earnest et al. 2012). As described previously, the nonlinear response of mosquito abundance to rainfall means that ENSO can drive outbreaks of mosquito-borne diseases through both severe drought and flood conditions (Anyamba et al. 2009, 2012; Gagnon et al. 2002). ENSO metrics are a particularly useful tool for forecasting disease outbreaks because they are constantly monitored around the world and conditions take several months to develop, providing substantial lead time.

Rainfall is also an important driver for vectors without aquatic larval stages. Rainfall is consistently negatively associated with tsetse fly populations and transmission risk (Grébaut et al. 2009; Rogers & Williams 1993), possibly due to negative effects of oversaturated soils on pupal survival. Conversely, rainfall is often positively associated with *Culicoides* midges and transmission of bluetongue virus (Gao et al. 2017; Guis et al. 2012; Walker & Davies 1971). However, rainfall is not always the strongest driver of midge seasonality (Mellor et al. 2000), and it has been posited that drought can increase breeding habitat for some species by exposing moist mud on the margins of lakes and streams (Berry et al. 2013). Different sand fly vectors for leishmaniasis live across a wide variety of humid, semi-arid, and arid habitats, and thus rainfall has positive (Furtado et al. 2016), negative (Gálvez et al. 2010; Miranda et al. 2015), and unimodal (Chaniotis et al. 1971) effects on their abundances across different species and locations. In the Australian Wheatbelt, rainfall influences the timing of aphid arrival on crops because it increases the growth of wild plants that serve as a 'green bridge' where aphids persist over the dry summer between growing seasons (Thackray et al. 2009). As with thermal effects on aphid flight phenology in the northern hemisphere, this timing has important consequences for viral transmission and disease severity (Thackray et al. 2009).

6.3.4 Other Factors

Even flying vectors have limited dispersal (Elbers et al. 2015; Thomas et al. 2013), making the spatial spread of disease over larger distances depend primarily on movement of infected hosts (Buckee et al. 2013; Stoddard et al. 2013). However, wind-based passive dispersal is important for vectors like aphids and leafhoppers (Parry 2013; Thresh et al. 1983), *Culicoides* midges (Elbers et al. 2015), and some species of mosquitoes (Elbers et al. 2015; Lapointe 2008). Long-distance wind dispersal of infected vectors likely initiated outbreaks of midge-borne bluetongue and African horse sickness (Durr et al. 2017; Elbers et al. 2015; Sedda et al. 2012), mosquito-borne Japanese encephalitis (Ritchie & Rochester 2001) and Rift Valley fever (Mapaco et al. 2012), and hemipteran-borne plant viruses (Thresh et al. 1983). Local wind patterns can also drive patterns of malaria at smaller spatial scales (Midega et al. 2012). Climate change is already shifting wind patterns at regional scales (Pryor et al. 2005), which may impact the patterns of disease spread over large spatial scales, as well as transmission at more local scales.

Although not climatic factors *per se*, sea level and the concentration of carbon dioxide are two related environmental factors that can impact transmission of vector-borne disease. Increased carbon dioxide in the atmosphere—the primary cause of anthropogenic climate change—directly influences plant physiology, which can cause cascading effects on vector traits and disease processes (Newman 2004; Trębicki et al. 2015). Sea level rise is a major predicted outcome of climate change (Rahmstorf 2007). Some mosquitoes reproduce in estuarine habitats rather than freshwater pools, including several that transmit Ross River virus. Accordingly, variation in tidal height drives abundances of these species and incidence of Ross River fever (Jacups et al. 2008; Kokkinn et al. 2009; Tong & Hu 2002). Thus, sea level rise may affect future transmission of Ross

River virus in Australia or other diseases transmitted by saltmarsh mosquitoes.

6.4 Habitat and Land Use

Habitat type and land use often drive variation in risk for vector-borne diseases, favoring infection by different pathogens in different environments (Fig. 6.4). For example, newly deforested areas on the edges of relatively undisturbed tropical forests are the main sites for transmission of *Plasmodium vivax* malaria by *Anopheles* spp. in South America (Chaves et al. 2018; MacDonald & Mordecai 2019; Santos & Almeida 2018; Vittor et al. 2006, 2009), while transmission of a suite of arboviruses (dengue, Zika, and chikungunya) by *Ae. aegypti* is maximized in urban areas (Jansen & Beebe 2010; Sheela et al. 2017). This system-specific variation in disease risk across gradients of ecosystem type and land use regime is driven by four major processes (Fig. 6.1). First, most vector species are only able to maintain high population abundances in a subset of habitats. Second, the host species that act as competent reservoirs for pathogens may be similarly restricted to a subset of habitats. Third, different habitats may affect vector competence or other traits that drive transmission independently from vector abundance and host community. Finally, the population density and contact rates of humans (or other focal host species) also varies across habitats. Hotspots for transmission may therefore reflect the distribution of humans on the landscape or increased exposure rather than risk stemming directly from the density of infectious vectors.

Habitat type and land use affect vector abundances through many underlying mechanisms. The availability of breeding sites is often critical. For instance, dam construction and agricultural irrigation has provided ideal breeding habitat for black-flies in Africa (Patz et al. 2000) and malaria vectors in Brazil (Tadei et al. 1998). Tsetse flies thrive in deforested areas that have been converted to agriculture but cannot survive in most urban areas because they require loose soil for depositing larvae (Franco et al. 2014b; Grébaut et al. 2009; Patz et al. 2000). Specific plant species and communities can also be critical habitat for certain vectors. For instance, palm trees are the preferred habitat for triatomine kissing bugs that spread Chagas disease (Abad-Franch et al. 2015), and a mosquito vector of West Nile virus was strongly associated with eastern red cedar stands across a varied landscape in the U.S. (O'Brien & Reiskind 2013). Ixodid ticks are typically associated with dense shrub-dominated or forested communities, as these habitats provide favorable temperature and humidity conditions for tick questing activity and survivorship (Eisen et al. 2006, 2010; MacDonald et al. 2017; Ostfeld et al. 1995; Padgett & Lane 2001; Swei et al. 2011). However, vectors and pathogens can adapt to new environments. For instance, South American sand flies are adrapting to urban environments and leishmaniasia has gone from a predominantly rural to a predominantly urban disease in Brazil (Jeronimo et al. 1994; Werneck et al. 2002). Similarly, *Ae. albopictua* mosquitoes are increasingly present in highly urban areas in Malaysia (Kwa 2008).

Deforestation in particular can have important effects on the abundances of vectors and disease transmission. In western Africa, deforestation has increased the relative proportion of savannah blackflies (versus forest blackflies), which carry the more severe form of onchocerciasis with higher blindness rates (Wilson et al. 2002). Deforestation can increase breeding habitat for mosquitoes (Leisnham et al. 2005; Norris 2004; Patz et al. 2004) and shift mosquito community composition toward species that vector human diseases. For instance, in the Amazon, sampling sites in deforested areas were more likely to contain malaria vectors, and deforested sites had significantly higher human biting rates (Tucker Lima et al. 2017; Vittor et al. 2006, 2009). Similarly, in Thailand mosquito diversity was highest and vector species abundance was lowest in intact forest habitat (Thongsripong et al. 2013). Some of these effects may be due to changing biotic interactions like the loss of predators (Hunt et al. 2017; Juliano & Lounibos 2005) or reduced competition from more sensitive species (Chase & Shulman 2009; Freed & Leisnham 2014). Additionally, several studies suggest that mosquito density increases at transitional zones between habitat types or in heterogeneous landscapes (Barros et al. 2011; Chaves et al. 2011; Despommier et al. 2006; Lothrop et al. 2002; Reiskind et al. 2017), and early stages of deforestation can contribute to this effect by increasing edge

habitat. However, the effects of deforestation may be temporary, as transmission risk changes again with further transitions in land use (Fig. 6.4) (Baeza et al. 2017).

Habitat type and land use can also drive disease transmission by influencing reservoir hosts. In some cases, the same habitat features that promote high vector densities also promote high densities or clustering of highly competent reservoir hosts, compounding transmission risk. For example, the same palm trees shelter both triatomine bug vectors and the mammal hosts that serve as reservoirs for the Chagas disease parasite (Abad-Franch et al. 2015). This effect is magnified in deforested areas where palm trees may be some of the only suitable habitat for mammals, and where highly competent possums are more common; together these effects on the host community are hypothesized to drive the observed increase in the infection prevalence of vectors (Gottdenker et al. 2012). Food resources are another important mechanism for habitat effects on reservoir host populations. For instance, the irrigation resulting from dam construction in Africa has increased resources, for gerbil populations that are important reservoirs for leishmaniasis parasites (Desjeux 2001). Additionally, increased acorn output from periodic oak tree masts temporarily increases the abundance of white-footed mice that are highly competent reservoir hosts for the Lyme pathogen (Jones 1998). This same food resource also attracts deer, which increases the abundance of blacklegged ticks, further increasing risk of Lyme transmission (Jones 1998).

The search for more general relationships between habitat conversion, reservoir community composition, and disease risk has been controversial (Civitello et al. 2015; Halsey 2019; Ostfeld & Keesing 2000; Randolph & Dobson 2012; Rohr et al. 2019; Salkeld et al. 2013). The 'dilution effect' hypothesis—in which increased biodiversity lowers average host competence and/or vector biting rate on competent hosts, and thus disease risk—was proposed to describe the impact of suburban development on the Lyme disease pathogen transmitted by blacklegged ticks in the northeastern U.S. In that system, smaller forest fragments had less diverse mammal communities with relatively higher proportions of highly-competent hosts

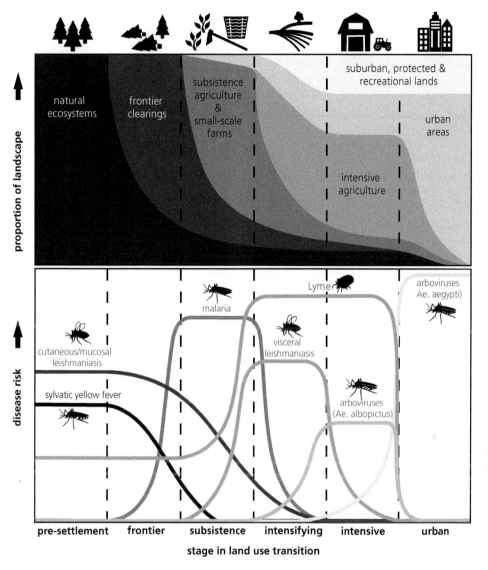

Figure 6.4 Different types of habitat and land use promote transmission of specific vector-borne diseases. Here, we show examples of different diseases and where their transmission is maximized across a land use gradient. Top panel is modified from Foley 2005.

(white-footed mice), resulting in increased disease risk (Allan et al. 2003; Brownstein et al. 2005b; LoGiudice et al. 2008; Ostfeld & Keesing 2000). However, more recent research found the opposite pattern in an area nearby: mammal communities were more diverse and less competent for transmitting Lyme disease in fragmented forests than in undisturbed forests (Linske et al. 2018). Further, some argue that 'amplification effects'—in which more diverse host communities are *more* likely to contain highly-competent hosts—are also likely to occur (Faust et al. 2017; Salkeld et al. 2013; Wood et al. 2014). Regardless, habitat and land use can drive the transmission of vector-borne pathogens by impacting the density of reservoir hosts that vary in their competence.

Habitat and land use can also influence vector traits that drive transmission independently from vector abundance. Deforestation and urban develop-ment often create 'heat islands' with higher average temperatures relative to nearby areas with more vegetation (Afrane et al. 2005; Imhoff et al. 2010; Kalnay & Cai 2003; Lindblade et al. 2000; Murdock et al. 2017), although urbanization and conversion to agriculture can instead lower temperatures in arid regions (Imhoff et al. 2010; Lobell & Bonfils 2008; Mahmood et al. 2006). In turn, these differing micro-climates can influence vector traits and predicted vectorial capacity (Afrane et al. 2005; LaDeau et al. 2015; Murdock et al. 2017). Food resources also vary across habitats and can influence vector traits. Laboratory studies show that food availability dur-ing the larval stage and the quantity and plant source of sugar resources during the adult stage can both alter adult mosquito survival, mating success, fecundity, transmission competency, and extrinsic incubation period duration (Alto et al. 2005; Ebrahimi et al. 2018; Gu et al. 2011; Hien et al. 2016; Lefèvre et al. 2013; Moller-Jacobs et al. 2014; Shapiro et al. 2016; Stone & Foster 2013; Stone et al. 2009; Yu et al. 2016). This research suggests that landscapes could be con-structed to minimize disease transmission. However, research in natural settings is limited, and studies have yielded conflicting results for the effects of simi-lar resource treatments (Alto et al. 2005; Gu et al. 2011; Lefèvre et al. 2013; Moller-Jacobs et al. 2014; Shapiro et al. 2016; Stone et al. 2012).

Habitat and land use can also determine the density of humans (or other focal hosts) and their contact with vectors. Newly deforested areas often bring humans into close proximity with tsetse fly, sand fly, and mosquito vectors (Confalonieri et al. 2014; Desjeux 2001; Patz et al. 2000; Van den Bossche et al. 2010). For tick-borne diseases, human movement and interaction with high risk land-scapes determines the rates of contact between infected ticks and human hosts (Eisen & Eisen 2016; Pepin et al. 2012), where human activity in specific types of habitat is much more likely to result in tick encounter and tick-borne pathogen transmission than others. In the specific case of Lyme disease in the northeastern US, suburban development in for-ested regions puts human populations in closer contact with vector ticks, elevating disease risk for humans in peri-domestic environments (Berry et al. 2018; Connally et al. 2009; Larsen et al. 2014; MacDonald et al. 2019). Zooprophylaxis is one way

that land use is already being manipulated to minimize disease transmission: the presence of live-stock can lower malaria incidence in humans by shifting bites to cattle in place of humans, even when vector densities increase in response to the new source of readily available blood meals (Franco et al. 2014a; Mutero et al. 2004).

6.5 Challenges and Approaches

Two major goals of vector-borne disease research are to infer the role of environmental variation from past transmission dynamics and to predict future transmission based on environmental factors. This research aims to influence environmental and health policy and to enable more proactive and effective strategies for control of vector-borne dis-eases. Three major challenges in inferring the role of environmental drivers and predicting transmission are: (1) nonlinearity, (2) interacting and correlated drivers, and (3) variation across temporal and spa-tial scales. Here, we review these challenges and how different approaches can be used to address them, depending on the specific goals at hand.

6.5.1 Challenges

6.5.1.1 Nonlinearity

The first challenge to inferring environmental drivers of vector transmission is that transmission is a nonlinear process at multiple scales. Vector and pathogen traits respond nonlinearly to environ-mental drivers like temperature and rainfall (Figs. 6.2 and 6.3, Section 6.3). Then, the force of infection (i.e. per capita transmission rate) depends nonlinearly on vector and pathogen traits (e.g. in the R_0 equation for mosquito-borne disease) (Dietz 1993) and vector densities (Khatchikian et al. 2012). Host infection dynamics are in turn a nonlinear function of the force of infection because transmission slows as susceptible hosts are depleted in the population (Smith et al. 2012). Together, nonlinearity and time lags in multiple processes can obscure the impact of environmental drivers in observed disease dynam-ics. Inferential approaches that look for linear responses in observed data can find different responses in different settings, making environmen-tal drivers appear highly context dependent. For example, if the true relationship between transmission and temperature or rainfall is non-monotonic (Fig. 6.2), these drivers may be positively, negatively, or not significantly correlated with cases of disease at different locations and times of year, even if the mechanistic influence of the driver is consistent.

6.5.1.2 Interacting or Correlated Drivers

The second challenge is that multiple environmental drivers of transmission may interact or be correlated in space or time. Different environmental drivers may not operate independently because they affect the same traits, and their impacts may have interactive effects on transmission (Guis et al. 2012; Shand et al. 2016; Vail & Smith 2002). For example, for vectors with aquatic larvae, temperature might be suitable for transmission throughout the year in some locations, but temperature suitability only translates into disease incidence when rainfall provides sufficient larval habitat to generate large vector populations (e.g. during the wet or dry season, as appropriate). Conversely, appropriate levels of rain will only generate large vector populations if they occur when temperatures are suitable. Further, multiple drivers often follow a seasonal pattern and thus co-vary, making it difficult to identify the causal driver(s) and to predict the outcome if drivers become decoupled in future climate scenarios. For diseases with strong seasonality, correlation analysis can suggest spurious relationships in the absence of a true causal link. Conversely, traditional time series analysis of seasonal data (SARIMA models) removes the seasonal pattern and measures anomalies from seasonal trends (Hu et al. 2004). These methods can underestimate the importance of a driver if it primarily affects transmission at the seasonal scale. Therefore, inferring causality from and making predictions for such interacting, co-varying, and seasonal environmental drivers is challenging with traditional statistical techniques, and requires system-specific knowledge of the ecology of the host-vector-pathogen interface.

6.5.1.3 Variation Across Temporal and Spatial Scales

The third challenge is that the relative impact of different environmental drivers may differ across temporal and spatial scales. Since transmission requires alignment of suitable ecological conditions (Fig. 6.1), the specific ecological process or environmental driver that is most limiting may vary by scale and setting (Plowright et al. 2017). For example, impacts of climate may be more apparent across larger geographic (e.g. degrees of latitude) and seasonal scales than at smaller scales. Some environmental drivers may vary more substantially over space than over time. For example, land use can vary over relatively small spatial scales (meters to kilometers) but usually over relatively longer time scales (years to decades). In contrast, weather can vary daily and weekly but usually over larger spatial scales (tens to hundreds of kilometers; though microclimate effects can also be important) (Murdock et al. 2017; Vanwambeke et al. 2007). Finally, human behavior and social determinants (e.g. housing construction, vector control, access to medical care, and water storage) can modify the influence of environmental drivers, and these complex interactions can be nonlinear, bidirectional, and operate at multiple scales (MacDonald et al. 2019; MacDonald & Mordecai 2019).

6.5.2 Approaches

Together, the challenges of nonlinearity, complexity of drivers, and scale can make inferring environmental drivers from observational data challenging. These challenges necessitate a suite of complementary methods for understanding environmental drivers of vector-borne disease transmission. The most appropriate type of inference depends on the specific question at hand. Further, combining multiple lines of evidence from different inference methods and data sources can provide the strongest support for environmental drivers of transmission (Metcalf et al. 2017; Munafò & Davey Smith 2018; Ostfeld & Brunner 2015; Tjaden et al. 2018).

6.5.2.1 Mechanistic and Statistical Modeling Approaches

Process-based, or mechanistic, approaches seek to understand environmental drivers of transmission by examining their impact on the traits and population properties that drive transmission (Rogers & Randolph 2006; Tjaden et al. 2018). Mechanistic approaches measure the impact of environmental drivers on traits and population processes directly through observations and experiments, then combine the trait responses into an overall effect on transmission or disease dynamics (e.g, Johansson et al. 2014; Mordecai et al. 2019; Ogden et al. 2005). They can incorporate multiple nonlinear impacts of a single driver as well as interacting environmental drivers. Mechanistic models are particularly important for predicting transmission in novel conditions or combinations of conditions for which observations do not currently exist. However, these approaches require detailed experimental work. Additionally, it can be difficult to connect experimental results with field observations in natural settings, making validation a challenge.

Statistical inference on observational data complements mechanistic approaches by allowing inference of environmental drivers from observed patterns of transmission (Tjaden et al. 2018; Larsen et al. 2019). This family of approaches uses observed vector abundance or human case data matched with data on hypothesized environmental covariates, and infers environmental drivers by matching patterns over space or time, including lagged and nonlinear relationships (e.g. Johansson et al. 2009; Pascual et al. 2008; Perkins et al. 2015; Stewart-Ibarra & Lowe 2013). Statistical inference from observational data assures field relevance. However, it is important to couple statistical approaches with biological knowledge of a system in order to anticipate nonlinear and complex relationships. Strong correlations between different environmental factors through time can mask true causal relationships, and it can be difficult to distinguish between proximate and ultimate drivers. Moreover, even when environment–transmission relationships can be inferred from observational data, application of results can be limited by a lack of mechanistic understanding (e.g. not knowing which traits are driving the response).

Many mechanistic and statistical approaches for estimating and predicting environmental impacts on vector-borne disease are available, and the best approach depends on the specific goals (Metcalf et al. 2017; Ostfeld & Brunner 2015; Tjaden et al. 2018). Accurate short-term forecasts for specific locations can often be constructed from past patterns of transmission alone (without mechanistic knowledge of a

system) if sufficient data are available (Johansson et al. 2016; Metcalf et al. 2017; Yamana et al. 2016). For example, Johnson et al. (2018) compared forecasting models for two dengue fever time series and found that mechanistic forecasts could capture the seasonality but failed to anticipate large outbreaks that the statistical forecasts predicted more successfully. The mechanistic approach required forecasting environmental predictors, which introduced error in anomalous years, while the statistical approach was able to phenomenologically match dynamical patterns from the disease data alone (Johnson et al. 2018; Yamana et al. 2016). Weighted averages of multiple forecast types (e.g. 'superensembles') can reduce error by smoothing over idiosyncrasies of individual forecasting approaches (Yamana et al. 2016). By contrast, understanding mechanism is important for predicting dynamics of disease in novel settings, such as for emerging pathogens, new or poorly monitored locations, or changing environmental conditions or transmission regimes, including the implementation of vector control or other interventions. Ultimately, mechanistic and statistical approaches both provide important insights and should be used to inform each other. For instance, mechanistic models can guide the design of statistical analyses by providing specific hypotheses (e.g. different types of nonlinearity). Additionally, if statistical models support predictions from a mechanistic model in a context-dependent manner, it may help reveal the relative importance of drivers over space and time or at different scales (Cohen et al. 2016; Mordecai et al. 2019).

6.5.2.2 Uncertainty, Causality, and Multiple Lines of Evidence

Uncertainty is an important but poorly characterized aspect of vector transmission dynamics. Process uncertainty emerges when the functional relationships between drivers and disease dynamics are unknown or poorly characterized, for example, due to interactive effects or time lags. Additionally, transmission is a stochastic process, so even an accurate mechanistic description of the transmission process cannot perfectly predict disease dynamics. Stochasticity is particularly important for rare diseases, which are harder to predict. For any given model, parameter uncertainty arises from measurement error in both the response variable (e.g. transmission probability) and environmental drivers, as well as unmeasured predictors. Vector population dynamics are extremely variable and challenging to model due to many sources of process and observation error; thus, analyses can be sensitive to the time scale of sample collection and aggregation (Jian et al. 2014, 2016). Finally, the direction and magnitude of environmental change is uncertain due to both human behavior (e.g. how much greenhouse gas emissions are released in the future) and complex responses of global systems to human activities (Intergovernmental Panel on Climate Change 2014). Accurately representing the true range of uncertainty requires propagating uncertainty due to measurement error, process error, true stochasticity through transmission models, incorporating multiple uncertain future environmental change scenarios, and communicating the results through credible intervals or scenario analyses (Dietze 2017; Johnson et al. 2015; Little et al. 2017; Metcalf et al. 2017; Ryan et al. 2019).

In the face of multiple, complex relationships between environmental drivers and transmission and a variety of sources of uncertainty, inferring causation is challenging. Support for environmental change causing changes in disease dynamics is strongest when statistical associations between environmental variation and disease transmission occur at the hypothesized time and place and in the hypothesized direction predicted by an independent mechanistic analysis (Metcalf et al. 2017; Randolph 2004; Rogers & Randolph 2006). Semi-mechanistic models are an alternative approach for inferring drivers in nonlinear transmission systems (Metcalf et al. 2017; Perkins et al. 2015). These models infer time-varying force of infection parameters while accounting for nonlinear dynamics; the parameters can then be statistically analyzed with the environmental drivers (Harris et al. 2019; Perkins et al. 2015). Another promising approach is fusing elements of experimental manipulation and natural observations (Murdock et al. 2017). Finally, detecting causality is an area of active philosophical and mathematical research, and new methods for inferring causation are still entering disease ecology (Sugihara et al. 2012; Xu et al. 2017).

No single method is sufficient for understanding environmental drivers across scales, systems, and settings. Multiple lines of evidence that support hypothesized mechanisms provide the strongest causal support for environmental drivers of disease (Metcalf et al. 2017; Munafò & Davey Smith 2018; Rogers & Randolph 2006). Using a complementary suite of approaches that includes exploring mechanisms, inferring drivers through spatial and temporal variation in transmission, and linking vector and case distributions can provide the deepest understanding and greatest predictive power across scales and settings (Munafò & Davey Smith 2018; Ostfeld & Brunner 2015; Tjaden et al. 2018).

6.5.2.3 Integrating Social Science, Economics, and Medicine

Interdisciplinary collaborations help identify when ecological mechanisms interact with socioeconomic mechanisms. For instance, political instability and war have led to the recent resurgence of leishmaniasis in the Middle East (Hotez 2018). Economic hardship following the fall of the Soviet Union led to increased tick-borne encephalitis in Eastern Europe when people increased foraging for food in wild areas and thus exposure to infected vectors (Šumilo et al. 2008). Frontier communities can experience a transient period with little access to healthcare and increased disease burden (de Castro et al. 2006). Further, established communities can get stuck in 'poverty traps', where there are reinforcing, positive feedbacks between low socioeconomic status and infectious disease burden (Bonds et al. 2010). Thus, the insights gained from an ecological perspective on vector-borne disease must be integrated with perspectives from other disciplines like sociology, economics, and medicine to correctly identify environmental drivers and achieve sustained reductions in disease. Interdisciplinary research is key for determining how human behavior influences disease risk and interacts with these environmental and socioeconomic drivers (Hammond et al. 2007; Koenraadt et al. 2006) and for designing interventions that are effective and accepted by the public (Brossard et al. 2019; Okamoto et al. 2016).

6.6 Future Directions

In this final section, we identify research gaps and frontiers in the technologies and analyses used for vector-borne disease research. Many of these new areas and approaches are interdisciplinary, as it is increasingly necessary to build links between human activity and technology, environmental change, social and economic systems, and the biology and ecology of vector-borne diseases. We break these future directions down into two broad categories: (1) fundamental biological questions and (2) emerging technologies, data, and analyses.

6.6.1 Fundamental Biological Questions

There are many gaps in our current knowledge of how environmental factors mechanistically drive variation in vector traits. First, while we know that climatic drivers interact in important ways (Shand et al. 2016), we lack a general or comparative framework for interactive effects of different drivers. Such a framework would also greatly enhance our understanding of land use change, as it often alters multiple environmental drivers at once. Second, our mechanistic understanding of how habitat and land use affects vector traits and populations is relatively poor compared to that of climatic drivers (Jones et al. 2008; Norris 2004). For most vector species we lack basic natural history connecting them to their preferred habitats and characterizing their niche breadths. Third, we need to develop methods to better quantify vector traits in the field in order to connect trait data from laboratory studies with field observations. For instance, mosquito lifespan—a trait that has a large influence on transmission—is typically shorter in nature than in the lab due to predation and other hazards (Brady et al. 2013; Clements & Paterson 1981; Macdonald 1952); however, it remains uncertain how effectively lab-based estimates of environmental impacts on lifespan reflect the survival of mosquitoes across environments in the field (Brady et al. 2013). Fourth, more empirical work is needed to determine the time scale at which environmental factors influence vector abundance and disease transmission (e.g. daily versus weekly average temperatures) and how to best account for fluctuations and time lags in these

factors. For instance, the current standard method to account for thermal variation (rate summation using thermal responses measured at constant temperatures: Bernhardt et al. 2018) has not been rigorously validated. Fifth, the importance of within- and among-population genetic variation for responses of vector and pathogen traits to environmental drivers is largely unknown. Moreover, higher order interactions (vector genotype x pathogen genotype x environment) are possible (Zouache et al. 2014). Finally, while vector microbiomes can impact their susceptibility to and competence for transmitting pathogens (Bonnet et al. 2017; Geiger et al. 2015; Jupatanakul et al. 2014; Narasimhan & Fikrig 2015), and microbiomes can vary by location (Coon et al. 2016), it is unclear how vector microbiomes and environmental variation may interact or be linked (Evans et al. 2020; see Chapter 13 this volume).

6.6.2 Emerging Technologies, Data, and Analyses

Consistent measurements of factors like temperature, rainfall, and land cover are critical for analyzing patterns of vector-borne disease driven by environmental variation and global change. Earth observations data (EO; i.e. satellite, airplane, and drone-based imagery), which provide globally-consistent measurements of the environment, have emerged as a key data source for mapping large-scale patterns of disease (Bhatt et al. 2013; Kraemer et al. 2015; Samy & Peterson 2016). EO data are often used to build species distribution models for vectors or pathogens, which search for patterns in observed distributions and relate them to environmental variables that best predict the presence, absence, or abundance of the species (Elith et al. 2006; Kalluri et al. 2007; Kraemer et al. 2015; Tjaden et al. 2018; White et al. 2012). These methods are powerful because they can take large, unstructured datasets and find relationships that may not be well characterized *a priori*, identifying or refining mechanistic relationships. The quantity and quality of EO data are dramatically increasing, as is access to analytical software (Gorelick et al. 2017). However, improved modeling approaches are needed to use these data more effectively.

Current challenges for using EO data include: scale mismatches, when the scale of field data does not match the scale of EO data (Anderson 2018); incomplete or unknown sampling effort (Phillips et al. 2009); and competing data typologies, such as the differences between categorical and continuous environmental patterns (Tucker Lima et al. 2017). Additionally, while the EO data itself is scalable, the current generation of models are not (Jansen & Beebe 2010). For instance, temperature-driven predictive transmission models are typically applied to coarse (5+ km spatial resolution), global-scale mean temperature data to predict national-scale health patterns (Bhatt et al. 2013; Ryan et al. 2015, 2019). This approach rarely captures fine-scale transmission dynamics, such as shifts in local vector habitat, microclimate, and human host susceptibility. Conversely, land use-driven predictive models are typically developed to represent local-scale phenomena and are often difficult to generalize to larger scales (Loveland et al. 2000; Sithiprasasna et al. 2005; Tadei et al. 1998). Further, maps often poorly characterize standardized land use patterns across highly heterogeneous environments (Gong et al. 2013) and classification typologies can be highly subjective and therefore inconsistent outside of specific contexts (Foody 2010; Tucker Lima et al. 2017; Verburg et al. 2011).

There is a growing movement to assemble large, publicly available vector and disease datasets to aid in the validation and development of models (e.g. Siraj et al. 2018, Johansson et al. 2019). Testing predictions from mechanistic models is a critical step in the scientific process; however, validating models with independent data is currently difficult for two reasons. First, the output of mechanistic models is often fundamentally different (e.g. relative transmission rates versus human incidence) or at different spatial and temporal scales than the data available for testing. Thus, better analytical methods are needed for comparing disparate types of data. Second, data on vector densities and disease incidence are sporadic, variable in format, and often not freely available (Cator et al. 2019). This is a missed opportunity given the extensive effort toward vector and disease surveillance undertaken by vector control and health agencies across the globe. To improve data availability and cross-compatibility, efforts are underway to collect, validate, and make data available for building, testing, and improving models. Examples of databases include VectorBASE, (bioinformatics databases of vector genomes, transcriptomes, proteomes, and insecticide resistance across vector populations: https://www.vectorbase.org/), Vector ByTE (ecoinformatics databases of vector traits and population dynamics [in development]: https://vectorbyte.org), the CDC Zika data repository (https://github.com/cdcepi/zika), and datasets provided for forecasting challenges for dengue cases and *Ae. aegypti* and *Ae. albopictus* (https://predict.cdc.gov/).

More synthetic analyses comparing environmental impacts across different vector-borne diseases and research methods would greatly enhance the field by sharing insights across systems. Currently, most research on vector-borne diseases is system-specific. While system-specific and local knowledge is often critical, much could be learned from identifying consistent trends and drivers, as well as important differences, across multiple diseases. Additionally, a wide variety of statistical and semi-mechanistic methods are used to study vector-borne disease time series, but most studies only use a single approach. Thus, it can be difficult to determine whether differing conclusions among studies are due to real differences in the underlying biology or simply reflect differences in methodology. Systematic analyses comparing results from multiple approaches and datasets would help to identify patterns and processes across systems, providing better information about the sensitivity and consistency of different methods for analyzing specific environmental drivers.

Finally, the biotechnology industry has a growing interest in developing solutions for vector-borne diseases. For instance, *Wolbachia*-infected *Ae. aegypti* mosquitoes—thought to be refractory for (i.e. not competent for transmitting) dengue, Zika, and other viruses—were released and successfully established in northeastern Australia (Hoffmann et al. 2011). More recently, Verily (Alphabet's life science subsidiary) released millions of sterile male *Ae. aegypti* mosquitoes—developed to temporarily suppress the mosquito population size—in Fresno, California during the summer of 2017 as a test for future releases in areas with Zika transmission

(Crawford 2017). Environmental drivers could interact with these biotechnology-based interventions and impact their efficacy (e.g. Ross et al. 2019). Thus, research is needed to explore these technologies in an ecological context and examine how the environment influences their success or failure. Similarly, Microsoft's Project Premonition seeks to use drones, robots, and other technological advances for mosquito and pathogen surveillance to improve disease detection (Microsoft 2019). Integrating these novel technologies with ecological knowledge is critical for achieving disease reduction and prevention.

6.6.3 Integrating Ecosystem Services and Management

The ecosystem services framework assigns value to the benefits that natural systems confer to humans. The framework aims to influence how land management policies are designed and implemented, shifting priorities to balance environmental quality, economic growth, and human well-being (Daily et al. 2009). This framework—which has traditionally been applied to services like clean water, food production via pollination, flood control, and regulating a stable climate—is increasingly being applied to infectious diseases (Foley 2005; Foley et al. 2007). However, it remains underutilized for these vector-borne diseases, given the strong links between land use and their transmission. Future research should incorporate collaborations with scientists from complementary disciplines to examine potential scenarios for economic development and land management. These predictions will help policymakers more effectively balance the needs for food, shelter, and economic prosperity while minimizing potential for vector-borne disease transmission. Crucially, this research would also engage local stakeholders that have often been omitted from discussions regarding their health and prosperity.

6.7 Summary

Transmission of vector-borne diseases emerges from the abundances, traits, and ecological interactions of arthropod vectors, hosts, and pathogens. These processes are sensitive to environmental factors like habitat and land use, temperature, humidity, and rainfall. Therefore, natural variation and anthropogenic changes in these factors have driven patterns of vector-borne diseases over the past decades, and will continue to impact them into the future. The current approaches and promising new directions will enable us to better understand and predict vector-borne disease transmission, and thus better protect human health and economic well-being.

References

Abad-Franch, F., Lima, M.M., Sarquis, O., et al. (2015). On palms, bugs, and Chagas disease in the Americas. *Acta Tropica*, 151, 126–41.

Abbasi, I., Trancoso Lopo de Queiroz, A., Kirstein, O.D., et al. (2018). Plant-feeding phlebotomine sand flies, vectors of leishmaniasis, prefer *Cannabis sativa*. *Proceedings of the National Academy of Sciences of the United States of America*, 115, 11790–5.

Adde, A., Roucou, P., Mangeas, M., et al. (2016). Predicting dengue fever outbreaks in French Guiana using climate indicators. *PLoS Neglected Tropical Diseases*, 10, e0004681.

Adeleke, M.A., Mafiana, C.F., Sam-Wobo, S.O., et al. (2010). Biting behaviour of *Simulium damnosum* complex and *Onchocerca volvulus* infection along the Osun River, Southwest Nigeria. *Parasites & Vectors*, 3, 93.

Afrane, Y.A., Lawson, B.W., Githeko, A.K., & Yan, G. (2005). Effects of microclimatic changes caused by land use and land cover on duration of gonotrophic cycles of *Anopheles Gambiae* (Diptera: Culicidae) in western Kenya highlands. *Journal of Medical Entomology*, 42, 974–80.

Alemu, A., Abebe, G., Tsegaye, W., & Golassa, L. (2011). Climatic variables and malaria transmission dynamics in Jimma town, south west Ethiopia. *Parasites & Vectors*, 4, 30.

Allan, B.F., Keesing, F., & Ostfeld, R.S. (2003). Effect of forest fragmentation on Lyme Disease risk. *Conservation Biology*, 17, 267–72.

Alsan, M. (2015). The effect of the tsetse fly on African development. *American Economic Review*, 105, 382–410.

Althouse, B.M. & Hanley, K.A. (2015). The tortoise or the hare? Impacts of within-host dynamics on transmission success of arthropod-borne viruses. *Philosophical Transactions of the Royal Society B: Biological Sciences*, 370, 20140299.

Altizer, S., Ostfeld, R.S., Johnson, P.T.J., Kutz, S., & Harvell, C.D. (2013). Climate change and infectious diseases: From evidence to a predictive framework. *Science*, 341, 514–19.

Alto, B.W., Lounibos, L.P., Higgs, S., & Juliano, S.A. (2005). Larval competition differentially affects arbovirus infection in *Aedes* mosquitoes. *Ecology*, 86, 3279–88.

Amarasekare, P. & Savage, V. (2012). A framework for elucidating the temperature dependence of fitness. *The American Naturalist*, 179, 178–91.

Anderson, C.B. (2018). Biodiversity monitoring, earth observations and the ecology of scale. *Ecology Letters*, 21, 1572–85.

Angilletta, M.J. (2006). Estimating and comparing thermal performance curves. *Journal of Thermal Biology*, 31, 541–5.

Anyamba, A., Chretien, J.-P., Small, J., et al. (2009). Prediction of a Rift Valley fever outbreak. *Proceedings of the National Academy of Sciences of the United States of America*, 106, 955–9.

Anyamba, A., Linthicum, K.J., Small, J.L., et al. (2012). Climate Teleconnections and Recent Patterns of Human and Animal Disease Outbreaks. *PLoS Neglected Tropical Diseases*, 6, e1465.

Arsnoe, I.M., Hickling, G.J., Ginsberg, H.S., McElreath, R., & Tsao, J.I. (2015). Different populations of blacklegged tick nymphs exhibit differences in questing behavior that have implications for human Lyme disease risk. *PLoS One*, 10, e0127450.

Baeza, A., Santos-Vega, M., Dobson, A.P., & Pascual, M. (2017). The rise and fall of malaria under land-use change in frontier regions. *Nature Ecology & Evolution*, 1, 108.

Barros, F.S.M., Arruda, M.E., Gurgel, H.C., & Honório, N.A. (2011). Spatial clustering and longitudinal variation of *Anopheles darlingi* (Diptera: Culicidae) larvae in a river of the Amazon: The importance of the forest fringe and of obstructions to flow in frontier malaria. *Bulletin of Entomological Research*, 101, 643–58.

Barton, P.S., Aberton, J.G., & Kay, B.H. (2004). Spatial and temporal definition of *Ochlerotatus camptorhynchus* (Thomson) (Diptera: Culicidae) in the Gippsland Lakes system of eastern Victoria. *Australian Journal of Entomology*, 43, 16–22.

Bayoh, M.N. (2001). Studies on the development and survival of *Anopheles gambiae* sensu stricto at various temperatures and relative humidities. Doctoral thesis. Durham: Durham University.

Bell, J.R., Alderson, L., Izera, D., et al. (2015). Long-term phenological trends, species accumulation rates, aphid traits and climate: Five decades of change in migrating aphids. *Journal of Animal Ecology*, 84, 21–34.

Benedum, C.M., Seidahmed, O.M.E., Eltahir, E.A.B., & Markuzon, N. (2018). Statistical modeling of the effect of rainfall flushing on dengue transmission in Singapore. *PLoS Neglected Tropical Diseases*, 12, e0006935.

Bernhardt, J.R., Sunday, J.M., Thompson, P.L., & O'Connor, M.I. (2018). Nonlinear averaging of thermal experience predicts population growth rates in a thermally variable environment. *Proceedings of the Royal Society B: Biological Sciences*, 285, 20181076.

Berry, B.S., Magori, K., Perofsky, A.C., Stallknecht, D.E., & Park, A.W. (2013). Wetland cover dynamics drive hemorrhagic disease patterns in white-tailed deer in the United States. *Journal of Wildlife Diseases*, 49, 501–9.

Berry, K., Bayham, J., Meyer, S.R., & Fenichel, E.P. (2018). The allocation of time and risk of Lyme: A case of ecosystem service income and substitution effects. *Environmental & Resource Economics*, 70, 631–50.

Bertrand, M.R. & Wilson, M.L. (1996). Microclimate-dependent survival of unfed adult *Ixodes scapularis* (Acari: Ixodidae) in nature: Life cycle and study design implications. *Journal of Medical Entomology*, 33, 619–27.

Bhatt, S., Gething, P.W., Brady, O.J., et al. (2013). The global distribution and burden of dengue. *Nature*, 496, 504–7.

Bi, P., Hiller, J.E., Cameron, a S., Zhang, Y., & Givney, R. (2009). Climate variability and Ross River virus infections in Riverland, South Australia, 1992–2004. *Epidemiology and Infection*, 137, 1486–93.

Bicout, D.J. & Sabatier, P. (2004). Mapping rift valley fever vectors and prevalence using rainfall variations. *Vector-Borne and Zoonotic Diseases*, 4, 33–42.

Bødker, R., Akida, J., Shayo, D., et al. (2003). Relationship between altitude and intensity of malaria transmission in the Usambara Mountains, Tanzania. *Journal of Medical Entomology*, 40, 706–17.

Boland, G.J., Melzer, M.S., Hopkin, A., Higgins, V., & Nassuth, A. (2004). Climate change and plant diseases in Ontario. *Canadian Journal of Plant Pathology*, 26, 335–50.

Bonds, M.H., Keenan, D.C., Rohani, P., & Sachs, J.D. (2010). Poverty trap formed by the ecology of infectious diseases. *Proceedings of the Royal Society B: Biological Sciences*, 277, 1185–92.

Bonnet, S.I., Binetruy, F., Hernández-Jarguín, A.M., & Duron, O. (2017). The tick microbiome: Why non-pathogenic microorganisms matter in tick biology and pathogen transmission. *Frontiers in Cellular and Infection Microbiology*, 7, 236.

Bowman, A.S. & Sauer, J.R. (2004). Tick salivary glands: Function, physiology and future. *Parasitology*, 129, S67–81.

Brady, O.J., Johansson, M.A., Guerra, C.A., et al. (2013). Modelling adult *Aedes aegypti* and *Aedes albopictus* survival at different temperatures in laboratory and field settings. *Parasites & Vectors*, 6, 351.

Brand, S.P.C. & Keeling, M.J. (2017). The impact of temperature changes on vector-borne disease transmission: *Culicoides* midges and bluetongue virus. *Journal of The Royal Society Interface*, 14, 20160481.

Brock, P.M., Fornace, K.M., Parmiter, M., et al. (2016). *Plasmodium knowlesi* transmission: Integrating quantitative approaches from epidemiology and ecology to understand malaria as a zoonosis. *Parasitology*, 143, 389–400.

Brossard, D., Belluck, P., Gould, F., & Wirz, C.D. (2019). Promises and perils of gene drives: Navigating the communication of complex, post-normal science. *Proceedings of the National Academy of Sciences of the United States of America*, 116, 7692–7.

Brownstein, J.S., Holford, T.R., & Fish, D. (2005a). Effect of climate change on Lyme disease risk in North America. *EcoHealth*, 2, 38–46.

Brownstein, J.S., Skelly, D.K., Holford, T.R., & Fish, D. (2005b). Forest fragmentation predicts local scale heterogeneity of Lyme disease risk. *Oecologia*, 146, 469–75.

Brunner, J.L., Killilea, M., & Ostfeld, R.S. (2012). Overwintering survival of nymphal *Ixodes scapularis* (Acari: Ixodidae) under natural conditions. *Journal of Medical Entomology*, 49, 981–7.

Buckee, C.O., Wesolowski, A., Eagle, N.N., Hansen, E., & Snow, R.W. (2013). Mobile phones and malaria: Modeling human and parasite travel. *Travel Medicine and Infectious Disease*, 11, 15–22.

Byrne, D.N. & Bellows, T.S. (1991). Whitefly biology. *Annual Review of Entomology*, 36, 431–57.

Cai, W., Borlace, S., Lengaigne, M., et al. (2014). Increasing frequency of extreme El Niño events due to greenhouse warming. *Nature Climate Change*, 4, 111–16.

Caillouët, K.A., Michaels, S.R., Xiong, X., Foppa, I., & Wesson, D.M. (2008). Increase in West Nile neuroinvasive disease after Hurricane Katrina. *Emerging Infectious Diseases*, 14, 804–7.

Caminade, C., Kovats, S., Rocklov, J., et al. (2014). Impact of climate change on global malaria distribution. *Proceedings of the National Academy of Sciences of the United States of America*, 111, 3286–91.

Campbell, K.M., Lin, C.D., Iamsirithaworn, S., & Scott, T.W. (2013). The complex relationship between weather and dengue virus transmission in Thailand. *The American Journal of Tropical Medicine and Hygiene*, 89, 1066–80.

Canto, T., Aranda, M.A., & Fereres, A. (2009). Climate change effects on physiology and population processes of hosts and vectors that influence the spread of hemipteran-borne plant viruses. *Global Change Biology*, 15, 1884–94.

de Castro, M.C., Monte-Mór, R.L., Sawyer, D.O., & Singer, B.H. (2006). Malaria risk on the Amazon frontier. *Proceedings of the National Academy of Sciences of the United States of America*, 103, 2452–7.

Cator, L.J., Johnson, L.R., Mordecai, E.A., et al. (2019). More than a flying syringe: Using functional traits in vector-borne disease research. *bioRxiv*.

Cazelles, B., Chavez, M., McMichael, A.J., & Hales, S. (2005). Nonstationary influence of El Niño on the synchronous dengue epidemics in Thailand. *PLoS Medicine*, 2, e106.

Chandran, R. & Azeez, P.A. (2015). Outbreak of dengue in Tamil Nadu, India. *Current Science*, 109, 171–6.

Chaniotis, B.N., Neely, J.M., Correa, M.A., Tesh, R.B., & Johnson, K.M. (1971). Natural population dynamics of phlebotomine sandflies in Panama. *Journal of Medical Entomology*, 8, 339–52.

Chase, J.M. & Shulman, R.S. (2009). Wetland isolation facilitates larval mosquito density through the reduction of predators. *Ecological Entomology*, 34, 741–7.

Chaves, L.F., Hamer, G.L., Walker, E.D., Brown, W.M., Ruiz, M.O., & Kitron, U.D. (2011). Climatic variability and landscape heterogeneity impact urban mosquito diversity and vector abundance and infection. *Ecosphere*, 2, 1–21.

Chaves, L.S.M., Conn, J.E., López, R.V.M., & Sallum, M.A.M. (2018). Abundance of impacted forest patches less than 5 km$_2$ is a key driver of the incidence of malaria in Amazonian Brazil. *Scientific Reports*, 8, 1–11.

Cheke, R.A., Basáñez, M.-G., Perry, M., et al. (2015). Potential effects of warmer worms and vectors on onchocerciasis transmission in West Africa. *Philosophical Transactions of the Royal Society B: Biological Sciences*, 370, 20130559.

Civitello, D.J., Cohen, J., Fatima, H., et al. (2015). Biodiversity inhibits parasites: Broad evidence for the dilution effect. *Proceedings of the National Academy of Sciences of the United States of America*, 112, 8667–71.

Clements, A.N. & Paterson, G.D. (1981). The Analysis of Mortality and Survival Rates in Wild Populations of Mosquitoes. *The Journal of Applied Ecology*, 18, 373.

Cohen, J.M., Civitello, D.J., Brace, A.J., et al. (2016). Spatial scale modulates the strength of ecological processes driving disease distributions. *Proceedings of the National Academy of Sciences of the United States of America*, 113, E3359–64.

Colón-González, F.J., Bentham, G., & Lake, I.R. (2011). Climate Variability and Dengue Fever in Warm and Humid Mexico. *The American Journal of Tropical Medicine and Hygiene*, 84, 757–63.

Confalonieri, U.E.C., Margonari, C., & Quintão, A.F. (2014). Environmental change and the dynamics of parasitic diseases in the Amazon. *Acta Tropica*, 129, 33–41.

Connally, N.P., Durante, A.J., Yousey-Hindes, K.M., Meek, J.I., Nelson, R.S., & Heimer, R. (2009). Peridomestic Lyme disease prevention: Results of a population-based case-control study. *American Journal of Preventive Medicine*, 37, 201–6.

Coon, K.L., Brown, M.R., & Strand, M.R. (2016). Mosquitoes host communities of bacteria that are essential for development but vary greatly between local habitats. *Molecular Ecology*, 25, 5806–26.

Coon, K.L. & Strand, M.R. (2020). Gut microbiome assembly and function in mosquitoes. In J.M. Drake, M.B. Bonsall & M.R. Strand, eds. *Population Biology of*

Vector-borne diseases, pp. 229-245. Oxford: Oxford University Press.

Costa, E.A.P. de A., Santos, E.M. de M., Correia, J.C., & Albuquerque, C.M.R. de (2010). Impact of small variations in temperature and humidity on the reproductive activity and survival of *Aedes aegypti* (Diptera, Culicidae). *Revista Brasileira de Entomologia*, 54, 488–93.

Crawford, J. (2017). Debug Fresno, our first U.S. field study. Verily Life Sciences. https://blog.verily.com/2017/07/debug-fresno-our-first-us-field-study.html.

Crump, A., Morel, C.M., & Omura, S. (2012). The onchocerciasis chronicle: from the beginning to the end? *Trends in Parasitology*, 28, 280–8.

Daily, G.C., Polasky, S., Goldstein, J., et al. (2009). Ecosystem services in decision making: Time to deliver. *Frontiers in Ecology and the Environment*, 7, 21–8.

Day, J. (2016). Mosquito oviposition behavior and vector control. *Insects*, 7, 65.

Day, J.F. & Shaman, J. (2009). Severe winter freezes enhance St. Louis encephalitis virus amplification and epidemic transmission in Peninsular Florida. *Journal of Medical Entomology*, 46, 1498–506.

Day, J.F., Curtis, G.A., & Edman, J.D. (1990). Rainfall-directed oviposition behavior of *Culex nigripalpus* (Diptera: Culicidae) and Its Influence on St. Louis Encephalitis Virus Transmission in Indian River County, Florida. *Journal of Medical Entomology*, 27, 43–50.

Dell, A.I., Pawar, S., & Savage, V.M. (2011). Systematic variation in the temperature dependence of physiological and ecological traits. *Proceedings of the National Academy of Sciences of the United States of America*, 108, 10591–6.

DeLong, D.M. (1971). The bionomics of leafhoppers. *Annual Review of Entomology*, 16, 179–210.

Desjeux, P. (2001). The increase in risk factors for leishmaniasis worldwide. *Transactions of the Royal Society of Tropical Medicine and Hygiene*, 95, 239–43.

Despommier, D., Ellis, B.R., & Wilcox, B.A. (2006). The Role of Ecotones in Emerging Infectious Diseases. *EcoHealth*, 3, 281–9.

Deutsch, C.A., Tewksbury, J.J., Huey, R.B., et al. (2008). Impacts of climate warming on terrestrial ectotherms across latitude. *Proceedings of the National Academy of Sciences of the United States of America*, 105, 6668–72.

Devictor, V., Julliard, R., & Jiguet, F. (2008). Distribution of specialist and generalist species along spatial gradients of habitat disturbance and fragmentation. *Oikos*, 117, 507–14.

Didham, R.K. (2010). Ecological Consequences of Habitat Fragmentation. In R. Jansson, ed. *Encyclopedia of Life Sciences*, pp. 1–11, John Wiley & Sons, Chichester, UK.

Dietz, K. (1993). The estimation of the basic reproduction number for infectious diseases. *Statistical Methods in Medical Research*, 2, 23–41.

Dietze, M.C. (2017). Ecological Forecasting. Princeton University Press, Princeton, New Jersey..

Durr, P.A., Graham, K., & van Klinken, R.D. (2017). Sellers' revisited: A big data reassessment of historical outbreaks of bluetongue and African horse sickness due to the long-distance wind dispersion of *Culicoides* midges. *Frontiers in Veterinary Science*, 4, 98.

Earnest, A., Tan, S.B., & Wilder-Smith, A. (2012). Meteorological factors and El Niño southern oscillation are independently associated with dengue infections. *Epidemiology and Infection*, 140, 1244–51.

Ebrahimi, B., Jackson, B.T., Guseman, J.L., Przybylowicz, C.M., Stone, C.M., & Foster, W.A. (2018). Alteration of plant species assemblages can decrease the transmission potential of malaria mosquitoes. *Journal of Applied Ecology*, 55, 841–51.

Eisen, L. & Eisen, R.J. (2016). Critical evaluation of the linkage between tick-based risk measures and the occurrence of Lyme disease cases: Table 1. *Journal of Medical Entomology*, 53, 1050–62.

Eisen, R.J., Eisen, L., & Lane, R.S. (2006). Predicting density of *Ixodes pacificus* nymphs in dense woodlands in Mendocino County, California, based on geographic information systems and remote sensing versus field-derived data. *American Journal of Tropical Medicine and Hygiene*, 74, 632–40.

Eisen, R.J., Eisen, L., Girard, Y.A., et al. (2010). A spatially-explicit model of acarological risk of exposure to *Borrelia burgdorferi*-infected *Ixodes pacificus* nymphs in north-western California based on woodland type, temperature, and water vapor. *Ticks and Tick-borne Diseases*, 1, 35–43.

Elbers, A.R.W., Koenraadt, C.J.M., & Meiswinkel, R. (2015). Mosquitoes and *Culicoides* biting midges: Vector range and the influence of climate change. *Scientific and Technical Review of the Office International des Epizooties*, 34, 123–37.

Elith, J.H. Graham, C. P. Anderson, R., et al. (2006). Novel methods improve prediction of species' distributions from occurrence data. *Ecography*, 29, 129–51.

Estrada-Peña, A. & de la Fuente, J. (2014). The ecology of ticks and epidemiology of tick-borne viral diseases. *Antiviral Research*, 108, 104–28.

Evans, M.V., Newberry, P.M., & Murdock, C.C. (2020). Carry-over effects of the larval environment in mosquito-borne disease systems. In J.M. Drake, M.R. Strand, & M. Bonsall, eds. *Population Biology of Vector-borne Diseases*, pp. 155–176. Oxford University Press, Oxford, UK.

Ezenwa, V.O., Godsey, M.S., King, R.J., & Guptill, S.C. (2006). Avian diversity and West Nile virus: Testing associations between biodiversity and infectious disease risk. *Proceedings of the Royal Society B: Biological Sciences*, 273, 109–17.

Faust, C.L., Dobson, A.P., Gottdenker, N., et al. (2017). Null expectations for disease dynamics in shrinking habitat: Dilution or amplification? *Philosophical Transactions of the Royal Society B: Biological Sciences*, 372, 20160173.

Feliciangeli, M.D. (2004). Natural breeding places of phlebotomine sandflies. *Medical and Veterinary Entomology*, 18, 71–80.

Ferraguti, M., Martínez-de la Puente, J., Roiz, D., Ruiz, S., Soriguer, R., & Figuerola, J. (2016). Effects of landscape anthropization on mosquito community composition and abundance. *Scientific Reports*, 6, 1–9.

Fischer, E.M. & Knutti, R. (2015). Anthropogenic contribution to global occurrence of heavy-precipitation and high-temperature extremes. *Nature Climate Change*, 5, 560–4.

Fischer, S., Alem, I.S., De Majo, M.S., Campos, R.E., & Schweigmann, N. (2011). Cold season mortality and hatching behavior of *Aedes aegypti* L. (Diptera: Culicidae) eggs in Buenos Aires City, Argentina. *Journal of Vector Ecology*, 36, 94–9.

Foley, J.A. (2005). Global consequences of land use. *Science*, 309, 570–4.

Foley, J.A., Asner, G.P., Costa, M.H., et al. (2007). Amazonia revealed: Forest degradation and loss of ecosystem goods and services in the Amazon basin. *Frontiers in Ecology and the Environment*, 5, 25–32.

Foody, G.M. (2010). Assessing the accuracy of land cover change with imperfect ground reference data. *Remote Sensing of Environment*, 114, 2271–85.

Forum on Microbial Threats (2008). Vector-Borne Diseases: Understanding the Environmental, Human Health, and Ecological Connections, Workshop Summary. National Academies Press, Washington, D.C.

Foster, W.A. (1995). Mosquito sugar feeding and reproductive energetics. *Annual Review of Entomology*, 40, 443–74.

Franco, A.O., Gomes, M.G.M., Rowland, M., Coleman, P.G., & Davies, C.R. (2014a). Controlling malaria using livestock-based interventions: A one health approach. *PLoS One*, 9, e101699.

Franco, J., Simarro, P., Diarra, A., & Jannin, J. (2014b). Epidemiology of human African trypanosomiasis. *Clinical Epidemiology*, 6, 257–275.

Freed, T.Z. & Leisnham, P.T. (2014). Roles of spatial partitioning, competition, and predation in the North American invasion of an exotic mosquito. *Oecologia*, 175, 601–11.

Fu, C. (2003). Potential impacts of human-induced land cover change on East Asia monsoon. *Global and Planetary Change*, 37: 219–229.

Furtado, N.V.R., Galardo, A.K.R., Galardo, C.D., Firmino, V.C., & Vasconcelos dos Santos, T. (2016). Phlebotomines (Diptera: Psychodidae) in a hydroelectric system affected area from northern Amazonian Brazil: Further insights into the effects of environmental changes on vector ecology. *Journal of Tropical Medicine*, 2016, 1–12.

Gagnon, A.S., Smoyer-Tomic, K.E., & Bush, A.B.G. (2002). The El Niño southern oscillation and malaria epidemics in South America. *International Journal of Biometeorology*, 46, 81–9.

Gálvez, R., Descalzo, M.A., Miró, G., et al. (2010). Seasonal trends and spatial relations between environmental/meteorological factors and leishmaniosis sand fly vector abundances in Central Spain. *Acta Tropica*, 115, 95–102.

Gao, X., Qin, H., Xiao, J., & Wang, H. (2017). Meteorological conditions and land cover as predictors for the prevalence of Bluetongue virus in the Inner Mongolia autonomous region of mainland China. *Preventive Veterinary Medicine*, 138, 88–93.

Garrett, K.A., Dendy, S.P., Frank, E.E., Rouse, M.N., & Travers, S.E. (2006). Climate change effects on plant disease: Genomes to ecosystems. *Annual Review of Phytopathology*, 44, 489–509.

Gatewood, A.G., Liebman, K.A., Vourc'h, G., et al. (2009). Climate and tick seasonality are predictors of *Borrelia burgdorferi* genotype distribution. *Applied and Environmental Microbiology*, 75, 2476–83.

Gatton, M.L., Kay, B.H., & Ryan, P.A. (2005). Environmental predictors of Ross River virus disease outbreaks in Queensland, Australia. *American Journal of Tropical Medicine and Hygiene*, 72, 792–9.

Geiger, A., Ponton, F., & Simo, G. (2015). Adult bloodfeeding tsetse flies, trypanosomes, microbiota and the fluctuating environment in sub-Saharan Africa. *The ISME Journal*, 9, 1496–507.

Geoghegan, J.L., Walker, P.J., Duchemin, J.-B., Jeanne, I., & Holmes, E.C. (2014). Seasonal drivers of the epidemiology of arthropod-borne viruses in Australia. *PLoS Neglected Tropical Diseases*, 8, e3325.

Ginsberg, H.S., Albert, M., Acevedo, L., et al. (2017). Environmental factors affecting survival of immature *Ixodes scapularis* and implications for geographical distribution of Lyme disease: The climate/behavior hypothesis. *PLoS One*, 12, e0168723.

Gong, P., Wang, J., Yu, L., et al. (2013). Finer resolution observation and monitoring of global land cover: First mapping results with Landsat TM and ETM+ data. *International Journal of Remote Sensing*, 34, 2607–54.

Gorelick, N., Hancher, M., Dixon, M., Ilyushchenko, S., Thau, D., & Moore, R. (2017). Google Earth Engine: Planetary-scale geospatial analysis for everyone. *Remote Sensing of Environment*, 202, 18–27.

Gottdenker, N.L., Chaves, L.F., Calzada, J.E., Saldaña, A., & Carroll, C.R. (2012). Host life history strategy, species diversity, and habitat influence *Trypanosoma cruzi* vector infection in changing landscapes. *PLoS Neglected Tropical Diseases*, 6, e1884.

Gottdenker, N.L., Streicker, D.G., Faust, C.L., & Carroll, C.R. (2014). Anthropogenic land use change and infectious diseases: A review of the evidence. *EcoHealth*, 11, 619–32.

Grébaut, P., Bena, J.-M., Manzambi, E.Z., et al. (2009). Characterization of sleeping sickness transmission sites in rural and Periurban areas of Kinshasa (République Démocratique du Congo). *Vector Borne Zoonotic Diseases*, 9, 631–6.

Gu, W., Müller, G., Schlein, Y., Novak, R.J., & Beier, J.C. (2011). Natural plant sugar sources of *Anopheles* mosquitoes strongly impact malaria transmission potential. *PLoS One*, 6, e15996.

Guis, H., Caminade, C., Calvete, C., Morse, A.P., Tran, A., & Baylis, M. (2012). Modelling the effects of past and future climate on the risk of bluetongue emergence in Europe. *Journal of The Royal Society Interface*, 9, 339–50.

Halsey, S. (2019). Defuse the dilution effect debate. *Nature Ecology & Evolution* 3, 145-6.

Hammond, S.N., Gordon, A.L., Lugo, E. del C., et al. (2007). Characterization of *Aedes aegypti* (Diptera: Culcidae) production sites in urban Nicaragua. *Journal of Medical Entomology*, 44, 851–60.

Harrigan, R.J., Thomassen, H.A., Buermann, W., & Smith, T.B. (2014). A continental risk assessment of West Nile virus under climate change. *Global Change Biology*, 20, 2417–25.

Harris, M., Caldwell, J.M., & Mordecai, E.A. (2019). Climate drives spatial variation in Zika epidemics in Latin America. *Proceedings of the Royal Society B: Biological Sciences*, 286, 20191578.

Hartley, D.M., Barker, C.M., Le Menach, A., Niu, T., Gaff, H.D., & Reisen, W.K. (2012). Effects of temperature on emergence and seasonality of West Nile virus in California. *American Journal of Tropical Medicine and Hygiene*, 86, 884–94.

Hien, D.F. d. S., Dabiré, K.R., Roche, B., et al. (2016). Plant-mediated effects on mosquito capacity to transmit human malaria. *PLoS Pathogens*, 12, e1005773.

Hoffmann, A.A., Montgomery, B.L., Popovici, J., et al. (2011). Successful establishment of *Wolbachia* in *Aedes* populations to suppress dengue transmission. *Nature*, 476, 454–7.

Hotez, P.J. (2018). Modern Sunni-Shia conflicts and their neglected tropical diseases. *PLoS Neglected Tropical Diseases*, 12, e0006008.

Houghton, R.A. (1999). The U.S. carbon budget: Contributions from land-use change. *Science*, 285, 574–8.

Hu, W., Nicholls, N., Lindsay, M., et al. (2004). Development of a predictive model for Ross River virus disease in Brisbane, Australia. *American Journal of Tropical Medicine and Hygiene*, 71, 129–37.

Hu, W., Tong, S., Mengersen, K., & Oldenburg, B. (2006). Rainfall, mosquito density and the transmission of Ross River virus: A time-series forecasting model. *Ecological Modelling*, 196, 505–14.

Hu, W., Clements, A., Williams, G., & Tong, S. (2010). Dengue fever and El Nino/Southern Oscillation in Queensland, Australia: A time series predictive model. *Occupational and Environmental Medicine*, 67, 307–11.

Hunt, S.K., Galatowitsch, M.L., & McIntosh, A.R. (2017). Interactive effects of land use, temperature, and predators determine native and invasive mosquito distributions. *Freshwater Biology*, 62, 1564–77.

Imhoff, M.L., Zhang, P., Wolfe, R.E., & Bounoua, L. (2010). Remote sensing of the urban heat island effect across biomes in the continental USA. *Remote Sensing of Environment*, 114, 504–13.

Intergovernmental Panel on Climate Change (2014). Climate Change 2014: Synthesis Report. Contribution of Working Groups I, II and III to the Fifth Assessment. Geneva, Switzerland: IPCC.

Jacups, S.P., Whelan, P.I., Markey, P.G., Cleland, S.J., Williamson, G.J., & Currie, B.J. (2008). Predictive indicators for Ross River virus infection in the Darwin area of tropical northern Australia, using long-term mosquito trapping data. *Tropical Medicine and International Health*, 13, 943–52.

Jansen, C.C., & Beebe, N.W. (2010). The dengue vector *Aedes aegypti*: What comes next. *Microbes and Infection*, 12, 272–9.

Jeronimo, S.M., Oliveira, R.M., Mackay, S., et al. (1994). An urban outbreak of visceral leishmaniasis in Natal, Brazil. *Transactions of the Royal Society of Tropical Medicine and Hygiene*, 88, 386–8.

Jian, Y., Silvestri, S., Brown, J., Hickman, R., & Marani, M. (2014). The temporal spectrum of adult mosquito population fluctuations: Conceptual and modeling implications. *PLoS One*, 9, e114301.

Jian, Y., Silvestri, S., Brown, J., Hickman, R., & Marani, M. (2016). The predictability of mosquito abundance from daily to monthly timescales. *Ecological Applications*, 26, 2611–22.

Johansson, M.A., Cummings, D.A.T., & Glass, G.E. (2009). Multiyear climate variability and dengue—El Niño southern oscillation, weather, and dengue incidence in Puerto Rico, Mexico, and Thailand: A Longitudinal Data Analysis. *PLoS Medicine*, 6, e1000168.

Johansson, M.A., Powers, A.M., Pesik, N., Cohen, N.J., & Staples, J.E. (2014). Nowcasting the Spread of Chikungunya Virus in the Americas. *PLoS One*, 9, e104915.

Johansson, M.A., Reich, N.G., Hota, A., Brownstein, J.S., & Santillana, M. (2016). Evaluating the performance of infectious disease forecasts: A comparison of climate-driven and seasonal dengue forecasts for Mexico. *Scientific Reports*, 6, 33707.

Johansson, M.A., Apfeldorf, K.M., Dobson, S., et al. (2019). An open challenge to advance probabilistic forecasting for dengue epidemics. *Proceedings of the National Academy of Sciences of the United States of America*, 116, 24268–74.

Johnson, L.R., Ben-Horin, T., Lafferty, K.D., et al. (2015). Understanding uncertainty in temperature effects on vector-borne disease: A Bayesian approach. *Ecology*, 96, 203–13.

Johnson, L.R., Gramacy, R.B., Cohen, J., et al. (2018). Phenomenological forecasting of disease incidence using heteroskedastic Gaussian processes: A dengue case study. *The Annals of Applied Statistics*, 12, 27–66.

Jones, C.G. (1998). Chain reactions linking acorns to gypsy moth outbreaks and Lyme disease risk. *Science*, 279, 1023–6.

Jones, A.E., Turner, J., Caminade, C., et al. (2019). Bluetongue risk under future climates. *Nature Climate Change*, 9, 153–7.

Jones, K.E., Patel, N.G., Levy, M.A., et al. (2008). Global trends in emerging infectious diseases. *Nature*, 451, 990–3.

Juliano, S.A. & Lounibos, L.P. (2005). Ecology of invasive mosquitoes: Effects on resident species and on human health. *Ecology Letters*, 8, 558–74.

Jupatanakul, N., Sim, S., & Dimopoulos, G. (2014). The insect microbiome modulates vector competence for arboviruses. *Viruses*, 6, 4294–313.

Kalluri, S., Gilruth, P., Rogers, D., & Szczur, M. (2007). Surveillance of arthropod vector-borne infectious diseases using remote sensing techniques: A review. *PLoS Pathogens*, 3, e116.

Kalnay, E. & Cai, M. (2003). Impact of urbanization and land-use change on climate. *Nature*, 423, 528–31.

Karim, M.N., Munshi, S.U., Anwar, N., & Alam, M.S. (2012). Climatic factors influencing dengue cases in Dhaka city: A model for dengue prediction. *The Indian Journal of Medical Research*, 136, 32–9.

Kaufmann, C., Mathis, A., & Vorburger, C. (2015). Sugar-feeding behaviour and longevity of European *Culicoides* biting midges: Sugar feeding of *Culicoides* biting midges. *Medical and Veterinary Entomology*, 29, 17–25.

Kelly-Hope, L.A., Purdie, D.M., & Kay, B.H. (2004). Ross River virus disease in Australia, 1886–1998, with analysis of risk factors associated with outbreaks. *Journal of Medical Entomology*, 41, 133–50.

Kerr, P., Liu, J., Cattadori, I., Ghedin, E., Read, A., & Holmes, E. (2015). Myxoma virus and the leporipoxviruses: An evolutionary paradigm. *Viruses*, 7, 1020–61.

Khatchikian, C.E., Prusinski, M., Stone, M., et al. (2012). Geographical and environmental factors driving the increase in the Lyme disease vector *Ixodes scapularis*. *Ecosphere*, 3, 1–18.

Kilham, L. & Woke, P.A. (1953). Laboratory transmission of fibromas (shope) in cottontail rabbits by means of fleas and mosquitoes. *Experimental Biology and Medicine*, 83, 296–301.

Kilpatrick, A.M. & Randolph, S.E. (2012). Drivers, dynamics, and control of emerging vector-borne zoonotic diseases. *The Lancet*, 380, 1946–55.

Knap, N., Durmisi, E., Saksida, A., Korva, M., Petrovec, M., & Avsic-Zupanc, T. (2009). Influence of climatic factors on dynamics of questing *Ixodes ricinus* ticks in Slovenia. *Veterinary Parasitology*, 164, 275–81.

Koenraadt, C.J.M. & Harrington, L.C. (2008). Flushing effect of rain on container-inhabiting mosquitoes *Aedes aegypti* and *Culex pipiens* (Diptera: Culicidae). *Journal of Medical Entomology*, 45, 28–35.

Koenraadt, C.J.M., Githeko, A.K., & Takken, W. (2004). The effects of rainfall and evapotranspiration on the temporal dynamics of *Anopheles gambiae s.s.* and *Anopheles arabiensis* in a Kenyan village. *Acta Tropica*, 90, 141–53.

Koenraadt, C.J.M., Tuiten, W., Sithiprasasna, R., Kijchalao, U., Jones, J.W., & Scott, T.W. (2006). Dengue knowledge and practices and their impact on *Aedes aegypti* populations in Kamphaeng Phet, Thailand. *American Journal of Tropical Medicine and Hygiene*, 74, 692–700.

Kokkinn, M.J., Duval, D.J., & Williams, C.R. (2009). Modelling the ecology of the coastal mosquitoes *Aedes vigilax* and *Aedes camptorhynchus* at Port Pirie, South Australia. *Medical and Veterinary Entomology*, 23, 85–91.

Kraemer, M.U., Sinka, M.E., Duda, K.A., et al. (2015). The global distribution of the arbovirus vectors *Aedes aegypti* and *Ae. albopictus*. *eLife*, 4, e08347.

Kraemer, M.U.G., Reiner, R.C., Brady, O.J., et al. (2019). Past and future spread of the arbovirus vectors *Aedes aegypti* and *Aedes albopictus*. *Nature Microbiology*, 4, 854–63.

Kurtenbach, K., Hanincová, K., Tsao, J.I., Margos, G., Fish, D., & Ogden, N.H. (2006). Fundamental processes in the evolutionary ecology of Lyme borreliosis. *Nature Reviews Microbiology*, 4, 660–9.

Kwa, B.H. (2008). Environmental change, development and vectorborne disease: Malaysia's experience with filariasis, scrub typhus and dengue. *Environment, Development and Sustainability*, 10, 209–17.

Labuda, M., Kozuch, O., Zuffová, E., Elecková, E., Hails, R.S., & Nuttall, P.A. (1997). Tick-borne encephalitis virus transmission between ticks cofeeding on specific immune natural rodent hosts. *Virology*, 235, 138–43.

LaDeau, S.L., Allan, B.F., Leisnham, P.T., & Levy, M.Z. (2015). The ecological foundations of transmission potential and vector-borne disease in urban landscapes. *Functional Ecology*, 29, 889–901.

Lafferty, K.D. (2009). The ecology of climate change and infectious diseases. *Ecology*, 90, 888–900.

Lafferty, K.D. & Mordecai, E.A. (2016). The rise and fall of infectious disease in a warmer world. *F1000Research*, 5, 2040.

Lambin, E.F., Turner, B.L., Geist, H.J., et al. (2001). The causes of land-use and land-cover change: Moving beyond the myths. *Global Environmental Change*, 11, 261–9.

Lambin, E.F., Tran, A., Vanwambeke, S.O., Linard, C., & Soti, V. (2010). Pathogenic landscapes: Interactions between land, people, disease vectors, and their animal hosts. *International Journal of Health Geographics*, 9, 54.

Lambrechts, L., Paaijmans, K.P., Fansiri, T., et al. (2011). Impact of daily temperature fluctuations on dengue virus transmission by *Aedes aegypti*. *Proceedings of the National Academy of Sciences of the United States of America*, 108, 7460–5.

Lane, R.S., Mun, J., Peribáñez, M.A., & Stubbs, H.A. (2007). Host-seeking behavior of *Ixodes pacificus* (Acari: Ixodidae) nymphs in relation to environmental parameters in dense-woodland and woodland-grass habitats. *Journal of Vector Ecology*, 32, 342.

Lapointe, D.A. (2008). Dispersal of *Culex quinquefasciatus* (Diptera: Culicidae) in a Hawaiian rain forest. *Journal of Medical Entomology*, 45, 600–9.

Larsen, A.E., Plantinga, A.J., & MacDonald, A.J. (2014). Lyme disease risk influences human settlement in the wildland–urban interface: Evidence from a longitudinal analysis of counties in the northeastern United States. *The American Journal of Tropical Medicine and Hygiene*, 91, 747–55.

Larsen, A.E., Meng, K., & Kendall, B.E. (2019). Causal analysis in control–impact ecological studies with observational data. *Methods in Ecology and Evolution*, 10, 924–34.

Lefèvre, T., Vantaux, A., Dabiré, K.R., Mouline, K., & Cohuet, A. (2013). Non-genetic determinants of mosquito competence for malaria parasites. *PLoS Pathogens*, 9, e1003365.

Leisnham, P.T., Slaney, D.P., Lester, P.J., & Weinstein, P. (2005). Increased larval mosquito densities from modified landuses in the Kapiti region, New Zealand: Vegetation, water quality, and predators as associated environmental factors. *EcoHealth*, 2, 313–22.

Li, C.F., Lim, T.W., Han, L.L., & Fang, R. (1985). Rainfall, abundance of *Aedes aegypti* and dengue infection in Selangor, Malaysia. *Southeast Asian Journal of Tropical Medicine and Public Health*, 16, 560–8.

Lindblade, K.A., Walker, E.D., Onapa, A.W., Katungu, J., & Wilson, M.L. (2000). Land use change alters malaria transmission parameters by modifying temperature in a highland area of Uganda. *Tropical Medicine and International Health*, 5, 263–74.

Lindgren, E. & Gustafson, R. (2001). Tick-borne encephalitis in Sweden and climate change. *Lancet*, 358, 16–18.

Lindgren, E., Tälleklint, L., & Polfeldt, T. (2000). Impact of climatic change on the northern latitude limit and population density of the disease-transmitting European tick *Ixodes ricinus*. *Environmental Health Perspectives*, 108, 119–23.

Linske, M.A., Williams, S.C., Stafford, K.C., & Ortega, I.M. (2018). *Ixodes scapularis* (Acari: Ixodidae) reservoir host diversity and abundance impacts on dilution of *Borrelia burgdorferi* (Spirochaetales: Spirochaetaceae) in residential and woodland habitats in Connecticut, United States. *Journal of Medical Entomology*, 55, 681–90.

Linthicum, K.J., Anyamba, A., Tucker, C.J., Kelley, P.W., Myers, M.F., & Peters, C.J. (1999). Climate and satellite indicators to forecast Rift Valley fever epidemics in Kenya. *Science*, 285, 397–400.

Little, E., Bajwa, W., & Shaman, J. (2017). Local environmental and meteorological conditions influencing the invasive mosquito *Ae. albopictus* and arbovirus transmission risk in New York City. *PLoS Neglected Tropical Diseases*, 11, e0005828.

Liu-Helmersson, J., Stenlund, H., Wilder-Smith, A., & Rocklöv, J. (2014). Vectorial capacity of *Aedes aegypti*: Effects of temperature and implications for global dengue epidemic potential. *PLoS One*, 9.

Lobell, D.B., & Bonfils, C. (2008). The effect of irrigation on regional temperatures: A spatial and temporal analysis of trends in California, 1934–2002. *Journal of Climate*, 21, 2063–71.

LoGiudice, K., Duerr, S.T.K., Newhouse, M.J., Schmidt, K.A., Killilea, M.E., & Ostfeld, R.S. (2008). Impact of host community composition on Lyme disease risk. *Ecology*, 89, 2841–9.

Lothrop, H.D., Lothrop, B., & Reisen, W.K. (2002). Nocturnal microhabitat distribution of adult *Culex tarsalis* (Diptera: Culicidae) impacts control effectiveness. *Journal of Medical Entomology*, 39, 574–82.

Loveland, T.R., Reed, B.C., Brown, J.F., et al. (2000). Development of a global land cover characteristics database and IGBP DISCover from 1 km AVHRR data. *International Journal of Remote Sensing*, 21, 1303–30.

Lyons, C.L., Coetzee, M., Terblanche, J.S., & Chown, S.L. (2014). Desiccation tolerance as a function of age, sex, humidity and temperature in adults of the African malaria vectors *Anopheles arabiensis* and *Anopheles funestus*. *Journal of Experimental Biology*, 217, 3823–33.

Mabaso, M.L.H., Craig, M., Vounatsou, P., & Smith, T. (2005). Towards empirical description of malaria seasonality in southern Africa: The example of Zimbabwe. *Tropical Medicine and International Health*, 10, 909–18.

Macdonald, G. (1952). The analysis of the sporozoite rate. *Tropical Diseases Bulletin*, 49, 569–86.

MacDonald, A.J., Hyon, D.W., Brewington, J.B., O'Connor, K.E., Swei, A., & Briggs, C.J. (2017). Lyme disease risk in southern California: abiotic and environmental drivers of *Ixodes pacificus* (Acari: Ixodidae) density and infection prevalence with *Borrelia burgdorferi*. *Parasites & Vectors*, 10, 7.

MacDonald, A.J., Larsen, A.E., & Plantinga, A.J. (2019). Missing the people for the trees: Identifying coupled natural–human system feedbacks driving the ecology of Lyme disease. *Journal of Applied Ecology*, 56, 354–64.

MacDonald, A.J. & Mordecai, E.A. (2019). Amazon deforestation drives malaria transmission, and malaria burden reduces forest clearing. *Proceedings of the National Academy of Sciences of the United States of America*, 116, 22212–18.

Mackenzie, J.S., Lindsay, M.D., & Broom, A.K. (2000). Effect of climate and weather on the transmission of Ross River and Murray Valley encephalitis viruses. *Microbiology Australia*, 21, 20–5.

Mahmood, R., Foster, S., Keeling, T., Hubbard, K., Carlson, C., & Leeper, R. (2006). Impacts of irrigation on 20th century temperature in the northern Great Plains. *Global and Planetary Change*, 54, 1–18.

Mapaco, L.P., Coetzer, J.A.W., Paweska, J.T., & Venter, E.H. (2012). An investigation into an outbreak of Rift Valley fever on a cattle farm in Bela-Bela, South Africa, in 2008. *Journal of the South African Veterinary Association*, 83, 47–54.

McCabe, G.J. & Bunnell, J.E. (2004). Precipitation and the occurrence of Lyme disease in the northeastern United States. *Vector Borne and Zoonotic Diseases*, 4, 143–8.

Medlock, J.M., Hansford, K.M., Bormane, A., et al. (2013). Driving forces for changes in geographical distribution of Ixodes ricinus ticks in Europe. *Parasites & Vectors*, 6, 1.

Medone, P., Ceccarelli, S., Parham, P.E., Figuera, A., & Rabinovich, J.E. (2015). The impact of climate change on the geographical distribution of two vectors of Chagas disease: Implications for the force of infection. *Philosophical Transactions of the Royal Society B: Biological Sciences*, 370, 20130560.

Mellor, P.S., Boorman, J., & Baylis, M. (2000). *Culicoides* biting midges: Their role as arbovirus vectors. *Annual Review of Entomology*, 45, 307–40.

Messina, J.P., Brady, O.J., Golding, N., et al. (2019). The current and future global distribution and population at risk of dengue. *Nature Microbiology*, 4, 1508–15.

Metcalf, C.J.E., Walter, K.S., Wesolowski, A., et al. (2017). Identifying climate drivers of infectious disease dynamics: Recent advances and challenges ahead. *Proceedings of the Royal Society B: Biological Sciences*, 284, 20170901.

Microsoft (2019). Project Premonition https://www.microsoft.com/en-us/research/project/project-premonition/.

Midega, J.T., Smith, D.L., Olotu, A., et al. (2012). Wind direction and proximity to larval sites determines malaria risk in Kilifi district in Kenya. *Nature Communications*, 3, 1–8.

Miranda, D.E. de O., Sales, K.G. da S., Faustino, M.A. da G., et al. (2015). Ecology of sand flies in a low-density residential rural area, with mixed forest/agricultural exploitation, in north-eastern Brazil. *Acta Tropica*, 146, 89–94.

Moller-Jacobs, L.L., Murdock, C.C., & Thomas, M.B. (2014). Capacity of mosquitoes to transmit malaria depends on larval environment. *Parasites & Vectors*, 7, 593.

Moncaz, A., Faiman, R., Kirstein, O., & Warburg, A. (2012). Breeding sites of *phlebotomus sergenti*, the sand fly vector of cutaneous leishmaniasis in the Judean Desert. *PLoS Neglected Tropical Diseases*, 6, e1725.

Moo-Llanes, D., Ibarra-Cerdeña, C.N., Rebollar-Téllez, E.A., Ibáñez-Bernal, S., González, C., & Ramsey, J.M. (2013). Current and future niche of North and Central American sand flies (Diptera: Psychodidae) in climate change scenarios. *PLoS Neglected Tropical Diseases*, 7, e2421.

Moore, S., Shrestha, S., Tomlinson, K.W., & Vuong, H. (2012). Predicting the effect of climate change on African trypanosomiasis: Integrating epidemiology with parasite and vector biology. *Journal of The Royal Society Interface*, 9, 817–30.

Mordecai, E.A., Paaijmans, K.P., Johnson, L.R., et al. (2013). Optimal temperature for malaria transmission is dramatically lower than previously predicted. *Ecology Letters*, 16, 22–30.

Mordecai, E.A., Caldwell, J.M., Grossman, M.K., et al. (2019). Thermal biology of mosquito-borne disease. *Ecology Letters*, 22, 1690–1708.

Morin, C.W., Monaghan, A.J., Hayden, M.H., Barrera, R., & Ernst, K. (2015). Meteorologically Driven Simulations of Dengue Epidemics in San Juan, PR. *PLoS Neglected Tropical Diseases*, 9, e0004002.

Müller, C.B., Williams, I.S., & Hardie, J. (2001). The role of nutrition, crowding and interspecific interactions in the development of winged aphids. *Ecological Entomology*, 26, 330–40.

Munafò, M.R. & Davey Smith, G. (2018). Robust research needs many lines of evidence. *Nature*, 553, 399–401.

Murdock, C., Evans, M., McClanahan, T., Miazgowicz, K., & Tesla, B. (2017). Fine-scale variation in microclimate across and urban landscape changes the capacity of *Aedes albopictus* to vector arboviruses. *PLoS Neglected Tropical Diseases*, 11, e0005640.

Mutero, C.M., Kabutha, C., Kimani, V., et al. (2004). A transdisciplinary perspective on the links between malaria and agroecosystems in Kenya. *Acta Tropica*, 89, 171–86.

Myburgh, E., Bezuidenhout, H., & Neville, E.M. (2001). The role of flowering plant species in the survival of blackflies (Diptera: Simuliidae) along the lower Orange River, South Africa. *Koedoe*, 44, 63–70.

Naish, S., Dale, P., Mackenzie, J.S., McBride, J., Mengersen, K., & Tong, S. (2014). Climate change and dengue: A critical and systematic review of quantitative modelling approaches. *BMC Infectious Diseases*, 14, 167.

Narasimhan, S. & Fikrig, E. (2015). Tick microbiome: The force within. *Trends in Parasitology*, 31, 315–23.

Needham, G.R. & Teel, P.D. (1991). Off-host physiological ecology of ixodid ticks. *Annual Review of Entomology*, 36, 659–81.

Newman, J.A. (2004). Climate change and cereal aphids: The relative effects of increasing CO_2 and temperature on aphid population dynamics. *Global Change Biology*, 10, 5–15.

Ng, J.C.K. & Falk, B.W. (2006). Virus-vector interactions mediating nonpersistent and semipersistent transmission of plant viruses. *Annual Review of Phytopathology*, 44, 183–212.

Nguyen, N.M., Thi Hue Kien, D., Tuan, T.V., et al. (2013). Host and viral features of human dengue cases shape the population of infected and infectious *Aedes aegypti* mosquitoes. *Proceedings of the National Academy of Sciences of the United States of America*, 110, 9072–7.

Nnko, H.J., Ngonyoka, A., Salekwa, L., et al. (2017). Seasonal variation of tsetse fly species abundance and prevalence of trypanosomes in the Maasai Steppe, Tanzania. *Journal of Vector Ecology*, 42, 24–33.

Norris, D.E. (2004). Mosquito-borne diseases as a consequence of land use change. *EcoHealth*, 1, 19–24.

Nwoke, B.E., Onwuliri, C.O., & Ufomadu, G.O. (1992). Onchocerciasis in Plateau State; Nigeria: Ecological background, local disease perception & treatment; and vector/parasite dynamics. *Journal of Hygiene, Epidemiology, Microbiology, and Immunology*, 36, 153–60.

O'Brien, V.A. & Reiskind, M.H. (2013). Host-seeking mosquito distribution in habitat mosaics of southern great plains cross-timbers. *Journal of Medical Entomology*, 50, 1231–9.

Ogden, N.H. & Lindsay, L.R. (2016). Effects of climate and climate change on vectors and vector-borne diseases: Ticks are different. *Trends in Parasitology*, 32, 646–56.

Ogden, N.H., Lindsay, L.R., Beauchamp, G., et al. (2004). Investigation of relationships between temperature and developmental rates of tick *Ixodes scapularis* (Acari: Ixodidae) in the laboratory and field. *Journal of Medical Entomology*, 41, 622–33.

Ogden, N.H., Bigras-Poulin, M., O'Callaghan, C.J., et al. (2005). A dynamic population model to investigate effects of climate on geographic range and seasonality of the tick *Ixodes scapularis*. *International Journal for Parasitology*, 35, 375–89.

Ogden, N.H., Maarouf, A., Barker, I.K., et al. (2006). Climate change and the potential for range expansion of the Lyme disease vector *Ixodes scapularis* in Canada. *International Journal for Parasitology*, 36, 63–70.

Ogden, N.H., Radojevic', M., Wu, X., Duvvuri, V.R., Leighton, P.A., & Wu, J. (2014). Estimated effects of projected climate change on the basic reproductive number of the Lyme disease vector *Ixodes scapularis*. *Environmental Health Perspectives*, 122, 631–8.

Okamoto, K.W., Gould, F., & Lloyd, A.L. (2016). Integrating transgenic vector manipulation with clinical interventions to manage vector-borne diseases. *PLoS Computational Biology*, 12, e1004695.

Oliver, J.H. (1989). Biology and systematics of ticks (Acari:Ixodida). *Annual Review of Ecology and Systematics*, 20, 397–430.

Olson, S.H., Gangnon, R., Elguero, E., et al. (2009). Links between climate, malaria, and wetlands in the Amazon Basin. *Emerging Infectious Diseases*, 15, 659–62.

Ostfeld, R.S. & Brunner, J.L. (2015). Climate change and *Ixodes* tick-borne diseases of humans. *Philosophical Transactions of the Royal Society B: Biological Sciences*, 370, 20140051.

Ostfeld, R.S. & Keesing, F. (2000). The function of biodiversity in the ecology of vector-borne zoonotic diseases. *Canadian Journal of Zoology*, 78, 2061–78.

Ostfeld, R.S., Cepeda, O.M., Hazler, K.R., & Miller, M.C. (1995). Ecology of Lyme disease: Habitat associations of ticks (*Ixodes Scapularis*) in a rural landscape. *Ecological Applications*, 5, 353–61.

Overgaard, J., Kearney, M.R., & Hoffmann, A.A. (2014). Sensitivity to thermal extremes in Australian *Drosophila* implies similar impacts of climate change on the distribution of widespread and tropical species. *Global Change Biology*, 20, 1738–50.

Paaijmans, K.P., Wandago, M.O., Githeko, A.K., & Takken, W. (2007). Unexpected High Losses of *Anopheles gambiae* Larvae Due to Rainfall. *PLoS One*, 2, e1146.

Paaijmans, K.P., Blanford, S., Bell, A.S., Blanford, J.I., Read, A.F., & Thomas, M.B. (2010). Influence of climate on malaria transmission depends on daily temperature variation. *Proceedings of the National Academy of Sciences of the United States of America*, 107, 15135–9.

Paaijmans, K.P., Heinig, R.L., Seliga, R.A., et al. (2013). Temperature variation makes ectotherms more sensitive to climate change. *Global Change Biology*, 19, 2373–80.

Padgett, K.A. & Lane, R.S. (2001). Life cycle of *Ixodes pacificus* (Acari: Ixodidae): Timing of developmental processes under field and laboratory conditions. *Journal of Medical Entomology*, 38, 684–93.

Padmanabha, H., Soto, E., Mosquera, M., Lord, C.C., & Lounibos, L.P. (2010). Ecological links between water storage behaviors and *Aedes aegypti* production: Implications for dengue vector control in variable climates. *EcoHealth*, 7, 78–90.

van Panhuis, W.G., Choisy, M., Xiong, X., et al. (2015). Region-wide synchrony and traveling waves of dengue across eight countries in Southeast Asia. *Proceedings of the National Academy of Sciences of the United States of America*, 112, 13069–74.

Parry, H.R. (2013). Cereal aphid movement: General principles and simulation modelling. *Movement Ecology*, 1, 14.

Pascual, M., Cazelles, B., Bouma, M.J., Chaves, L.F., & Koelle, K. (2008). Shifting patterns: Malaria dynamics and rainfall variability in an African highland. *Proceedings of the Royal Society B: Biological Sciences*, 275, 123–32.

Patz, J.A., Graczyk, T.K., Geller, N., & Vittor, A.Y. (2000). Effects of environmental change on emerging parasitic diseases. *International Journal for Parasitology*, 30, 1395–405.

Patz, J.A., Daszak, P., Tabor, G.M., et al. (2004). Unhealthy landscapes: Policy recommendations on land use change and infectious disease emergence. *Environmental Health Perspectives*, 112, 1092–8.

Paull, S.H., Horton, D.E., Ashfaq, M., et al. (2017). Drought and immunity determine the intensity of West Nile virus epidemics and climate change impacts. *Proceedings of the Royal Society B: Biological Sciences*, 284, 20162078.

Paz, S. (2015). Climate change impacts on West Nile virus transmission in a global context. *Philosophical Transactions of the Royal Society B: Biological Sciences*, 370, 20130561.

Peña-García, V.H., Triana-Chávez, O., & Arboleda-Sánchez, S. (2017). Estimating effects of temperature on dengue transmission in Colombian Cities. *Annals of Global Health*, 83, 509–18.

Pepin, K.M., Eisen, R.J., Mead, P.S., et al. (2012). Geographic variation in the relationship between human Lyme disease incidence and density of infected host-seeking *Ixodes scapularis* nymphs in the eastern United States. *American Journal of Tropical Medicine and Hygiene*, 86, 1062–71.

Perkins, T.A., Metcalf, C.J.E., Grenfell, B.T., & Tatem, A.J. (2015). Estimating drivers of autochthonous transmission of chikungunya virus in its invasion of the Americas. *PLoS Currents*, 7.

Petersen, L.R., Brault, A.C., & Nasci, R.S. (2013). West Nile virus: Review of the literature. *Journal of the American Medical Association*, 310, 308–15.

Phillips, S.J., Dudík, M., Elith, J., et al. (2009). Sample selection bias and presence-only distribution models: Implications for background and pseudo-absence data. *Ecological Applications*, 19, 181–97.

Platonov, A.E., Fedorova, M.V., Karan, L.S., Shopenskaya, T.A., Platonova, O.V., & Zhuravlev, V.I. (2008). Epidemiology of West Nile infection in Volgograd, Russia, in relation to climate change and mosquito (Diptera: Culicidae) bionomics. *Parasitology Research*, 103, 45–53.

Plowright, R.K., Parrish, C.R., McCallum, H., et al. (2017). Pathways to zoonotic spillover. *Nature Reviews Microbiology.*, 15, 502–10.

Pontes, R.J., Spielman, A., Oliveira-Lima, J.W., Hodgson, J.C., & Freeman, J. (2000). Vector densities that potentiate dengue outbreaks in a Brazilian city. *The American Journal of Tropical Medicine and Hygiene*, 62, 378–83.

Pryor, S.C., Barthelmie, R.J., & Kjellström, E. (2005). Potential climate change impact on wind energy resources in northern Europe: Analyses using a regional climate model. *Climate Dynamics*, 25, 815–35.

Purse, B.V., Carpenter, S., Venter, G.J., Bellis, G., & Mullens, B.A. (2015). Bionomics of temperate and tropical *Culicoides* midges: Knowledge gaps and consequences for transmission of *Culicoides*-borne viruses. *Annual Review of Entomology*, 60, 373–92.

Rahmstorf, S. (2007). A semi-empirical approach to projecting future sea-level rise. *Science*, 315, 368–70.

Randolph, S.E. (2004). Evidence that climate change has caused 'emergence' of tick-borne diseases in Europe? *International Journal of Medical Microbiology*, 293 Suppl 37, 5–15.

Randolph, S.E. & Dobson, A.D.M. (2012). Pangloss revisited: A critique of the dilution effect and the biodiversity-buffers-disease paradigm. *Parasitology*, 139, 847–63.

Randolph, S.E., Green, R.M., Peacey, M.F., & Rogers, D.J. (2000). Seasonal synchrony: The key to tick-borne encephalitis foci identified by satellite data. *Parasitology*, 121 (Pt 1), 15–23.

Reisen, W.K., Cayan, D., Tyree, M., Barker, C.M., Eldridge, B., & Dettinger, M. (2008). Impact of climate variation on mosquito abundance in California. *Journal of Vector Ecology*, 33, 89–98.

Reiskind, M.H., Griffin, R.H., Janairo, M.S., & Hopperstad, K.A. (2017). Mosquitoes of field and forest: The scale of habitat segregation in a diverse mosquito assemblage: Scale of mosquito habitat segregation. *Medical and Veterinary Entomology*, 31, 44–54.

Riou, J., Poletto, C., & Boëlle, P.-Y. (2017). A comparative analysis of Chikungunya and Zika transmission. *Epidemics*, 19, 43–52.

Ritchie, S.A. & Rochester, W. (2001). Wind-blown mosquitoes and introduction of Japanese encephalitis into Australia. *Emerging Infectious Diseases*, 7, 900–8.

Robert, Y., Woodford, J.A.T., & Ducray-Bourdin, D.G. (2000). Some epidemiological approaches to the control of aphid-borne virus diseases in seed potato crops in northern Europe. *Virus Research*, 71, 33–47.

Rodgers, S.E., Zolnik, C.P., & Mather, T.N. (2007). Duration of exposure to suboptimal atmospheric moisture affects nymphal blacklegged tick survival. *Journal of Medical Entomology*, 44, 372–5.

Rogers, D.J. & Randolph, S.E. (2000). The global spread of malaria in a future, warmer world. *Science*, 289, 1763–6.

Rogers, D.J. & Randolph, S.E. (2006). Climate change and vector-borne diseases. *Advances in Parasitology*, 62, 345–81.

Rogers, D.J. & Williams, B.G. (1993). Monitoring trypanosomiasis in space and time. *Parasitology*, 106 Suppl, S77–92.

Rohr, J.R., Civitello, D.J., Halliday, F.W., et al. (2019). Towards common ground in the biodiversity–disease debate. *Nature Ecology & Evolution*, 4, 24–33.

Ross, P.A., Ritchie, S.A., Axford, J.K., & Hoffmann, A.A. (2019). Loss of cytoplasmic incompatibility in Wolbachia-infected *Aedes aegypti* under field conditions. *PLoS Neglected Tropical Diseases*, 13, e0007357.

Rozendaal, J.A. (1992). Relations between *Anopheles darlingi* breeding habitats, rainfall, river level and malaria transmission rates in the rain forest of Suriname. *Medical and Veterinary Entomology*, 6, 16–22.

Ruel, J.J. & Ayres, M.P. (1999). Jensen's inequality predicts effects of environmental variation. *Trends in Ecology & Evolution*, 14, 361–6.

Ryan, S.J., McNally, A., Johnson, L.R., et al. (2015). Mapping physiological suitability limits for malaria in Africa under climate change. *Vector Borne and Zoonotic Diseases*, 15, 718–25.

Ryan, S.J., Carlson, C.J., Mordecai, E.A., & Johnson, L.R. (2019). Global expansion and redistribution of Aedes-borne virus transmission risk with climate change. *PLoS Neglected Tropical Diseases*, 13, e0007213.

Salkeld, D.J., Padgett, K.A., & Jones, J.H. (2013). A meta-analysis suggesting that the relationship between biodiversity and risk of zoonotic pathogen transmission is idiosyncratic. *Ecology Letters*, 16, 679–86.

Samy, A.M. & Peterson, A.T. (2016). Climate change influences on the global potential distribution of bluetongue virus. *PLoS One*, 11, e0150489.

Sang, R., Arum, S., Chepkorir, E., et al. (2017). Distribution and abundance of key vectors of Rift Valley fever and other arboviruses in two ecologically distinct counties in Kenya. *PLoS Neglected Tropical Diseases*, 11, e0005341.

Santos, A.S. & Almeida, A.N. (2018). The impact of deforestation on malaria infections in the Brazilian Amazon. *Ecological Economics*, 154, 247–56.

Sedda, L., Brown, H.E., Purse, B.V., Burgin, L., Gloster, J., & Rogers, D.J. (2012). A new algorithm quantifies the roles of wind and midge flight activity in the bluetongue epizootic in northwest Europe. *Proceedings of the Royal Society B: Biological Sciences*, 279, 2354–62.

Shah, M.M., Krystosik, A.R., Ndenga, B.A., et al. (2019). Malaria smear positivity among Kenyan children peaks at intermediate temperatures as predicted by ecological models. *Parasites & Vectors*, 12, 288.

Shaman, J., Day, J.F., & Stieglitz, M. (2005). Drought-induced amplification and epidemic transmission of west Nile virus in southern Florida. *Journal of Medical Entomology*, 42, 134–41.

Shand, L., Brown, W.M., Chaves, L.F., et al. (2016). Predicting West Nile virus infection risk from the synergistic effects of rainfall and temperature. *Journal of Medical Entomology*, 53, 935–44.

Shapiro, L.L.M., Murdock, C.C., Jacobs, G.R., Thomas, R.J., & Thomas, M.B. (2016). Larval food quantity affects the capacity of adult mosquitoes to transmit human malaria. *Proceedings of the Royal Society B: Biological Sciences*, 283, 20160298.

Sheela, A.M., Ghermandi, A., Vineetha, P., Sheeja, R.V., Justus, J., & Ajayakrishna, K. (2017). Assessment of relation of land use characteristics with vector-borne diseases in tropical areas. *Land Use Policy*, 63, 369–80.

Sheppard, L.W., Bell, J.R., Harrington, R., & Reuman, D.C. (2016). Changes in large-scale climate alter spatial synchrony of aphid pests. *Nature Climate Change*, 6, 610–13.

Shocket, M.S., Ryan, S.J., & Mordecai, E.A. (2018). Temperature explains broad patterns of Ross River virus transmission. *eLife*, 7, e37762.

Shocket, M.S., Verwillow, A.B., Numazu, M.G., et al. (2020). Transmission of West Nile virus and other temperate mosquito-borne viruses peaks at intermediate environmental temperatures. *bioRxiv*, https://www.biorxiv.org/content/10.1101/597898v3.full.

Shultz, J.M., Russell, J., & Espinel, Z. (2005). Epidemiology of tropical cyclones: The dynamics of disaster, disease, and development. *Epidemiologic Reviews*, 27, 21–35.

Siraj, A.S., Santos-Vega, M., Bouma, M. J., Yadeta, D., Ruiz Carrascal, D., & Pascual, M. (2014). Altitudinal changes in malaria incidence in highlands of Ethiopia and Colombia. *Science*, 343, 1154–8.

Siraj, A.S., Bouma, M.J., Santos-Vega, M., et al. (2015). Temperature and population density determine reservoir regions of seasonal persistence in highland malaria. *Proceedings of the Royal Society B: Biological Sciences*, 282, 20151383.

Siraj, A.S., Rodriguez-Barraquer, I., Barker, C.M., et al. (2018). Spatiotemporal incidence of Zika and associated environmental drivers for the 2015–2016 epidemic in Colombia. *Scientific Data*, 5, 180073.

Sithiprasasna, R., Lee, W.J., Ugsang, D.M., & Linthicum, K.J. (2005). Identification and characterization of larval and adult anopheline mosquito habitats in the Republic of Korea: Potential use of remotely sensed data to estimate mosquito distributions. *International Journal of Health Geographics*, 4, 17.

Smith, D.L., Battle, K.E., Hay, S.I., Barker, C.M., Scott, T.W., & McKenzie, F.E. (2012). Ross, Macdonald, and a theory for the dynamics and control of mosquito-transmitted pathogens. *PLoS Pathogens*, 8, e1002588.

Solano, P., Salou, E., Rayaisse, J.-B., et al. (2015). Do tsetse flies only feed on blood? *Infection, Genetics and Evolution*, 36, 184–9.

Sorensen, C.J., Borbor-Cordova, M.J., Calvello-Hynes, E., Diaz, A., Lemery, J., & Stewart-Ibarra, A.M. (2017). Climate variability, vulnerability, and natural disasters: A case study of Zika Virus in Manabi, Ecuador following the 2016 earthquake: Climate, vulnerability, and disasters. *GeoHealth*, 1, 298–304.

Stafford, K.C. (1994). Survival of immature *Ixodes scapularis* (Acari: Ixodidae) at different relative humidities. *Journal of Medical Entomology*, 31, 310–14.

Sterling, S.M., Ducharne, A., & Polcher, J. (2013). The impact of global land-cover change on the terrestrial water cycle. *Nature Climate Change*, 3, 385–90.

Stewart Ibarra, A.M., Ryan, S.J., Beltrán, E., Mejía, R., Silva, M., & Muñoz, Á. (2013). Dengue vector dynamics (*Aedes aegypti*) influenced by climate and social factors in Ecuador: Implications for targeted control. *PLoS One*, 8, e78263.

Stewart Ibarra, A.M., & Lowe, R. (2013). Climate and non-climate drivers of dengue epidemics in southern coastal Ecuador. *The American Journal of Tropical Medicine and Hygiene*, 88, 971–81.

Stoddard, S.T., Forshey, B.M., Morrison, A.C., et al. (2013). House-to-house human movement drives dengue virus transmission. *Proceedings of the National Academy of Sciences of the United States of America*, 110, 994–9.

Stone, C.M. & Foster, W.A. (2013). Plant-sugar feeding and vectorial capacity. In *Ecology Of Parasite-Vector Interactions*, C.J.M. Koenraadt, & W. Takken, eds. Wageningen Acad Publ: Postbus 220, 6700 Ae Wageningen, Netherlands, pp. 35–79.

Stone, C.M., Taylor, R.M., Roitberg, B.D., & Foster, W.A. (2009). Sugar deprivation reduces insemination of *Anopheles gambiae* (Diptera: Culicidae), despite daily recruitment of adults, and predicts decline in model populations. *J. Med. Entomol.*, 46, 1327–37.

Stone, C.M., Jackson, B.T., & Foster, W.A. (2012). Effects of plant-community composition on the vectorial capacity and fitness of the malaria mosquito *Anopheles gambiae*. *The American Journal of Tropical Medicine and Hygiene*, 87, 727–36.

Subak, S. (2003). Effects of climate on variability in Lyme disease incidence in the northeastern United States. *American Journal of Epidemiology*, 157, 531–8.

Sugihara, G., May, R., Ye, H., et al. (2012). Detecting causality in complex ecosystems. *Science*, 338, 496–500.

Šumilo, D., Bormane, A., Asokliene, L., et al. (2008). Socio-economic factors in the differential upsurge of tick-borne encephalitis in central and Eastern Europe. *Reviews in Medical Virology*, 18, 81–95.

Sundararaman, S.A., Liu, W., Keele, B.F., et al. (2013). *Plasmodium falciparum*-like parasites infecting wild apes in southern Cameroon do not represent a recurrent source of human malaria. *Proceedings of the National Academy of Sciences of the United States of America*, 110, 7020–5.

Swei, A., Meentemeyer, R., & Briggs, C.J. (2011). Influence of abiotic and environmental factors on the density and infection prevalence of *Ixodes pacificus* (Acari: Ixodidae) with *Borrelia burgdorferi*. *Journal of Medical Entomology*, 48, 20–8.

Tadei, W.P., Thatcher, B.D., Santos, J.M., Scarpassa, V.M., Rodrigues, I.B., & Rafael, M.S. (1998). Ecologic observations on anopheline vectors of malaria in the Brazilian Amazon. *American Journal of Tropical Medicine and Hygiene*, 59, 325–35.

Takaoka, H. (2015). Review of the biology and ecology of adult blackflies in relation to the transmission of onchocerciasis in Guatemala. *Tropical Medicine and Health*, 43, 71–85.

Tall, J.A. & Gatton, M.L. (2019). Flooding and arboviral disease: Predicting Ross river virus disease outbreaks across inland regions of south-eastern Australia. *Journal of Medical Entomology*, 57, 241–51.

Thackray, D.J., Diggle, A.J., & Jones, R.A.C. (2009). BYDV predictor: A simulation model to predict aphid arrival, epidemics of *Barley yellow dwarf virus* and yield losses in wheat crops in a Mediterranean-type environment. *Plant Pathology*, 58, 186–202.

Thomas, M.B. & Blanford, S. (2003). Thermal biology in insect-parasite interactions. *Trends in Ecology & Evolution*, 18, 344–50.

Thomas, C.J., Cross, D.E., & Bøgh, C. (2013). Landscape movements of *Anopheles gambiae* malaria vector mosquitoes in rural Gambia. *PLoS One*, 8, e68679.

Thomas, S.M., Obermayr, U., Fischer, D., Kreyling, J., & Beierkuhnlein, C. (2012). Low-temperature threshold for egg survival of a post-diapause and non-diapause European aedine strain, *Aedes albopictus* (Diptera: Culicidae). *Parasites & Vectors*, 5, 100.

Thongsripong, P., Green, A., Kittayapong, P., Kapan, D., Wilcox, B., & Bennett, S. (2013). Mosquito vector diversity across habitats in central Thailand endemic for dengue and other arthropod-borne diseases. *PLoS Neglected Tropical Diseases*, 7, e2507.

Thresh, J.M., Scorer, R.S., Harrington, R., Pedgley, D.E., Nuttall, P.A., & Sellers, R.F. (1983). The long-range dispersal of plant viruses by arthropod vectors [and discussion]. *Philosophical Transactions of the Royal Society B: Biological Sciences*, 302, 497–528.

Tian, H.Y., Bi, P., Cazelles, B., et al. (2015). How environmental conditions impact mosquito ecology and Japanese encephalitis: An eco-epidemiological approach. *Environment international*, 79, 17–24.

Tjaden, N.B., Caminade, C., Beierkuhnlein, C., & Thomas, S.M. (2018). Mosquito-borne diseases: Advances in modelling climate-change impacts. *Trends in Parasitology*, 34, 227–45.

Tokarz, R.E. & Smith, R.C. (2020). Crossover dynamics of *Culex* (Diptera: Culicidae) vector populations determine WNV transmission intensity. *Journal of Medical Entomology*, 57, 289–96.

Tolba, M.K. & El-Kholy, O.A. (1992). Land degradation. In M.K. Tolba, & O.A. El-Kholy, eds. The World Environment 1972–1992, pp. 131–155. Springer, Dordrecht. 55.

Tong, S. & Hu, W. (2002). Different responses of Ross River virus to climate variability between coastline and inland cities in Queensland, Australia. *Occupational and Environmental Medicine*, 59, 739–44.

Trębicki, P., Nancarrow, N., Cole, E., et al. (2015). Virus disease in wheat predicted to increase with a changing climate. *Global Change Biology*, 21, 3511–19.

Trewin, B.J., Kay, B.H., Darbro, J.M., & Hurst, T.P. (2013). Increased container-breeding mosquito risk owing to drought-induced changes in water harvesting and storage in Brisbane, Australia. *International Health*, 5, 251–8.

Tucker Lima, J.M., Vittor, A., Rifai, S., & Valle, D. (2017). Does deforestation promote or inhibit malaria transmission in the Amazon? A systematic literature review and critical appraisal of current evidence. *Philosophical Transactions of the Royal Society B: Biological Sciences*, 372, 20160125.

Ukawuba, I. & Shaman, J. (2018). Association of spring-summer hydrology and meteorology with human West Nile virus infection in West Texas, USA, 2002–2016. *Parasites & Vectors*, 11, 224.

Vail, S.G. & Smith, G. (2002). Vertical movement and posture of blacklegged tick (Acari: Ixodidae) nymphs as a function of temperature and relative humidity in laboratory experiments. *Journal of Medical Entomology*, 39, 842–6.

Van den Bossche, P., Rocque, S. de L., Hendrickx, G., & Bouyer, J. (2010). A changing environment and the epidemiology of tsetse-transmitted livestock trypanosomiasis. *Trends in Parasitology*, 26, 236–43.

Vanwambeke, S.O., Lambin, E.F., Eichhorn, M.P., et al. (2007). Impact of land-use change on dengue and malaria in northern Thailand. *EcoHealth*, 4, 37–51.

Vasilakis, N., Cardosa, J., Hanley, K.A., Holmes, E.C., & Weaver, S.C. (2011). Fever from the forest: Prospects for the continued emergence of sylvatic dengue virus and its impact on public health. *Nature Reviews Microbiology*, 9, 532–41.

Verburg, P.H., Neumann, K., & Nol, L. (2011). Challenges in using land use and land cover data for global change studies: Land use and land cover data for global change studies. *Global Change Biology*, 17, 974–89.

Vinogradova, E.B. (2000). Culex Pipiens Pipiens *Mosquitoes Taxonomy, Distribution, Ecology, Physiology, Genetics, Applied Importance and Control*. Sofia: Pensoft.

Vittor, A.Y., Gilman, R.H., Tielsch, J., et al. (2006). The effect of deforestation on the human-biting rate of *Anopheles darlingi*, the primary vector of Falciparum malaria in the Peruvian Amazon. *American Journal of Tropical Medicine and Hygiene*, 74, 3–11.

Vittor, A.Y., Pan, W., Gilman, R.H., et al. (2009). Linking deforestation to malaria in the Amazon: Characterization of the breeding habitat of the principal malaria vector, *Anopheles darlingi*. *The American Journal of Tropical Medicine and Hygiene*, 81, 5–12.

Voordouw, M.J. (2015). Co-feeding transmission in Lyme disease pathogens. *Parasitology*, 142, 290–302.

Walker, A.R. & Davies, F.G. (1971). A preliminary survey of the epidemiology of bluetongue in Kenya. *Journal of Hygiene*, 69, 47–60.

Werneck, G.L., Rodrigues, L., Santos, M.V., et al. (2002). The burden of *Leishmania Chagasi* infection during an urban outbreak of visceral leishmaniasis in Brazil. *Acta Tropica*, 83, 13–18.

West, J.S., Townsend, J.A., Stevens, M., & Fitt, B.D.L. (2012). Comparative biology of different plant pathogens to estimate effects of climate change on crop diseases in Europe. *European Journal of Plant Pathology*, 133, 315–31.

Whelan, P.I., Jacups, S.P., Melville, L., et al. (2003). Rainfall and vector mosquito numbers as risk indicators for mosquito-borne disease in central Australia, 27, 110–16.

White, E.P., Thibault, K.M., & Xiao, X. (2012). Characterizing species abundance distributions across taxa and ecosystems using a simple maximum entropy model. *Ecology*, 93, 1772–8.

Wilson, M.D., Cheke, R.A., Fiasse, S.P.J., et al. (2002). Deforestation and the spatio-temporal distribution of savannah and forest members of the *Simulium damnosum* complex in southern Ghana and south-western Togo. *Transactions of the Royal Society of Tropical Medicine and Hygiene*, 96, 632–9.

Wittmann, E.J. & Baylis, M. (2000). Climate change: Effects on *Culicoides*-transmitted viruses and implications for the UK. *The Veterinary Journal*, 160, 107–17.

Wittmann, E.J., Mellor, P.S., & Baylis, M. (2001). Using climate data to map the potential distribution of *Culicoides imicola* (Diptera: Ceratopogonidae) in Europe. *Scientific and Technical Review of the Office International des Epizooties*, 20, 731–40.

Wood, C.L., Lafferty, K.D., DeLeo, G., Young, H.S., Hudson, P.J., & Kuris, A.M. (2014). Does biodiversity protect humans against infectious disease? *Ecology*, 95, 817–32.

Woodruff, R.E., Guest, C.S., Garner, M.G., et al. (2002). Predicting Ross River virus epidemics from regional weather data. *Epidemiology*, 13, 384–93.

World Health Organization (2014). A global brief on vector-borne diseases. Geneva: WHO.

Wu, X., Lu, Y., Zhou, S., Chen, L., & Xu, B. (2016). Impact of climate change on human infectious diseases:

Empirical evidence and human adaptation. *Environment International*, 86, 14–23.

Xu, H.-Y., Fu, X., Lee, L.K.H., et al. (2014). Statistical modeling reveals the effect of absolute humidity on dengue in Singapore. *PLoS Neglected Tropical Diseases*, 8, e2805.

Xu, L., Stige, L.C., Chan, K.-S., et al. (2017). Climate variation drives dengue dynamics. *Proceedings of the National Academy of Sciences of the United States of America*, 114, 113–18.

Yamana, T.K., Kandula, S., & Shaman, J. (2016). Superensemble forecasts of dengue outbreaks. *Journal of The Royal Society Interface*, 13, 20160410.

Yu, B.-T., Ding, Y.-M., Mo, X.-C., Liu, N., Li, H.-J., & Mo, J.-C. (2016). Survivorship and fecundity of *Culex pipiens pallens* feeding on flowering plants and seed pods with differential preferences. *Acta Tropica*, 155, 51–7.

Zacarias, O.P. & Andersson, M. (2011). Spatial and temporal patterns of malaria incidence in Mozambique. *Malaria Journal*, 10, 189.

Zahler, M. & Gothe, R. (1995). Effect of temperature and humidity on longevity of unfed adults and on oviposition of engorged females of *Dermacentor reticulatus* (Ixodidae). *Applied Parasitology*, 36, 200–11.

Zamora-vilchis, I., Williams, S.E., & Johnson, C.N. (2012). Environmental temperature affects prevalence of blood parasites of birds on an elevation gradient: Implications for disease in a warming climate. *PLoS One*, 7, e39208.

Zhou, G., Minakawa, N., Githeko, A.K., & Yan, G. (2004). Association between climate variability and malaria epidemics in the East African highlands. *Proceedings of the National Academy of Sciences of the United States of America*, 101, 2375–80.

Zouache, K., Fontaine, A., Vega-Rua, A., et al. (2014). Three-way interactions between mosquito population, viral strain and temperature underlying chikungunya virus transmission potential. *Proceedings of the Royal Society B: Biological Sciences.*, 281, 20141078.

CHAPTER 7

Population Biology of *Culicoides-*Borne Viruses of Livestock in Europe

Simon Gubbins

7.1 Introduction

Culicoides biting midges (Diptera: Ceratopogonidae) are small blood-sucking insects that occur on all large landmasses except Antarctica and New Zealand. Their importance lies principally in the ability of a small number of species in the genus to act as vectors of viruses affecting humans and other animal species (Mellor et al. 2000; Carpenter et al. 2013; Purse et al. 2015). Of those viruses affecting humans, only Oropouche virus has been identified as being primarily transmitted by *Culicoides*, though they may also play a less important, poorly defined role in the transmission of other human arboviruses (Carpenter et al. 2013). By contrast, *Culicoides* species are widely recognized as the principal vectors for a number of internationally important arboviruses that infect animals, notably bluetongue virus, epizootic haemorrhagic disease virus, African horse sickness virus, Akabane virus, and Schmallenberg virus (Purse et al. 2015). Here the focus is on bluetongue virus and Schmallenberg virus, both of which have emerged to pose a threat to European livestock in the past two decades.

Bluetongue virus (BTV) is a segmented, double-stranded RNA virus of the genus *Orbivirus* in the family *Reoviridae*. It is the causative agent of blue-tongue, a disease of ruminants, including cattle, sheep, goats, and deer, which can be particularly severe in sheep and deer. Multiple strains of BTV exist and, currently, the main means of strain classification is serotype, of which twenty-seven have been characterized to date (Nomikou et al. 2015). Prior to the 1990s, BTV occurred only sporadically in Europe (Mellor et al. 2008; Wilson & Mellor 2009). Between 1998 and 2005 multiple strains (serotypes 1, 2, 4, 9, and 16) of BTV emerged in southern and eastern Europe in a series of incursions into many countries that had never previously recorded the virus (Mellor et al. 2008; Wilson & Mellor 2009). Despite the potential for expansion further into Europe, most BTV serotypes have failed to do so (Wilson & Mellor 2009). However, in 2006 BTV serotype 8 (BTV-8) was detected for the first time in the Netherlands and subsequently spread to Belgium, Germany, France, and Luxembourg in 2006–2007 (Wilson & Mellor 2009). Further spread occurred in 2007 to England, Denmark, Norway, and Sweden, which were previously considered to be at the very limits of potential BTV transmission (Carpenter et al. 2009a). Since 2006, outbreaks of BTV have occurred throughout Europe, though none has been as far-reaching or damaging as the 2006 BTV-8 outbreak.

Schmallenberg virus (SBV) is a single stranded RNA virus of the Simbu serogroup of the genus *Orthobunyavirus* in the family *Bunyaviridae*. It was first identified in 2011 (Hoffmann et al. 2012) after dairy cattle in Germany and the Netherlands were reported to be affected by an unknown disease causing a short period of clinical signs such as fever,

Simon Gubbins, *Population Biology of* Culicoides-*Borne Viruses of Livestock in Europe*. In: *Population Biology of Vector-Borne Diseases* In: *Population Biology of Vector-Borne Diseases*. Edited by: John M. Drake, Michael B. Bonsall, and Michael R. Strand: Oxford University Press (2021). © Simon Gubbins.
DOI: 10.1093/oso/9780198853244.003.0007

diarrhea, and reduced milk production. From November 2011 onwards malformations in newborn lambs and calves associated with SBV were reported in Germany, The Netherlands, Belgium, France, Luxemburg, United Kingdom, Italy, and Spain (European Food Safety Authority 2012). By spring of 2013 SBV had been reported across much of Europe (European Food Safety Authority 2013). Based on detection of viral RNA in insects caught in light traps, it was confirmed that SBV is transmitted by *Culicoides* biting midges (Elbers et al. 2013; Balenghien et al. 2014), in common with other related viruses (St George and Standfast 1989). Since 2013 SBV has continued to circulate in Europe, with virus repeatedly detected in the blood of adult ruminants and malformed calves and lambs (Wernike and Beer 2017).

A number of *Culicoides* species have been implicated as vectors of BTV or SBV in Europe (Purse et al. 2015): one Afro-tropical species, *C. imicola*; and six Palearctic species, *C. obsoletus*, *C. scoticus*, *C. dewulfi*, and *C. chiopterus* (sometimes referred to as the *Obsoletus* group), and *C. pulicaris* and *C. punctatus* (sometimes referred to as the *Pulicaris* group). In southern Europe, *C. imicola* tends to be the main vector species, while in northern Europe one or more of the six Palearctic species tend to predominate, though the ranges of the species do overlap (Purse et al. 2007). Finally, most laboratory studies of virus and vector interactions (including for European viruses) involve *C. sonorensis*, a North American vector species, because the availability of colony populations of this species make it amenable to experimentation.

7.2 Transmission of *Culicoides*-borne Viruses

The transmission cycle for *Culicoides*-borne viruses comprises two phases. The first relates to normal transmission while vectors are active[1] (i.e. within a season) (Fig. 7.1). This phase is well understood and, indeed, is common to many other vector-borne diseases. An infectious midge bites a susceptible

host and transmits the virus to it in its saliva. The virus replicates in the host and, after completing the latent (or intrinsic incubation) period, the host becomes infectious. Susceptible midges feeding on the infectious host take up the virus with their blood meal. The ingested virus replicates in the infected vector until it reaches the salivary glands (the extrinsic incubation period or EIP) at which point the midge becomes infectious and the cycle is completed.

The second phase relates to transmission between seasons (Fig 7.1), when vectors are inactive and viruses escape detection for periods of time that are longer than either the lifespan of an adult midge (typically <20 days; Mellor et al. 2000) or the duration of viremia in the mammalian host (typically <60 days; Singer et al. 2001). This second phase is often referred to as overwintering, and is much less well understood (Wilson et al. 2008; Mayo et al. 2016). A number of mechanisms by which *Culicoides*-borne viruses may overwinter have been proposed (Wilson et al. 2008; Mayo et al. 2016), including transovarial transmission in the vector, long-lived adult midges, persistent infection in the host or transplacental transmission in the host (Fig. 7.1).

Current evidence supports two of these routes. First, BTV-infected female *C. sonorensis* were caught on farms in California during the inter-seasonal period (Mayo et al. 2014, 2016). This most likely reflects overwintering in long-lived female midges infected in the preceding seasonal period of BTV transmission. It could, however, also represent low level circulation of infection during the interseasonal period. Second, field and experimental evidence shows that SBV (Wernike and Beer 2017) and some strains of BTV (van der Sluijs et al. 2016) can be transmitted transplacentally in both cattle and sheep.

At larger scales, that is between farms or between regions, transmission occurs principally via two routes: movement of infected animals or dispersal of infected vectors. Animal movements have been implicated in the introduction of BTV to Europe via the Eurasian 'ruminant street', a contiguous region of high ruminant density running from India and Pakistan to south-eastern Europe (Wilson and Mellor 2009). The role of animal movements in the

---

[1] Other (i.e. non-vector) transmission routes have been identified for some *Culicoides*-borne viruses (e.g. oral, needle or direct contact), but these appear to be of limited importance in their epidemiology (van der Sluijs et al. 2016).

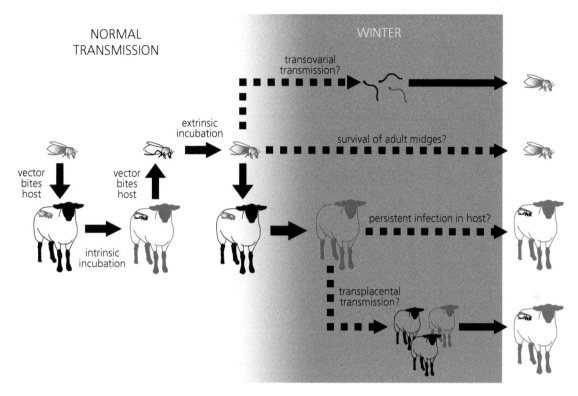

Figure 7.1 The transmission cycle for *Culicoides*-borne viruses during summer (left) and winter (right).

Source: Reproduced from Wilson et al. (2008), distributed under a Creative Commons Attribution License.

spread of BTV and SBV between farms has also been explored (Turner et al. 2012, 2019; Ensoy et al. 2013; Gubbins et al. 2014; Sumner et al. 2017). For example, analysis of the BTV-8 and SBV outbreaks in United Kingdom suggested that while movement of infected animals accounts for only a small proportion (<10 percent) of transmission between farms, it is important for introducing viruses to new areas (Sumner et al. 2017).

Long-distance dispersal of midges over water has been linked to incursions of *Culicoides*-borne viruses to islands hundreds of kilometers from the nearest source of virus, where introduction via animal movements had been discounted (Sellers et al. 1978, 1979; Alba et al. 2004; Ågren et al. 2010; Burgin et al. 2013). By contrast, dispersal over land, as inferred from outbreak data, typically follows a stepping-stone pattern, with limited evidence of single, long-distance dispersal events (Hendrickx et al. 2008; Sedda et al. 2012; Sedda and Rogers 2013). Moreover, analysis of the BTV-8 outbreaks in north-western

Europe indicated that dispersal of infected vectors accounts for a majority (>90 percent) of transmission between farms (Sedda et al. 2012; Sumner et al. 2017).

7.3 Factors Influencing Transmission of *Culicoides*-borne Viruses

The factors influencing the normal transmission cycle (Fig. 7.1) can be usefully characterized using the basic reproductive ratio (R_0). This quantity is defined as 'the average number of secondary cases caused by an average primary case in an entirely susceptible population' (Keeling & Rohani 2008). The basic reproductive ratio is a threshold quantity, with $R_0 > 1$ required for an infection to be able to invade a host population. A number of studies have estimated R_0 for BTV (Courtejoie et al. 2018) or SBV (Gubbins et al. 2014) and, for the case of a single host and single vector, R_0 is given by,

$$R_0 = \sqrt{\frac{ba}{\mu}\left(\frac{kv}{kv+\mu}\right)^k \times \frac{\beta ma}{r}}. \qquad (7.1)$$

This expression can be extended to consider multiple host or vector species (Gubbins et al. 2008; Turner et al. 2013).

The expression for the basic reproductive ratio, (7.1), can be understood heuristically in relation to the transmission cycle (Fig. 7.1). After a vector takes an infected blood meal, it must complete the EIP. Assuming that the duration of the EIP follows a gamma distribution with mean $1/v$ and variance $1/kv^2$ (Carpenter et al. 2011), the probability that the midge will survive the EIP is $(kv/(kv+\mu))^k$ where μ is the vector mortality rate. Following completion of the EIP the midge remains infectious for the rest of its lifespan, which will be $1/\mu$ days on average. During this time, the midge will bite susceptible hosts a times per day (where a is the reciprocal of the time interval between blood meals, assumed to be equal to the biting rate) and a proportion, b, of these bites will result in a newly infected host. Once

infected, a host will remain infectious for the duration of viremia, which lasts $1/r$ days on average.[2] During this time the host will be bitten by susceptible midges on average $m \times a$ times per day (here $m = N/H$ is the vector to host ratio and N and H are the number of vectors and hosts, respectively), a proportion, β, of which will result in a newly infected vector.

The basic reproductive ratio, (7.1), involves eight constituent parameters (or ten if there is appreciable disease-associated mortality). These parameters are influenced by factors related to the host, virus, vector, and environment and, indeed, all parameters are affected by a combination of two or three of these factors (Table 7.1). For the two best quantified viruses, namely BTV-8 and SBV, param-

[2] When there is disease-associated mortality, the mean duration for which a host remains infectious becomes a more complex expression. If the duration of viremia is assumed to follow a gamma distribution with mean $1/r$ and variance $1/nr^2$, the appropriate expression is $1/d \times (1-(nr/(nr+d))^n)$ where d is the disease-associated mortality rate (Gubbins et al. 2008, 2012).

Table 7.1 Factors influencing the transmission of *Culicoides*-borne viruses.

Parameter	H	Vi	Ve	E	examples and comments
probability of transmission from vector to host	✓	✓			can be a very efficient process (Baylis et al. 2008); saliva proteins play a role in the efficiency with which a midge transmits a virus (Darpel et al. 2011)
probability of transmission from host to vector	✓	✓	✓		differs by *Culicoides* species and virus; other factors such as rearing temperature or co-infection with bacterial endosymbionts may also influence whether or not a midge becomes infected after taking a blood meal (see review by Carpenter et al. 2015)
reciprocal of time interval between blood meals			✓	✓	assumed equal to the biting rate; depends on temperature, with blood meals taken more frequently at warmer temperatures (Mullens and Holbrook 1991; Lysyk and Danyk 2007; Veronesi et al. 2009)
vector-to-host ratio	✓		✓	✓	numerous host, vector and environmental factors influence vector abundance and, hence, the vector to host ratio (see review by Purse et al. 2015)
duration of viremia	✓	✓			differs between host species (cattle vs sheep) and viruses (e.g. BTV-8 vs SBV) (see estimates in Table 7.2)
disease-associated mortality rate	✓	✓			bluetongue is typically more severe in sheep and is often mild or asymptomatic in cattle (Maclachlan et al. 2009) (see estimates in Table 7.2); SBV causes no or only mild, transient clinical signs (Wernike and Beer 2017)
extrinsic incubation period (EIP)	✓	✓	✓		depends on temperature, with shorter EIP at higher temperatures (Mullens et al. 2004; Carpenter et al. 2011); threshold temperature for replication varies among BTV strains and vector species (Carpenter et al. 2011)
vector mortality rate			✓	✓	depends on temperature, with shorter lifespan at warmer temperatures (Gerry and Mullens 2000; Wittmann et al. 2002); also affected by relative humidity (Wittmann et al. 2002)

† H - host; Vi—virus; Ve—vector; E - environment

eter values and functions for those parameters which depend on temperature are presented in Table 7.2. These have been estimated using laboratory and field data to construct informative prior distributions for the parameters, which were then updated by fitting the models to epidemic data to generate posterior inferences (Gubbins et al. 2014; Sumner et al. 2017). Comparing the estimates for the two viruses highlights the dependence of the parameters influencing R_0 on factors related to the host, virus, vector, and environment (in particular, temperature).

Table 7.2 Parameters influencing the transmission of *Culicoides*-borne viruses with quantitative estimates.

Source: Bluetongue virus serotype 8 (Sumner et al. 2017) and Schmallenberg virus (Gubbins et al. 2014)

Description		symbol	bluetongue virus serotype 8 (BTV-8)		Schmallenberg virus (SBV)		comments
			estimate[†]	95% CI[‡]	estimate	95% CI	
probability of transmission from vector to host		b	0.84	(0.68, 0.96)	0.76	(0.46, 0.95)	-
probability of transmission from host to vector		β	0.022	(0.0073, 0.042)	0.14	(0.07, 0.26)	-
vector to host ratio		m	Gamma($s_v, \mu_v/s_v$)		-	-	varies amongst farms
mean vector to host ratio		μ_v	1774	(688, 3141)	-	-	
shape parameter for vector to host ratio		s_v	1.62	(0.54, 3.17)	-	-	
reciprocal of the time interval between blood meals		a	$a(T) = 0.0002T(T - 3.7)(41.9 - T)^{1/2.7}$				assumed equal to the biting rate; function taken from Mullens et al. (2004)
duration of viremia (cattle)*	mean	$1/r$	20.5	(18.7, 22.3)	3.04	(1.63, 5.91)	-
	shape	n	4.8	(3.6, 5.8)	11	(2, 20)	
disease-associated mortality rate (cattle)		d	1.1×10^{-3}	$(1.2 \times 10^{-4}, 2.9 \times 10^{-3})$	0	-	-
duration of viremia (sheep)*	mean	$1/r$	16.1	(14.1, 18.2)	4.37	(2.24, 9.02)	-
	shape	n	11.8	(5.1, 19.2)	11	(1, 20)	
disease-associated mortality rate (sheep)		d	6.2×10^{-3}	$(5.9 \times 10^{-4}, 1.7 \times 10^{-2})$	0	-	-
extrinsic incubation period (EIP)*	mean	$1/\nu$	$\nu(T) = \alpha(T - T_{min})$				reciprocal of mean EIP depends on temperature (Carpenter et al. 2011)
	shape	k	9.7	(2.3, 21.9)	6	(2, 30)	
virus replication rate		α	0.020	(0.016, 0.024)	0.030	(0.016, 0.045)	-
threshold temperature for virus replication		T_{min}	13.2	(12.8, 13.7)	12.4	(10.5, 14.0)	-
vector mortality rate		μ	$\mu(T) = 0.009 \exp(0.16T)$				depends on temperature; function taken from Gerry and Mullens (2000)

* the duration of viremia in cattle and sheep and the duration of the EIP were assumed to follow gamma distributions; the mean and shape parameterise the distribution

† posterior median

‡ 95% credible interval

Although the basic reproductive ratio, (7.1), is useful for identifying factors which influence the transmission of *Culicoides*-borne viruses, it ignores several important aspects. First, it only reflects transmission at a local scale, which in the case of livestock viruses, such as BTV and SBV, is typically a single farm. Second, it potentially obscures the effects of seasonally-varying factors, such as temperature or vector abundance, particularly when used quantitatively (i.e. to determine if and when the threshold at $R_0 = 1$ is exceeded) (Grassly & Fraser 2006; Brand & Keeling 2017).

One way of addressing these limitations is to consider the transmission of *Culicoides*-borne viruses across a landscape. Several mathematical models have been developed to do this, largely in relation to the spread of BTV-8 (Courtejoie et al. 2018). Those parameters and factors that influence R_0 still play a role in spread across a landscape, principally through their influence on transmission within a farm. In addition, spatial heterogeneities in livestock and vector distributions and in environmental factors will further affect patterns of spread. For example, analysis of outbreak data for BTV-8 in France (Durand et al. 2010) and Belgium (Ensoy et al. 2013) and for BTV-1 in France (Pioz et al. 2014) indicated that farm density, cattle and sheep densities, temperature, precipitation, elevation, and land cover (urban, pasture and forest) influence the rate of spread.

7.4 BTV-8 and SBV in Northwestern Europe

BTV-8 (introduced in 2006) and SBV (introduced in 2011) emerged in the same region of northwestern Europe and are transmitted by the same constellation of *Culicoides* vector species (Purse et al. 2015). In addition, the host species are the same and the host populations were entirely susceptible. Despite this, SBV spread faster and further than BTV-8 did over a comparable time span (Table 7.3; see maps in Carpenter et al. 2009b for BTV-8 and in European Food Safety Authority 2012 for SBV). Furthermore, a greater proportion of animals became infected with SBV compared with BTV-8 (Table 7.3) as shown by seroprevalence of the two viruses in cattle and sheep in Belgium and the Netherlands (Méroc

et al. 2008, 2013, 2014; van Schaik et al. 2008; Veldhuis et al. 2013).

In terms of transmission parameters BTV-8 and SBV are known to differ in three ways (Table 7.2). First, the duration of viremia in cattle and sheep is much shorter for SBV than for BTV-8. Second, the probability of transmission from host to vector is much higher for SBV than for BTV-8. Third, the duration of the EIP may be different, with faster viral replication for SBV than for BTV-8. The question is how these differences in transmission parameters combine to create the different rates of spread.

The basic reproductive ratio for both viruses exceeds one for a wide range of temperatures, but R_0 is consistently higher for SBV than for BTV-8. For a given temperature, it is around 25 percent higher in cattle (Fig. 7.2a) and 75 percent higher in sheep (Fig. 7.2b). In addition, the temperature at which R_0 exceeds one is lower by 1°C for SBV, though the optimum temperature (around 20°C) is similar for both viruses (Fig. 7.2). This overall effect is a combination of the differences in the underlying parameters (Fig. 7.2). The differences in R_0 between the viruses are consistent with greater transmission of SBV compared with BTV-8 at a local scale, as shown by the different within-herd seroprevalences for the two viruses (Table 7.3).

Scaling-up to consider spread across a landscape (in this case, United Kingdom) the difference in

Table 7.3 Comparison of bluetongue virus serotype 8 (BTV-8) and Schmallenberg (SBV) virus in north-western Europe.

		BTV-8	SBV
countries affected†		Belgium France Germany The Netherlands Luxemburg	Belgium France Germany The Netherlands Luxemburg Italy Spain United Kingdom
within-herd seroprevalence (%)	Cattle	Belgium: 23.8 The Netherlands: 39.3	Belgium: 86.3 The Netherlands: 98.5
	Sheep	-	Belgium: 84.3 The Netherlands: 89.0

† at the end of the first year of the epidemic (2006 for BTV-8, 2011 for SBV)

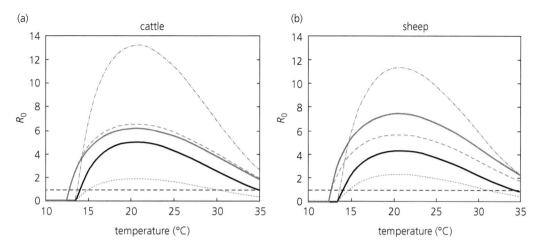

Figure 7.2 Basic reproductive ratio (R_0) for bluetongue virus serotype 8 (BTV-8) and Schmallenberg virus (SBV) in (*a*) cattle and (*b*) sheep and its dependence on temperature. The curves show R_0 for: BTV-8 (solid black line); BTV-8, but with duration of viremia as for SBV (grey dotted line); BTV-8, but with probability of transmission from host to vector as for SBV (grey dash-dotted line); BTV-8, but with duration of the EIP as for SBV (grey dashed line); and SBV (solid grey line). Each curve is the posterior median for R_0 using the parameters for BTV-8 and SBV in Table 7.2. The horizontal black dashed line indicates the threshold at $R_0 = 1$.

spread between BTV-8 and SBV is unlikely to have been a simple consequence of different temperatures in 2006 compared with 2011: in every scenario SBV in 2011 spreads further than BTV-8 in 2006 or, indeed, in 2011 (Fig. 7.3). The shorter duration of viremia for SBV compared with BTV-8 has the expected effect of reducing spread (Fig. 7.3). The higher probability of transmission from host to vector and the shorter EIP both increase spread, but when the increases in these two parameters are combined with the shorter duration of viremia, there is not a subsequent reduction in spread (Fig. 7.3).

Consequently, the increased spread of SBV compared with BTV-8 can be attributed to the increased probability of transmission from host to vector and of the increased viral replication rate for SBV, both of which have the effect of increasing the prevalence of infected vectors (which is consistent with field observations; Elbers et al. 2013). The higher prevalence increases the force of infection between farms due to dispersal of infected vectors (Sumner et al. 2017), which enables SBV to spread further and faster than BTV-8.

7.5 BTV Strains in Europe

By contrast to BTV-8 and SBV in northwestern Europe, the situation with the pattern of spread of

the different BTV strains is much more complex. This reflects the invasion of multiple strains of BTV over the past decades, the different vector species and their ecologies and the impact of control measures, especially vaccination.

Identifying the reasons underlying the emergence of BTV in Europe between 1998 and 2006 is challenging because of the lack of longitudinal surveillance data on both BTV and its vectors. There is circumstantial evidence that emergence was linked to regional warming over the corresponding time period because: (i) temperatures increased in outbreak areas; (ii) different strains spread simultaneously from multiple sources; and (iii) there were negligible changes to livestock management and land-cover (Purse et al. 2005, 2015). This warming was hypothesised to have influenced the distribution of BTV in two ways. First, it allowed range expansion of *C. imicola* (Purse et al. 2005), though more recent genetic evidence suggests that *C. imicola* was already well established prior to the first reported BTV outbreaks in Cyprus in 1924 (Jacquet et al. 2015). Second, it increased the ability of *C. imicola* or of Palearctic *Culicoides* vector species to transmit the virus (Purse et al. 2005, 2007; Guis et al. 2012). Interestingly, recent phylogeographic analyses did not identify a significant role of temperature in the

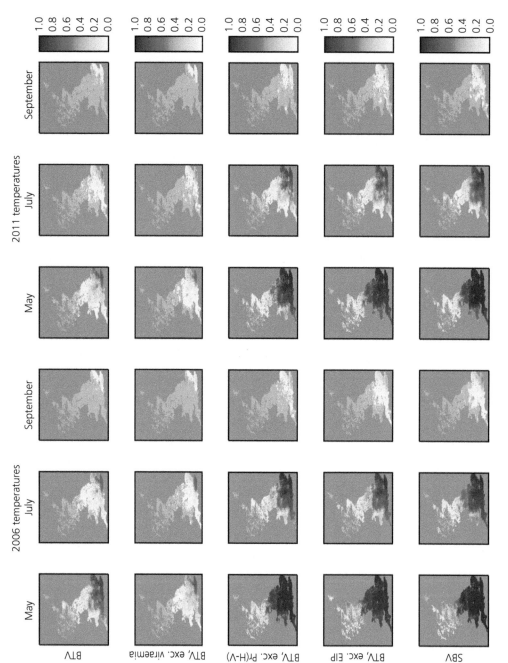

Figure 7.3 Predicted spatial spread of bluetongue virus serotype 8 (BTV-8) and Schmallenberg virus (SBV) in the United Kingdom. The maps show the cumulative probability of infection (see scale bar) expressed as the proportion of simulated outbreaks for which at least one farm was affected within each 5 km grid square. Columns differ in the time of the initial incursion (1 May, 1 July, 1 September) and the temperature data used (2006 or 2011). Rows differ in the parameters used for the simulations (from top to bottom): BTV-8; BTV-8, but with duration of viremia as for SBV (BTV, exc. viremia); BTV-8, but with probability of transmission from host to vector as for SBV (BTV, exc. Pr(H–V)); BTV-8, but with duration of the EIP as for SBV (BTV, exc. EIP); and SBV. Results are based on one hundred replicates of the model presented in Sumner et al. (2017) for an incursion to south-east England with parameters sampled from their joint posterior distribution (see Table 7.2). The simulated epidemics assume no measures are implemented to control spread of the virus.

expansion of BTV (Jacquot et al. 2017). However, the authors suggest this could be a consequence of reasons either methodological (e.g. a linear response to temperature was assumed; cf. Fig. 7.3) or biological (e.g. strain variation in the response to temperature, which was not accounted for).

Two studies have explored the emergence of BTV in Europe through the impact of changing temperatures on the basic reproductive ratio (Guis et al. 2012; Brand & Keeling 2017). At the European scale, statistical modelling of trap catch data predicted that the vector to host ratio for 2000 to 2008 (compared to the 1961-1999 mean) increased for *C. imicola* in much of southern Europe, but decreased for the *C. obsoletus* group across Europe (Guis et al. 2012). However, the approach used in the study means it is not clear if these changes resulted in the basic reproductive ratio exceeding the threshold at $R_0 = 1$ (in the case of *C. imicola*), or merely a change in R_0 that was already above the threshold. Analysis of the basic reproductive ratio for BTV in United Kingdom between 1950 and 2015 suggested that there were times of the year during which R_0 exceeded one throughout this period, but that the length of the transmission season has increased in recent years (Brand & Keeling 2017). However, neither of these studies considered possible changes in underlying parameters, for example, due to strain variation.

Indeed, teasing out the roles of the underlying host, viral, vector and environmental factors in more detail is challenging because of limited quantitative data. Life history parameters are not generally available for most European *Culicoides* vector species (Purse et al. 2015; White et al. 2017). This makes predicting vector population dynamics difficult, at least mechanistically (White et al. 2017) and predicting their response to changes in climate even more so. It also means that those parameters directly influencing BTV transmission, such as biting and mortality rates, have to be assumed to be the same as for *C. sonorensis*, a North American vector species (Table 7.2).

A further complication is that incursions of multiple BTV strains into Europe has resulted in a complex strain landscape (Nomikou et al. 2015). This is a consequence of the number of incursions, but also

of the segmented nature of the BTV genome and, hence, its ability to reassort (i.e. for viruses infecting the same cell to exchange segments). A detailed understanding of how this strain diversity influences transmission is lacking. However, the series of incursions of BTV into Europe (and, indeed, of SBV) and subsequent spread demonstrate that substantial co-evolution of vector and viruses is not a pre-requisite for epidemics to occur. This contrasts with the view of BTV existing as a number of stable episystems that was prevalent prior to the 2000s (Tabachnick 2004; Purse et al. 2015).

Nonetheless, there are vector-virus interactions which influence the transmission of BTV strains. Studies using colony-reared midges found no significant variation in the probability of transmission from host to vector for three BTV strains (serotypes 9, 10, and 16) (Carpenter et al. 2011). By contrast, field-caught *C. obsoletus* and *C. pulicaris* group midges in the United Kingdom showed geographic variation in oral susceptibility to BTV-9 infection (Carpenter et al. 2008), though it is not clear whether this reflects variation between sites in competence or in species composition. Outbreaks in Sicily provide tentative evidence that there are differences between the *C. obsoletus* and *C. pulicaris* groups in terms of the level of trap catches required for transmission to occur (Torina et al. 2004), indicative of underlying differences between these species in terms of, for example, vector competence.

The extrinsic incubation period can also differ between BTV strains and between vector species (Carpenter et al. 2011). These differences arise both through differences in the viral replication rate and in the threshold temperature required for replication. Analysis of outbreak data has demonstrated a threshold temperature required for transmission of BTV between farms (de Koeijer et al. 2011; Boender et al. 2014; Napp et al. 2016). This threshold differed between outbreaks, estimated at 15°C for BTV-8 in north-western Europe (de Koeijer et al. 2011; Boender et al. 2014) and 20°C for BTV-1 in southern Spain (Napp et al. 2016), presumably reflecting differences between the circulating strains, but also between vector species involved in transmission in the two regions (Palearctic *Culicoides* species in north-western Europe and *C. imicola* in southern Spain).

7.6 Overwintering of *Culicoides*-borne Viruses

Evidence from the field shows that *Culicoides*-borne viruses are able to persist during periods when vectors are inactive (Wilson et al. 2008; Mayo et al. 2016) and normal transmission does not occur (Fig. 7.1). The ability of these viruses to overwinter was spectacularly demonstrated when BTV-8 re-emerged in France in 2015 following an

apparent absence of five years. Because the virus was almost identical to that circulating up to 2010 (Bréard et al. 2016), this was most likely the re-emergence of an existing strain, possibly following low level circulation in livestock or wildlife, rather than the introduction of a new one (Sailleau et al. 2017).

Only limited modeling work has examined the overwintering of *Culicoides*-borne viruses. One study has explored the possibility of overwintering of BTV via long-lived vectors or low-level circulation (Napp et al. 2011). In the context of overwintering of BTV-8 in Germany, the authors concluded that while these are possible mechan-

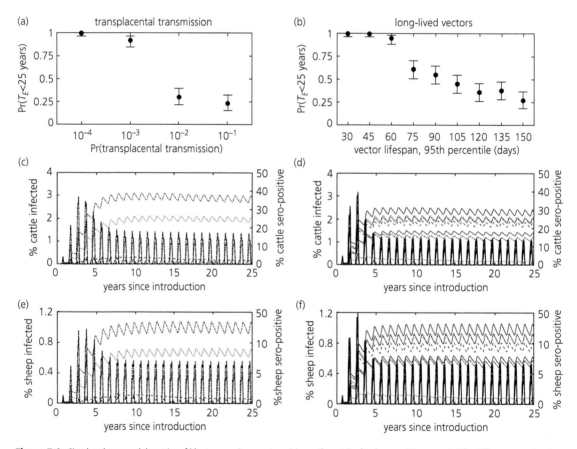

Figure 7.4 Simulated temporal dynamics of bluetongue virus serotype 8 in south east England over a 25-year period for different overwintering scenarios. (*a,b*) Proportion of replicates in which the extinction time (*T$_E$*) is less than 25 years for different (*a*) levels of transplacental transmission or (*b*) vector lifespans (95th percentile at the minimum mortality rate). (*c-f*) Mean proportion (%) of (*c,d*) cattle or (*e,f*) sheep that are infected (black lines; left axis) or seropositive (grey lines; right axis) when overwintering occurs via (*c,e*) transplacental transmission or (*d,f*) long-lived vectors. Results are based on 100 replicates of the model.

isms, the probability of their occurrence was too low for them to have been the principal means by which the virus overwintered. A further study has considered the potential for persistence of BTV via transplacental transmission, particularly in the context of the re-emergence of BTV-8 in France (European Food Safety Authority 2017). Simulation results demonstrated that BTV could persist for at least 25 years (i.e. the duration of the simulations) and that, after around five years, the dynamics of BTV settled into a stable pattern of seasonal outbreaks (cf. Fig. 7.4). In addition, the model showed that while vaccination suppressed the prevalence of infection, it did not eradicate the virus within five years (even at 95 percent coverage) and, if vaccination was stopped, it could re-emerge. This was the case for simulated outbreaks in the United Kingdom, France, Andalusia (Spain) and Sardinia (Italy).

More detailed analysis using the same model (European Food Safety Authority 2017) suggests that both transplacental transmission and long-lived vectors[3] are potentially efficient overwintering mechanisms (Fig. 7.4). Moreover, results suggest that there are threshold levels below which persistence is unlikely to occur: $<10^{-2}$ for the probability of transplacental transmission and <75 days for the vector lifespan (Fig. 7.4a,b). For comparison, the probability of transplacental transmission has been estimated from field data to be >18 percent for BTV-8 (De Clercq et al. 2008; Darpel et al. 2009), while the lifespan for a *Culicoides* biting midge is typically up to 20 days, but may exceptionally be up to 90 days (Mellor et al. 2000).

When the virus does persist, the dynamics of BTV settles down into a stable cycle of annual epidemics with the number of cases reaching its peak in late August–early September (Fig. 7.4c-f). The long-term prevalence of infection increases as either the probability of transplacental transmission or vector lifespan increases (Fig. 7.4).

---

[3] In this case, the lifespan for a vector was constrained by assuming the vector mortality rate must equal or exceed a value such that only 5 percent of midges live for longer than a specified lifespan.

7.7 Key Knowledge Gaps

A number of models have been developed to understand the population biology of *Culicoides*-borne viruses in Europe. These are able to capture the dynamics of the viruses, at least during a single vector season, and to generate insights into the factors influencing their transmission. They are also able to address practical questions regarding the design of control strategies, especially vaccination campaigns. That said, three points should be kept in mind. First, the models have largely focused on two viruses, BTV-8 and SBV. Second, the potential complexities of disease dynamics where multiple *Culicoides* species are involved in transmission have largely been ignored by assuming there is only a single vector (i.e. all the potential vector species behave in the same way). Third, certain key parameters (vector biting and mortality rates) are assumed to be the same as for the North American vector, *C. sonorensis*.

The focus on BTV-8 and SBV is largely a consequence of the scale and the unexpected nature of these epidemics. This resulted in an urgent need from policy makers for practical advice on how the viruses are likely to spread and the potential impact of different control measures. The scale of the outbreaks also meant that large data sets were generated which could be used to investigate the population dynamics of the viruses in some detail. However, the generic structure of the models can readily be adapted for other strains of BTV or for other *Culicoides*-borne viruses.

The simplified treatment of vector ecology and use of other species as proxies reflects the lack of available data on European *Culicoides* species (Purse et al. 2015). Although averaging across species (i.e. assuming there is a single vector) has allowed predictions to be made, more detailed understanding of the transmission of *Culicoides*-borne viruses will require separating out the species and the roles they play. For example, there is considerable uncertainty in estimates for several key parameters, notably the probability of transmission from host to vector and the vector to host ratio (Table 7.2). Considering *Culicoides* species individually may help explain this variation.

Although overwintering of *Culicoides*-borne viruses clearly occurs, the precise mechanisms are still poorly understood. Long-lived vectors have been shown to play a role in overwintering of BTV in California (Mayo et al. 2014, 2016), but it is not clear if this could also be the case in Europe. Trapping on farms during winter would give some indication of whether or not long-lived midges are found on farms and, hence, could play a role in maintaining viruses between seasons. Transplacental transmission has been demonstrated in the field for SBV (Wernike & Beer 2017) and BTV-8 (De Clercq et al. 2008; Darpel et al. 2009), but its occurrence has not been demonstrated for other strains of BTV (van der Sluijs et al. 2016) or for other *Culicoides*-borne viruses.

Acknowledgements

SG acknowledges funding from the Biotechnology and Biological Sciences Research Council (BBSRC) [grant codes: BBS/E/I/00007033, BBS/E/I/00007036 and BBS/E/I/00007037].

References

Ågren, E.C.C., Burgin, L., Sternberg Lewerin, S., Gloster, J., & Elvander, M. (2010). Possible means of introduction of bluetongue virus serotype 8 (BTV-8) to Sweden in August 2008: Comparison of results from two models for atmospheric transport of the *Culicoides* vector. *Veterinary Record*, 167, 484–8.

Alba, A., Casal, J., & Domingo, M. (2004). Possible introduction of bluetongue into the Balearic Islands, Spain, via air streams. *Veterinary Record*, 155, 460–31.

Balenghien, T., Pagès, N., Goffredo, M. et al. (2014). The emergence of Schmallenberg virus across *Culicoides* communities and ecosystems in Europe. *Preventive Veterinary Medicine*, 116, 36–369.

Baylis, M., O'Connell, L., & Mellor, P.S. (2008). Rates of bluetongue virus transmission between *Culicoides sonorensis* and sheep. *Medical and Veterinary Entomology*, 22, 228–37.

Boender, G.J., Hagenaars, T.J., Elbers, A.R.W. et al. (2014). Confirmation of spatial patterns and temperature effects in bluetongue virus serotype-8 transmission in NW-Europe from the 2007 reported case data. *Veterinary Research*, 45, 75.

Brand, S.P. & Keeling, M.J. (2017). The impact of temperature on vector-borne disease transmission: *Culicoides* midges and bluetongue virus. *Journal of the Royal Society Interface*, 14, 20160481.

Bréard, E., Sailleau, C., Quenault, H. et al. (2016). Complete genome sequence of bluetongue virus serotype 8, which reemerged in France in August 2015. *Genome Announcements*, 4, e00163–16.

Burgin, L., Gloster, J., Sanders, C., Mellor, P.S., Gubbins, S., & Carpenter, S.J. (2013). Investigating incursions of bluetongue virus using a model of long-distance *Culicoides* biting midge dispersal. *Transboundary and Emerging Diseases*, 60, 263–72.

Carpenter, S., Lunt, H.L., Arav, D., Venter, G.J., & Mellor, P.S. (2008). Oral susceptibility to bluetongue virus of *Culicoides* (Diptera: Ceratopogonidae) from the United Kingdom. *Journal of Medical Entomology*, 43, 73–8.

Carpenter, S., Wilson, A., and Mellor, P.S. (2009a). *Culicoides* and the emergence of bluetongue virus in northern Europe. *Trends in Microbiology*, 17, 172–8.

Carpenter, S., Wilson, A., & Mellor, P.S. (2009b). Bluetongue virus and *Culicoides* in the UK: The impact of research on policy. *Outlooks in Pest Management*, 20, 161–4.

Carpenter, S., Wilson, A., Barber, J. et al. (2011). Temperature dependence of the extrinsic incubation period of orbiviruses in *Culicoides* biting midges. *PLoS One*, 6, e27987.

Carpenter, S., Groschup, M.H., Garros, C., Felippe-Bauer, M.L., & Purse, B.V. (2013). *Culicoides* biting midges, arboviruses and public health in Europe. *Antiviral Research*, 100, 102–13.

Carpenter, S., Veronesi, E., Mullens, B., & Venter, G. (2015). Vector competence of *Culicoides* for arboviruses: Three major periods of research, their influence on current studies and future directions. *Revue Scientifique et Technique de l'O.I.E.*, 34, 97–112.

Courtejoie, N., Zanella, G., & Durand, B. (2018). Bluetongue transmission and control in Europe: A systematic review of compartmental mathematical models. *Preventive Veterinary Medicine*, 156, 113–25.

Darpel, K.E., Batten C.A., Veronesi, E. et al. (2009). Transplacental transmission of bluetongue virus 8 in cattle, GB. *Emerging Infectious Diseases*, 15, 2025–8.

Darpel, K.E., Langner, K.F.A., Nimtz, M. et al. (2011). Saliva proteins of vector *Culicoides* modify structure and infectivity of bluetongue virus particles. *PLoS One*, 6, e17545.

De Clercq, K., De Leeuw, I., Verheyden, B. et al. (2008). Transplacental infection and apparent immunotolerance induced by a wild-type bluetongue virus serotype 8 natural infection. *Transboundary and Emerging Diseases*, 55, 352–9.

de Koeijer, A.A., Boender, G.J., Nodelijk, G., Staubach, C., Méroc, E., & Elbers, A.R.W. (2011). Quantitative analysis of transmission parameters for bluetongue virus serotype 8 in Western Europe in 2006. *Veterinary Research*, 42, 53.

Durand, B., Zanella, G., Biteau-Coroller, F. et al. (2010). Anatomy of bluetongue virus serotype 8 epizootic wave, France, 2007–2008. *Emerging Infectious Diseases*, 16, 1861–8.

Elbers, A.R.W., Meiswinkel, R., van Weezep, E., Sloet van Oldruitenborgh-Oosterbaan, M.M. & Kooi, E.A. (2013). Schmallenberg virus in *Culicoides* spp. biting midges, the Netherlands, 2011. *Emerging Infectious Diseases*, 19, 106–9.

Ensoy, C., Aerts, M., Welby, S., van der Stede, Y. & Faes, C. (2013). A dynamic spatio-temporal model to investigate the effect of cattle movements on the spread of bluetongue BTV-8 in Belgium. *PLoS One*, 8, e78591.

European Food Safety Authority (2012). 'Schmallenberg' virus: Analysis of the epidemiological data and assessment of impact. *The EFSA Journal*, 10, 2768.

European Food Safety Authority (2013). *'Schmallenberg' Virus: Analysis of the Epidemiological Data. Supporting Publications 2013:EN-429*. Parma, Italy: European Food Safety Authority.

European Food Safety Authority (2017). Bluetongue: Control, surveillance and safe movement of animals. *The EFSA Journal*, 15, 4698.

Gerry, A.C. & Mullens, B.A. (2000). Seasonal abundance and survivorship of *Culicoides sonorensis* (Diptera: Ceratopogonidae) at a southern Californian dairy, with reference to potential bluetongue virus transmission and persistence. *Journal of Medical Entomology*, 37, 675–88.

Grassly, N.C. & Fraser, C. (2006). Seasonal infectious disease epidemiology. *Proceedings of the Royal Society B: Biological Sciences*, 273, 2541–50.

Gubbins, S., Carpenter, S., Baylis, M., Wood, J.L.N. & Mellor, P.S. (2008). Assessing the risk of bluetongue to UK livestock: Uncertainty and sensitivity analysis of a temperature dependent model for the basic reproduction number. *Journal of the Royal Society Interface*, 5, 363–71.

Gubbins, S., Hartemink, N.A., Wilson, A.J. et al. (2012). Scaling from challenge experiments to the field: Quantifying the impact of vaccination on the transmission of bluetongue virus serotype 8. *Preventive Veterinary Medicine*, 105, 297–308.

Gubbins, S., Turner, J., Baylis, M. et al. (2014). Inferences about the transmission of Schmallenberg virus within and between farms. *Preventive Veterinary Medicine*, 116, 380–90.

Guis, H., Caminade, C., Calvete, C., Morse, A.P., Tran, A., & Baylis, M. (2012). Modelling the effects of past and future climate change on the risk of bluetongue emergence in Europe. *Journal of the Royal Society Interface*, 9, 339–50.

Hendrickx, G., Gilbert, M., Staubach, C. et al. (2008). A wind density model to quantify the airborne spread of *Culicoides* species during north-western Europe bluetongue epidemic, 2006. *Preventive Veterinary Medicine*, 87, 162–81.

Hoffmann, B., Scheuch, M., Höper, D. et al. (2012). Novel orthobunyavirus in cattle, Europe, 2012. *Emerging Infectious Diseases*, 18, 469–72.

Jacquet, S., Garros, C., Lombaert, E. et al. (2015). Colonization of the Mediterranean basin by the vector biting midge species *Culicoides imicola*: An old story. *Molecular Ecology*, 24, 5707–25.

Jacquot, M., Nomikou, K., Palmarini, M., Mertens, P., & Biek, R. (2017). Bluetongue virus spread in Europe is a consequence of climatic, landscape and vertebrate host factors as revealed by phylogenetic inference. *Proceedings of the Royal Society B: Biological Sciences*, 284, 20170919.

Keeling, M.J. & Rohani, P. (2008). *Modeling Infectious Diseases in Animals and Humans*. Princeton, U.S.A.: Princeton University Press.

Lysyk, T. & Danyk, T. (2007). Effects of temperature on life history parameters of adult *Culicoides sonorensis* (Diptera: Ceratopogonidae) in relation to geographical origin and vectorial capacity for bluetongue virus. *Journal of Medical Entomology*, 44, 741–51.

Maclachlan, N.J., Drew, C.P., Darpel, K.E., & Worwa, G. (2009). The pathology and pathogenesis of bluetongue. *Journal of Comparative Pathology*, 141, 1–16.

Mayo, C.E., Mullens, B.A., Reisen, W.K. et al. (2014). Seasonal and inter-seasonal dynamics of bluetongue virus infection of dairy cattle and Culicoides sonorensis midges in northern California—implications for virus overwintering in temperate zones. *PLoS One*, 9, e0106975.

Mayo, C.E., Mullens, B.A., Gibbs, E.P.J., & Maclachlan, N.J. (2016). Overwintering of bluetongue virus in temperate zones. *Veterinaria Italiana*, 52, 243–6.

Mellor, P.S., Boorman, J., & Baylis, M. (2000). *Culicoides* biting midges: Their role as arbovirus vectors. *Annual Review of Entomology*, 45, 307–40.

Mellor, P.S., Carpenter, S., Harrup, L., Baylis, M., & Mertens, P.P.C. (2008). Bluetongue in Europe and the Mediterranean basin: History of occurrence prior to 2006. *Preventive Veterinary Medicine*, 87, 4–20.

Méroc, E., Faes, C., Herr, C. et al. (2008). Establishing the spread of bluetongue virus at the end of the 2006 epidemic in Belgium. *Veterinary Microbiology*, 131, 133–44.

Méroc, E., Poskin, A., Van Loo, H. et al. (2013). Large-scale cross-sectional survey of Schmallenberg virus in Belgian cattle at the end of the first vector season. *Transboundary and Emerging Diseases*, 60, 4–8.

Méroc, E., De Regge, N., Riocreux, F., Caij, A.B., van den Berg, T. & van der Stede, Y. (2014). Distribution of Schmallenberg virus and seroprevalence in Belgian sheep and goats. *Transboundary and Emerging Diseases*, 61, 425–31.

Mullens, BA. & Holbrook, F.R. (1991). Temperature effects on the gonotrophic cycle of *Culicoides variipennis* (Diptera: Ceratopogonidae). *Journal of the American Mosquito Control Association*, 7, 588–91.

Mullens, B.A., Gerry, A.C., Lysyk, T.J., & Schmidtmann, E.T. (2004). Environmental effects on vector competence and virogenesis of bluetongue virus in *Culicoides*: Interpreting laboratory data in a field context. *Veterinaria Italiana*, 40, 160–6.

Napp, S., Gubbins, S., Calistri, P. et al. (2011). Quantitative assessment of the probability of bluetongue virus overwintering by horizontal transmission: application to Germany. *Veterinary Research*, 42, 4.

Napp, S., Allepuz, A., Purse, B.V. et al. (2016). Understanding spatio-temporal variability in the reproduction ratio of the bluetongue (BTV-1) epidemic in southern Spain (Andalusia) in 2007 using epidemic trees. *PLoS One*, 11, e0151151.

Nomikou, K., Hughes, J., Wash, R. et al. (2015). Widespread reassortment shapes the evolution and epidemiology of bluetongue virus following European invasion. *PLoS Pathogens*, 11, e1005056.

Pioz, M., Guis, H., Pleydell, D. et al. (2014). Did vaccination slow the spread of bluetongue in France? *PLoS One*, 9, e85444.

Purse, B.V., Mellor, P.S., Rogers, D.J., Samuel, A.R., Mertens, P.P.C., & Baylis, M. (2005). Climate change and the recent emergence of bluetongue in Europe. *Nature Reviews Microbiology*, 3, 171–81.

Purse, B.V., McCormick, B.J.J., Mellor, P.S. et al. (2007). Incriminating bluetongue virus vectors with climate envelope models. *Journal of Applied Ecology*, 44, 1231–42.

Purse, B.V., Carpenter, S., Venter, G.J., Bellis, G., & Mullens, B.A. (2015). Bionomics of temperate and tropical *Culicoides* midges: Knowledge gaps and consequences for transmission of *Culicoides*-borne viruses. *Annual Review of Entomology*, 60, 373–92.

Sailleau, C., Bréard, E., Viarouge, C. et al. (2017). Re-emergence of bluetongue virus serotype 8 in France, 2015. *Transboundary and Emerging Diseases*, 64, 998–1000.

Sedda, L. & Rogers, D.J. (2013). The influence of wind in the Schmallenberg virus outbreak in Europe. *Scientific Reports*, 3, 3361.

Sedda, L., Brown, H.E., Purse, B.V., Burgin, L., Gloster, J., & Rogers, D.J. (2012). A new algorithm quantifies the roles of wind and midge flight activity in the bluetongue epizootic in northwest Europe. *Proceedings of the Royal Society of London Series B: Biological Sciences*, 279, 2354–62.

Sellers, R.F., Pedgley, D.E., & Tucker, M.R. (1978). Possible windborne spread of bluetongue to Portugal, June-July 1956. *Journal of Hygiene*, 81, 189–96.

Sellers, R.F., Gibbs, E.P.J., Herniman, K.A.J., Pedgley, D.E., & Tucker, M.R. (1979). Possible origin of the bluetongue epidemic in Cyprus, August 1977. *Journal of Hygiene*, 83, 547–55.

Singer, R.S., MacLachlan, N.J., & Carpenter, T.E. (2001). Maximal predicted duration of viremia in bluetongue virus-infected cattle. *Journal of Veterinary Diagnostic Investigation*, 13, 43–9.

St George, T.D. & Standfast, H.A. (1989). Simbu group viruses with teratogenic potential. In: T.P. Monath, ed. *The Arboviruses: Epidemiology and Ecology* (Volume IV). Boca Raton, Florida, U.S.A.: RC Press, pp. 145–66.

Sumner, T., Orton, R.J., Green, D.M., Kao, R.R., & Gubbins, S. (2017). Quantifying the roles of host movement and vector dispersal in the transmission of vector-borne diseases of livestock. *PLoS Computational Biology*, 13, e1005470.

Tabachnick, W.J. (2004). *Culicoides* and the global epidemiology of bluetongue virus infection. *Veterinaria Italia*, 40, 145–50.

Torina, A., Caracappa, S., Mellor, P.S., Baylis, M., & Purse, B.V. (2004). Spatial distribution of bluetongue virus and its Culicoides vectors in Sicily. *Medical and Veterinary Entomology*, 18, 81–9.

Turner, J., Bowers, R.G., & Baylis, M. (2012). Modelling bluetongue virus transmission between farms using animal and vector movements. *Scientific Reports*, 2, 319.

Turner, J., Bowers, R.G., & Baylis, M. (2013). Two-host, two-vector basic reproduction ratio (R_0) for bluetongue. *PLoS One*, 8, e53128.

Turner, J., Jones, A.E., Heath, A.E. et al. (2019) The effect of temperature, farm density and foot-and-mouth disease restrictions on the 2007 UK bluetongue outbreak. *Scientific Reports*, 9, 112.

van der Sluijs, M.T.W., de Smit, A.J., & Moormann, R.J. (2016). Vector independent transmission of the vector-borne bluetongue virus. *Critical Reviews in Microbiology*, 42, 57–64.

van Schaik, G., Berends, I.M.G.A., van Langen, H., Elbers, A.R.W., & Vellema, P. (2008). Seroprevalence of bluetongue serotype 8 in cattle in the Netherlands in spring 2007, and its consequences. *Veterinary Record*, 163, 441–4.

Veldhuis, A.M.B., van Schaik, G., Vellema, P. et al. (2013). Schmallenberg virus epidemic in the Netherlands: Spatiotemporal introduction in 2011 and seroprevalence in ruminants. *Preventive Veterinary Medicine*, 112, 35–47.

Veronesi, E., Venter, G.J., Labuschagne, K., Mellor, P.S., & Carpenter, S. (2009). Life-history parameters of *Culicoides* (Avarita) *imicola* Kieffer in the laboratory at different rearing temperatures. *Veterinary Parasitology*, 163, 370–3.

Wernike, K. & Beer, M. (2017). Schmallenberg virus: A novel virus of veterinary importance. *Advances in Virus Research*, 99, 39–60.

White, S.M., Sanders, C.J., Shortall, C.R., & Purse, B.V. (2017). Mechanistic modelling for predicting seasonal abundance of Culicoides biting midges and the impacts of insecticide control. *Parasites & Vectors*, 10, 162.

Wilson, A.J. & Mellor, P.S. (2009). Bluetongue in Europe: Past, present and future. *Philosophical Transactions of the Royal Society of London Series B*, 364, 2669–81.

Wilson, A.J., Darpel, K., & Mellor, P.S. (2008). Where does bluetongue virus sleep in the winter? *PLoS Biology*, 6, e210.

Wittmann, E.J., Mellor, P.S., & Baylis, M. (2002). Effect of temperature on the transmission of orbiviruses by the biting midge, *Culicoides sonorensis*. *Medical and Veterinary Entomology*, 16, 147–56.

CHAPTER 8

Ecological Interactions Influencing the Emergence, Abundance, and Human Exposure to Tick-Borne Pathogens

Maria A. Diuk-Wasser, Maria del Pilar Fernandez, and Stephen Davis

8.1. Introduction

Tick-borne diseases involve complex interactions between multiple pathogens, tick vectors, and a wide range of vertebrate hosts. Understanding the population biology of each component and the interactions between them is essential to controlling disease vectors and reducing human health impact. Tick-borne diseases have emerged in the last 50 years in temperate areas in North America, Europe, and Asia, where they constitute the most significant vector-borne diseases. Ticks in the *Ixodes ricinus* species complex have a Holarctic distribution but are restricted to regions with temperate climates (Diuk-Wasser et al. 2016). Four species within this complex account for most of the transmission of human *Ixodes*-borne pathogens: *Ixodes pacificus* along the Pacific coast of the United States; *Ixodes scapularis* in the Northeast, upper Midwest, and South of the United States; *Ixodes ricinus* across Europe; and *Ixodes persulcatus* in Asia. (Diuk-Wasser et al. 2016). Lyme disease (LD) (Lyme borreliosis in Europe), caused by the spirochete *Borrelia burgdorferi*, is the most commonly reported tick-borne disease worldwide, with an estimate of more than 300,000 cases per year that continues to increase annually (Schwartz et al. 2015). The disease was first described in Lyme, Connecticut, in 1976, and has expanded throughout much of the Northeast and Midwest United States as well as Canada, at a mean estimated rate of ~11km/year in New England (Walter et al. 2016). This geographical expansion of

LD endemic areas has been attributed to range expansion of *I. scapularis* in the eastern United States and evidence for this association is based on the spatial patterns of invasion, with *B. burgdorferi* infection or LD establishment following the invasion of ticks with a few years lag (Ogden et al. 2013). The rarity of *B. burgdorferi*-free populations of *I. scapularis*, only four out of 54 (7.4 percent) in a nation-wide study (Diuk-Wasser et al. 2012), indicates that the main limiting factor for the spread of *B. burgdorferi* and possibly other tick-borne pathogens is the distribution and abundance of their shared vector. In turn, tick and pathogen persistence and spread are dependent on the abundance and distribution of natural hosts through complex demographic and movement dynamics.

In this chapter, we synthesize how tick life history traits influence tick-host-pathogen interactions in human-modified landscapes, driving pathogen emergence, persistence, expansion and impacts on human health. We discuss how land use change—and climate change to a limited extent—mediate these interactions. In Section 8.2, we investigate key interactions required for the emergence of tick-borne pathogens as represented by the basic reproductive number, R_0. We present early models that introduce the concept of a 'dilution effect' by considering two joint abundance thresholds for pathogen emergence, one for the tick and the other for the pathogen. We then examine advances in R_0 modeling for emerging tick-borne pathogens that assess the influence of alternative transmission pathways (Nonaka et al.

Maria A. Diuk-Wasser, Maria del Pilar Fernandez, and Stephen Davis, *Ecological Interactions Influencing the Emergence, Abundance, and Human Exposure to Tick-Borne Pathogens* In: *Population Biology of Vector-Borne Diseases*. Edited by: John M. Drake, Michael B. Bonsall, and Michael R. Strand: Oxford University Press (2021). © Maria A. Diuk-Wasser, Maria del Pilar Fernandez, and Stephen Davis.
DOI: 10.1093/oso/9780198853244.003.0008

2010, Davis & Bent 2011) and tick seasonality (Ogden & Lindsay 2016, Gatewood et al. 2009, Dunn et al. 2013) on the persistence of tick-borne pathogens with short infectious periods. In Section 8.3, we explore how land use and host community composition impact the abundance and distribution of infected ticks as well as the probability of human exposure to ticks in endemic areas. We discuss the multiple interactions involved in hypothesized linkages between biodiversity, habitat fragmentation and human health, and propose a coupled natural-human system (CNH) framework for tick-borne diseases. By accounting for nonlinearities and feedbacks in the system, the CNH framework may help explain some paradoxical or contradictory findings reported in the literature. We close in Section 8.4 with a call for defining the boundaries of the complex system of tick-borne diseases to facilitate discussions and empirical-theoretical integration. Throughout, we focus mainly on *Ixodes scapularis* and the LD system in the United States but draw on other hard bodied (Ixodid) tick-borne disease systems as needed to illustrate patterns or concepts.

8.2. Emergence of Tick-Borne Pathogens: Developments in R_0 Models

The more recent emergence of new or previously rare *I. scapularis*-borne pathogens has brought renewed interest in understanding the potential role for multiple transmission pathways on persistence. Among these pathogens, *Babesia microti*, *Anaplasma phagocytophilum* and Powassan virus (POWV) have been recognized as human pathogens for almost as long as *B. burgdorferi*, but have emerged at a slower pace (Walter et al. 2016). *Borrelia miyamotoi* (Krause et al. 2013), *Borrelia mayonii* (Pritt et al. 2016) and *Ehrlichia muris eauclairensis* (Pritt et al. 2011) have only been recognized as human pathogens in the last ~10 years. The latter two pathogens, *Borrelia mayonii* and *Ehrlichia muris eauclairensis*, have so far exclusively detected in the Midwest United States (Pritt et al. 2016; Pritt et al. 2011). It is unknown to what extent enhanced pathogen discovery has played a role in the observed emergence and, in contrast with the emergence of mosquito-borne viruses, evolutionary changes have not been implicated in tick-borne disease emergence. Genomic studies of *B. burgdorferi sensu stricto*, the dominant *Borrelia* species in North America, and *Babesia microti*, indicate that their diversity is ancient and geographically widespread, well pre-dating the LD epidemic of the past ~40 years, as well as the Last Glacial Maximum ~20,000 years ago (Walter et al. 2017; Carpi et al. 2016). This indicates that the recent emergence of human LD and babesiosis likely reflects ecological change—climate change and land use changes over twwhe past century—affecting tick-host-pathogen interactions rather than evolutionary change of the bacterium (Walter et al. 2017; Wood & Lafferty 2013, Barbour & Fish 1993).

In this section, we describe key tick life history traits that influence transmission of tick-borne pathogens (8.2.1); we investigate how the minimum requirement of two hosts results in nonlinear thresholds for emergence (8.2.2) and examine how alternative pathways in different ecological settings can facilitate the emergence of pathogens of short infectious period. We discuss advances in R_0 model development that incorporate these traits and discuss future directions in spatially-explicit R_0 modeling (8.2.3).

8.2.1. *Ixodes scapularis* Life Cycle: Key Life History Traits

A distinguishing trait of hard-bodied (Ixodid) ticks is the two-year (or longer) life cycle, which generally exceeds the longevity of the small mammals they infest (less than a year). Ticks may actually be considered as the reservoirs of infection, with small mammals passaging the infection to subsequent cohorts of ticks. For simplicity, however, we will adhere to the convention of calling ticks 'vectors'. *Ixodes scapularis* feeds one time during each of three life stages (larva, nymph, adult), at intervals of weeks, months or even years. One-host feeding lasts from three to five days depending on the life stage. The remaining time is spent in the environment undergoing development and molting to the next stage. The immature stages (larvae and nymphs) feed on a wide range of mammalian and bird hosts; adult *I. scapularis* depend almost exclusively on white-tailed deer (*Odocoileus virginianus*) or other large hosts to mate and for females to obtain a final blood meal. *I. scapularis* only moves a few meters during each life stage so host movement and host habitat use determine its dispersal patterns. Vertebrate hosts also vary in the quality of the bloodmeal for the ticks as well as their competence as reservoirs for *B. burgdorferi*: white-footed mouse (*Peromyscus leucopus*) is the most competent reservoir host in the Northeast United States, while other hosts exhibit various levels of host competence, including eastern chipmunks (*Tamias striatus*), Northern short tailed shrews (*Blarina brevicauda*), northern raccoons (*Procyon lotor*), Virginia opossums (*Didelphis virginiana*), eastern grey squirrels (*Sciurus carolinensis*), and some ground-foraging birds, among others (Linske et al. 2017). White-tailed deer cannot sustain *B. burgdorferi* infections, i.e. they are virtually incompetent (Telford et al. 1988). Here we explore how these tick-host-pathogen interactions influence the basic reproduction number, R_0, for tick-borne pathogens, particularly how R_0 varies with the abundance of competent and non-competent hosts in a simplified system (8.2.2), as well as with the timing of immature tick feeding and different transmission pathways (8.2.3).

8.2.2. Joint Thresholds for Emergence: Tick and Pathogen

A key feature of Ixodid tick-borne pathogens is the requirement of (at least) two types of hosts: one for the tick and one for the pathogen, which are typically not the same species. Here we review the consequences of a two-host system for pathogen emergence. Using the louping ill virus as a model, Norman et al. (1999) considered a system in which one species is host for the tick only ('non-viremic' host), while the other species is host for both the tick and the pathogen ('viremic host'). Using simple deterministic models, they derived joint threshold abundance curves to illustrate the basic reproductive number, R_0, of both the ticks and the virus. They found a nonlinear relationship between the non-viremic host density and the pathogen R_0. Increasing

non-viremic host abundance first amplifies pathogen transmission by increasing tick abundance to a critical density-dependent saturation point after which these hosts predominantly divert larval tick bites away from the amplifying hosts (Dobson & Foufopoulos 2001; Hudson et al. 1995) (Fig. 8.1). Norman et al. (1999) and Rosa & Pugliese (2007) referred to this reduction in infection at high non-viremic host abundances as a 'dilution effect'. Limited studies empirically evaluated the role of non-viremic hosts—usually deer, in diverting immature tick bites from pathogen-amplifying hosts. In an experimental enclosure study, Perkins et al. (2006) observed greater tick feeding intensity on rodents and TBEV rodent seropositivity within deer exclosures than outside of them, presumably due to increased larval feeding on highly competent rodents where deer were removed. In the LD system in the United States, Williams et al. (2017)

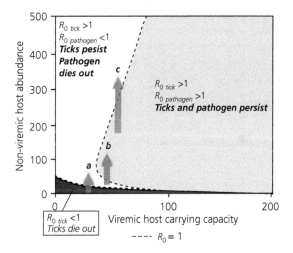

Figure 8.1 Pathogen and tick joint threshold abundance curves (R_0 pathogen=1 and R_0 ticks=1, respectively) in non-viremic/viremic host space, derived from a simple deterministic model using the louping ill virus as a model system. These curves define areas where ticks and pathogen persist (grey area), where pathogen dies out but ticks persist (white area), and where the density of the non-viremic host cannot sustain the tick population (dark gray area). The three arrows indicate how increases in non-viremic host abundance (e.g. deer) would first enable tick persistence (from $R_{0\,tick}<1$ to $R_{0\,tick}>1$) (a), then pathogen persistence (from $R_{0\,pathogen}<1$ to $R_{0\,pathogen}>1$) (b) but then 'dilute' infection and drive the pathogen to extinction (from $R_{0\,pathogen}>1$ to $R_{0\,pathogen}<1$) (c).
Adapted from Norman et al. (1999).

found a significant increase in nymphal infection prevalence when deer were removed, suggesting larvae shifted to feeding on highly competent small mammal hosts given deer absence. Ogden & Tsao (2009) concluded that either amplification or dilution may occur with increasing deer abundance, with the outcome depending on the precise mechanisms of competition, host contact rates with ticks, and acquired host resistance to ticks. This nonlinear 'dilution-by-deer' effect is particularly problematic in the context of deer reduction efforts, since control efforts would first need to cross the threshold point before a reduction in R_0 can be achieved.

8.2.3 Tick Phenology and Transmission Pathways

8.2.3.1 The 'Canonical' Seasonal Feeding Inversion in the Northeast United States

The typical two-year life cycle of *I. scapularis* in the Northeast United States results in a seasonal inversion of nymphal and larval activity (Yuval & Spielman 1990). During their first year, larval ticks take one blood meal during the late summer months, molt to nymphs, diapause over the winter and take another bloodmeal during their second year in late spring and early summer (Yuval & Spielman 1990, Piesman & Spielman 1979, Wilson & Spielman 1985). Because the seasonal peak of nymphal activity (May-June) precedes the majority of larval activity (August) in the Northeast United States, infected nymphs can transmit the pathogen to a large proportion of small rodent hosts before the larvae acquire the pathogen by feeding on the same hosts in late summer. The efficiency of this horizontal transmission pathway is dependent, however, on the survival of the hosts and the persistence of a transmissible infection between the times of nymphal and larval feeding (Gatewood et al. 2009, Hamer et al. 2012). While many *B. burgdorferi* strains are considered persistent in white-footed mice, other strains, as well as most other *I. scapularis*-borne pathogens, have short infectious periods (Hanincova et al. 2008, Dunn et al. 2014, States et al. 2017; Ogden et al. 2007), posing the question of how these pathogens/strains persist in nature.

8.2.3.2. Alternative Transmission Pathways

Horizontal transmission is less efficient for pathogens with short infectious periods than for pathogens with longer or chronic infections such as *B. burgdorferi*. For this reason, these pathogens are often partially maintained through alternative transmission pathways. One such pathway is cofeeding or nonsystemic transmission, which is a form of horizontal transmission between infected and uninfected ticks feeding in close proximity on any hosts, including reservoir incompetent and immune hosts. Cofeeding is key for the maintenance of tick-borne encephalitis virus (TBEV)—a flavivirus causing tick-borne encephalitis in Europe (Labuda et al. 1997). Modeling studies postulate it plays an important role in the maintenance of Powassan virus in the United States—which is genetically and antigenically linked to TBEV (Nonaka et al. 2010), although this has not been experimentally demonstrated. Cofeeding has also been demonstrated for *B. burgdorferi* in Europe (Gern & Rais 1996) and the United States (States et al. 2017). Finally, cofeeding transmission has been demonstrated in *Ehrlichia muris eauclairensis*, a recently discovered pathogen with a short infectious period (Karpathy et al. 2016), which induces high mortality rates in *P. leucopus* (Lynn et al. 2017). Another transmission pathway is vertical transmission from female adult ticks to eggs, which has been demonstrated for TBEV, POWV, and for *Borrelia miyamotoi* (Rollend et al. 2013; Costero & Grayson 1996; Rehacek 1962). In addition, vertical transmission from female rodents to offspring has been demonstrated for *B. microti* in three *Microtus* spp in Poland (Tolkacz et al. 2017) and in *P. leucopus* and *Microtus pennsylvanicus* in the northeastern United States (Tufts & Diuk-Wasser 2018). The relative importance of these pathways in nature remains to be determined, though modeling studies indicate that multiple pathways are likely to play a role in maintaining these pathogens (Nonaka et al. 2010).

8.2.3.3 Alternative Phenologies

For pathogens/strains with short infectious periods that can be transmitted either horizontally or through cofeeding, larvae and nymphs need to host-seek synchronously. Immature feeding synchrony has

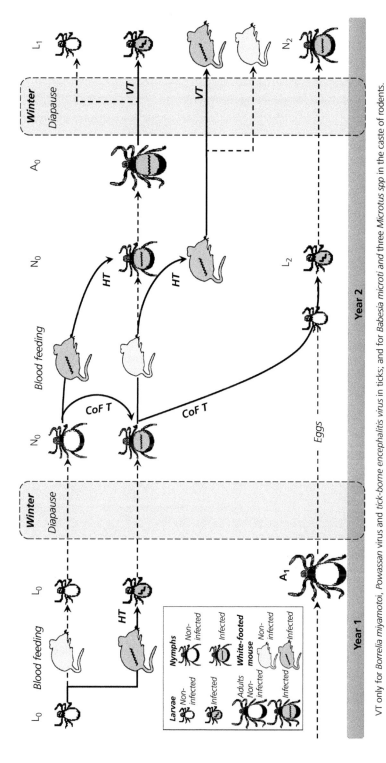

Figure 8.2 Transmission pathways for pathogens transmitted by ticks during the typical a two-year life cycle. Ticks can acquire infection by feeding on an infected host (horizontal transmission, HT), while cofeeding on the same host (CoF T) or vertically (VT). Tick cohorts are indicated by the stage and number (e.g., L_0 are larvae of cohort '0'). The dashed arrows indicate stage transitions by development or reproduction.

VT only for *Borrelia miyamotoi*, *Powassan virus* and *tick-borne encephalitis virus* in ticks; and for *Babesia microti* and three *Microtus spp* in the caste of rodents.

been found to be an essential determinant of TBEV distribution in Europe (Randolph et al. 2000; Randolph et al. 1999; Randolph & Rogers 2000). In the United States, more synchronous feeding in the Midwest than the Northeast may explain why short-lived *B. burgdorferi* strains are relatively more common in the former (Gatewood et al. 2009). Furthermore, the fact that recently discovered pathogens with short infectious periods, *Borrelia mayonii* and *Ehrlichia muris eauclairensis*, have so far been found only in the Midwest, is consistent with a key role of feeding synchrony in transmission. Climate likely plays a role in tick phenology: increased rates of fall cooling (Randolph et al. 2000) and the amplitude of the annual temperature cycle (Gatewood et al. 2009) have been found to be associated with increased feeding synchrony. The proposed mechanism is that a faster rate of fall cooling in the Midwest may cause unfed larvae to pass the winter in quiescence or behavioral diapause, from which they emerge and host-seek synchronously with nymphs in the spring (Randolph & Rogers 2000). Tick phenology, in particular immature feeding synchrony, thus interacts with multiple transmission pathways to facilitate persistence of strains or pathogens of short infectious periods.

8.2.3.4 Integration through R_0 Models

Models for the basic reproduction number were first developed from quantities derived from stability analyses of systems of differential equations (e.g. Rosa et al. 2003; Norman et al. 1999; Rosa & Pugliese 2007). A drawback of R_0 expressions arising from systems of differential equation models has been that they assume a tick biting rate, as in models for mosquito-borne diseases, whereas ticks bite only once per life stage and so an increase or decrease in biting rate would affect the duration of each life stage but will not necessarily affect R_0 (Hartemink et al. 2008). Stage-structured R_0 models have since been developed that incorporate both the complex life cycle of ticks and the multiple transmission routes that allow tick-borne pathogens to spread and persist (Hartemink et al. 2008; Matser et al. 2009; Davis & Bent 2011; Dunn et al. 2013). The key technique in such models is to classify infectious hosts/ticks into 'host types' defined by the stage at which they were infected (note the term 'host type' here is used to refer to both ticks and reservoir hosts). For example, ticks infected from birth (via vertical transmission) are considered a different host type to ticks infected while feeding as larvae.

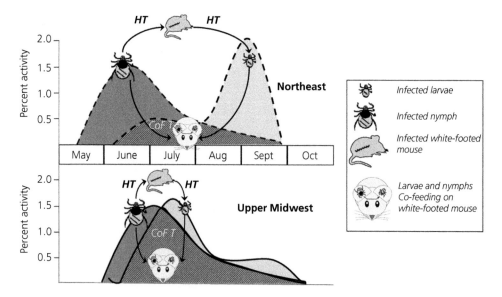

Figure 8.3 Influence of tick feeding phenology on the dominant transmission pathway between immature stages of Ixodes scapularis (larvae and nymphs) in the Northeast and Upper Midwest United States. While horizontal transmission (HT) can occur in both phenologies depending on the duration of pathogen infection, cofeeding transmission (coF T) is highly dependent on the overlap between larval and nymphal feeding, so it is more likely in the Upper Midwest.

The stage-structured approach to R_0 has been validated by Harrison et al. (2011) and extended further by Nonaka et al. (2010); Davis & Bent (2011); and Dunn et al. (2013) to explicitly include the phenologies of immature ticks and their interaction with the duration of the infectious period (see Fig. 8.3). Such models for R_0 have generated insight into how shifts in the seasonality of tick feeding might affect the persistence of tick-borne pathogens. For example, Davis & Bent (2011) found that the greater overlap of larva and nymph feeding activity in the Upper Midwest of the United States substantially increased R_0, especially for those pathogens with shorter infectious periods, such as *B. microti*, *A. phagocytophilum* and POWV. Nonaka et al. (2010) demonstrated that the consistent prevalence of POWV observed in tick populations could be maintained by a combination of low vertical, intermediate cofeeding and high systemic transmission rates. They also extended the model to patho-

gens that cause chronic infections in hosts and found that cofeeding transmission could contribute to elevating prevalence even in these systems, which is consistent with empirically-based models by States et al. (2017).

8.2.3.5 Insights from R_0 to Understand the Ecological Determinants and Spatial Distribution of Tick-Borne Pathogens

A global sensitivity analysis of R_0 for tick-borne pathogens of *I. scapularis* was performed by Dunn *et al.* (2013) to rank the importance of the parameters in terms of their contribution to the observed variation in R_0. This analysis concluded that the establishment of *B. burgdorferi* is most sensitive to the transmission efficiency from the vertebrate host to *I. scapularis* larvae, larval survivorship from fed larva to feeding nymphs and the probability of finding a competent host. Notable in these models is the weak dependence of R_0 on tick abundance, which

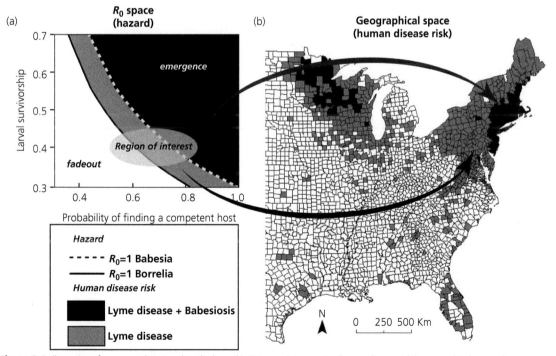

Figure 8.4 Illustration of an approach to translate the hazard in R0 space into mapping human disease risk in geographical space. a) Hypothetical threshold curves ($R_0=1$) for *Babesia microti* and *Borrelia burgdorferi* plotted as a function of two influential variables: larval survivorship and the probability of finding a competent host. b) If information on these ecological variables were available in geographic space, the hazard could be mapped to predict the geographic distribution of human cases of Lyme disease and babesiosis per county (as reported here from 2011 to 2013).
Source: Adapted from Diuk-Wasser et al. 2016.

may be exacerbated by the co-aggregation of ticks on a subset of vertebrate hosts (Harrison & Bennett 2012; Johnstone-Robertson et al. 2019) and may explain why some tick-borne pathogens are found everywhere ticks are.

A model for R_0 provides a method to make site-specific or region-specific predictions about whether the pathogen is likely to emerge. Such a model can also be used to explicitly identify threshold targets to curb emergence. As our understanding of what the key drivers of R_0 are in tick-borne pathogens, a next step would be to develop spatially-explicit maps of R_0 to validate the models and inform the public about emergence 'hot spots'. As an illustration of the approach, Fig. 8.4 shows hypothetical threshold curves for *B. burgdorferi* and *B. microti* for a range of values of the two most biologically relevant parameters. If these parameters could be assessed at an appropriate spatial scale, they could be used to populate a habitat suitability map for these pathogens that would inform predictions of human risk.

8.3 Host Community Impacts on Tick-Borne Pathogens in Anthropogenic Landscapes: Links to Human Disease Risk

Tick-borne diseases—like most zoonotic diseases, involve many host species, each of which can play multiple roles in maintaining tick populations and contributing to pathogen transmission. Hosts vary in the quality of the blood meal they provide to ticks, their reservoir competence for different pathogens and strains, and their capacity as tick and pathogen dispersers. In Section 8.2.2, we described the consequences of the required minimum of two host species for the emergence and persistence of most Ixodid tick-borne pathogens; host-associated model terms combined all hosts that fed and/or infected ticks. Here we shift the focus away from R_0 and the process of emergence to consider the implications of the diverse community of hosts for human risk of infection in endemic settings such as *B. burgdorferi* in the Northeast United States.

The process of anthropogenic land use change leading to forest fragmentation can affect the enzootic transmission cycle of *B. burgdorferi* and the risk of human disease by directly or indirectly affecting the abundance, demography, behavior, movement, immune response, and contact between host species and vectors (Gottdenker et al. 2014). However, a disproportionate share of the research linking land use change with tick-borne disease risk has concerned the role of host community diversity on disease risk. Spurred by an interest in the role of biodiversity as a buffer against infectious diseases, the dilution effect defined by Norman et al. (1999) with an emphasis on deer was expanded to include dilution by other hosts with varying levels of pathogen competence. As more diverse communities are assumed to have a higher proportion of low competence (dilution) hosts, the dilution effect was re-cast as a mechanism by which biodiversity can serve as an ecosystem service to human health (Ostfeld & Keesing 2000). Accordingly, the new meaning of the dilution effect has been called the 'biodiversity-buffers-disease' hypothesis by Randolph & Dobson (2012) and Randolph (2013). As applied to the LD system in the United States, the dilution effect hypothesis proposes that habitat fragmentation results in a shift in species assemblages to one dominated by white-footed mice and other competent small rodents because of competitor and predator release in small forest patches (Nupp & Swihart 1996). Because white footed mice can feed and infect a larger number of immature ticks compared to other vertebrate hosts, the density of infected nymphs, as well as human risk of acquiring LD are predicted to increase as forests become fragmented, compared to contiguous parcels of undisturbed land presumably occupied by more ecologically diverse host communities (LoGiudice et al. 2003). Debate ensued after this formulation of the dilution effect hypothesis, both in support (Ostfeld & Keesing 2013; Civitello et al. 2015) and questioning its theoretical basis, empirical evidence, or generality (Randolph & Dobson 2012, Randolph, 2013; Wood & Lafferty, 2013). Additionally, two recent syntheses propose that the association between biodiversity and LD may vary depending on the spatial scale of analysis (Wood & Lafferty 2013) and along a fragmentation gradient (Kilpatrick et al. 2017). These studies postulated a positive link between biodiversity and LD at broad spatial scales, as new areas became forested or when urban areas transition to suburban/exurban land cover types and equivocal evidence for a negative link between biodiversity and LD risk at varying levels of biodiversity within forests.

A limitation of empirical and theoretical studies assessing the effect of fragmentation on host diversity and the LD enzootic transmission cycle is the focus on forest patch size and isolation with disregard to the composition and functional role of the matrix habitat (Allan et al. 2003, Logiudice et al. 2008). Moreover, the process of fragmentation is only one of the components of anthropogenic land use change, which can be defined as 'all components of change in the quantity and quality of land cover types as habitat for organisms and productive land for humans' (Didham 2010). In the

Northeast United States, urbanization is the main driver of changes in land cover types, with anthropogenic landscapes spanning high, medium and low intensity developed areas (i.e., urbanization gradient) as defined by the United States Geological Survey Multi Resolution Land Characteristics Consortium classification (Homer et al. 2002). In this section, we discuss the shortcomings of these simpler views of the fragmentation process (8.3.1) and synthesize the multiple processes involved as the system transitions from forest to low intensity developed areas (8.3.2), and from low to medium and high intensity developed areas (8.3.3) (Fig. 8.5). In Section 8.3.4, we explore the links between the enzootic hazard and LD incidence.

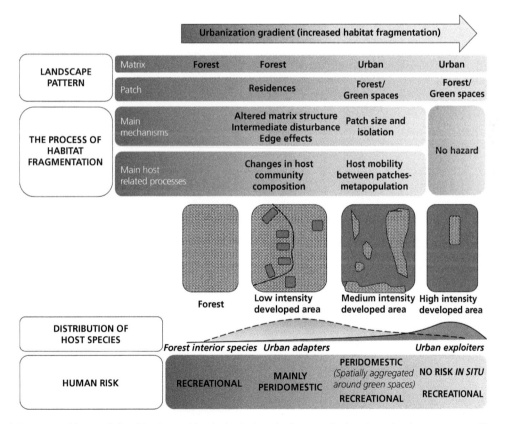

Figure 8.5 Conceptual framework describing the transitions in the dominant landscape mechanisms, host related processes, types of hosts and types of human exposure across a forest to urban matrix transition.

Box 8.1 Synthetic framework: Coupled natural-human systems

Zoonotic diseases occur at the interface of human and ecosystem health, which are both interrelated and affected by human activities (Ellis & Wilcox 2009; Wilcox & Jessop 2010). However, the biophysical, ecological, human behavioral and social components of disease transmission are often studied separately, either as human systems with input/output to natural systems or as natural systems subject to human disturbances (An 2012). Therefore, the complexity of the ecological, biophysical and human interaction fails to be captured. In this context, an integrative framework of Coupled Natural and Human systems (CNH) has emerged (Liu et al. 2016), acknowledging that natural and human system dynamics involve reciprocal interactions and complex feedback loops across multiple scales, featuring nonlinearity, thresholds and time lags (Liu et al. 2007; An 2012), giving rise to emerging properties that cannot be explained in a reductionist paradigm (Arthur 1993; Gell-Mann 1995). Here, we apply the CNH framework to the transmission of Lyme disease in the Northeast United States (Fig. 8.6). We propose that the enzootic transmission cycle of *B. burgdorferi* (i.e. the natural system) and the determinants of human exposure to *I. scapularis* (i.e. the human system), are linked by changes in land use, which will directly and indirectly affect transmission dynamics and, in turn, influence human-tick interactions and the risk of humans to acquire LD at different temporal and spatial scales.

Defining the Outcomes

Diverse outcomes have been used to evaluate the different components of the LD system, mainly the density of infected nymphs (DIN) and Lyme disease incidence in humans (LDI), sometimes assuming a direct relationship between these outcomes and failing to account for the complexity of their interaction and confounding effects. Nymphal infection prevalence (NIP) is sometimes used as an outcome, although it is limited by not accounting for tick density. To help clarify potential sources of controversy, we follow the suggestion of Hosseini et al. (2017), to separate the concepts of hazard and risk caused by exposure to a hazard.

Hazard: potential sources of harm from microbes, such as viruses, bacteria and other pathogens (e.g. pathogen abundance). In the case of the LD system, DIN constitutes a measure of hazard and is a direct result of the processes intervening in the natural system.

Risk: the likelihood of an adverse effect (e.g. disease) caused by exposure to a hazard, including the processes that lead to exposure and the vulnerability of individuals. In the

CNH framework we consider the LD risk as the human risk of acquiring the disease given human-tick encounters (i.e. exposure) and their vulnerability.

Exposure: the likelihood of contact, including vector-borne transmission, between humans and the hazards. In the LD system, exposure is determined by the multiple processes underlying human-tick interactions.

Vulnerability: the possibility given exposure that the hazard can cause harm. This term could be expanded to consider the social vulnerability as the 'defenselessness, insecurity, and exposure to risks, shocks and stress' experienced by the individuals and their ability to cope with them and overcome the negative outcomes (Ratnapradipa et al. 2017; Chambers 1989).

Clear delineation of the multiple components of risk can facilitate comparisons across studies and systems and advance the theoretical and applied development of the field.

Components of the coupled natural and human system of tick-borne diseases in the Northeast

Similarly, we need to define the components of the coupled system:

Processes of the natural system: the dynamics of the enzootic cycle are driven by host and vector population dynamics, host-vector encounter rates and other biophysical factors associated with landscape composition and configuration. Changes in host community composition will have a direct effect on the natural system, although these are inherently confounded with changes in land use (Section 8.2).

Processes of the human system: Human exposure to ticks will mainly depend on processes that pertain to the human system: for example, certain outdoor activities and mobility patterns, the use of personal protection measures and environmental interventions at the household and neighborhood levels. Prior exposures to ticks and LD can also influence human exposure to ticks by changing individual risk perception and adaptive behaviors (Section 3).

Coupling processes: Anthropogenically-driven habitat fragmentation—as a result of changes in land use, will have direct and indirect effects on the enzootic hazard—the natural system (Section 8.3). Changes in hazard, in turn, will affect risk perception and alter human interactions with the environment, leading to changes in human-tick interactions. Thus, habitat fragmentation will:

1. Directly affect enzootic transmission dynamics within and between forest patches by acting on host movement and tick dispersal. The degree of fragmentation and landscape configuration will influence the host metacommunity dynamics and tick persistence in endemic areas.

2. Indirectly affect the assemblage of the host community in forest patches, which will in turn affect host-tick encounter frequency, tick abundance, and DIN.

3. At a regional scale, it may also affect human exposure to ticks by determining settling decisions in the wildland-urban interface (Larsen et al. 2014).

In addition, at a smaller spatial scale, humans have direct effects on the enzootic cycle by carrying out interventions at the household level (e.g. landscape design or spraying) and neighborhood levels (e.g., deer culling), which may affect habitat suitability for tick and small mammals hosts and thus influence the density of ticks at the edge of residences.

Ultimately, human-tick interactions and LDI emerge as a new property of this complex system, as a result of the dynamics of the enzootic cycle of LD (natural system) shaped by the landscape and land use configuration, and the human responses to the hazard (human system). Thus, understanding the mechanisms coupling the natural and human systems across multiple scales can facilitate targeting control efforts and overcome potential barriers and unintended consequences.

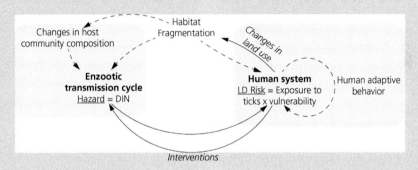

Figure 8.6 Coupled natural-human system proposed for Lyme disease. The natural system are factors and processes involved in the enzootic transmission cycle, and the human system are factors and processes determining human exposure to ticks and the pathogen.

8.3.1 Fragmentation as a Landscape-level Rather than a Patch-level Phenomenon

A limitation of studies evaluating the effects of host community composition on the enzootic hazard and LD risk, is the underlying assumption that habitat fragmentation will have a direct negative effect on host diversity via habitat loss, ignoring other plausible effects of fragmentation on the host community (Box 8.1). Habitat fragmentation is an umbrella term that refers to the process by which habitat loss results in the division of large, continuous habitats into a greater number of smaller patches with varying degrees of isolation from each other by a matrix of dissimilar habitats (Didham 2010). Habitat fragmentation integrates the temporal dynamics as well as the spatial pattern of remaining habitat by considering it as a landscape-level phenomenon, i.e. a patch-level process (patch area, edge effects and patch shape complexity) that can only be understood within a landscape context (isolation and matrix structure) (Fahrig 2003). The mechanistic basis for the effect of habitat fragmentation on host population and community dynamics, includes not only decreased patch area and increased patch isolation, but also increased edge effects, altered patch shape and altered matrix structure (Didham 2010). The relative importance of each of these mechanisms on the enzootic hazard will vary depending on the context, for instance along an urbanization gradient (McClure & Diuk-Wasser 2018). It is thus not surprising that fragmentation studies focusing on individual forest patches of different sizes have yielded conflicting results: regions where patches are large often correspond to overall larger habitat amount (Fahrig 2003) and these patches are potentially more connected to potential source patches in a metapopulation (Estrada-Peña 2002). In the following sections, we discuss the mechanisms influencing changes in hazard across land use transitions along an urbanization gradient, including the most likely host species responses, as determined by their human-adaptive traits (Fig. 8.5).

8.3.2 Transition from Forest to Low Intensity Residential

8.3.2.1 Patch Metrics: An Incomplete Measure of Fragmentation for Tick-Borne Diseases

An assumption of the dilution effect hypothesis is that habitat fragmentation indirectly affects the hazard by creating smaller forest fragments that host a greater proportion of highly competent white-footed mice (Nupp & Swihart 1996; Berl et al. 2018; Anderson et al. 2003). However, smaller fragments are expected to undergo stochastic extinction or near-extinction events for *B. burgdorferi* or *I. scapularis*, potentially leading to a non-linear relationship between patch size and tick density or infection. This nonlinearity may explain why Allan et al. (2003) found increased densities of *I. scapularis* nymphs in patches < 2ha, but LoGiudice et al. (2008) did not find enough ticks to estimate infection in the smallest patches and did not find a significant linear relationship between nymphal infection prevalence (NIP) and species richness or Shannon diversity index. LoGiudice et al. (2008) postulated that these inconsistent observations concerning the relationship between fragment size and tick infection prevalence suggest an intriguing paradox. Small fragments that support high densities of reservoir hosts sometimes support high NIP, but these same fragments would be most likely to undergo stochastic extinction or near-extinction events for *B. burgdorferi* or *I. scapularis*, leading to extreme temporal variability in LD risk. Differences in patch connectivity across the matrix may account for the differences in patch occupancy found in Allan et al. (2003) vs. LoGiudice et al. (2008). Characterization of the matrix is thus key to identifying the mechanisms dominating the process of habitat fragmentation in each study, depending on where they are situated in an urbanization gradient (Fig. 8.5).

8.3.2.2 The Intermediate Disturbance Hypothesis: Increased Biodiversity in Fragmented Landscapes

Habitat fragmentation can have either negative or positive effects on host population abundance. A positive effect on host diversity has been called the 'intermediate disturbance hypothesis', which postulates that initial human impacts of suburban sprawl are sometimes relatively mild and promote environmental heterogeneity because different habitats occur alongside each other (McKinney 2008). For many species, human-modified land-use types provide supplementary or complementary resources that may compensate for limited resource availability in habitat patches (Ries et al. 2004). Strictly speaking, then, there is no dividing line between what is a 'patch' and what is a 'matrix' in these cases, and some species may well perceive the whole landscape as 'habitat' (Haila 2002). Consistently with the intermediate disturbance hypothesis, Linske et al. (2017) compared clusters of residences with typical peridomestic heterogeneity in Connecticut, United States, with continuous forest patches and found higher mammalian diversity around the residences. Consistent with one postulate of the the dilution effect hypothesis, host infection with *B. burgdorferi* in these more diverse residential areas was lower. However, this study contradicts the proposed inverse relationship between habitat fragmentation and biodiversity often associated with the dilution effect hypothesis and questions the use of the former as a proxy for the latter.

8.3.2.3 Edge Effects

Human-tick interactions occur mostly on edges between residential and forested areas. Edges are three-dimensional zones of transition between habitats (Cadenasso et al. 2003), and have no absolute quantifiable dimensions unless comparisons are made relative to both the adjacent patch and matrix interiors (Fig. 8.5). Most studies on *Ixodes* spp risk have compared differently sized forest fragments—ignoring the intervening matrix (a 'one-sided' approach sensu Fonseca & Joner 2007), rather than sampling ticks across the whole gradient from the interior of the forest to the matrix habitat (a 'two-sided' approach). This tendency is particularly surprising as most drivers of edge effects originate externally to the forest patch and humans are often exposed in open areas adjacent to forests. Moreover, because of properties such as shading, wind turbulence and spillover effects, the observed edge response for some variables may actually occur well outside the structural vegetation edge of the patch, and therefore be missed entirely if sampling only occurs within the patch. A study examining the role of vegetation on the abundance of *P. leucopus* in relation to forest-old field edges found that capture probability was enhanced by increased vegetation structural complexity and by proximity to the forest edge (Manson et al. 1999). In contrast, tick abundance was found to be greater in interior forest habitats than in edge habitats (Finch et al. 2014; Horobik et al. 2006). In a modeling study, Li et al. (2012) postulated that the presence of grasslands adjacent to woodlands could act as sinks for ticks, since the hosts readily use this land cover type, but ticks are more likely to desiccate there. An integrated understanding of landscape dynamics of tick-borne pathogens should include assessment of patch, matrix and edge dynamics, as well as patch connectivity.

8.3.3 Transition from Low to Medium/High Intensity Residential

8.3.3.1 The Trade-Off of Ecosystem Services Provided by Green Spaces in an Urban Matrix

In highly urbanized areas, parks and other green spaces provide suitable habitat for ticks infected with various tick-borne pathogens. These areas may facilitate the invasion of ticks and pathogens that can then act as sources of tick-borne pathogens for visitors or neighborhoods highly connected to urban parks. Although the positive effect of green spaces on human well-being (ecosystem service) is well established, this benefit needs to be weighed against potential negative effects (ecosystem disservices), which are often overlooked (Escobedo et al. 2011). With high human densities in cities, these emerging tick-borne infections can cause a significant public health burden.

8.3.3.2 Metapopulation/Metacommunity Dynamics: Patch Connectivity is Key

The importance of patch connectivity for tick-borne pathogens in urban areas was highlighted in two recent publications (VanAcker et al. 2019; Heylen et al. 2019). These studies found that patch connectivity, in addition to other patch characteristics, was critical for the persistence of ticks and pathogens in a metapopulation. Matrix properties can affect the dispersal and movement of individuals between patches (Gascon et al. 1999), and the degree of structural contrast between patch and matrix determines the permeability of habitat edges to propagule movement (Collinge & Palmer 2002), which taken together can be the prime determinants of colonization-extinction dynamics (Kupfer et al. 2006) and species loss. Future studies of *I. scapularis*-borne pathogens in fragmented landscapes should thus consider metapopulation/metacommunity dynamics. Patch connectivity is expected to increase in importance as the landscape becomes more fragmented and patch isolation increases along an urbanization gradient (Fig 8.5).

8.3.4 Human-tick Interactions: The Conversion of Hazard into Risk

Humans are incidental hosts for *B. burgdorferi*, and human exposure to ticks is mainly peridomestic, although recreational exposure also occurs (Finch et al. 2014; Connally et al. 2009). Peridomestic exposure to *I. scapularis* ticks occurs at the edge of residences, where they intersect with adjacent forested areas. Therefore, human-tick interaction is expected to occur more frequently as habitat fragmentation increases because of edge effects (Larsen et al. 2014). However, the relationship is non-linear: along an urbanization gradient, human exposure to ticks may initially increase in suburban neighborhoods (i.e., low intensity developed areas) but eventually decline at very high fragmentation levels (i.e., high intensity developed areas), as humans are less likely to come into contact with forests and forest edges (Wood & Lafferty 2013). Consistently, three studies found a nonlinear effect of fragmentation on Lyme disease incidence (LDI). In the region surrounding Lyme, Connecticut. Cromley et al. (1998) found that medium-intensity development, as compared to low-intensity development, was protective against LD. An analysis of twelve Maryland counties found that the percentage of forest cover had a quadratic relationship to LDI, with half-forested landscapes with a large percentage of forest-herbaceous edge statistically associated with the highest LDI (Jackson et al. 2006). Similarly, McClure & Diuk-Wasser (2018) found a unimodal relationship between percentage of forest cover and mean patch area and LDI, which was strongly modified by landscape connectivity. Moreover, Brownstein et al. (2005), who replicated the inverse relationship between mean forest patch size and DIN in suburban residential areas, found the opposite relationship between LDI and mean forest patch size and isolation; LDI increased with patch size and decreased with patch isolation. These studies indicate that, even if fragmentation results in smaller patches with increased DIN, human exposure and disease incidence may decline as fewer residences contain or adjoin forested areas and fewer people interact with greenspaces.

The term 'Lyme disease risk' is often used to refer solely to DIN (Frank et al. 1998)—a human's risk of encountering a LD-infected nymph, under the assumption that tick density is directly related to the likelihood of tick encounter. However even direct measurements of DIN do not always yield linear associations to human LDI in the region, implying significant variations in tick exposure mediated by human behavior (Connally et al. 2006). Although LDI has been found to increase with DIN at town and county levels (Kitron & Kazmierczak 1997; Stafford et al. 1998; Falco et al. 1999; Diuk-Wasser et al. 2012; Pepin et al. 2012), a local scale empirical study found a positive relationship between landscape configuration and LDI in humans only after adjusting for some control measures employed by residents (Finch et al. 2014). Tick exposure may be reduced by modifying the household peridomestic environment or adopting personal protection measures.

To understand how the enzootic hazard translates to LD risk, we need to further understand how human behavior acts to convert hazard into risk, which depends on risk perception in different ecological and social contexts. These behaviors may result in positive or negative feedbacks between the enzootic cycle and human risk at different scales (see coupled nature and human system in Box 8.1), and may partially account for the different stages of LD epidemic spread and endemic equilibrium (Burtis et al. 2016). At a regional-scale, a study found a behavioral response to LD risk via settlement decisions at the wildland-urban interface (Larsen et al. 2014) that could enhance changes in land use and affect the enzootic hazard. At a local scale, interventions to reduce the enzootic hazard (via reducing the density of ticks or deer targeted interventions for example) may have a direct effect on the enzootic hazard which will, in turn, indirectly affect LD incidence in humans by changing their behavior regarding LD prevention. The use of preventive behaviors such as wearing protective clothing and using tick repellent has been found to increase at high levels of awareness of tick-borne diseases as a health threat (Aenishaenslin et al. 2017; Valente et al. 2015).

8.4 Towards a Unifying Approach: Defining the Boundaries of the System

The inherent complexity of the LD system, which results from the multiplicity of interconnected relationships and levels requires an integrative approach that not only recognizes the complex LD ecology involving multiple hosts, transmission pathways and tick phenologies, but also the interwoven nature of natural and human components (Box). The apparently contradictory findings of different studies evaluating the fragmentation effects on the enzootic hazard and human disease risk may be partly due to differences in the spatial and temporal scales, the components of the system considered, and the measured outcome. Therefore, in order to reach a consensus among studies, their boundaries need to be explicitly defined.

To this end, Pickett et al. (2005), proposed a three-dimensional framework to facilitate the biocomplexity concept in coupled human–natural systems. Each study should be placed in a bounded dimensional space determined by the spatial, organizational and temporal dimensions and include descriptions of the (a) degree of system integration,

(b) resolution of system components, and (c) system boundedness (Pickett et al. 2005; Ellis & Wilcox 2009). The integration of studies focusing on different aspects of the LD system under a unifying framework will deepen our understanding of the eco-bio-social determinants of LD transmission (and other tick-borne diseases) by acknowledging the complexity of the LD system while simultaneously aiming to achieve the simplest models capable of explaining its dynamics depending on the context.

References

Aenishaenslin, C., Bouchard, C., Koffi, J.K., & Ogden, N.H. (2017). Exposure and preventive behaviours toward ticks and Lyme disease in Canada: Results from a first national survey. *Ticks and Tick-borne Diseases*, 8, 112–18.

Allan, B.F., Keesing, F., & Ostfeld, R.S. (2003). Effect of forest fragmentation on Lyme disease risk. *Conservation Biology*, 17, 267–72.

An, L. (2012). Modeling human decisions in coupled human and natural systems: Review of agent-based models. *Ecological Modelling*, 229, 25–36.

Anderson, C. S., Cady, A.B., & Meikle, D. B. (2003). Effects of vegetation structure and edge habitat on the density and distribution of white-footed mice (*Peromyscus leucopus*) in small and large forest patches. *Canadian Journal of Zoology*, 81, 897–904.

Arthur, W.B. (1993). Why do things become more complex? *Scientific American*, 268, 144–4.

Barbour, A.G. & Fish, D. (1993). The biological and social phenomenon of Lyme disease. *Science*, 260, 1610–6.

Berl, J.L., Kellner, K.F., Flaherty, E.A,. & Swihart, R.K. (2018). Spatial variation in diversity of white-footed mice along edges in fragmented habitat. *American Midland Naturalist*, 179, 38–50.

Brownstein, J.S., Skelly, D.K., Holford, T.R., & Fish, D. (2005). Forest fragmentation predicts local scale heterogeneity of Lyme disease risk. *Oecologia*, 146, 469–75.

Burtis, J.C., Sullivan, P., Levi, T., Oggenfuss, K., Fahey, T.J., Ostfeld, R.S. (2016). The impact of temperature and precipitation on blacklegged tick activity and Lyme disease incidence in endemic and emerging regions. *Parasites & Vectors*, 9, 606.

Cadenasso, M.L., Pickett, S.T.A., Weathers, K.C., & Jones, C.G. (2003). A Framework for a theory of ecological boundaries. *BioScience*, 53, 750–8.

Carpi, G., Walter, K. S., Mamoun, C. B., *et al.* (2016). *Babesia microti* from humans and ticks hold a genomic signature of strong population structure in the United States. *BMC Genomics*, 17, 888.

Schwartz, A.M., Hinckley, A.F., Mead, P.S., Hook, S.A., Kugeler, K.J. (2017). Surveillance for Lyme disease—United States, 2008–2015. Morbidity and Mortality Weekly Report *Surveillance Summaries*, 66:1–12.

Chambers, R. (1989). Vulnerability, coping and policy. *IDS Bulletin*, 37(4), 33–40.

Civitello, D.J., Cohen, J., Fatima, H., et al. (2015). Biodiversity inhibits parasites: Broad evidence for the dilution effect. *Proceedings of the National Academy of Sciences of the United States of America*, 112, 8667–71.

Collinge, S.K. & Palmer, T.M. (2002). The influences of patch shape and boundary contrast on insect response to fragmentation in California grasslands. *Landscape Ecology*, 17, 647–56.

Connally, N.P., Durante, A.J., Yousey-Hindes, K.M., Meek, J.I., Nelson, R.S., & Heimer, R. (2009). Peridomestic Lyme disease prevention: results of a population-based case-control study. *American Journal of Preventative Medicine*, 37, 201–6.

Connally, N.P., Ginsberg, H.S., & Mather, T.N. (2006). Assessing peridomestic entomological factors as predictors for Lyme disease. *Journal of Vector Ecology*, 31, 364–70.

Costero, A. & Grayson, M.A. (1996). Experimental transmission of Powassan virus (Flaviviridae) by *Ixodes scapularis* ticks (Acari: Ixodidae). *American Journal of Tropical Medicine and Hygiene*, 55, 536–46.

Cromley, E.K., Cartter, M.L., Mrozinski, R.D., & Ertel, S.H. (1998). Residential setting as a risk factor for Lyme disease in a hyperendemic region. *American Journal of Epidemiology*, 147, 472–7.

Davis, S. & Bent, S.J. (2011). Loop analysis for pathogens: niche partitioning in the transmission graph for pathogens of the North American tick Ixodes scapularis. *Journal of Theoretical Biology*, 269, 96–103.

Didham, R. K. (2010). Ecological consequences of habitat fragmentation. In: *Encyclopedia of Life Sciences (ELS)*. Chichester, UK: John Wiley & Sons, Ltd.

Diuk-Wasser, M.A., Hoen, A.G., Cislo, P., et al. (2012). Human risk of infection with *Borrelia burgdorferi*, the Lyme disease agent, in eastern United States. *The American Journal of Tropical Medicine and Hygiene*, 86, 320–7.

Diuk-Wasser, M.A., Vannier, E., & Krause, P.J. (2016). Coinfection by ixodes tick-borne pathogens: Ecological, epidemiological, and clinical consequences. *Trends in Parasitology*, 32, 30–42.

Dobson, A. & Foufopoulos, J. (2001). Emerging infectious pathogens of wildlife. *Philosophical Transactions of the Royal Society B: Biological Sciences*, 356, 1001–12.

Dunn, J.M., Davis, S., Stacey, A., & Diuk-Wasser, M.A. (2013). A simple model for the establishment of tick-borne pathogens of *Ixodes scapularis*: A global sensitivity analysis of R0. *Journal of Theoretical Biology*, 335, 213–21.

Dunn, J.M., Krause, P.J., Davis, D., et al. (2014). *Borrelia burgdorferi* promotes the establishment of *babesia microti* in the northeastern United States. *PloS One*, 9, e115494.

Ellis, B.R. & Wilcox, B.A. (2009). The ecological dimensions of vector-borne disease research and control. *Cadernos de Saude Publica/ministerio da saude, fundacao oswaldo cruz, escola nacional de saude publica*, 25 suppl 1, s155–67.

Escobedo, F.J., Kroeger, T., & Wagner, J.E. (2011). Urban forests and pollution mitigation: Analyzing ecosystem services and disservices. *Environmental Pollution*, 159, 2078–87.

Estrada-Peña, A. (2002). Understanding the relationships between landscape connectivity and abundance of *Ixodes ricinus* ticks. *Experimental and Applied Acarology*, 28, 239–48.

Fahrig, L. (2003). Effects of habitat fragmentation on biodiversity. *Annual Review of Ecology, Evolution and Systematics*, 34, 487–515.

Falco, R.C., Mckenna, D.F., Daniels, T.J., et al. (1999). Temporal relation between *Ixodes scapularis* abundance and risk for Lyme disease associated with *erythema migrans*. *American Journal of Epidemiology*, 149, 771–6.

Finch, C., Al-damluji, M.S., Krause, P. et al. (2014). Integrated assessment of behavioral and environmental risk factors for Lyme disease infection on Block Island, Rhode Island. *PloS One*, 9, e84758–8.

Fonseca, C. & Joner, F. (2007). Two-sided edge effect studies and the restoration of endangered ecosystems. *Restoration Ecology*, 15, 613–19.

Frank, D.H., Fish, D., & Moy, F.H. (1998). Landscape features associated with lyme disease risk in a suburban residential environment. *Landscape Ecology*, 13, 27–36.

Gascon, C., Lovejoy, T.E., Bierregaard, R.O., et al. (1999). Matrix habitat and species richness in tropical forest remnants. *Biological Conservation*, 91, 223–9.

Gatewood, A.G., Liebman, K.A., Vourc'h, G., et al. (2009). Climate and tick seasonality are predictors of *borrelia burgdorferi* genotype distribution. *Applied Environmental Microbiology*, 75, 2476–83.

Gell-Mann, M. (1995). What is complexity? *Complexity*, 1, 16–19.

Gern, L. & Rais, O. (1996). Efficient transmission of *borrelia burgdorferi* between cofeeding *Ixodes ricinus* ticks (acari: Ixodidae). *Journal of Medical Entomology*, 33, 189–92.

Gottdenker, N.L., Streicker, D.G., Faust, C.L., & Carroll, C. R. (2014). Anthropogenic land use change and infectious diseases: A review of the evidence. *Ecohealth*, 11, 619–32.

Haila, Y. (2002). A Conceptual genealogy of fragmentation research: From island biogeography to landscape ecology. *Ecological Applications*, 12, 321–34.

Hamer, S.A., Hickling, G.J., Sidge, J.L., Walker, E.D., & Tsao, J.I. (2012). Synchronous phenology of juvenile Ixodes scapularis, vertebrate host relationships, and associated patterns of *Borrelia burgdorferi* ribotypes in the midwestern United States. *Ticks and Tick-Borne Diseases*, 3, 65–74.

Hanincova, K., Ogden, N. H., Diuk-Wasser, M. et al. (2008). Fitness variation of *Borrelia burgdorferi* sensu stricto strains in mice. *Applied Environmental Microbiology*, 74, 153–7.

Harrison, A. & Bennett, N.C. (2012). The importance of the aggregation of ticks on small mammal hosts for the establishment and persistence of tick-borne pathogens: an investigation using the R0 model. *Parasitology*, 139, 1605–13.

Harrison, A., Montgomery, W.I., & Bown, K.J. (2011). Investigating the persistence of tick-borne pathogens via the R0 model. *Parasitology*, 138, 896–905.

Hartemink, N.A., Randolph, S.E., Davis, S.A., & Heesterbeek, J.A. (2008). The basic reproduction number for complex disease systems: Defining R(0) for tick-borne infections. *American Naturalist*, 171, 743–54.

Heylen, D., Lasters, R., Adriaensen, F., Fonville, M., Sprong, H., & Matthysen, E. (2019). Ticks and tick-borne diseases in the city: Role of landscape connectivity and green space characteristics in a metropolitan areas. *Science of the Total Environment*, 670, 941–9.

Homer, C., Huang, C. Yang, L., & Wylie, B.K. (2002). Development of a circa 2000 landcover database for the United States. Publications of the US Geological Survey, 108. https://digitalcommons.unl.edu/usgspubs/108.

Horobik, V., Keesing, F., & Ostfeld, R.S. (2006). Abundance and *Borrelia burgdorferi*-infection prevalence of nymphal *Ixodes scapularis* ticks along forest–field edges. *EcoHealth*, 3, 262–8.

Hosseini, P.R., Mills, J.N., Prieur-Richard, A.-H., et al. (2017). Does the impact of biodiversity differ between emerging and endemic pathogens? The need to separate the concepts of hazard and risk. *Philosophical Transactions of the Royal Society B: Biological Sciences*, 372, 9–20160129.

Hudson, P., Norman, R., Laurenson, M., et al. (1995). Persistence and transmission of tick-borne viruses: *Ixodes ricinus* and louping-ill virus in red grouse populations. *Parasitology*, 111, S49–58.

Jackson, L.E., Hilborn, E.D., & Thomas, J.C. (2006). Towards landscape design guidelines for reducing Lyme disease risk. *International Journal of Epidemiology*, 35, 315–22.

Johnstone-Robertson, S., Diuk-Wasser, & M.A., Davis, S. (2019). Incorporating tick feeding into R_0 for tick-borne pathogens. *Theoretical Population Biology*, 131, 25–37.

Karpathy, S.E., Allerdice, M.E., Sheth, M., Dasch, G.A., & Levin, M.L. (2016). Co-feeding transmission of the

Ehrlichia muris-like agent to mice (*Mus musculus*). *Vector Borne and Zoonotic Diseases*, 16, 145–50.

Kilpatrick, A.M., Dobson, A.D.M., Levi, T. et al. (2017). Lyme disease ecology in a changing world: Consensus, uncertainty and critical gaps for improving control. *Philosophical Transactions of the Royal Society of London B: Biological Sciences*, 372.

Kitron, U.D. & Kazmierczak, J.J. (1997). Spatial analysis of the distribution of Lyme disease in Wisconsin. *American Journal of Epidemiology*, 145, 558–66.

Krause, P.J., Narasimhan, S., Wormser, G.P. et al. (2013). Human *Borrelia miyamotoi* infection in the United States. *New England Journal of Medicine*, 368, 290–1.

Kupfer, J.A., Malanson, G.P., & Franklin, S.B. (2006). Not seeing the ocean for the islands: The mediating influence of matrix-based processes on forest fragmentation effects. *Global Ecology and Biogeography*, 15, 8–20.

Labuda, M., Kozuch, O., Zuffova, E., Eleckova, E., Rosie, H., & Nuttall, P.A. (1997). Tick-borne encephalitis virus transmission between ticks cofeeding on specific immune natural rodent hosts. *Virology*, 235, 138–43.

Larsen, A.E., Plantinga, A.J., & Macdonald, A.J. (2014). Lyme disease risk influences human settlement in the wildland–urban interface: Evidence from a longitudinal analysis of counties in the Northeastern United States. *The American Journal of Tropical Medicine and Hygiene*, 91, 747–55.

Li, S.,Hartemink, N., Speybroeck, N. & Vanwambeke, S.O. (2012). Consequences of landscape fragmentation on Lyme disease risk: A cellular automata approach. *PLoS One*, 7, e39612.

Linske, M.A., Williams, S.C., K.C., 3rd, & Ortega, I.M. (2017). *Ixodes scapularis* (Acari: Ixodidae) reservoir host diversity and abundance impacts on dilution of *Borrelia burgdorferi* (Spirochaetales: Spirochaetaceae) in residential and woodland habitats in Connecticut, United *Journal of Medical Entomology*, 55(3), 681–90.

Liu, J., Dietz, T., Carpenter, S.R. et al. (2007). Coupled human and natural systems. *AMBIO: A Journal of the Human Environment*, 36, 639–49.

Liu, J., Hull, V., Carter, N., Viña, A. & Yang, W. (2016). *Framing Sustainability of Coupled Human and Natural Systems*. Oxford Univeristy Press, Oxford.

Logiudice, K., Duerr, S.T.K., Newhouse, M. J. et al. (2008). Impact of host community composition on Lyme disease risk. *Ecology*, 89, 2841–9.

Logiudice, K., Ostfeld, R.S., Schmidt, K.A., & Keesing, F. (2003). The ecology of infectious disease: Effects of host diversity and community composition on Lyme disease risk. *Proceedings of the National Academy of Sciences of the United States of America*, 100, 567–71.

Lynn, G.E., Oliver, J.D., Cornax, I., O'Sullivan, M.G., & Munderloh, U.G. (2017). Experimental evaluation of *Peromyscus leucopus* as a reservoir host of the *Ehrlichia muris*-like agent. *Parasites & Vectors*, 10, 48.

McClure, M. & Diuk-Wasser, M.A. (2018). Reconciling the entomological hazard and disease risk in the Lyme disease system. *International Journal of Environmental Research and Public Health*, 15:1048

Manson, R.H., Ostfeld, R.S., & Canham, C.D. (1999). Responses of a small mammal community to heterogeneity along forest-old-field edges. *Landscape Ecology*, 14, 355–67.

Matser, A., Hartemink, N., Heesterbeek, H., Galvani, A., & Davis, S. (2009). Elasticity analysis in epidemiology: An application to tick-borne infections. *Ecology Letters*, 12, 1298–305.

Mckinney, M.L. (2008). Effects of urbanization on species richness: A review of plants and animals. *Urban Ecosystems*, 11, 161–76.

Nonaka, E., Ebel, G.D., & Wearing, H.J. (2010). Persistence of pathogens with short infectious periods in seasonal tick populations: The relative importance of three transmission routes. *PLoS One*, 5, e11745.

Norman, R., Bowers, R.G., Begon, M., & Hudson, P.J. (1999). Persistence of tick-borne virus in the presence of multiple host species: Tick reservoirs and parasite mediated competition. *Journal of Theoretical Biology*, 200, 111–18.

Nupp, T.E. & Swihart, R.K. (1996). Effect of forest patch area on population attributes of white-footed mice (*Peromyscus leucopus*) in fragmented landscapes. *Canadian Journal of Zoology*, 74, 467–72.

Ogden, N., Bigras-Poulin, M., O'Callaghan, C.J. et al. (2007). Vector seasonality, host infection dynamics and fitness of pathogens transmitted by the tick Ixodes scapularis. *Parasitology*, 134, 209–27.

Ogden, N.H. & Lindsay, L.R. (2016). Effects of climate and climate change on vectors and vector-borne diseases: Ticks are different. *Trends in Parasitology*, 32, 646–56.

Ogden, N.H., Lindsay, L.R. & Leighton, P.A. (2013). Predicting the rate of invasion of the agent of Lyme disease *Borrelia burgdorferi*. *Journal of Applied Ecology*, 50, 510–18.

Ogden, N.H. & Tsao, J.I. (2009). Biodiversity and Lyme disease: Dilution or amplification? *Epidemics*, 1, 196–206.

Ostfeld R. S. & Keesing F. (2000). Biodiversity and disease risk: The case of Lyme Disease. *Conservation Biology*, 14: 722–728.

Ostfeld, R. S. & Keesing, F. (2013). Straw men don't get Lyme disease: Response to Wood and Lafferty. *Trends in Ecology & Evolution*, 28, 502–3.

Pepin, K.M., Eisen, R.J., Mead, P.S., et al. (2012). Geographic variation in the relationship between human Lyme disease incidence and density of infected host-seeking *Ixodes scapularis* nymphs in the Eastern United States,

American Journal of Tropical Medicine and Hygiene, 86, 1062–71.

Perkins, S.E., Cattadori, I.M., Tagliapietra, V., Rizzoli, A.P., & Hudson, P.J. (2006). Localized deer absence leads to tick amplification. *Ecology*, 87, 1981–6.

Pickett, S.T.A., Cadenasso, M.L., & Grove, J.M. (2005). Biocomplexity in coupled natural–human systems: A multidimensional framework. *Ecosystems*, 8, 225–32.

Piesman, J. & Spielman, A. (1979). Host-associations and seasonal abundance of immature ixodes Dammini in Southeastern Massachusetts. *Annals of the Entomological Society of America*, 72, 829–832.

Pritt, B.S., Mead, P.S., Johnson, D.K.H. et al. (2016). Identification of a novel pathogenic Borrelia species causing Lyme borreliosis with unusually high spirochaetaemia: A descriptive study. *The Lancet Infectious Diseases*, 16, 556–64.

Pritt, B.S., Sloan, L.M., Johnson, D.K. et al. (2011). Emergence of a new pathogenic *Ehrlichia* species, Wisconsin and Minnesota, 2009. *New England Journal of Medicine*, 365, 422–9.

Randolph, S.E. (2013). Commentary on 'A Candide response to Panglossian accusations by Randolph and Dobson: Biodiversity buffers disease'. *Parasitology*, 140, 1199–200.

Randolph, S.E. & Dobson, A.D.M. (2012). Pangloss revisited: A critique of the dilution effect and the biodiversity-buffers-disease paradigm. *Parasitology*, 139, 847–63.

Randolph, S.E., Green, R.M., Peacey, M.F. & Rogers, D.J. (2000). Seasonal synchrony: The key to tick-borne encephalits foci idendtified by satellite data. *Parasitology*, 121, 15–23.

Randolph, S.E., Miklisová, D., Lysy, J., Rogers, D.J., & Labuda, M. (1999). Incidence from coincidence: Patterns of tick infestations on rodents facilitate transmission of tick-borne encephalitis virus. *Parasitology*, 118, 177–86.

Randolph, S.E. & Rogers, D.J. (2000). Fragile transmission cycles of tick-borne encephalitis virus may be disrupted by predicted climate change, *Proceedings of the Royal Society B: Biological Sciences*, 267, 1741–4.

Ratnapradipa, D., McDaniel, J.T., & Barger, A. (2017). Social vulnerability and Lyme disease incidence: A regional analysis of the United States, 2000–2014. *Epidemiology Biostatistics and Public Health*, 14, e12158.

Rehacek, J. (1962). Transovarial transmission of tick-borne encephalitis virus by ticks. *Acta Virologica*, 6, 220–6.

Ries, L., Fletcher, R.J., Battin, J., & Sisk, T.D. (2004). Ecological responses to habitate edges: Mechanisms, models, and variability explained. *Annual Reviews in Ecology, Evolution and Systematics*, 35, 491–522.

Rollend, L., Fish, D., & Childs, J.E. (2013). Transovarial transmission of Borrelia spirochetes by *Ixodes scapularis*:

A summary of the literature and recent observations. *Ticks and Tick-Borne Diseases*, 4, 46–51.

Rosà, R. & Pugliese, A. (2007). Effects of tick population dynamics and host densities on the persistence of tick-borne infections. *Mathematical Biosciences*, 208, 216–40.

Rosà, R., Pugliese, A., Norman, R., & Hudson, P.J. (2003). Thresholds for disease persistence in models for tick-borne infections including non-viraemic transmission, extended feeding and tick aggregation. *Journal of Theoretical Biology*, 224, 359–76.

Stafford, K.C., Cartter, M.L., Magnarelli, L.A., Ertel, S.H., & Mshar, P.A. (1998). Temporal correlations between tick abundance and prevalence of ticks infected with Borrelia burgdorferi and increasing incidence of Lyme disease. *Journal of Clinical Microbiology*, 36, 1240–4.

States, S.L., Huang, C.I., Davis, S., Tufts, D.M., & Diuk-Wasser, M.A. (2017). Co-feeding transmission facilitates strain coexistence in *Borrelia burgdorferi*, the Lyme disease agent. *Epidemics*, 19, 33–42.

Telford, S.R., Moore, S.I., Wilson, M.L, Spielman, A., Mather, T.N. (1988). Incompetence of deer as reservoirs of the Lyme disease spirochete. *American Journal of Tropical Medicine and Hygiene*, 39, 105–9.

Tolkacz, K., Bednarska, M., Alsarraf, M. et al. (2017). Prevalence, genetic identity and vertical transmission of *Babesia microti* in three naturally infected species of vole, *Microtus* spp. (Cricetidae). *Parasites & Vectors*, 10, 66.

Tufts, D. & Diuk-Wasser, M.A. (2018). Transplacental transmission of tick-borne *Babesia microti* in its natural host *Peromyscus leucopus*. *Parasites & Vectors*, 11, 286.

Valente, S.L., Wemple, D., Ramos, S., Cashman, S.B., & Savageau, J.A. (2015). Preventive behaviors and knowledge of tick-borne illnesses. *Journal of Public Health Management and Practice*, 21, E16-23.

VanAcker, M.C., Little, E., Molaei, G., Bajwa, W. & Diuk-Wasser, M.A. (2019). Enhancement of risk for Lyme disease by landscape connectivity, New York, New York, USA. *Emerging Infectious Diseases* 25, 1036–43.

Walter, K.S., Carpi, G., Caccone, A., & Diuk-Wasser, M.A. (2017). Genomic insights into the ancient spread of Lyme disease across North America. *Nature Ecology and Evolution*, 1, 1569–76.

Walter, K.S., Pepin, K.M., Webb, C.T., Gaff, H.D., Krause, P.J., Pitzer, V.E., & Diuk-Wasser, M.A. (2016). Invasion of two tick-borne diseases across New England: Harnessing human surveillance data to capture underlying ecological invasion processes. *Proceedings of the Royal Society B: Biological Sciences*, 283: 20160834.

Wilcox, B. & Jessop, H. (2016). Ecology and human health. In: Frumkin, H. (ed.). *Environmental Health: From Global to Local*. Josset-Bass, San Francisco.

Williams, S.C., Stafford, K.C., Molaei, G., & Linske, M.A. (2017). Integrated control of nymphal *Ixodes scapularis*:

Effectiveness of white-tailed deer reduction, the entomopathogenic fungus *Metarhizium anisopliae*, and fipronil-based rodent bait boxes. *Vector-Borne and Zoonotic Diseases*, 18, 55–64.

Wilson, M.L. & Spielman, A. (1985). Seasonal activity of immature *Ixodes Dammini* (Acari: Ixodidae). *Journal of Medical Entomology*, 22, 408–14.

Wood, C.L. & Lafferty, K.D. (2013). Biodiversity and disease: A synthesis of ecological perspectives on Lyme disease transmission. *Trends in Ecology & Evolution*, 28, 239–47.

Yuval, B. & Spielman, A. (1990). Duration and regulation of the developmental cycle of *Ixodes dammini* (Acari: Ixodidae). *Journal of Medical Entomology*, 27, 196–201.

CHAPTER 9

Carry-over Effects of the Larval Environment in Mosquito-Borne Disease Systems

Michelle V. Evans, Philip M. Newberry, and Courtney C. Murdock

The potential for the environment to shape mosquito dynamics is an important foundation for vector control efforts and our understanding of mosquito-borne disease. Mosquitoes have a complex life-cycle involving multiple life-stages that experience environments separated by both time and space (Fig. 9.1). For example, eggs may be oviposited prior to a dry season, remain dormant during the dry season, hatch months later into an aquatic immature stage within a confined aquatic habitat, and eventually emerge as an adult in the terrestrial environment. Like other organisms that experience ontogenetic niche-shifts, mosquito phenotypes can be shaped indirectly by the environment experienced by earlier life stages, a phenomenon known as carry-over effects (Harrison et al. 2011; Benard & McCauley 2008). Carry-over effects are found across a diversity of organisms with complex life-cycles or life-cycles that span multiple environments, such as migratory birds (Norris & Taylor 2006) and amphibians (Vonesh 2005), and are common in invertebrates, like mosquitoes, that undergo complete metamorphosis. Carry-over effects in mosquitoes can occur at multiple life-stage transitions or across generations. Mosquitoes that enter diapause can have significantly reduced fitness post-diapause (Bradshaw et al. 1998) and egg quiescence can alter the larval and adult phenotypes (Perez & Noriega 2013). Transgenerational effects occur when the maternal or paternal environment impacts offspring fitness (e.g. Zirbel & Alto 2018), a well-known phenomenon in insects (Mousseau & Dingle 1991). Carry-over effects of the larval environment on adult phenotypes are especially important given the focus on larval habitats in vector control efforts, and are the primary focus of this chapter (Fig. 9.1).

Manifestations of carry-over effects differ by organism, as each is influenced by specific abiotic and biotic factors. For small ectothermic invertebrates such as mosquitoes, temperature can play an especially important role through its effect on metabolic processes (Angilleta et al. 2004). Temperature generally has a unimodal relationship with metabolic processes and fitness in invertebrates, with cool temperatures slowing metabolism and high temperatures destroying proteins or causing mortality (Huey & Stevenson 1979; Angilleta et al. 2004). Larvae that take longer to develop are able to assimilate more resources during the immature stage, and generally emerge as larger adults (Reiskind & Zarrabi 2012). The nutritional resources of an earlier life stage have lasting effects on later phenotypes. Resource levels indirectly impact later life-stages by controlling the growth rate of the current life stage or by limiting future energy stores following metamorphosis. In resource-rich larval environments, a "silver spoon effect" can occur, whereby individuals experience high fit-

Michelle V. Evans, Philip M. Newberry, and Courtney C. Murdock, *Carry-over Effects of the Larval Environment in Mosquito-Borne Disease Systems* In: *Population Biology of Vector-Borne Diseases.* Edited by: John M. Drake, Michael B. Bonsall, and Michael R. Strand: Oxford University Press (2021).
DOI: 10.1093/oso/9780198853244.003.0009

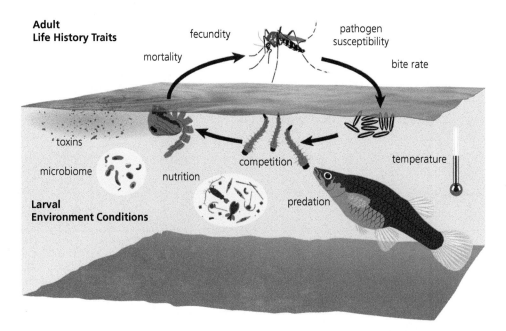

Figure 9.1 Life-cycle of a mosquito, from egg to larva to pupa to adult. Environmental carry-over effects of the larval environment and the relevant adult life history traits discussed in this chapter are labeled.
Source: Eric Marty.

ness, even if the current environment is of low quality, due to lasting effects of the high-quality larval environment (Monaghan 2008). Species interactions, such as competition or microbial mutualisms, at one life-stage may also carry-over to influence later life-stages. This could be in direct response to an interaction such as crowding, or indirectly through another organism's impact on resources (Beckerman et al. 2002). Larval mosquitoes require a microbial community to develop into adults (Coon et al. 2016a), and the microbial community can also persist across life stages and interact with pathogens within the mosquito vector (Carrington et al. 2018).

Carry-over effects influence disease transmission dynamics through their effects on adult phenotypes or life-history traits. A unifying concept to quantify disease transmission is vectorial capacity (VC), the number of infectious mosquito bites that arise due to the introduction of one infectious person into a population (Klempner et al. 2007; Smith et al. 2012). VC is determined by mosquito and pathogen dynamics, and can be summarized by the equation:

$$VC = \frac{ma^2bp^n}{-\ln(p)}$$

Mosquito density (m) is determined by fecundity and survival rates, with a higher mosquito density leading to a higher VC. Biting rate (a) is squared to include the bite needed for the mosquito to become infected and the bite needed to transmit the infection to the next host, leading to a non-linear, positive relationship with VC. Vector competence, the probability that a mosquito feeding on an infectious host becomes infectious, is represented by b, and is dependent on the mosquito-pathogen combination. A pair with a higher vector competence will have an overall higher VC. The probability that a mosquito will live long enough to transmit the pathogen depends on mosquito survival rate (p) and the extrinsic incubation period (n), the number of days required for an ingested pathogen to reach the salivary glands for a mosquito to become infectious. An increase in mosquito survival or a decrease in the extrinsic incubation period would increase VC. Sensitivity analyses have found that adult survival disproportionately affects pathogen transmission because it both impacts mosquito population density through its effect on adult mortality rates and controls pathogen transmission by limiting the time-period a pathogen has to reach the salivary glands and be transmitted before the mosquito dies (Brady et al. 2015). The length of the gonotrophic cycle of a mosquito similarly can influence the frequency of fecundity events and the human biting rate. Carry-over effects on mosquitoes can affect all of the parameters included in calculations of *VC* (for a review focused on vector competence, see Alto & Lounibos 2013), and therefore have important consequences for disease transmission.

In this chapter, we review the prior work on carry-over effects in mosquito systems to summarize current knowledge of the impacts of larval temperature, resource levels, competition and the microbiome on mosquito-borne disease transmission via the four mosquito life-history traits in the *VC* equation. We explore past efforts at including these effects in models of mosquito-borne diseases and identify future directions for modeling. Finally, we identify future research directions in the field of carry-over effects and mosquito-borne diseases.

9.1 Overview of Carry-Over Effects

9.1.1 Temperature

Temperature is the best studied environmental factor with regards to mosquito-borne disease. Natural variation in temperature across latitudinal and altitudinal gradients allows for statistical analyses of disease incidence data (e.g. Pascual et al. 2006, Siraj et al. 2014), and the development of programmable incubators has enabled empirical work exploring how environmental temperatures impact both larval and adult mosquitoes. Temperature at the larval stage can influence adult phenotypes by exposing individuals to thermal stress at low or high extremes. Increasing temperatures also quicken the development of larval mosquitoes, resulting in shorter larval stages. Following classic ectothermic theory, adult mosquito life-history traits have a uni-modal relationship with temperature, although the exact shape of this curve varies by trait and mosquito species (Mordecai et al. 2013, 2017) and may differ with fluctuating temperatures (Carrington et

al. 2013; Colinet et al. 2015). Most studies, however, do not separate the carry-over effects of temperature in the larval environment from the direct effects of temperature during the adult life-stage, which makes it difficult to infer if the effect of temperature is due to an indirect carry-over effect or to the direct effects of the present temperature. In addition to the direct effects of temperature on mosquito metabolic rates, temperature can indirectly influence mosquitoes through its effect on other trophic levels. For example, the growth of microbial populations that mosquitoes feed on is itself temperature dependent (Ratkowsky et al. 1982), and changes to the density of this resource due to temperature can cause bottom-up trophic effects on mosquito larvae.

9.1.1.1 Fecundity

Fecundity has a well-known positive relationship with female adult body-size and mass (Lounibos et al. 2002, Armbruster & Hutchinson 2002). This enables studies to predict adult female fecundity without directly measuring an oviposition event. Although this size-fecundity relationship is assumed to be fixed, there is variation in the magnitude and strength of this relationship across temperature, resource availability, and genotypes (Costanzo et al. 2018). At low temperatures near species' thermal minima, this relationship becomes weaker and can disappear (Costanzo et al. 2018, Westby & Juliano 2015). This variation in allometric relationships is discussed later in this chapter. Following the general rule that ectotherms are smaller in hotter environments (Atkinson 1994), females reared in higher temperature aquatic environments are smaller than those from cooler environments and generally have lower fecundity (Lounibos et al. 2002, Ezeakacha & Yee 2019). In *Anopheles* mosquitoes, higher larval temperatures also decrease the probability of an oviposition event, reducing lifetime fecundity, although this is also highly dependent on temperatures at the adult stage (Christiansen-Jucht et al. 2015). No studies have directly measured the effects of acute heat or cold stress in the larval stage on adult fecundity. However, studies of *Drosophila* have found that heat-induced expression of heat shock proteins can depress fecundity (Sørensen et al. 1999, Silbermann & Tatar 2000), and a similar fitness trade-off may exist in mosquitoes. Seasonal variations in temperature often coincide with shifts in photoperiod, and longer daylengths can result in larger (Yee et al. 2017: *Ae. albopictus*) or smaller females (Costanzo et al. 2015: *Ae. albopictus* & *Ae. aegypti*). Given that longer daylengths often correspond with warmer temperatures, the effect of photoperiod can potentially mediate or amplify some of the temperature-fecundity trends found in the lab when applied to the field.

9.1.1.2 Pathogen Susceptibility

Carry-over effects of larval temperature on the susceptibility of adults to pathogens have been primarily studied in *Aedes*-virus systems, with a focus on arboviruses of human health concern, such as dengue (DENV), West Nile (WNV), and chikungunya (CHIKV) viruses. There are two primary ways by which temperature is predicted to influence adult susceptibility to pathogens. One is through its impact on adult immunity, reasoning that adults from thermally stressful larval environments will have lower immune function. While the carry-over effects of temperature on immunity have not been directly studied, Murdock et al. (2012) found temperature to determine expression of genes relevant for immune functioning in adults, identifying the potential for temperature in the larval stage to impact this via carry-over effects. The second pathway is through an indirect effect on blood meal size. Smaller adults may have lower teneral reserves and in response will imbibe a larger blood meal, ingesting more virus particles in the process. Most studies do not investigate specific mechanisms, but rather focus on the overall change in virus susceptibility in adults from different larval rearing temperatures. Of these studies, results are mixed, with some finding that cooler larval temperatures increase infection and dissemination (Westbrook et al. 2010: *Ae. albopictus* & CHIKV, Evans et al. 2018: *Ae. albopictus* & DENV, Adelman et al. 2013: *Ae. aegypti* & CHIKV / YFV) and others finding that cooler temperatures decrease susceptibility (Alto & Bettinardi 2013: *Ae. albopictus* & DENV). Some of these conflicting results may be due to differences in temperature treatments across studies, with Westbrook et al. (2010) including a wide range of thermal extremes. Additionally, the genotype of the mosquito and the

virus could alter these results, as populations are adapted to their respective temperature regimes. These genotype x genotype x temperature interactions have been observed with ambient adult temperatures (Zouache et al. 2014, Gloria-Soria et al. 2017) and may exist for carry-over effects due to larval temperatures as well.

9.1.1.3 Adult Mortality

Larval temperature seems to have less of an effect on adult mortality rates than other life-history traits. Studies have found that the larval rearing temperature does not impact *Aedes* adults' mortality under normal, non-stressed conditions (Alto & Bettinardi 2013; Ezeakacha & Yee 2019), however this may be dependent on whether larval temperatures are constant vs. fluctuating (Westby & Juliano 2015). The lack of carry-over effects on mortality may be due to the strong filtering effects of the larval environment. Individuals that experience increased mortality due to temperature may die during the larval stage, and only those that are more thermally tolerant survive to emergence. Indeed, temperature has strong effects on survival during the larval stage (Couret et al. 2014: *Ae aegypti*) and the impact on mortality may be most strongly felt during this stage. On the other hand, Christiansen-Jucht et al. (2014) found that increases in larval temperature increased adult mortality in *An. gambiae*, but only for increases of 8°C and not 4°C. This could suggest genus-level differences in these carry-over effects, but more study is needed to investigate this.

9.1.1.4 Biting Rate

We know very little about how larval temperatures impact adult biting rates. One study found that larval temperature only impacted the probability of taking a blood meal after the third gonotrophic cycle (Christiansen-Jucht et al. 2014: *An gambiae*), suggesting that this carry-over effect may also be mediated by the age of the adult. Scott et al. (2000) found that wing length explained 18 percent of the variation in the frequency of blood meals in wild-caught *Ae. aegypti*, with smaller mosquitoes from warmer environments feeding more frequently. Similarly, smaller *An. gambiae* require multiple bloodmeals to complete their gonotrophic cycle

(Lyimo & Takken 1993). Therefore, mosquitoes reared in warmer environments, which emerge at a smaller size, may have a higher biting rate than those from cooler environments. When measured empirically, larval environmental temperature did not influence the willingness of *Ae. triseriatus* mosquitoes to bloodfeed during their first gonotrophic cycle, in spite of a significant difference in wing length amongst temperature treatments (Westby & Juliano 2015). However, whether this willingness to feed when offered a first bloodmeal translates directly to biting rate across a lifespan has yet to be examined.

9.1.2 Nutrition

Nutrition during the larval stage can have a lasting impact on adults by impeding or facilitating larval development rates and by impacting the quantity of teneral reserves. Teneral reserves are the lipids, proteins, and carbohydrates in a mosquito's body that are available following eclosion. During the brief teneral phase following eclosion, these reserves serve as the primary source of energy for adults as they undergo the final stages of maturation and development (Briegel 2003). Naturally, these reserves are directly impacted by the quality and quantity of nutrients found in the larval environment. The nutrition of the larval environment can be assessed in terms of quantity, or availability, of resources or by the quality of specific resources (e.g. plant vs. animal based) found in the environment. Quantity and quality of resources will likely have differential impacts on adult fitness, as they operate on different developmental mechanisms, and should be considered separately.

9.1.2.1 Fecundity

As with temperature, much of the knowledge regarding larval nutrition and adult fecundity is based on fecundity's relationship with mosquito body size. Generally, higher resource availability or diet quality results in larger bodied mosquitoes that lay more eggs, although the extent of this does differ by species (Buckner et al. 2016: *Ae. aegypti* & *Ae. albopictus*). Further, this size-fecundity relationship can be modulated by temperature (Buckner et al. 2016), and may be weaker at higher levels of resource availability (Costanzo et al. 2018). Larval nutrition can also impact other aspects of fecundity. *An. stephensi* from low nutrition larval environments not only lay fewer eggs per oviposition event, but also are less likely to mate and have a longer gonotrophic cycle than individuals from high nutrition larval environments (Moller-Jacobs et al. 2014). Over a mosquito's lifetime, these effects can compound to result in an overall lower lifetime fecundity than expected through an effect on egg number only.

9.1.2.2 Pathogen Susceptibility

An individual's susceptibility to a pathogen, such as a virus or *Plasmodium*, is also influenced by larval nutrition. The availability of nutrients during mosquito development can influence an adult's innate and adaptive immune responses. Mosquito innate immunity consists partly of physical barriers to infection, such as the midgut epithelial lining, which prevents pathogens from escaping the midgut into the hemolymph (Hillyer 2016). It is hypothesized that resource-poor larval environments result in smaller mosquitoes with thinner epithelial linings, which would increase susceptibility to infection, however, studies find mixed evidence for this in *Ae. aegypti* (Grimstad & Walker 1991, Telang et al. 2012). Mosquito adaptive immune responses include mechanisms such as phagocytosis, melanization, lysis, or RNA interference (Hillyer 2016). Innate immune responses are higher and more efficient in mosquitoes from higher nutrition larval environments, resulting in higher melanization rates (Suwanchaichinda & Paskewitz 1998: *An. gambiae*) and fat body derived immune factors (Telang et al. 2012: *Ae. aegypti*).

The mechanisms cited previously suggest that mosquitoes from low quality larval environments will be more susceptible to pathogen infection, however the reality is much more complex. In the *Anopheles*-malaria system, higher quality diets often result in higher infection rates, as measured by the prevalence and intensity of oocyst formation (Takken et al. 2013, Moller-Jacobs et al. 2014, Vantaux et al. 2016a). In the *Aedes*-virus systems, higher quality diets often result in lower virus susceptibility, as measured by virus dissemination and mosquito infectiousness (Takahashi 1976, Grimstad &

Haramis 1984, Paige et al. 2019), although some studies have found no effect of larval nutrition on infection dynamics (Jennings & Kay 1999). This difference across systems may be due to the differences in host-pathogen ecology. Lower larval nutrition can limit infection rates by limiting resource availability for the parasite inside the host, while it can increase infection rates by suppressing the hosts' immune system. Because infection rates tend to be lower in adults from lower quality larval environments, the *Plasmodium-Anopheles* interaction may approximate a consumer-resource interaction within the mosquito (Costa et al. 2018), rather than a top-down control by the mosquito's immune system.

9.1.2.3 Adult Mortality

The effect of larval nutrition on adult longevity has been primarily studied in *Anopheles* mosquitoes. The majority of studies find that increased larval resources either decrease adult mortality rates or have no effect (Aboagye-Antwi & Tripet 2010: *An. gambiae*, Araújo et al. 2012: *An. darlingi*, Takken et al. 2013: *An. gambiae & An. Stephensi*, Barreaux et al. 2018: *An. gambiae*, Chandrasegaran et al. 2018: *Ae. aegypti*). Moller-Jacobs et al. (2014) found mortality was higher for malaria-infected *An. stephensi* during the oocyst development stage of the parasite for adults that were reared in low nutrition larval environments. The stress of infection may interact with past larval environments to cause differential adult mortality in this instance. Interestingly, Aboagye-Antwi & Tripet (2010) found no evidence for a carry-over effect even in the face of desiccation stress on adults, which would be expected to increase reliance on teneral reserves. Mosquitoes reared in low-nutrition environments may be able to offset a poor larval environment by increased consumption of water, nectar, and blood as adults, negating any increase in mortality risk. Comparatively, Vantaux et al. (2016a) found that *An. coluzzi* adults from low nutrition larval environments live nearly one day longer than those from higher nutrition larval environments. Further, this effect was magnified in *Plasmodium*-infected females. In this case, a poor larval environment may have resulted in smaller adults that require less energy than larger adults emerging from high-quality larval environments, and so have a relatively longer lifespan.

9.1.2.4 Biting Rate

Mosquitoes from higher nutrition environments land on hosts more frequently (Klowden et al. 1988: *Ae. aegypti*, Nasci 1991: *Ae. aegypti*) and have a higher bite rate (Araújo et al. 2012: *An. darlingi*). However, smaller *Anopheles* from low quality environments are more likely to feed multiple times within their gonotrophic cycle (Takken et al. 1998: *An. gambiae*). If the allometric relationship discussed previously holds true, small mosquitoes from low nutrition larval environments will likely require multiple blood meals to complete their gonotrophic cycles, resulting in higher bite rates.

9.1.3 Competition

Interactions, such as competition, facilitation, or predation, can also cause carry-over effects. For species that oviposit in containers, especially, larvae often cohabit a container with conspecifics or other mosquito species, leading to intra- and inter-specific competition over resources. Competition can cause resource limitation, resulting in carry-over effects similar to those of low nutrition in the larval environment. Notably, this would more closely represent a change to nutrition quantity than to nutrition quality. However, effects of species interactions do not necessarily need to be resource-mediated. For example, Moore & Fisher (1969) found that *Ae. aegypti* larvae release a chemical compound referred to as a growth retardant factor in water that can slow development of *Ae. albopictus* larvae in the same water body. Recent work provides conflicting evidence for the existence of chemical interference in mosquitoes (Dye 1982, Bédhomme et al. 2005; Silberbush et al. 2014) and suggests that conventional interference and resource competition are the primary drivers of changes in mosquito development rates due to species interactions (Dye 1984, Roberts & Kokkinn 2010).

9.1.3.1 Fecundity

Competition impacts adult fecundity in ways similar to nutrition. Individuals emerging from habitats with higher intra- or inter-specific densities likely had less access to resources as larvae and may be smaller, which is associated with lower fecundity.

Indeed, the majority of studies find that increasing the density of individuals results in adult females with shorter wing-lengths (Armistead et al. 2008: *Ae. japonicus* & *Ae. atropalpus,* Muturi et al. 2011: *Ae. aegypti* & *Ae. albopictus,* Noden et al. 2016: *Ae. albopictus,* but not *Ae. aegypti*) and body mass (Chandrasegaran et al. 2018: *Ae. aegypti,* Ezeakacha & Yee 2019: *Ae. albopictus*). These results hold true whether the increase in density is due to conspecifics or heterospecifics (Armistead et al. 2008: *Ae. japonicus* & *Ae. atropalpus*; Paaijmans et al. 2009: *An. gambiae* & *An arabiensis*), suggesting it is density-mediated and does not differ across intra- and interspecific competition. A study that directly measured fecundity found a similarly negative effect of increasing larval density on adult fecundity (Moore & Fisher 1969: *Ae. aegypti*), lending support for the findings from indirect measurements. However, a more recent study of competition followed cohorts of mosquitoes throughout their life and found competition to have no effect on *Ae. albopictus* fecundity, but that high levels of interspecific competition did decrease *Ae. aegypti* fecundity relative to low levels of intraspecific competition (Noden et al. 2016). A negative impact of increased competition on adult fecundity represents a density-dependent negative feedback on population growth and is a pathway by which carry-over effects impact population-scale dynamics (Hawley 1985a).

9.1.3.2 Pathogen Susceptibility

The carry-over effects of competition on pathogen susceptibility have only been studied in the *Aedes*-virus systems. This may be because *Aedes* species larvae are often found in containers, and competition has been studied in these mosquitoes more than in species that inhabit natural water bodies (but see Roux et al. 2015 for a discussion of the effect of predation on *P. falciparum* in *An. coluzzii*). Results on competition differ across *Aedes* species and virus, with little clear trend. Increased inter-specific competition with *Ae. aegypti* leads to increased dissemination rates of dengue-2 virus and Sindbis virus in *Ae. albopictus,* but inter-specific competition with *Ae. albopictus* had no effect on dissemination rates of these viruses in *Ae. aegypti* (Alto et al. 2005, 2008). Similarly, *Ae. triseriatus* infection rates with LaCrosse Virus increase with an increasing propor-

tion of *Ae. albopictus* competitors, even if the overall larval density remains the same (Bevins 2008). The effect of interspecific competition on *Ae. aegypti* pathogen susceptibility can be influenced by other environmental factors, which may interact with competition to increase pathogen susceptibility. For example, intraspecific competition increases Sindbis virus infection and dissemination rates at 20°C, but decreases it at 30°C (Muturi et al. 2012). Similarly, the presence of an insecticide can lead to a positive effect of intra-specific competition on Sindbis virus infection and dissemination rates (Muturi et al. 2011). While most studies find either a positive or null effect of competition on pathogen susceptibility, one study found that mosquitoes reared in crowded larval environments had lower rates of DENV2 infection and dissemination (Kang et al. 2017). In this instance, stressful larval conditions due to competition may "prime" the immune system, thereby decreasing susceptibility to pathogens in the adult stage. Additionally, a study of filarial worms in *Ae. aegypti* found that individuals from high-density larval environments had lower infection prevalence of *Brugia pahangi* (Breaux et al. 2014). While this may be a similar case of "priming", it could also mirror dynamics seen in *Anopheles*-malaria systems, with higher competition amongst parasites within smaller mosquitoes originating from high-density environments. Further study should investigate differences in vector-pathogen dynamics between micro- and macro-parasite systems.

9.1.3.3 Adult Mortality

Adult mortality is generally predicted to increase with increasing competition in the larval environment, due to increased resource limitation at high larval densities. Studies of the effects of larval intraspecific densities on adult mortality support this hypothesis (Reiskind & Lounibos 2009: *Ae. aegypti,* Alto et al. 2012: *Cx. pipiens,* Breaux et al. 2014: *Ae. aegypti*). A high resource environment can mediate the positive effects of high intraspecific densities on adult mortality, which suggests that this effect is in fact resource-mediated (Alto et al. 2012). There is some evidence that the relationship is non-linear, with larger adults from low species densities experiencing higher mortality rates than those from intermediate densities (Hawley 1985b: *Ae. sierrensis*, Juliano et al. 2014: *Ae. aegypti*). Large adults may require more energy than intermediate-sized adults, resulting in higher mortality rates, however this is simple speculation. Empirical work on energy requirements and models of dynamic energy budgets (as has been done for *Schistosomiasis* (Civitello et al. 2018)) are needed to confirm this hypothesis. Adult mortality is also highly dependent on the adult environment, which can amplify or mediate carry-over effects. For example, higher mortality due to competition may be more pronounced in a low humidity adult environment where mosquitoes are exposed to desiccation stress (Reiskind & Lounibos 2009).

9.1.3.4 Biting Rate

To our knowledge, no studies have directly measured the effect of larval competition on adult biting rates in mosquitoes. Given that carry-over effects due to competition are likely resource mediated, we hypothesize that adult biting rates would respond to increased larval competition similarly as they respond to low nutrition larval environments. That is, smaller mosquitoes from environments with higher interspecific or intraspecific densities would have a higher biting rate.

9.1.4 Microbiome

Current research indicates the importance of carry-over effects mediated by the mosquito microbiome. The effects of larval inoculation with select bacterial taxa may be most evident in the resulting impact of larval health on adult mosquito fitness (Souza et al. 2019), or even simply the ability to successfully develop to the adult stage (Coon et al. 2016a, 2017, Chapter 13 in current volume). Beyond these carry-over effects from the larval stage, the midgut microbiome of adult mosquitoes does seem to impact mosquito life history traits that influence the next generation, like egg production (Coon et al. 2016b). While the microbial communities of the mosquito midgut are dynamic across life stages, there is notably a dramatic turnover in bacterial taxa occurring between the larval and adult stages in the case of field derived *Anopheles* (Wang et al. 2011). The mechanism separating the direct transfer of larval and adult microbiomes is in the expulsion of the

bolus and peritrophic matrix before pupation (Moll et al. 2001; Moncayo et al. 2005; Duguma et al. 2015). The initial inoculation of the mosquito larva's midgut is predominantly from the microbial community that the larvae emerge into (Lindh et al. 2008; Coon et al. 2016a). However, there is overlap with the nutritional acquisition of microbes, as many medically significant mosquito genera (*Anopheles*, *Aedes*, and *Culex*) are known to feed on microorganisms and detritus through a mix of feeding modes (Merrit et al. 1992). Some microbes are digested or otherwise do not persist in the larval gut, but environmental taxa do survive and contribute to the larval microbiome (Duguma et al. 2013).

Given the limitations for direct vertical transfer of symbionts that this mechanism presents, the carry-over effects of the larval microbiome are predominantly driven by the effects of larval health on adult developmental success (Chouaia et al. 2012), lifespan, and fecundity (Coon et al. 2016b). Still, vertical transmission of the microbiome can occur from parent to offspring in the case of some endosymbionts which can enter somatic cells and reproductive organs in the case of the bacteria taxa *Wolbachia* (Dorigatti et al. 2018) or *Asaia* (Mitraka et al. 2013). These vertical transmission events are of great interest for pathogen control efforts, given the microbe's ability to block infection of DENV in *Ae. aegypti* (Carrington et al. 2018).

9.1.4.1 Fecundity

The pathogen *Elizabethkingia meningoseptica* can spread from ovary to eggs during the gonotrophic cycle (Akhouayri et al. 2013), and the potential endosymbiont vector control agent *Wolbachia* demonstrates the potential for adult to offspring transmission. Some evidence suggests these cases of vertical transmission can outcompete the environmental inoculation of the midgut in cases where the egg tissue itself contains symbiotic bacterial taxa (Akhouayri et al. 2013). Larval inoculation occurs predominantly through bacterial taxa present in the larval environment (Lindh et al. 2008, Coon et al. 2016a), some of which persists to the adult stage, possibly through some degree of survival of peritrophic matrix loss (Duguma et al. 2015) or transstadially as individuals emerge from the aquatic environment (Lindh et al. 2008). This indirect path can allow the carry over of the larval microbial community into the adult stage in spite of meconial peritrophic matrix egestion.

9.1.4.2 Pathogen Susceptibility

The microbiome-pathogen susceptibility relationship depends highly on the mosquito host, the bacterial taxa, and the pathogen in question. While understanding of the relationship between host microbiome and pathogen susceptibility in mosquitoes is still developing, the endosymbiont *Wolbachia* already deployed in arbovirus control efforts demonstrates the potential impacts microbes can have on mosquito-pathogen interactions and potentially disease transmission. These carry-over effects from inoculated larvae to adult mosquitoes can show positive associations with West Nile Virus in *Culex* (Dodson et al. 2014) and negative relationships with dengue, chikungunya, and avian *Plasmodium gallinaceum* in *Aedes* (Moreira et al. 2009).

Empirical work investigating *Anopheles* and the rodent malaria *Plasmodium berghei* further shows variable interactions resulting from different strains of *Wolbachia* (Hughes et al. 2012). While this evidence may not necessarily reflect natural interactions that would happen beyond the rodent model into human systems, it is suggestive of how different microbial taxa can have divergent impacts on pathogen susceptibility. Further, the effect of microbial diversity on pathogen susceptibility (e.g. blocks, does nothing, or even enhances) can depend on prevailing environmental conditions (Murdock et al. 2014: *An stephensi* & *P. yoelli*). The association of mosquitoes with different microbial communities can also result in negative associations with vector-borne pathogens, likely due to immune system priming by bacteria or direct interference by bacteria with viral pathogens (Ramirez et al. 2012, 2014, Dickson et al. 2017). This has implications for the transmission of arboviruses impacting human health, like West Nile virus (Novakova et al. 2017), demonstrating the need to understand the carry-over of the larval microbiome into the adult mosquito. The nature of these interactions is species-pathogen system specific, and this warrants in-depth investigation to understand the breadth of interactions between the within host microbiome, the mosquito-pathogen interaction, and vector-borne disease transmission.

9.1.4.3 Adult Mortality

Bacterial symbionts are generally believed to be essential for mosquito larval development through the creation of an anoxic environment (Coon et al. 2017). Hypoxic signaling is the strongest candidate for pupation triggering in some species (Coon et al. 2017). The ultimate carry-over effects of the larval microbiome on adult fitness are known to include pupation rates, adult size, and immune function (Dickson et al. 2017, Souza et al. 2019). A further effect is that some bacteria, such as some species of *Asaia* and strains of *Escherichia coli*, increase development rates and adult survival with increased microbe population numbers (Souza et al. 2019, Mitraka et al. 2013). Additionally, the resident microbiome can decrease development rates when the midgut bacterial population levels are experimentally decreased while also resulting in smaller adults (Chouaia et al. 2012). The primary carry-over effects in these cases are at the population level, influencing the dynamics of larval to adult mortality. Microbial communities in larvae are linked to the characteristics of the adult vector, and the interplay of adult mosquito pathogens and the larval microbiome has implications for understanding the disease dynamics of vector borne disease systems.

9.1.4.4 Biting Rate

Currently the relationship between the carry-over effects of the larval microbiome and adult biting rates is not well understood. The adult mosquito is sensitive to the microbiome of its hosts and of potential oviposition sites, but whether larval microbial communities impact later feeding behavior is unknown. It can be speculated that biting rates are indirectly influenced by carry-over effects that alter adult traits such as body size, which may influence female blood meal seeking as discussed previously. Future research could investigate this question.

9.1.5 Additional Drivers of Carry-Over Effects

In addition to temperature, resources, competition, and the microbiome, there are many other biotic and abiotic factors that can lead to carry-over effects in mosquito systems. Species interactions besides competition, such as predation or parasitism, can cause carry-over effects. Non-consumptive, or trait-mediated, effects of predators have been well studied in mosquitoes, and often result in carry-over effects across life stages. Given the role of mosquito predators such as *Gambusia* in some vector control programs, it is especially important to understand the non-consumptive effects of predation to predict how vector control may impact disease transmission in unintended ways. Predation in the larval stage can decrease the size of adult females (van Uitregt et al. 2012, Roux et al. 2015, Beketov & Liess 2007), increase adult mortality (Bellamy & Alto 2018, Ower & Juliano 2019), increase blood feeding willingness (Ower & Juliano 2019), and decrease the prevalence of gravid females (Roux et al. 2015). As with all carry-over effects, the effect of predation is dependent on other aspects of the larval environment, such as larval density on nutrient availability, and is strongest in highly competitive or low resource environments (Beketov & Liess 2007, Chandrasegan et al. 2018, Ower & Juliano 2019). For mosquito species in natural water bodies, such as tree holes or ponds, other invertebrates can facilitate larval growth through increased leaf decomposition rates and bacterial and fungal activity, which serves as a resource for mosquito larvae (Pelz-Stelinski et al. 2011). This increased resource availability could lead to nutritionally-mediated carry-over effects, as discussed.

Toxins in the larval environment, including pollutants and vector control agents such as larvicides, can also lead to carry-over effects. Toxins that cause high larval mortality, such as malathion or *Bti*, can indirectly lead to carry-over effects through their direct effects on larval densities (Muturi et al. 2011, 2010). This results in carry-over effects similar to those driven by competition or nutrition. Additionally, toxins can directly impact adult life history traits through developmental mechanisms (Naresh et al. 2013, Op de Beeck et al. 2016, Vantaux et al. 2016b). Certain toxins have been specifically investigated for their ability to control disease transmission, such as metal nanoparticles. The presence of silver nanoparticles (AgNP), which are derived from plants, in the larval environment act as a larvicide, and has been shown to reduce transmission of *Plasmodium* in *vitro* and *in vivo* in mammals (Murugan et al. 2016). Whether these non-lethal effects occur in the mosquito vector are unknown (Benelli et al. 2017), and their ability to induce transmission blocking via carry-over effects is an avenue for further study.

9.2 Modeling carry-over effects

The inclusion of carry-over effects further complicates the modeling of mosquito-borne diseases. For compartmental models, this may involve adding compartments for adults that are dependent on the larval environment from which they came. For correlative or predictive models, information on the larval environment will need to be included as predictor variables. As with models that only include the adult environment, both types of models can be used to predict disease risk across space or time. Similarly, modeling can lend insight to the mechanics of the system. Compartmental models are best suited to identify mechanisms by which the larval environment influences disease risk by explicitly including parameters for each mechanism influenced by the larval environment (e.g. effect of nutrition on each life history traits). Correlative models could identify which larval environment (e.g. temperature vs. competition) is leading to the largest changes in mosquito abundance or disease cases through methods that quantify variable importance.

If the larval and adult environments are highly correlated, it may be possible to simplify models and not explicitly account for both environments. In special cases, this is possible with temperature, as air temperature may be used to estimate aquatic temperatures. However, in heterogeneous landscapes such as cities, exogenous factors such as the amount of impervious surface or housing types may alter the relationship between the temperature of larval and adult environments across space (Cator et al. 2013, Murdock et al. 2017). Aquatic temperatures experience smaller fluctuations in temperature than air temperatures due to the higher heat capacity of water, resulting in lower maximum temperatures in the field (Kumar et al. 2018). Changes in the daily air temperature range can alter adult susceptibility to pathogens (Lambrechts et al. 2011: *Ae. aegypti* & DENV), and the daily aquatic temperature range may be equally important for carry-over effects (Bradshaw 1980). Approaches that assume aquatic and air temperatures to be identical, or simply offset, may over or under-estimate disease risk by misestimating larval environmental temperatures.

Further, resources and competition in the larval environment may themselves be influenced by mosquitoes, creating feedbacks that must be accounted for. For example, in order to estimate the equilibrium population size of a mosquito species that experiences density-dependent population growth due to intraspecific competition, carry-over effects on fecundity and longevity will need to be taken into account (e.g. Hawley 1985a). Once the equilibrium size is reached, the equilibrium density could then impact adult susceptibility to pathogens and biting rates. Some empirical work (reviewed previously) has found that the strength of density-dependence is temperature-dependent. Models that include temperature-dependence will then need to account for seasonal forcing as temperatures change throughout the season.

As a first step, theoretical modeling should begin to include additional complexities due to carry-over effects. While these models cannot predict disease risk for a specific time and place, they help characterize relationships between the larval and adult environments and mosquito-borne disease risk. Theoretical models can also identify which changes to the larval environment or life history traits could have the largest impact on adult mosquito fitness. This question could be addressed with multi-faceted empirical studies that co-vary or vary individually (e.g. comparing a combination of treatments of temperature, nutrition, competition, and microbial communities), but these would be unrealistically large experiments. Parameterized mechanistic

models, or even comprehensive meta-analyses across larval environments, could help pare down the number of treatments to include. Some work has been done in this regard, using sensitivity analyses to identify the importance of adult longevity on vectorial capacity, given its exponential term in the mathematical equation (Johnson et al. 2015). In order to validate assumptions of theoretical models and create accurate predictive models, however, more empirical studies are needed to parameterize carry-over effects. Several studies have conducted literature searches to collate life history traits from empirical work (e.g. Mordecai et al. 2013; Shocket et al. 2019), and some of this information has begun to be organized into open access databases such as VecTraits (https://www.vectorbyte.org/). However, given the relatively few studies that focus on carry-over effects, only the direct effects of the current environments are included in these studies. As more empirical work is conducted and traits are parameterized, it will be possible to include realistic carry-over effects in models. Doing so will allow us to identify important carry-over effects through sensitivity analyses, and guide future empirical research.

9.3 Synthesis and Future Directions

9.3.1 Inferring Life-History Traits through Allometric Relationships

Many studies use allometric relationships to estimate life history traits rather than measuring them directly. This is especially common when measuring fecundity, which has a close, linear relationship with mosquito body size (Armbruster & Hutchinson

---

Box 9.1 Carry-over effects and the Ross-Macdonald model

Mosquito-borne diseases are often characterized by their basic reproductive number (R_0), defined as the number of secondary cases in a completely susceptible population resulting from one infectious individual. While there are many different derivations of R_0, the majority derive from the Ross-McDonald equation and include the life history traits discussed in this

chapter. Because traits interact in multiplicative and additive ways and vary in their response to the larval environment, as discussed in this chapter, it is difficult to intuit how changes to the larval environment will impact adult R_0. A potential approach is to define traits in the R_0 equation as a function of the larval environmental temperature and parameterize these

continued

Box 9.1 *Continued*

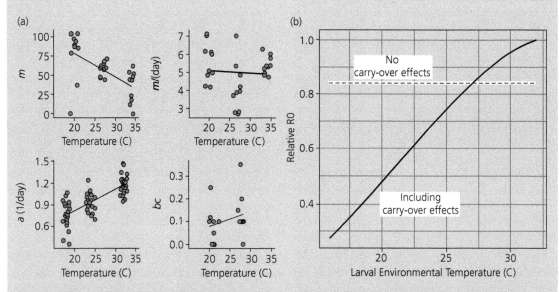

Figure 9.2 The results of an example R_0 model including the carry-over effects of larval temperature on relevant mosquito life history traits. Panel A depicts example carry-over effects of larval temperature on fecundity (*m*), lifespan (*l*), biting rate (*a*), and vector competence (*bc*) with data from the literature and linear regressions plotted. Panel B compares R_0 resulting from models that do and do not include the carry-over effects of larval temperatures for ambient adult temperatures of 27°C.

relationships with empirical data. As a case-study, we present the results of this approach for *Ae. albopictus* and DENV across differing larval temperatures.

We parameterized the Ross-Macdonald R_0 equation (Smith et al. 2012), where parameters that include carry-over effects as described in this chapter are denoted as a function of larval temperature, *LT*:

$$R_0 = \frac{m(LT)a(LT)^2 \, bc(LT)l(LT)}{r} e^{-\frac{v}{l(LT)}}$$

Where *m* is the mosquito density, approximated following Parham & Micheal (2010) as the ratio between the birth and death rates, *a* is the biting rate, *bc* is the vector competence, *l* is the adult mosquito lifespan, *v* is the extrinsic incubation rate, and *r* is the daily rate each human recovers from infection. All parameters that are a function of carry-over effects are parameterized from the literature for a constant adult temperature of 27°C: *m* from Ezeakacha & Yee 2019, *bc* from Evans et al. 2018, *l* from Ezeakacha & Yee 2019 (Fig. 9.2A). To approximate biting rate, we applied the equation for *Ae. aegypti* biting rate as a function of wing length from Scott et al. (2000) to wing lengths of *Ae. albopictus* reported in Westbrook et al. (2010) (Fig. 9.2A). The extrinsic incubation

rate and the human recovery rate were held constant at values of 0.05 day⁻¹ and 0.1 day⁻¹, respectively. We then scaled R_0 to its maximum value, resulting in a relative value of R_0. To examine how carry-over effects potentially influence R_0, we compared this model to one parameterized for a 27°C larval environment, matching the 27°C adult environment (Fig. 9.2B). Code to reproduce this example is located on the figshare repository (doi: 10.6084/m9.figshare.9487895).

This simple example demonstrates the effect of larval environmental temperature on disease transmission, decreasing R_0 below the ambient adult temperature of 27°C and increasing it above this temperature. The relationship with R_0 is not linear, particularly due to the quadratic biting rate term in the equation corresponding to the two bites needed for a mosquito to acquire and transmit a pathogen. These non-linearities would be magnified as further complexity is added to this model, such as fitting non-linear regressions to parameters and including interactions between the adult and larval temperatures on life-history traits. Similar models could be constructed for all of the larval environmental characteristics discussed here, depending on the availability of empirical data, and could lend insight into how multiple carry-over effects may impact disease transmission without the need for large factorial experiments.

2002) and is time and labor intensive to measure directly. Other traits are often compared to body size as well, and could be inferred through a similar method, although the relationships are not quite as clear. Biting rates, for example, are higher in smaller *Aedes* mosquitoes (Scott et al. 2000, Maciel-De-Freitas et al. 2007), particularly for the first gonotrophic cycle, when they may take multiple blood meals to counteract low teneral reserves (Takken et al. 1998). Similarly, body size is assumed to relate to the thickness of the midgut epithelial layer, which serves as a barrier to pathogen dissemination (Telang et al. 2012). Body size can also impact mosquito movement, with smaller sizes allowing access to indoor hosts through small gaps in eaves and screens that would otherwise be protected and increasing dispersal ability (Maciel-De-Freitas et al. 2007). However, these allometric relationships are not always stable, and certain combinations of larval environments can alter the usually positive wing length-body mass relationship behind these relationships (Costanzo et al. 2018; Costanzo et al. 2015; Reiskind & Zarrabi 2012; Siegel et al. 1992; Zeller & Koella 2016). This occurs as body size is assumed to be a predictor of these traits, regardless of how it is attained. An unexplored area of research is examining the relative contribution of environmental and genetic determinants of body size, and how each alters assumed allometric relationships. Therefore, while these relationships can drive hypotheses, it is preferable to test life history traits directly through empirical work. If the allometric relationships do hold true regardless of the environmental driver, however, this could greatly simplify predictions of disease risk as they could be inferred directly from the body size of field-caught mosquitoes.

9.3.2 Temperature Mismatch Across Life-stages

The majority of studies exploring the effects of the environment on mosquito fitness do not separate the effects of the larval and adult environments. In the case of larval diet and competition, this may be because those environments do not translate as well to the adult stage, where the diet consists of nectar and, in the case of females, blood, and adults are not likely to be competing for hosts. Temperature, however, directly impacts the adult phenotype and may amplify or reduce carry-over effects of temperature from the larval environment. Further, in the field, aquatic and ambient temperatures are rarely the same, with aquatic environments characterized by lower temperatures with smaller daily fluctuations than the neighboring ambient environment (Kumar et al. 2018). This can cause a mismatch in the aquatic and ambient environments that is rarely accounted for in laboratory studies. In order to better understand these effects, experiments that cross the larval and adult temperature treatments are needed to tease apart indirect carry-over effects, direct effects from the adult environment, and changes to adult fitness caused by the shift in environmental temperature from the larval to adult environment. Unfortunately, these experiments are rare, but see Ezeakacha and Yee (2019) for such a study in the *Aedes* system. Crossed experiments can also be expanded to explore transgenerational effects of temperature and other characteristics of the larval environment (e.g. Tran et al. 2018).

Laboratory studies exploring the carry-over effects of temperature implement constant temperatures that are maintained throughout the larval stage. In the field, however, temperature fluctuates on a daily and seasonal basis, particularly for container species. These fluctuations allow for differing minima and maxima while maintaining the same mean, which can expose mosquitoes to temperature stressors. Heat stress as short as ten minutes can increase a mosquito's susceptibility to virus infection through its effect on thermally-sensitive proteins that play a role in RNA interference (Mourya et al. 2004). While some studies have been conducted comparing other mosquito life-history traits in constant and fluctuating thermal environments (Paaijmans et al. 2010; Carrington et al. 2013), they do not separate the effects of the larval and adult environment and do not consider carry-over effects explicitly. Just as studies on adult mosquitoes are beginning to incorporate more realistic thermal profiles, future studies on carry-over effects should expand from simple constant temperature treatments.

9.3.3 Increased Focus on Mechanism

Much of the work conducted to date fails to explore the mechanisms behind the observed carry-over effects. Some studies have focused on changes to different immune mechanisms (e.g. iRNA, epithelial lining) in response to the larval environment (Telang et al. 2012), but the majority only measure the changes in parasite prevalence or intensity rates. Understanding how the larval environment leads to changes in specific immune function could lead to larval habitat management strategies that increase refractoriness to pathogens in adults, in addition to simply reducing the abundance of habitats. Similarly, studies on larval nutrition use broad categories of resources, such as high or low, rather than exploring a gradient of resource levels or varying certain nutrients specifically. Studies that do consider nutrient stoichiometry are relatively recent, but have found that the percent nitrogen and carbon in the larval environment results in changes to adult nutrient stoichiometry, particularly for *Aedes* species (Yee et al. 2015). Adult nutrient stoichiometry, in turn, may determine adult life history traits, as has been shown for *Ae. aegypti* susceptibility to and dissemination of Zika virus (Yee et al. 2019; Paige et al. 2019). The existence of these stoichiometrically-driven carry-over effects highlights changes in water quality and eutrophication as an otherwise understudied global change that could have implications for mosquito-borne disease transmission.

9.3.4 Including Multiple Carry-over Effects

Future work on carry-over effects should study them in combination with other carry-over effects of the larval environment, as well as consider the direct effects of the larval environment on population dynamics. For example, the direct effects of temperature on the mosquito development rate may be a much more important factor in determining mosquito population growth rate than changes in adult fecundity resulting from changes to adult female body size. This can be done through the use of realistically parameterized models and sensitivity analyses to identify which life-history traits have the biggest impact on mosquito population growth or disease transmission. Sensitivity analyses could similarly compare the strength of carry-over effects and direct effects of the adult environment on disease transmission. These modeling exercises can inform empirical work by identifying which larval environments and traits are likely to have significant impacts on mosquito-borne disease transmission. The relative contribution of carry-over effects to disease transmission can also be assessed through a combination of laboratory and field work. For example, Juliano et al. (2014) found that frequency of DENV infection in field caught *Ae. aegypti* increased with increasing female body size, in spite of lower infection rates at higher body sizes, suggesting that longer lifespans of larger mosquitoes may be more important than pathogen susceptibility in determining DENV transmission in this population.

9.4 Conclusions

As with other organisms that experience ontogenetic niche shifts, the adult fitness of mosquitoes is impacted by the environment of previous life stages, with consequences for life-history traits relevant to mosquito-borne disease transmission. Much of this is mediated through allometric relationships with body size or mass. In general, studies have found that the assumed allometric relationships hold true when the life-history traits are directly measured, however they differ by species, and likely population, so should be constructed at the finest level possible. The majority of work has been conducted in the laboratory (but see Evans et al. 2018) or with laboratory-adapted strains, such as the Rockefeller strain of *Ae. aegypti*, that no longer resemble their wild counterparts. Experimental designs should strive to include more environmental variation, including covarying environmental factors that cause carry-over effects, to test hypotheses in more realistic field conditions. There is also a clear bias toward *Aedes*-virus research, with very little work exploring how species interactions such as competition or microbial diversity impact *Anopheles* species and their malaria parasites. Several *Anopheles* species are becoming urbanized and adapting to oviposit in artificial containers like the more domesticated *Aedes* species (Kamdem et al. 2012, Surendran et al. 2019), and there is a clear need to understand how their introduction to the aquatic container food web will impact expanding *Anopheles* and existing mosquito species in containers. Future work should further explore the mechanisms behind carry-over effects, which would allow us to predict the magnitude and direction of carry-over effects in novel environments. Most of the past work has been conducted in a laboratory setting, often exploring one carry-over effect at a time. The use of realistic larval environments, via semi-field studies, for example, is needed to ensure that these laboratory studies translate to the field, and to investigate any interactions between characteristics of the larval environment, such as between temperature and larval resources.

Modeling can inform this work by identifying which larval environments and life-history traits are most likely to have significant impacts on adult transmission potential. Understanding the role of the microbiome will also inform the creation of realistic larval environments. Future research directions should seek a better understanding of how different pathogens interact with gut microbial communities while also untangling the resulting effects of the larval microbiome on vector borne disease transmission. The relationship between mosquito ecology and the surrounding environment is a foundation for vector control programs, and the inclusion of prior environments and their associated carry-over effects will allow for more targeted and efficient public health interventions to control mosquito-borne diseases.

Acknowledgements

We thank J.M. Drake for the invitation to write this chapter, and L.P. Lounibos and one anonymous referee for their very helpful comments and suggestions. We also thank E. Marty for the creation of our conceptual figure.

References

Aboagye-Antwi, F. & Tripet, F. (2010). Effects of larval growth condition and water availability on desiccation resistance and its physiological basis in adult *Anopheles gambiae* sensu stricto. *Malaria Journal*, 9, 225.

Adelman, Z.N., Anderson M.A.E., Wiley M.R., et al. (2013). Cooler temperatures destabilize RNA interference and increase susceptibility of disease vector mosquitoes to viral infection. *PLoS Neglected Tropical Diseases*, 7, e2239.

Akhouayri, I.G., Habtewold T., & Christophides G.K. (2013). Melanotic pathology and vertical transmission of the gut commensal *Elizabethkingia meningoseptica* in the major malaria vector *Anopheles gambiae*. *PLoS One*, 8, e77619.

Alto, B.W. & Bettinardi, D. (2013). Temperature and dengue virus infection in mosquitoes: Independent effects on the immature and adult stages. *American Journal of Tropical Medicine and Hygiene*, 88, 497–505.

Alto, B W., Lounibos, L.P., Higgs, S., & Juliano, S.A. (2005). Larval competition differentially affects arbovirus infection in *Aedes* mosquitoes. *Ecology*, 86, 3279–88.

Alto, B.W., Lounibos, L.P., Mores, C, & Reiskind, M.H. (2008). Larval competition alters susceptibility of adult *Aedes* mosquitoes to dengue infection. *Proceedings of the Royal Society B: Biological Sciences*, 275, 463–71.

Alto, B.W. & L.P. Lounibos. (2013). Vector competence for arboviruses in relation to the larval environment of mosquitoes. In W. Takken & C.J.M. Koenraadt, eds. *Ecology of Parasite-Vector Interactions*, pp. 81–101. Wageningen Academic Publishers, Wageningen, The Netherlands.

Alto, B.W., Muturi E.J., & Lampman R.L. (2012). Effects of nutrition and density in *Culex pipiens*. *Medical and Veterinary Entomology*, 26, 396–406.

Angilleta, M.J., Steury, T.D., & Sears, M.W. (2004). Temperature, growth rate, and body size in ectotherms: Fitting piece of a life-history puzzle. *Integrative and Comparative Biology*, 44, 498–509.

Araújo, da-Silva, M. Gil, L.H.S., & de-Almeida e-Silva, A. (2012). Larval food quantity affects development time, survival and adult biological traits that influence the vectorial capacity of *Anopheles darlingi* under laboratory conditions. *Malaria Journal*, 11, 261.

Armbruster, P. & Hutchinson R.A. (2002). Pupal mass and wing length as indicators of fecundity in *Aedes albopictus* and *Aedes geniculatus* (Diptera: Culicidae). *Journal of Medical Entomology*, 39, 699–704.

Armistead, J.S., Arias, J.R., Nishimura, N., & Lounibos, L.P. (2008). Interspecific larval competition between *Aedes albopictus* and *Aedes japonicus* (Diptera: Culicidae) in northern Virginia. *Journal of Medical Entomology*, 45, 629–37.

Atkinson, D. (1994). Temperature and organism size—A biological law for ectotherms? *Advances in Ecological Research*, 25, 1–58

Barreaux, A.M.G., Stone, C.M., Barreaux, P., & Koella, J.C. (2018). The relationship between size and longevity of the malaria vector *Anopheles gambiae* (s.s.) depends on the larval environment. *Parasites & Vectors*, 11, 485.

Bédhomme, S., Agnew, P., Sidobre, C., & Michalakis Y. (2005). Pollution by conspecifics as a component of intraspecific competition among *Aedes aegypti* larvae. *Ecological Entomology*, 30, 1–7.

Beckerman, A., Benton, T.G., Ranta, E., Kaitala, V., & Lundberg, P. (2002). Population dynamic consequences of delayed life-history effects. *Trends in Ecology & Evolution*, 17, 263–9.

Beketov, M.A. & Liess, M. (2007). Predation risk perception and food scarcity induce alterations of life-cycle traits of the mosquito *Culex pipiens*. *Ecological Entomology*, 32, 405–10.

Bellamy, S.K. & Alto, B.W. (2018). Mosquito responses to trait- and density-mediated interactions of predation. *Oecologia*, 187, 233–43.

Benelli, G., Caselli, A., & Canale, A. (2017). Nanoparticles for mosquito control: Challenges and constraints. *Journal of King Saud University—Science*, 29, 424–35.

Benard, M.F. & McCauley S.J. (2008). Integrating across life-history stages: Consequences of natal habitat effects on dispersal. *The American Naturalist* 171, 553–67.

Bevins, S.N. (2008). Invasive mosquitoes, larval competition, and indirect effects on the vector competence of native mosquito species (Diptera: Culicidae). *Biological Invasions*, 10, 1109–17.

Bradshaw, W.E. (1980). Thermoperiodism and the thermal environment of the pitcher-plant mosquito, *Wyeomyia smithii*. *Oecologia*, 46, 13–17.

Bradshaw, W.E., Armbruster, P.A., & Holzapfel, C.M. (1998). Fitness consequences of hibernal diapause in the pitcher-plant mosquito, *Wyeomyia smithii*. *Ecology*, 79, 1458–62.

Brady, O.J., Godfray, H.J.C., Tatem, A.J. et al. (2015). Adult vector control, mosquito ecology and malaria transmission. *International Health*, 7, 121–9.

Breaux, J.A., Schumacher, M.K., & Juliano, S.A. (2014). What does not kill them makes them stronger: Larval environment and infectious dose alter mosquito potential to transmit filarial worms. *Proceedings of the Royal Society B: Biological Sciences*, 281, 20140459.

Briegel, H. (2003). Physiological bases of mosquito ecology. *Journal of Vector Ecology*, 28, 1–11.

Buckner, E.A., Alto, B.W., & Lounibos, L.P. 2016. Larval temperature–food effects on adult mosquito infection and vertical transmission of dengue-1 virus. *Journal of Medical Entomology*, 53, 91–8.

Carrington, L.B., Armijos, M.V., Lambrechts, L., Barker, C.M., & Scott, T.W. (2013). Effects of fluctuating daily temperatures at critical thermal extremes on *Aedes aegypti* life-history traits. *PLoS One*, 8, e58824.

Carrington, L.B., Tran, B.C.N, Le, N.T.H. et al. (2018). Field- and clinically derived estimates of Wolbachia-mediated blocking of dengue virus transmission potential in *Aedes aegypti* mosquitoes. *Proceedings of the National Academy of Sciences of the United States of America*, 115, 361–6.

Cator, L.J., Thomas, S., Paaijmans, K.P. et al. (2013). Characterizing microclimate in urban malaria transmission settings: A case study from Chennai, India. *Malaria Journal*, 12, 1–1.

Chandrasegaran, K., Kandregula, S.R., Quader, S., & Juliano, S.A. (2018). Context-dependent interactive effects of non-lethal predation on larvae impact adult longevity and body composition. *PloS One*, 13, e0192104.

Colinet, H., Sinclair, B.J., P. Vernon, P., & Renault, D. (2015). Insects in fluctuating thermal environments. *Annual Review of Entomology*, 60, 123–40.

Chouaia, B., Rossi, P., Epis, S. et al. (2012). Delayed larval development in *Anopheles* mosquitoes deprived of *Asaia* bacterial symbionts. *BMC Microbiology*, 12, S2.

Christiansen-Jucht, C.D., P.E. Parham, P.E., Saddler, A., Koella, J.C., & Basáñez M.-G. (2015). Larval and adult environmental temperatures influence the adult reproductive traits of *Anopheles gambiae s.s. Parasites & Vectors*, 8, 456.

Christiansen-Jucht, C., Parham, P.E., Saddler, A., Koella, J.C., & Basáñez, M.-G. (2014). Temperature during larval development and adult maintenance influences the survival of *Anopheles gambiae s.s. Parasites & Vectors*, 7, 489.

Civitello, D.J., Hiba, F., Johnson, L.R., Nisbet, R., & Rohr, J.R. (2018). Bioenergetic theory predicts infection dynamics of human schistosomes in intermediate host snails across ecological gradients. *Ecology Letters*, 21(5), 692–701.

Coon, K.L., Brown, M R., & Strand M.R. (2016a). Mosquitoes host communities of bacteria that are essential for development but vary greatly between local habitats. *Molecular Ecology*, 25, 5806–26.

Coon, K.L., Brown, M.R., & Strand, M.R. (2016b). Gut bacteria differentially affect egg production in the anautogenous mosquito *Aedes aegypti* and facultatively autogenous mosquito *Aedes atropalpus* (Diptera: Culicidae). *Parasites & Vectors*, 9, 375.

Coon, K.L., Valzania, L., McKinney, D.A., Vogel, K.J., Brown, M.R., & Strand, M.R. (2017). Bacteria-mediated hypoxia functions as a signal for mosquito development. *Proceedings of the National Academy of Sciences of the United States of America*, 114, E5362–9.

Costa, G., Gildenhard, M., Eldering, M., et al. (2018). Non-competitive resource exploitation within mosquito shapes within-host malaria infectivity and virulence. *Nature Communications*, 9, 3474.

Costanzo, K.S., Schelble, S.S., Jerz, K.K., & Keenan M. (2015). The effect of photoperiod on life history and blood-feeding activity in *Aedes albopictus* and *Aedes aegypti* (Diptera: Culicidae). *Journal of Vector Ecology*, 40, 164–71.

Costanzo, K.S., Westby, K.M., & Medley K.A. (2018). Genetic and environmental influences on the size-fecundity relationship in *Aedes albopictus* (Diptera: Culicidae): Impacts on population growth estimates? *PLoS One*, 13, e0201465.

Couret, J., Dotson, E., & Benedict, M.Q. (2014). Temperature, larval diet, and density effects on development rate and survival of *Aedes aegypti* (Diptera: Culicidae). *PLoS One*, 9, e87468.

Dickson, L.B., Jiolle, D., Minard, G. et al. (2017). Carryover effects of larval exposure to different environmental bacteria drive adult trait variation in a mosquito vector. *Science Advances* 3, e1700585.

Dodson, B.L., Hughes, G.L., Paul, O., Matacchiero, A.C., Kramer, L.D., & Rasgon, J.L. (2014). Wolbachia enhances West Nile virus (WNV) infection in the mosquito *Culex tarsalis*. *PLoS Neglected Tropical Diseases*, 8, e2965.

Dorigatti, I., McCormack, C., Nedjati-Gilani, G., & Ferguson, N.M. (2018). Using *Wolbachia* for dengue control: Insights from modelling. *Trends in Parasitology*, 34, 102–13.

Duguma D., Rugman-Jones P., Kaufman M.G., et al. (2013). Bacterial communities associated with *Culex* mosquito larvae and two emergent aquatic plants of bioremediation importance. *PLoS One*, 8(8), e72522.

Duguma, D., Hall, M.W., Rugman-Jones, P., Stouthamer R., Terenius, O., Neufeld, J.D., & Walton, W.E. (2015). Developmental succession of the microbiome of *Culex* mosquitoes. *BMC Microbiology*, 15, 140.

Dye, C. (1982). Intraspecific competition amongst larval *Aedes aegypti*: Food exploitation or chemical interference? *Ecological Entomology*, 7, 39–46.

Dye, C. (1984). Competition amongst larval *Aedes aegypti*: The role of interference. *Ecological Entomology*, 9, 355–7.

Evans, M.V., Shiau, J.C., Solano, N., Brindley, M.A., Drake, J.M., & Murdock, C.C. (2018). Carry-over effects of urban larval environments on the transmission potential of dengue-2 virus. *Parasites & Vectors* 11, 426.

Ezeakacha, N.F. & Yee, D.A. (2019). The role of temperature in affecting carry-over effects and larval competition in the globally invasive mosquito *Aedes albopictus*. *Parasites & Vectors*, 12, 123.

Gloria-Soria, A., Armstrong, P.M., Powell, J.R., & P.E. Turner, P.E. (2017). Infection rate of *Aedes aegypti* mosquitoes with dengue virus depends on the interaction between temperature and mosquito genotype. *Proceedings of the Royal Society B: Biological Sciences*, 284, 20171506.

Grimstad, P.R. & Haramis, L.D. (1984). *Aedes Triseriatus* (Diptera: Culicidae) and La Crosse Virus III. Enhanced oral transmission by nutrition-deprived mosquitoes. *Journal of Medical Entomology*, 21, 249–56.

Grimstad, P.R. & Walker, E.D. (1991). *Aedes triseriatus* (Diptera: Culicidae) and La Crosse Virus. IV. Nutritional deprivation of larvae affects the adult barriers to infection and transmission. *Journal of Medical Entomology*, 28, 378–86.

Harrison, X.A., Blount, J.D., Inger, R., Norris, D.R., & Bearhop, S. (2011). Carry-over effects as drivers of fitness differences in animals. *Journal of Animal Ecology*, 80, 4–18.

Hawley, W. (1985a). Population dynamics of *Aedes sierrensis*. In L. Lounibos, J. Rey, & J. Frank, eds. Ecology of Mosquitoes Proceedings of a Workshop. Florida Medical Entomology Laboratory, University of Florida, Vero Beach FL, USA.

Hawley, W.A. (1985b). The effect of larval density on adult longevity of a mosquito, *Aedes sierrensis*: Epidemiological consequences. *Journal of Animal Ecology*, 54, 955.

Hillyer, J.F. (2016). Insect immunology and hematopoiesis. *Developmental and Comparative Immunology*, 58, 102–18.

Huey, R.B. & Stevenson, R. (1979). Integrating thermal physiology and ecology of ectotherms: A discussion of approaches. *American Zoologist*, 19, 357–66.

Hughes, G.L., Vega-Rodriguez, J., Xue, P. & Rasgon, J.L. (2012). *Wolbachia* strain wAlbB enhances infection by the rodent malaria parasite *Plasmodium berghei* in *Anopheles gambiae* mosquitoes. *Applied and Environmental Microbiology*, 78, 1491–5.

Jennings, C.D. & Kay, B.H. (1999). Dissemination barriers to Ross River virus in *Aedes vigilax* and the effects of larval nutrition on their expression. *Medical and Veterinary Entomology*, 13, 431–8.

Johnson, L.R., Ben-Horin, T., Lafferty, K.D. et al. (2015). Understanding uncertainty in temperature effects on vector-borne disease: A Bayesian approach. *Ecology*, 96, 203–13.

Juliano, S.A., Ribeiro, G.S., Maciel-de-Freitas, R. et al. (2014). She's a femme fatale: Low-density larval development produces good disease vectors. *Memórias do Instituto Oswaldo Cruz*, 109, 1070–7.

Kamdem, C., Fossog, B.T., Simard, F. et al. (2012). Anthropogenic habitat disturbance and ecological divergence between incipient species of the malaria mosquito *Anopheles gambiae*. *PLoS One*, 7, e39453.

Kang, D.S., Alcalay, Y., Lovin, D.D. et al. (2017). Larval stress alters dengue virus susceptibility in *Aedes aegypti* (L.) adult females. *Acta Tropica*, 174, 97–101.

Klempner, M.S., Unnasch, T.R., & Hu, L.T. (2007). Taking a bite out of vector-transmitted infectious diseases. *New England Journal of Medicine*, 356, 2567–9.

Klowden, M.J., Blackmer, J.L., & Chambers, G.M. (1988). Effects of larval nutrition on the host-seeking behavior of adult *Aedes aegypti* mosquitoes. *Journal of the American Mosquito Control Association*, 4, 73–5.

Kumar, G., Pande, V., Pasi, S., Ojha, V.P., & Dhiman, R.C. (2018). Air versus water temperature of aquatic habitats in Delhi: Implications for transmission dynamics of *Aedes aegypti*. *Geospatial Health*, 13, 707.

Lambrechts, L., Paaijmans, K.P., Fansiri, T. et al. (2011). Impact of daily temperature fluctuations on dengue virus transmission by *Aedes aegypti*. *Proceedings of the National Academy of Sciences of the United States of America*, 108, 7460–5.

Lindh, J.M., Borg-Karlson, A.-K., & Faye, I. (2008). Transstadial and horizontal transfer of bacteria within a colony of *Anopheles gambiae* (Diptera: Culicidae) and oviposition response to bacteria-containing water. *Acta Tropica*, 107, 242–50.

Lounibos, L.P., Suarez, S., Menendez, Z. et al. (2002). Does temperature affect the outcome of larval competition between *Aedes aegypti* and *Aedes albopictus*? *Journal of Vector Ecology*, 27, 86–95.

Lyimo, E.O., & Takken, W. (1993). Effects of adult body size on fecundity and the pre-gravid rate of *Anopheles gambiae* females in Tanzania. *Medical and Veterinary Entomology*, 7, 328–32.

Maciel-De-Freitas, R., Codeço, C.T., & De-Oliveira, L. (2007). Body size-associated survival and dispersal rates of *Aedes aegypti* in Rio de Janeiro. *Medical and Veterinary Entomology*, 21, 284–92.

Merrit, R.W., Dadd, R.H., & Walker, E.D. (1992). Feeding behavior, natural food, and nutritional relationships of larval mosquitoes. *Annual Review of Entomology*, 37, 349–76.

Mitraka, E., Stathopoulos, S., Siden-Kiamos, I., Christophides, G.K., & Louis, C. (2013). *Asaia* accelerates larval development of *Anopheles gambiae*. *Pathogens and Global Health*, 107, 305–11.

Moll, R.M., Romoser, W.S., Modrakowski, M.C., Moncayo, A.C., & Lerdthusnee, K. (2001). Meconial peritrophic nembranes and the fate of midgut bacteria during mosquito (Diptera: Culicidae) metamorphosis. *Journal of Medical Entomology*, 38, 29–32.

Moller-Jacobs, L.L., Murdock, C.C., & Thomas, M.B. (2014). Capacity of mosquitoes to transmit malaria depends on larval environment. *Parasites & Vectors*, 7, 593.

Mousseau, T.A., & Dingle, H. (1991). Maternal effects in insect life histories. *Annual Review of Entomology*, 36, 511–34.

Monaghan, P. (2008). Early growth conditions, phenotypic development and environmental change. *Philosophical Transactions of the Royal Society of London B: Biological Sciences*, 363, 1635–45.

Moncayo, A.C., Lerdthusnee, K., Leon, R., Robich, R.M., & Romoser, W.S. (2005). Meconial peritrophic matrix structure, formation, and meconial degeneration in mosquito pupae/pharate adults: Histological and ultrastructural aspects. *Journal of Medical Entomology*, 42, 939–44.

Moore, C.G. & Fisher, B.R. (1969). Competition in mosquitoes: Density and species ratio effects on growth, mortality, fecundity, and production of growth retardant. *Annals of the Entomological Society of America*, 62, 1325–31.

Mordecai, E.A., Cohen, J.M., Evans, M.V. et al. (2017). Detecting the impact of temperature on transmission of Zika, dengue, and chikungunya using mechanistic models. *PLoS Neglected Tropical Diseases*, 11, e0005568.

Mordecai, E.A., Paaijmans, K.P., Johnson, L.R. et al. (2013). Optimal temperature for malaria transmission is dramatically lower than previously predicted. *Ecology Letters*, 16, 22–30.

Moreira, L.A., Iturbe-Ormaetxe, I., Jeffery, J.A. et al. (2009). A *Wolbachia* symbiont in *Aedes aegypti* limits infection with dengue, chikungunya, and Plasmodium. *Cell*, 139, 1268–78.

Mourya, D.T., Yadav, P., & Mishra, A.C. (2004). Effect of temperature stress on immature stages and susceptibility of *Aedes aegypti* mosquitoes to chikungunya virus. *American Journal of Tropical Medicine and Hygiene*, 70, 346–50.

Murdock, C., Paaijmans, K., Bell, A. et al. (2012). Complex effects of temperature on mosquito immune function., 279, 3357–66.

Murdock, C.C., Blanford, S., Hughes, G.L., Rasgon, J.L., & Thomas, M.B. (2014). Temperature alters *Plasmodium* blocking by *Wolbachia*. *Scientific Reports*, 4, 3932.

Murdock, C.C., Evans, M.V., McClanahan, T.D., Miazgowicz, K.L., & Tesla, B. (2017). Fine-scale variation in microclimate across an urban landscape shapes variation in mosquito population dynamics and the potential of *Aedes albopictus* to transmit arboviral disease. *PLoS Neglected Tropical Diseases*, 11, e0005640.

Murugan, K., Panneerselvam, C., Samidoss, P. et al. (2016). *In vivo* and *in vitro* effectiveness of *Azadirachta indica*-synthesized silver nanocrystals against *Plasmodium berghei* and *Plasmodium falciparum*, and their potential against malaria mosquitoes. *Research in Veterinary Science*, 106, 14–22.

Muturi, E.J., Blackshear, M., & Montgomery, A. (2012). Temperature and density-dependent effects of larval environment on *Aedes aegypti* competence for an alphavirus. *Journal of Vector Ecology*, 37, 154–61.

Muturi, E.J., Costanzo, K., Kesavaraju, B., & Alto, B.W. (2011). Can pesticides and larval competition alter susceptibility of *Aedes* mosquitoes (Diptera: Culicidae) to arbovirus infection? *Journal of Medical Entomology*, 48, 429–36.

Naresh, K.A., Murugan, K., Shobana, K., & Abirami, D. (2013). Isolation of *Bacillus sphaericus* screening larvicidal, fecundity, and longevity effects on malaria vector *Anopheles stephensi*. *Scientific Research and Essays*, 8, 425–31.

Nasci, R.S. (1991). Influence of larval and adult nutrition on biting persistence in *Aedes aegypti* (Diptera: Culicidae). *Journal of Medical Entomology*, 28, 522–6.

Noden, B.H., O'Neal, P.A., J.Fader, E., & Juliano, S.A. (2016). Impact of inter- and intra-specific competition among larvae on larval, adult, and life-table traits of *Aedes aegypti* and *Aedes albopictus* females. *Ecological Entomology*, 41, 192–200.

Norris, D.R. & Taylor, C.M. (2006). Predicting the consequences of carry-over effects for migratory populations. *Biology Letters*, 2, 148–51.

Novakova, E., Woodhams, D.C., Rodríguez-Ruano, S.M. et al. (2017). Mosquito microbiome dynamics, a background for prevalence and seasonality of West Nile virus. *Frontiers in Microbiology*, 8, 256.

Op de Beeck, L., Janssens, L., & Stoks, R. (2016). Synthetic predator cues impair immune function and make the biological pesticide *Bti* more lethal for vector mosquitoes. *Ecological Applications*, 26, 355–66.

Ower, G.D. & Juliano, S.A. (2019). The demographic and life-history costs of fear: Trait-mediated effects of threat of predation on *Aedes triseriatus*. *Ecology and Evolution*, 9, 3794–806.

Paaijmans, K.P., Blanford, S., Bell, A.S., Blanford, J.I., Read, A.F., & Thomas, M.B. (2010). Influence of climate on malaria transmission depends on daily temperature variation. *Proceedings of the National Academy of Sciences of the United States of America*, 107, 15135–9.

Paaijmans, K.P., Huijben, S., Githeko, A.K., and Takken W. (2009). Competitive interactions between larvae of the malaria mosquitoes *Anopheles arabiensis* & *Anopheles gambiae* under semi-field conditions in western Kenya. *Acta Tropica*, 109, 124–30.

Paige, A.S., Bellamy, S.K., Alto, B.W., Dean, C.L., & Yee, D.A. (2019). Linking nutrient stoichiometry to Zika virus transmission in a mosquito. *Oecologia*, 191, 1–10.

Parham, P.E. & Michael, E. (2010). Modeling the effects of weather and climate change on malaria transmission. *Environmental Health Perspectives*, 118, 620–6.

Pascual, M., Ahumada, J.A., Chaves, L.F., Rodó, X., & Bouma, M. (2006). Malaria resurgence in the East African highlands: Temperature trends revisited. *Proceedings of the National Academy of Sciences of the United States of America*, 103, 5829–34.

Pelz-Stelinski, K., Kaufman, M.G. & Walker, E.D. (2011). Beetle (Coleoptera: Scirtidae) facilitation of larval mosquito growth in tree hole habitats is linked to multitrophic microbial interactions. *Microbial Ecology*, 62, 690.

Perez, M.H. & Noriega, F.G. (2013). *Aedes aegypti* pharate 1st instar quiescence: A case for anticipatory reproductive plasticity. *Journal of Insect Physiology*, 59, 318–24.

Ramirez, J.L., Souza-Neto, J., Torres Cosme, R., et al. (2012). Reciprocal tripartite interactions between the *Aedes aegypti* midgut microbiota, innate immune system and dengue virus influences vector competence. *PLoS Neglected Tropical Diseases*, 6, e1561.

Ratkowsky, D.A., Olley, J., McMeekin, T. A., & Ball, A. (1982). Relationship between temperature and growth rate of bacterial cultures. *Journal of Bacteriology*, 149, 1–5.

Reiskind, M.H. & Lounibos, L.P. (2009). Effects of intraspecific larval competition on adult longevity in the mosquitoes *Aedes aegypti* and *Aedes albopictus*. *Medical and Veterinary Entomology*, 23, 62–8.

Reiskind, M.H. & Zarrabi, A.A. (2012). Is bigger really bigger? Differential responses to temperature in measures of body size of the mosquito, Aedes albopictus. *Journal of Insect Physiology*, 58, 911–17.

Roberts, D. & Kokkinn, M. (2010). Larval crowding effects on the mosquito *Culex quinquefasciatus*: physical or chemical?. *Entomologia Experimentalis et Applicata*, 135, 271–5.

Roux, O., Vantaux, A., Roche, B. et al. (2015). Evidence for carry-over effects of predator exposure on pathogen transmission potential. *Proceedings of the Royal Society B: Biological Sciences*, 282, 20152430.

Scott, T.W., Amerasinghe, P.H., Morrison, A.C. et al. (2000). Longitudinal studies of *Aedes aegypti* (Diptera: Culicidae) in Thailand and Puerto Rico: Blood feeding frequency. *Journal of Medical Entomology*, 37, 89–101.

Shocket, M.S., Verwillow, A.B., Numazu, M.G. et al. (2019). Transmission of West Nile virus and other temperate mosquito-borne viruses occurs at lower environmental temperatures than tropical mosquito-borne diseases. *bioRxiv*, 597898.

Siegel, J.P., Novak, Lampman, R.L., & Steinly, B.A. (1992). Statistical appraisal of the weight–wing length relationship of mosquitoes. *Journal of Medical Entomology*, 29, 711–14.

Silberbush, A., Tsurim, I., Rosen, R., Margalith, Y., & Ovadia, O. (2014). Species-specific non-physical interference competition among mosquito larvae. *PLoS One*, 9, e88650.

Silbermann, R., & Tatar, M. (2000). Reproductive costs of heat shock protein in transgenic *Drosophila melanogaster*. *Evolution*. 54, 2038–45.

Siraj, A.S., Santos-Vega, M., Bouma, M.J. Yadeta, D., Carrascal, D.R., & Pascual, M. (2014). Altitudinal changes in malaria incidence in highlands of Ethiopia and Colombia. *Science*, 343, 1154–8.

Smith, D.L., Battle, K.E., Hay, S.I., Barker, C.M., Scott, T.W., & McKenzie, F.E. (2012). Ross, Macdonald, and a theory for the dynamics and control of mosquito-transmitted pathogens. *PLoS Pathogens*, 8, e1002588.

Sørensen, J.G., Michalak, P., Justesen, J., & Loeschcke, V. (1999). Expression of the heat-shock protein HSP70 in *Drosophila buzzatii* lines selected for thermal resistance. *Hereditas*, 131, 155–64.

Souza, R.S., Virginio, F., Riback, T.I.S., Suesdek, L., Barufi, J.B., & Genta, F.A. (2019). Microorganism-based larval diets affect mosquito development, size and nutritional reserves in the yellow fever mosquito *Aedes aegypti* (Diptera: Culicidae). *Frontiers in Physiology*, 10, 152.

Surendran, S.N., Sivabalakrishnan, K., Sivasingham, A. et al. (2019). Anthropogenic factors driving recent range expansion of the malaria vector *Anopheles stephensi*. *Frontiers in Public Health*, 7, 53.

Suwanchaichinda, C. & Paskewitz, S.M. (1998). Effects of larval nutrition, adult body size, and adult temperature on the ability of *Anopheles gambiae* (Diptera: Culicidae) to melanize sephadex beads. *Journal of Medical Entomology*, 35, 157–61.

Takahashi, M. (1976). The effects of environmental and physiological conditions of *Culex tritaeniorhynchus* on the pattern of transmission of Japanese encephalitis virus. *Journal of Medical Entomology*, 13, 275–84.

Takken, W., Klowden, M.J., & Chambers, G.M. (1998). Effect of body size on host seeking and blood meal utilization in *Anopheles gambiae sensu stricto* (Diptera: Culicidae): The disadvantage of being small. *Journal of Medical Entomology*, 35, 639–45.

Takken, W., Smallegange, R.C., Vigneau, A.J. et al. (2013). Larval nutrition differentially affects adult fitness and *Plasmodium* development in the malaria vectors *Anopheles gambiae* and *Anopheles stephensi*. *Parasites & Vectors*, 6, 345.

Telang, A., Qayum, A.A., Parker, A., Sacchetta, B.R., & Byrnes G.R. (2012). Larval nutritional stress affects vector immune traits in adult yellow fever mosquito *Aedes aegypti* (Stegomyia aegypti). *Medical and Veterinary Entomology*, 26, 271–81.

Tran, T.T., Janssens, L., Dinh, K.V., & Stoks, R. (2018). Transgenerational interactions between pesticide exposure and warming in a vector mosquito. *Evolutionary Applications*, 11, 906–17.

van Uitregt, V.O., Hurst, T.P., & Wilson R.S. (2012). Reduced size and starvation resistance in adult mosquitoes, *Aedes notoscriptus*, exposed to predation cues as larvae. *Journal of Animal Ecology*, 81, 108–15.

Vantaux, A., Lefèvre, T., Cohuet, A. Dabiré, K.R., Roche, B., & Roux, O. (2016a). Larval nutritional stress affects vector life history traits and human malaria transmission. *Scientific Reports*, 6, 36778.

Vantaux, A., Ouattarra, I., Lefèvre, T., & Dabiré, K.R. (2016b). Effects of larvicidal and larval nutritional stresses on *Anopheles gambiae* development, survival and competence for Plasmodium falciparum. *Parasites & Vectors*, 9, 226.

Vonesh, J.R. (2005). Sequential predator effects across three life stages of the African tree frog, *Hyperolius spinigularis*. *Oecologia*, 143, 280–90.

Wang, Y., Gilbreath III, T.M., Kukutla, P., Yan, G.& Xu J. (2011). Dynamic gut microbiome across life history of the malaria mosquito *Anopheles gambiae* in Kenya. *PLoS One*, 6, e24767.

Westbrook, C.J., Reiskind,, M.H.Pesko,, K.N.Greene, K.E., & Lounibos, L.P. (2010). Larval environmental temperature and the susceptibility of *Aedes albopictus* Skuse (Diptera: Culicidae) to chikungunya virus. *Vector-Borne and Zoonotic Diseases*, 10, 241–7.

Westby, K.M. & Juliano, S.A. (2015). Simulated seasonal photoperiods and fluctuating temperatures have limited effects on blood feeding and life history in *Aedes triseriatus* (Diptera: Culicidae). *Journal of Medical Entomology*, 52, 896–906.

Yee, D.A., Kaufman, M.G., & Ezeakacha, N F. (2015). How diverse detrital environments influence nutrient stoichiometry between males and females of the co-occurring container mosquitoes *Aedes albopictus*, *Ae. aegypti*, and *Culex quinquefasciatus*. *PLoS One*, 10, e0133734-18.

Yee, D.A., Ezeakacha, N.F., & Abbott, K.C. (2017). The interactive effects of photoperiod and future climate change may have negative consequences for a widespread invasive insect. *Oikos*, 126, 40–51.

Yee, S.H., Yee, D.A., de Jesus Crespo, R., Oczkowski, A., Bai, F., & Friedman S. (2019). Linking water quality to *Aedes aegypti* and Zika in flood-prone neighborhoods. *EcoHealth*, 16, 191–209.

Zeller, M. & Koella, J.C. (2016). Effects of food variability on growth and reproduction of *Aedes aegypti*. *Ecology and Evolution*, 6, 552–9.

Zirbel, K.E. & Alto, B.W. (2018). Maternal and paternal nutrition in a mosquito influences offspring life histories but not infection with an arbovirus. *Ecosphere*, 9, e02469.

Zouache, K., Fontaine, A., Vega-Rua, A. et al. (2014). Three-way interactions between mosquito population, viral strain and temperature underlying chikungunya virus transmission potential. *Proceedings of the Royal Society B: Biological Sciences*, 281, 20141078.

CHAPTER 10

Incorporating Vector Ecology and Life History into Disease Transmission Models: Insights from Tsetse (*Glossina* spp.)

Sinead English, Antoine M.G. Barreaux, Michael B. Bonsall,
John W. Hargrove, Matt J. Keeling, Kat S. Rock, and Glyn A. Vale

10.1 Introduction

Epidemiological models of vector-borne diseases are important quantitative tools for assessing disease risk and predicting the efficacy of various options for disease control. For example, models can tell us the speed and extent to which certain changes in vector dynamics, such as an enhanced death rate due to trapping or climate change, will affect vector abundance and disease incidence in various situations (reviewed in Parham et al. 2015; see example of Rock et al. 2017). The reliability of models for such purposes depends on a solid understanding of the biology of the vectors, parasites and hosts, but pertinent information is often sparse, especially from the field (Cator et al. 2019).

All such models make assumptions about aspects of vector biology and disease transmission. In many cases, as in the various formulations of the Ross-MacDonald model for the malaria-mosquito system (Smith et al. 2012, Chapter 2 in current volume), it is assumed that crucial parameters have a fixed value, often derived from mean values measured in laboratory experiments. Reality is invariably more complex, however, involving parameter values that differ substantially within and among individuals, strains, species and situations. While such detailed

data on host (humans or other animals) biology—for example in terms of variation in immunity—is often incorporated into models, similar detail on vector and parasite biology has received less attention. For example, the death rate of vectors is often assumed to be constant throughout life, yet it is increasingly apparent that individuals may suffer enhanced mortality when very young or very old (Hargrove et al. 2011; Harrington et al. 2008). Incorporating such complexity on vector biology into epidemiological models can have implications for the outcomes of disease transmission dynamics (Bellan 2010; Rock et al. 2015).

In this chapter, we first describe the classic approach to modeling vector-borne disease. We give several examples of how building on this approach with more complicated assumptions about vector and parasite traits can significantly change the predicted disease dynamics. We discuss the importance of detailed studies in natural vector populations to parameterize these models. We focus largely on tsetse (*Glossina* spp.), vectors of protozoan trypanosome parasites (Leak 1999). Animal African trypanosomiasis (AAT) causes extensive morbidity and mortality in livestock across Africa, resulting in an estimated economic loss in livestock and crop production of at least 1.3

Sinead English, Antoine M.G. Barreaux, Michael B. Bonsall, John W. Hargrove, Matt J. Keeling, Kat S. Rock, and Glyn A. Vale, *Incorporating Vector Ecology and Life History into Disease Transmission Models: Insights from Tsetse* (Glossina *spp.)* In: *Population Biology of Vector-Borne Diseases.* Edited by: John M. Drake, Michael B. Bonsall, and Michael R. Strand: Oxford University Press (2021). © Sinead English, Antoine M.G. Barreaux, Michael B. Bonsall, John W. Hargrove, Matt J. Keeling, Kat S. Rock, and Glyn A. Vale.
DOI: 10.1093/oso/9780198853244.003.0010

billion USD per year (Kristjanson et al. 1999). Some trypanosome species can infect humans, causing different forms of human African trypanosomiasis (HAT), which is extremely debilitating, and often fatal if untreated. While HAT incidence has fallen in recent years, it remains endemic in certain areas (Büscher et al. 2017).

In addition to the animal and public health importance of understanding tsetse, this vector is especially interesting since we have much field and laboratory data indicating the substantial within-population variation in traits used to parameterize classic epidemiological models. Moreover, tsetse flies provide a striking case study of the importance of considering species-specific variation in vector ecology for understanding disease transmission. The flies have an unusual life history, with a low rate of daily mortality, slow rate of reproduction and extreme maternal investment, which has important implications for population dynamics, the spread of disease, and the success of control strategies (Hargrove 2004; Benoit et al. 2015).

10.2 Disease Transmission Dynamics

10.2.1 Developing the Ross-Macdonald Framework

Across all vector-borne diseases, the foundation of epidemiological modeling is the Ross-Macdonald model. Smith et al. (2012) outline many variations on this modeling framework which have been applied to mosquito-borne infections. These classic models of vector-borne disease require only a few simple inputs to define the rate of growth of a disease, or reproductive ratio, which can be characterized as R_0, the average number of secondary infections arising in a totally susceptible population through the introduction of a single infectious case. Using the example of malaria and referring to Fig. 10.1, we have the full cycle, host-vector-host reproduction number given by:

$$R_0^2 = \frac{ma^2bce^{-gn}}{rg}.$$

The disease will persist if, and only if, $R_0 > 1$. R_0 increases with increasing values of the host suscepti-

bility (b), host infectiousness to the vector (c) and the vector to host density (m). It also increases with decreasing values of the rate of recovery in hosts (r), the adult vector mortality (g) and the incubation period of the parasite in the vector (n). While aspects of host immunity are clearly central to this model, we focus on the vector and parasite dynamics. Of particular importance is the vector biting rate (a): this occurs as a quadratic term, reflecting the fact that two bites are needed to complete the transmission cycle: one bite to acquire the infection and another to transmit it. There is considerable sensitivity to the value of a: for example, halving the biting rate reduces the disease growth rate by a factor of four. Similarly, the infection can be driven to extinction by reducing m, the density of vectors relative to the hosts. This is because, if there are relatively few vectors per host, it is unlikely that each host will be bitten the necessary two times to acquire and transmit the infection.

As with all models, the Ross–Macdonald model and its variants must involve some degree of simplification of reality. Thus, the art of model formulation is to judge whether the mathematical convenience of the Ross–Macdonald model is truly fit-for-purpose, or whether entomological and epidemiological realities demand more complex modeling. In the latter case, the leading questions are whether data are sufficient to identify the appropriate structure and parameterize the model, and what extra empirical work might be needed to inform model development. Moreover, attempting to produce and incorporate all of the additional complexity to represent the many nuances of disease transmission, from the vector perspective, requires not just additional information about the vector's biology, but also fuller details of pathogen development within the vector—the literature on which is generally sparse (Cator et al. 2019; Ohm et al. 2018). In order to determine the necessary level of complexity, more detailed models have to be created, and the level of detail required depends on the question being posed: are we trying to learn more about the biology, for example, or to make predictions for certain control interventions? Data can be used to validate models, estimate performance, or aid in selecting among alternative model formulations.

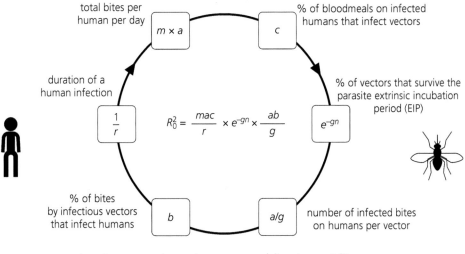

Figure 10.1 The full cycle basic reproductive ratio (R_0^2) of a vector-borne disease, based on the Ross–Macdonald model, which describes the number of vertebrate hosts (in this case, humans) infected after the parasite has established and matured in a single species of arthropod vector (in this case, tsetse).

Here we outline different approaches that have been used to incorporate biological complexity about vectors into classic disease models. In particular, we focus on models that include a more realistic representation of several key aspects of vector biology and ecology, including age-dependent processes, feeding behavior, the extrinsic incubation period (EIP), and the impact of climate. These aspects of vector biology and ecology are important for the disease transmission cycle (Fig. 10.1). In particular, age-dependent mortality and the relationship between vector lifespan and EIP are key to determining the distribution of vectors in a population. As explained previously, bite rate plays a key role with its squared term in the R_0 equation. Finally, ectothermic vectors and parasites are all strongly influenced by climatic factors, particularly temperature, as explained in more detail next.

10.2.2 Life Expectancies and Biting Behavior

The standard Ross–Macdonald model assumes a constant death rate for each vector, which in turn leads to an exponentially distributed life-expectancy,

such that most adult vectors in the population have a short lifespan but a few are much longer lived. It is the latter group that are of prime interest since they will have survived long enough for an infection to be picked up, incubated and passed on. However, the assumption that death rate is constant across the lifespan is unlikely to be realistic for most vectors. Across many animals, including arthropod vectors, individuals often face increasing risks of death as they get older (Nussey et al. 2012), due for example to physiological damage accrued across the lifespan, and many species also pass through a vulnerable early-life period with higher mortality. Such age-dependent mortality has been shown in *Aedes* mosquitoes (Styer et al. 2007; Harrington et al. 2008), and has consequences for models of disease spread (Bellan 2010; Ben-Ami 2019). In Section 10.3, we describe evidence for such age-dependent effects in tsetse (Hargrove 1990; Hargrove et al. 2011).

The Ross–Macdonald model also assumes a constant biting rate although, in reality, blood-feeding vectors follow a more cyclic blood-meal interval, going several days without taking a risky meal before gorging themselves (e.g., tsetse: Randolph et al. 1992). Finally, the EIP of the parasite is unlikely to be constant in a population but, in mosquitoes for example, can vary with temperature, larval conditions and vector or parasite genetic background (Ohm et al. 2018). For trypanosomes developing in tsetse, the EIP will vary depending on the species of trypanosome and tsetse as well as the parasite isolates (Dale et al. 1995), and whether the fly has experienced nutritional stress (Akoda et al. 2009).

Without first developing models that account for this biological complexity—on mortality, feeding and EIP—it is not possible to know how important they are to transmission (Fig. 10.2). Modeling all these factors is made complicated by mechanistic interactions amongst them (Cator et al. 2019), and it is potentially misleading that most mathematical investigations usually focus on just one or two of these features.

Rock et al. (2015) provide a deterministic, partial differential equation framework for incorporating more details on the distributions of vector life expectancies and feeding intervals in understanding disease transmission. Their model captures the status of each vector by two interdependent variables: the age of the adult vector and the time since the last bite. Considering the age of the vector explicitly allows the model to capture such patterns as mortality rates increasing with age, or senescence, as well as other age-dependent processes. Moreover, allowance for age effects can be particularly pertinent during control campaigns that increase the death rate of vectors, or reduce the birth rate, since such operations can drastically change the population's age structure (Van Sickle & Phelps 1988). Similarly, a more realistic approach to modeling bite rate is considered. The time since last bite increases over time, but is reset on each occasion that the vector feeds. In this way the gonotrophic cycle or the time between feeds can be explicitly modeled, which in turn sets the risk of acquiring and transmitting infection.

Including age and biting behavior in models significantly alters infection prevalence in both host and vector populations (Fig. 10.3). Compared to an exponentially distributed life-expectancy, with constant death rate, a more realistic life expectancy, whereby mortality increases with age, leads to a reduction in the number of infected vectors. Intuitively, this pattern occurs because of the reduc-

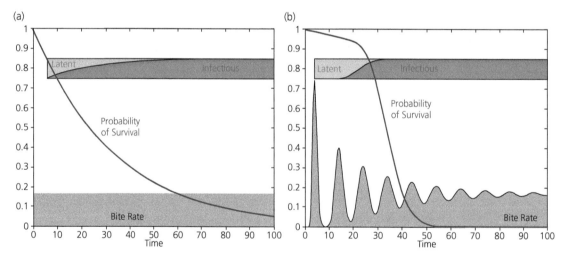

Figure 10.2 Comparison of (a) a simple model, where mortality, bite rates and the rate of passing though the extrinsic incubation period are constant, with (b) a more complex model, with more realistic distributions but the same mean values as the simpler model. The highly flexible model, based on midge biology, involves drawing a random waiting time from a distribution that can take any shape depending on assumptions (Brand et al. 2016). In this example of the complex model, EIP and time between blood-meals are gamma distributed (k=50 and k=20 respectively); while survival probability is a mixture of a gamma distributed senescence (k=30) and a constant mortality rate. In both figures, infection is shown as if starting at a single time-point, whereas in reality the time of infection is also a distribution related to the feeding behavior.
Source: Figure adapted from Brand et al. (2016).

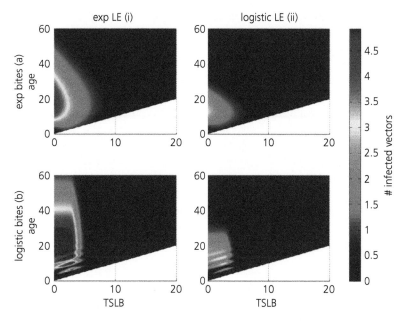

Figure 10.3 Distribution of infected vectors for various ages and times since the last bloodmeal (TSLB), based on the model of, and reproduced from, Rock et al. (2015) taking into account the distribution of life expectancy (exponential on the left, logistic on the right) and whether the time between bloodmeals has an exponential (upper panels) or logistic (lower panels) distribution.

Source: Figure adapted from Rock et al. (2015).

tion in the abundance of long-lived vectors which can infect multiple times. In contrast to the age effects, the effect of different feeding patterns on infection transmission is more subtle and depends on the assumed distribution of life expectancy. When life expectancy is exponential, a lower variance in the time between feeds leads to an increase in transmission: long-lived vectors are guaranteed to take more blood-meals. Against this, when life expectancy is tightly distributed, a lower variance in the time between feeds leads to a decrease in transmission as there is a far lower chance of a vector taking multiple feeds. These generic findings highlight that knowing the distribution (not just the mean) of traits such as feeding rate is critically important for modeling the transmission of infection (Rock et al. 2015).

In principle, a third dimension—in addition to vector age and time since last bloodmeal—could be included in the formulation of Rock et al. (2015) to capture the time since infection, and therefore model the impact of the EIP distribution. This would be a more challenging computational under-

taking, but nonetheless would be interesting, given increasing appreciation that variation in the EIP can have substantial consequences for disease transmission and can itself be influenced by characteristics of the vector and parasite (Ohm et al. 2018).

Such an approach to considering the distribution of the EIP and the time between blood-meals has been described by Brand et al. (2016). In their model, the simplifying assumption is made that the life expectancy is exponentially distributed, which is a reasonable approximation for vectors such as midges and sandflies. The choice of an exponential life expectancy also allows for a more analytic consideration. Results show that, compared to the standard assumption of the simple Ross–Macdonald model, a less variable feeding cycle with a more fixed time between blood-meals leads to a decrease in R_0; a similar trend occurs for a less variable EIP. Intuitively, the higher variances (and exponential distributions) that arise from simple ODE models predict that—in the extremes of the distributions—many more transmission cycles can be achieved, so increasing R_0.

Whether such distributions make significant contributions to the models depends on the key life history traits of the vector or parasite as well as, of course, the vertebrate hosts. For example, one might only need to consider the distribution of the EIP if this is relatively long compared to the adult vector lifespan, as is the case for midges and sandflies with their relatively short lifespans. In contrast, for longer-lived species such as tsetse, variation in the EIP is likely to have less effect on infection transmission since most individual flies will live long enough to transmit the infection regardless of whether the EIP is short or long (Randolph 1998).

10.2.3 Environmental Factors, Vector Population and Disease Dynamics

Vector populations are often highly sensitive to their local environment. This might include habitat features, such as water bodies needed for mosquito breeding (Godfray 2013), or the riverine woodland preferred by some tsetse species (Leak 1999). Climatic factors tend to be especially important in determining the abundance of vectors—which in turn increases transmission, as illustrated by the occurrence of vector to host ratio, m, that appears in the calculation of R_0 (Fig. 10.1). For example, outbreaks of Rift Valley Fever in Africa have been linked to periods of high rainfall, which in turn increases the vegetation cover and creates a higher abundance of the mosquito vectors (Métras et al. 2015).

Climate, especially temperature, can have effects beyond the impacts on vector abundance. Arthropod vectors are ectotherms and thus particularly vulnerable to changes in temperature. In general, warmer temperatures speed up vector life cycles, such that the rate of senescence increases and EIP and feeding intervals decrease. The variance of the EIP may also vary with temperature, for example, malaria EIP has higher variance at lower temperatures (Ohm et al. 2018). Moreover, seasonal conditions in early life can shape later schedules of survival, reproduction and the ability to transmit infections. In the malaria mosquito, *Anopheles gambiae*, for example, larvae reared at higher temperature develop into smaller adults (Barreaux et al. 2018) and larval temperature affects the ability of adults to transmit the malaria parasite, although

whether there is an increase or decrease with higher temperatures depends on larval food stress (Barreaux et al. 2016).

Each of these effects—development rate, EIP, feeding intervals—may respond differently to changes in temperature, in ways that are non-linear and that differ between vector species (Shocket et al. 2018; Tesla et al. 2018; Mordecai et al. 2019). As such, temperature can impact R_0 through any of these mechanisms, from vector survival through to physiology and behavior (Brand & Keeling 2017). Predicting the impacts of climate change is thus far from simple, although it is clear that many diseases normally confined to warm regions will spread to temperate locations as the climate becomes more suited to vector survival (Wilson et al. 2008; Campbell et al. 2015).

10.3 Vector Ecology and Life History

Taken together, these considerations indicate that a more complete picture of trait variations within and between vector populations is important for developing accurate models of vector abundance and disease incidence, and greater reliability in our predictions and explanations for the effects of control programs and climate change. Realistic representation of vector ecology and life history in models of disease transmission will require reliable data for such matters under a range of conditions. These requirements are now considered largely in the context of the tsetse.

10.3.1 Peculiarities of Tsetse Biology

While most insects deposit hundreds of eggs directly into the environment, each female tsetse matures a single egg once every nine days and retains it in her uterus (Leak 1999; Benoit et al. 2015). There it hatches and the larva feeds on a milk-like substance until, when grown to about the size of its mother, it is extruded onto shady soil. It then burrows from sight, to pupate and emerge a few weeks later as a new adult, whereupon it soon searches for its first bloodmeal (Leak 1999, Fig.10.4). This slow but relatively secure method of reproduction requires a great investment in each individual off-

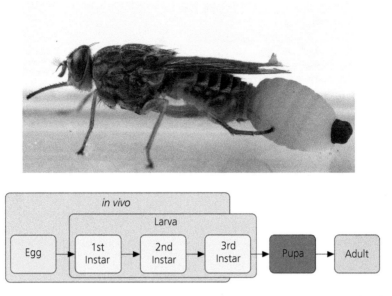

Figure 10.4 Upper panel: tsetse female giving birth to a larva (photo credit: Daniel Hargrove). Lower panel: tsetse life cycle.

spring, amounting to about 25 percent of the total energy budget (Bursell & Taylor 1980) and a huge transfer of fat reserves to a larva born with as much as 50 percent more fat than its mother (Hargrove & Muzari 2015).

In mosquitoes and many other haematophagous insects, only females feed on blood and, even for this sex, energy is also derived from feeding on nectar (Godfray 2013). In contrast, both sexes of tsetse feed exclusively on blood (Leak 1999), which almost doubles the proportion of effective vectors in the population, given an approximate 50:50 sex ratio in emerging flies (Buxton 1955)—before sex differences in mortality affect distributions—relative to those species, like mosquitoes, where only one sex can transmit disease. Moreover, this aspect of tsetse biology also maximizes the proportion of feeds potentially capable of transmitting infection. A further distinction is that tsetse gain all of their water requirements from the blood of their host: thus, unlike mosquitoes, they do not need water bodies for the development of the immature stages (Kleynhans & Terblanche 2009). Consequently, although the abundance of tsetse can change during the dry season (Hargrove 2004), tsetse are unusual in being able to breed, feed and transmit infection throughout the year. Indeed, given the slow rate of reproduction of tsetse, it is essential that the adults live a long time, with no aestivation (Hargrove et al. 2011).

10.3.2 Age Effects

10.3.2.1 Age-Related Changes in Survival

The most direct way of investigating age-dependent survival of tsetse in the field involves serial recaptures of individually marked adult flies released into the wild shortly after emergence (Hargrove 1990). A mark-release-recapture experiment conducted on Redcliff Island, Lake Kariba, Zimbabwe showed that tsetse (*G. morsitans morsitans*) suffered particularly high mortality for the first week of adult life. Thereafter, however, mortality rates among females fell to 1 percent per day by the age of three weeks, and remained below 2 percent per day until the age of 12 weeks. In contrast, male mortality increased with age more rapidly (Hargrove et al. 2011) (Fig. 10.5).

Other, indirect, methods rely on assessing the age distribution of individuals captured in the field, either by measuring the degree of wing fray (Jackson 1946), or, in the case of females, by examining the

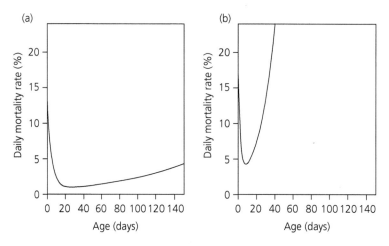

Figure 10.5 Estimated daily mortality in (A) female and (B) male *G. m. morsitans* based on a mark-recapture study on Redcliff Island, Lake Kariba, Zimbabwe.

Source: Data from Hargrove et al. 2011.

ovarian state (Hargrove 1991), as females follow a predictable order of ovulation. These methods are difficult to interpret due to biases in flies attracted to traps based on their age, nutritional and pregnancy status (Hargrove 1991). Moreover, they require an assumption that the age structure of a population is stable over time, which is unrealistic given, for example, seasonal effects on birth and death rates (Hargrove 2004). Dynamic models, which do not require the last assumption—and which have accounted for sampling biases—suggest that pupae and young adult flies have higher mortality than older adults, and are more sensitive to hot temperatures (Ackley & Hargrove 2017, section 10.3.3).

10.3.2.2 Age-Related Changes in Reproduction

Tsetse are unique among arthropod vectors in the significant role that the mother plays in determining offspring size and fat reserves at emergence (English et al. 2016; Hargrove et al. 2018). As there is no free-living larval stage, the only source of energy for the offspring is its mother—from the time it is a newly fertilized egg to the point it takes its first blood meal as a teneral adult. Given this high maternal investment, and the relatively long lifespan of tsetse, one might expect females to show reproductive decline in late life as has been found in many wild animal studies (Nussey et al. 2012). Laboratory studies on the size of offspring produced by females of known age have found mixed results (Langley & Clutton-Brock 1998; English et al. unpublished data; Fig. 10.6A,B). Ongoing research by the co-authors and colleagues is currently investigating these patterns and how they depend on maternal nutrition and the costs of reproduction.

While ovarian age estimation—described previously—allows for the estimation of female age in the wild, field studies of the size of offspring produced by female tsetse of various estimated ages can be exceedingly difficult because larval deposition sites are usually distributed widely and are visited by a fly infrequently. This makes it impracticable to catch females and their offspring at the time of larval deposition. During the hot dry season in Zimbabwe, however, larval deposition is largely concentrated into disused aardvark burrows for which artificial variants can be devised as traps, allowing the capture of mother-pupa pairs (Muzari & Hargrove 2005). Work with such artificial burrows suggests that females produce significantly smaller offspring in their first reproductive cycle, although there appears to be little change thereafter (English et al. 2016; Hargrove et al. 2018, Fig. 10.6 C,D).

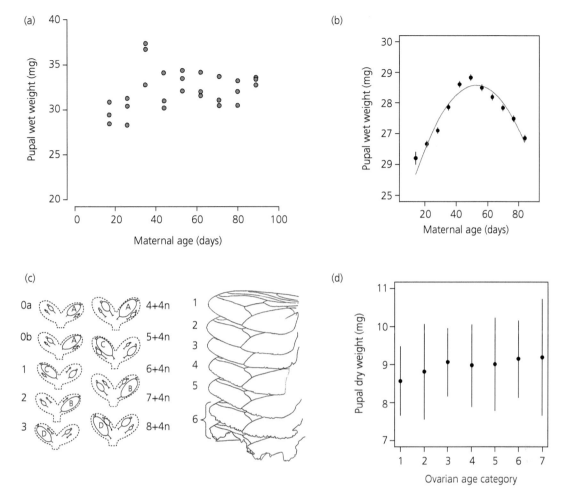

Figure 10.6 A. Pupal wet weight against maternal age in *G m. morsitans*, data from Langley & Clutton-Brock 1998, B. Pupal wet weight against maternal age from the breeding colony of *G. m. morsitans* at Liverpool School of Tropical Medicine (English et al., unpublished data). C. Means of assessing fly age in the wild using ovarian dissection (left), modified from (Hargrove 2013) and wing fray (right, modified from Jackson 1946). D. Pupal dry weight against maternal age as estimated by ovarian category from *G. m. moristans* in the field, data from English et al. 2016.

10.3.3. Effects of Temperature

10.3.3.1. Temperature-Related Size of Flies

The seasonal effects in tsetse have been studied mostly by assessing the body size of samples caught at standard host-like baits throughout the year, with wing length usually taken as an index of body size. There are well-marked annual cycles in wing length for female tsetse captured in Zimbabwe. Length is maximal in the cool season, and minimal in the hot dry season (Hargrove et al. 2019, Fig. 10.7). Moreover, data from artificial burrows also show that, as tem-

peratures increase, mothers produce progressively smaller offspring (English et al. 2016; Hargrove et al. 2018). As temperatures rise during the hot dry season, offspring become smaller and less viable: offspring emerging with low reserves of fat are likely to die quickly, and smaller flies have low survival prospects particularly at extremes of temperature (Phelps & Clarke 1974; Dransfield et al. 1989). Neither of these studies found any evidence for size-specific losses in older flies. Overall, it is the mortalities of pupae and young adults that are most affected by temperature (Ackley & Hargrove 2017).

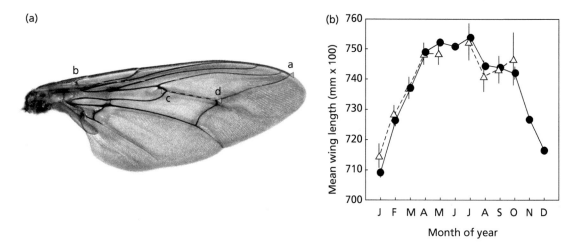

Figure 10.7 A. Wing of a tsetse fly and landmarks used to assess size (Photo credit: Lee Haines). B. Changes in size of tsetse (determined by mean wing length) across months of the year, for tsetse caught using stationary, odor-baited traps (solid points and line), or a vehicle-mounted electric trap (open triangles and dashed line).

Source: Photo and figure (modified) from Hargrove et al. (2019).

10.3.2. Direct Impact of Temperature on Birth and Death

While temperature has an indirect effect on adult mortality, by governing fly size, the main effects of temperature on tsetse abundance are likely to be via more direct impacts on birth and death rate. In this regard, laboratory studies have led to seemingly reliable formulae for the way that temperature governs the rates of development of eggs, larvae and pupae (Hargrove 2004). The equally important understanding of the effect of temperature on the death rate has been produced by a combination of field and laboratory work (Hargrove 2004).

The effect of temperature on pupal death is likely to dominate any other temperature effect on mortality. The most notable aspect of this is how tightly the viability of tsetse pupae is confined to 16–31°C, with drastic increases in mortality outside this range (Hargrove & Vale 2019). This is partly due to patterns of fat metabolism, as the proportion of original fat used during pupal development is a quadratic function of temperature, taking a minimum at about 25°C (Hargrove & Vale 2019). Fat reserves at the point of emergence are crucial for allowing the young adult fly time to acquire its first blood meal.

10.4. Summary and Outlook

The past century or so has seen a rich history of research into tsetse biology, ecology and physiology, all of which have provided parameters that could inform and improve modeled predictions of disease dynamics beyond what is possible by using only the simple parameters included in a classic Ross–Macdonald formulation. We are now at a point where we can better incorporate the biological trait variation (Section 10.3) to the epidemiological models that can deal with such variation (Section 10.2) to develop detailed models of trypanosome transmission dynamics and the efficacy of different control methods.

To improve the reliability of our models of the dynamics of vector populations and infection transmission, continued efforts are required on two fronts. First, classic modeling frameworks can be extended to incorporate more biological complexity. Second, more studies of vector populations under natural and experimental conditions can provide the data required for these models. To assist in both the development of models and the design of empirical studies, continued exchange of methodology and ideas across fields is necessary: facilitated by collaborative grants, workshops and research networks.

10.4.1 Data from the Laboratory and Field

In this chapter, we have emphasized the need for detailed data on vector and pathogen biology to extend classic epidemiological models. In all of this, it is the biological work that is the main limiting factor, since the biology of the vectors, parasites and hosts dictates the required structure of the models, and also provides the appropriate values of parameters. How are we to produce the sort of biological information required?

In tsetse, laboratory studies of the biology of life histories (Phelps & Burrows 1969; Langley & Clutton-Brock 1998), and of the course of infections (Welburn & Maudlin 1999) have produced exceedingly useful data. Further laboratory work could also be important, for example in extending our knowledge of the way that starvation and temperature extremes interact to influence mortality and reproduction of the vectors (Terblanche et al. 2008). Such experiments are more powerful when done in close collaboration with studies in the field. For example, it is only field work that can indicate the actual temperatures to which vectors are naturally exposed. In that particular matter we need to extend our knowledge of temperatures in micro-habitats, rather than in meteorological screens alone.

A key theme of this chapter is the importance of considering distributions of traits in epidemiological models. We have discussed how such traits—for examples adult size—can vary across season. Much of the empirical research described in Section 10.3, in the context of tsetse, has been focused on the savannah tsetse (*G. m. morsitans* and *G. pallidipes*) which have been the subject of extensive study in the Zambezi Valley, Zimbabwe (Leak 1999). There are at least thirty-one species and subspecies of tsetse, largely grouped according to their distribution in different habitat types (Leak 1999), and more comparative work on trait variation both in different populations and across different species will be extremely insightful for modeling tsetse population dynamics and disease transmission.

10.4.2 Future Directions for Models

While our examples in Section 10.2 highlight recent advances in expanding epidemiological models to incorporate aspects of vector and pathogen complexity to disease transmission models, there is still scope for future work. For example, Cator et al. (2019) have recently made a call for more explicit mechanistic models of functional traits when modeling vector populations and vector-borne disease. The framework they propose makes explicit links between vector traits, model parameters, population dynamics and disease transmission. Such models will be particularly informative when predicting how vector populations will shift in distribution, and competence at transmitting disease, under different scenarios of climate change, given the complex mechanistic basis underlying temperature variation, physiological traits, life histories and population dynamics.

10.4.3 Modeling as a Tool to Make Policy Decisions

The previous limitations aside, we now highlight the benefits of using detailed models to make predictions to inform policy on vector control; and the benefits of interactions between modelers and empiricists to improve said models. We end with an example of how such modeling has been applied, in the context of tsetse. Rock et al. (2017) modeled the effects of different intervention strategies—disease surveillance and vector control—in achieving the goal of zero transmission of Gambian HAT by 2030. In order to develop their model, they incorporated the complexity of tsetse biology including senescence (higher mortality at later ages) and explicit consideration of the tsetse life cycle. Their model highlights the dual importance of both active screening and vector control in achieving the elimination goals of Gambian HAT.

10.5. Concluding Remarks

In our rapidly warming world, we need predictive models more than ever to understand and prepare for changes in the distribution of disease-carrying arthropods and to develop strategies to mitigate risks to humans, crops and livestock, and wildlife. Such models, in turn, require a solid understanding of the biology not only of the vertebrate hosts but also of the arthropod vectors and the pathogens they transmit. In this chapter, we have considered how understanding certain key aspects of vector biology—such as how mortality changes with age, and how survival is affected by temperature—can improve these models. For such models on savannah tsetse, we are fortunate to have access to a long history of intensive field and laboratory work, although there is always more to be done. For other species of tsetse, and other vectors with other life histories, we may be faced with more substantial gaps in our understanding. An awareness of how this uncertainty affects model predictions, and the most efficient means to reduce this uncertainty, is key to advancing our ability to deal with the challenge of vector-borne diseases and the burden they impose on global health and biodiversity.

Acknowledgements

This work was funded by the Biotechnology and Biological Sciences Research Council (BBSRC, grant no. BB/P006159/1). SE is supported by a Royal Society Dorothy Hodgkin Fellowship.We thank Lee Haines, Jennifer Lord and Steve Torr for helpful discussion.

References

Ackley, S.F. & Hargrove, J.W. (2017). A dynamic model for estimating adult female mortality from ovarian dissection data for the tsetse fly Glossina pallidipes Austen sampled in Zimbabwe. *PLoS Neglected Tropical Diseases*, 11, e0005813.

Akoda, K., Van den Abbeele, J., Marcotty, T., De Deken, R., Sidibe, I., & Van den Bossche, P. (2009). Nutritional stress of adult female tsetse flies (Diptera: Glossinidae) affects the susceptibility of their offspring to trypanosomal infections. *Acta Tropica*, 111, 263–267.

Barreaux, A.M.G., Barreaux, P., Thiévent, K., Koella, J.C. (2016). Larval environment influences vector competence of the malaria mosquito Anopheles gambiae. *MalariaWorld Journal*, 7, 1–6.

Barreaux, A.M.G., Stone, C.M., Barreaux, P., Koella, J.C. (2018). The relationship between size and longevity of the malaria vector Anopheles gambiae (s.s.) depends on the larval environment. *Parasites & Vectors*, 11, 485.

Bellan, S.E. (2010). The importance of age dependent mortality and the extrinsic incubation period in models of mosquito-borne disease transmission and control. *PloS One*, 5, e10165.

Ben-Ami, F. (2019). Host age effects in invertebrates: Epidemiological, ecological, and evolutionary implications. *Trends in Parasitology*, 35, 466–80.

Benoit, J.B., Attardo, G.M., Baumann, A.A., Michalkova, V., & Aksoy, S. (2015). Adenotrophic viviparity in tsetse flies: Potential for population control and as an insect model for lactation. *Annual Review of Entomology*, 60, 351–71.

Brand, S.P.C. & Keeling, M.J. (2017). The impact of temperature changes on vector-borne disease transmission: Culicoides midges and bluetongue virus. *Journal of the Royal Society Interface*, 14, 20160481.

Brand, S.P.C., Rock, K.S., & Keeling, M.J. (2016). The interaction between vector life history and short vector life in vector-borne disease transmission and control. *PLoS Computational Biology*, 12, e1004837.

Bursell, E. & Taylor, P. (1980). An energy budget for Glossina (Diptera: Glossinidae). *Bulletin of Entomological Research*, 70, 187–96.

Büscher, P., Cecchi, G., Jamonneau, V., & Priotto, G. (2017). Human African trypanosomiasis. *The Lancet*, 390, 2397–409.

Buxton, P.A. (1955). *The Natural History of Tsetse Flies. An Account of the Biology of the Genus Glossina (Diptera)*. Lewis & Co, London.

Campbell, L.P., Luther, C., Moo-Llanes, D., Ramsey, J.M., Danis-Lozano, R., & Peterson, A.T. (2015). Climate change influences on global distributions of dengue and chikungunya virus vectors. *Philosophical Transactions of the Royal Society B*, 370, 20140135.

Cator, L., Johnson, L.R., Mordecai, E.A., El Moustaid, F., Smallwood, T., La Deau, S., Johansson, M., Hudson, P.J., Boots, M., Thomas, M.B. et al. (2019). More than a flying syringe: Using functional traits in vector borne disease research. *bioRxiv*, 501320.

Dale, C., Welburn, S.C., Maudlin, I., & Milligan, P.J.M. (1995). The kinetics of maturation of trypanosome infections in tsetse. *Parasitology*, 111, 187–191.

Dransfield, R.D., Brightwell, R., Kiilu, J., Chaudhury, M.F., & Abie, D.A.A. (1989). Size and mortality rates of Glossina pallidipes in the semi-arid zone of southwestern Kenya. *Medical and Veterinary Entomology*, 3, 83–95.

English, S., Cowen, H., Garnett, E., & Hargrove, J.W. (2016). Maternal effects on offspring size in a natural population of the viviparous tsetse fly. *Ecolological Entomology*, 41, 618–26.

Godfray, H.C.J. (2013). Mosquito ecology and control of malaria. *Journal of Animal Ecology*, 82, 15–25.

Hargrove, J., English, S., Torr, S.J., Lord, J., Haines, L.R., van Schalkwyk, C., Patterson, J., & Vale, G. (2019). Wing length and host location in tsetse (Glossina spp.): Implications for control using stationary baits. *Parasites & Vectors*, 12, 24.

Hargrove, J.H. (2013). An example from the world of tsetse flies. *Mathematical Biosciences and Engineering*, 10, 691–704.

Hargrove, J.W. (1990). Age-dependent changes in the probabilities of survival and capture of the tsetse, Glossina morsitans morsitans Westwood. *International Journal of Tropical Insect Science*, 11(3), 323–30.

Hargrove, J.W. (1991). Ovarian ages of tsetse flies (Diptera: Glossinidae) caught from mobile and stationary baits in the presence and absence of humans. *Bulletin of Entomological Research*, 81, 43–50.

Hargrove, J.W. (2004). Tsetse population dynamics. In: Maudlin I, Holmes P, Miles M, eds. The Trypanosomiases, p. 113–37. CABI Publishing, Wallingford, UK.

Hargrove, J.W. & Muzari, M.O. (2015). Nutritional levels of pregnant and postpartum tsetse Glossina pallidipes Austen captured in artificial warthog burrows in the Zambezi Valley of Zimbabwe. *Physiological Entomology*, 40, 138–48.

Hargrove, J.W., Muzari M.O., & English, S. (2018). How maternal investment varies with environmental factors and the age and physiological state of wild tsetse Glossina pallidipes and Glossina morsitans morsitans. *Royal Society Open Science*, 5, 171739.

Hargrove, J.W., Ouifki, R., & Ameh, H. (2011). A general model for mortality in adult tsetse (Glossina spp.). *Medical and Veterinary Entomology*, 25, 385–94.

Hargrove, J.W. & Vale, G.A. (2019). Models for the rates of pupal development, fat consumption and mortality in tsetse (Glossina spp). *Bulletin of Entomological Research*, June 13, 1–13.

Harrington, L.C, Françoisevermeylen, Jones, J.J., Kitthawee, S., Sithiprasasna, R., Edman, J.D., & Scott, T.W. (2008). Age-dependent survival of the dengue vector Aedes aegypti (Diptera: Culicidae) demonstrated by simultaneous release–recapture of different age cohorts. *Journal of Medical Entomology*, 45, 307–13.

Jackson, C.H.N. (1946). An artificially isolated Generation of Tsetse Flies (Diptera). *Bulletin of Entomological Research*, 37, 291–9.

Kleynhans, E. & Terblanche, J.S. (2009). The evolution of water balance in Glossina (Diptera: Glossinidae): Correlations with climate. *Biology Letters*, 5, 93–6.

Kristjanson, P.M., Swallow, B.M., Rowlands, G.J., Kruska, R.L., & de Leeuw, P.N. (1999). Measuring the costs of African animal trypanosomosis, the potential benefits of control and returns to research. *Agricultural Systems*, 59, 79–98.

Langley, P.A. & Clutton-Brock, T.H. (1998). Does reproductive investment change with age in tsetse flies, *Glossina morsitans morsitans* (Diptera: Glossinidae)? *Functional Ecology*, 12, 866–70.

Leak, S.G.A. (1999). Tsetse biology and ecology: Their role in the epidemiology and control of trypanosomosis. CABI Publishing. Oxford and New York. 568pp.

Métras, R., Jewell, C., Porphyre, T., Thompson, P.N., Pfeiffer, D.U., Collins, L.M., White, R.G. (2015). Risk factors associated with Rift Valley fever epidemics in South Africa in 2008–11. *Science Reports*, 5, 9492.

Mordecai, E.A., Caldwell, J.M., Grossman, M.K., Lippi, C.A., Johnson, L.R., Neira, M., Rohr, J.R., Ryan, S.J., Savage, V., Shocket, M.S., et al. (2019). Thermal biology of mosquito-borne disease. *Ecology Letters*, 22, 1690–708.

Muzari, M.O. & Hargrove, J.W. (2005). Artificial larviposition sites for field collections of the puparia of tsetse flies Glossina pallidipes and G. m. morsitans (Diptera: Glossinidae). *Bulletin of Entomological Research*, 95, 221–9.

Nussey, D.H., Froy, H., Lemaitre, J.-F., Gaillard, J.-M., & Austad, S.N. (2012). Senescence in natural populations of animals: Widespread evidence and its implications for bio-gerontology. *Ageing Research Reviews*, 12, 214–25.

Ohm, J.R., Baldini, F., Barreaux, P., et al. (2018). Rethinking the extrinsic incubation period of malaria parasites. *Parasites & Vectors*, 11, 178.

Parham, P.E., Waldock, J., Christophides, G.K., et al. (2015). Climate, environmental and socio-economic change: Weighing up the balance in vector-borne disease transmission. *Philosophical Transactions of the Royal Society B*, 370, 20130551.

Phelps, R. & Clarke, G. (1974). Seasonal elimination of some size classes in males of *Glossina morsitans morsitans* Westw. (Diptera, Glossinidae). *Bulletin of Entomological Research*, 64, 313–24.

Phelps, R.J. & Burrows, P.M. (1969). Puparial Duration in Glossina Morsitans Orientalis Under Conditions of Constant Temperature. *Entomologia Experimentalis et Applicata*, 12, 33–43.

Randolph, S.E. (1998). Ticks are not Insects: Consequences of Contrasting Vector Biology for Transmission Potential. *Parasitology Today*, 14, 186–92.

Randolph, S.E., Williams, B.G., Rogers, D.J., & Connor, H. (1992). Modeling the effect of feeding-related mortality on the feeding strategy of tsetse (Diptera: Glossinidae). *Medical and Veterinary Entomology* 6(3), 231–40.

Rock, K.S., Torr, S.J., Lumbala, C., & Keeling, M.J. (2017). Predicting the Impact of Intervention Strategies for Sleeping Sickness in Two High-Endemicity Health Zones of the Democratic Republic of Congo. *PLoS Neglected Tropical Diseases*, 11, e0005162.

Rock, K.S., Wood, D.A., & Keeling, M.J. (2015). Age- and bite-structured models for vector-borne diseases. *Epidemics*, 12, 20–9.

Shocket, M.S., Ryan, S.J., & Mordecai, E.A. (2018). Temperature explains broad patterns of Ross River virus transmission. *eLife*, 7, 37762.

Smith, D.L., Battle, K.E., Hay, S.I., Barker, C.M., Scott, T.W., & McKenzie, F.E. (2012). Ross, Macdonald, and a theory for the dynamics and control of mosquito-transmitted pathogens. *PLoS Pathogens*, 8, e1002588.

Styer, L.M., Carey, J.R., Wang, J.-L., & Scott, T.W. (2007). Mosquitoes do senesce: Departure from the paradigm of constant mortality. *American Journal of Tropical Medicine and Hygiene*, 76, 111–17.

Terblanche J.S., Clusella-Trullas, S., Deere, J.A., & Chown, S.L. (2008). Thermal tolerance in a south-east African population of the tsetse fly Glossina pallidipes (Diptera, Glossinidae): Implications for forecasting climate change impacts. *Journal of Insect Physiology*, 54, 114–27.

Tesla, B., Demakovsky, L.R., Mordecai, E.A., et al. (2018). Temperature drives Zika virus transmission: Evidence from empirical and mathematical models. *Proceedings of the Royal Society B Biological Sciences*, 285, 20180795.

Van Sickle, J. & Phelps, R.J. (1988). Age distributions and reproductive status of declining and stationary populations of Glossina pallidipes Austen (Diptera: Glossinidae) in Zimbabwe. *Bulletin of Entomological Research*, 78, 51–61.

Welburn, S.C. & Maudlin, I. 1999. Tsetse–trypanosome interactions: Rites of passage. *Parasitology Today*, 15, 399–403.

Wilson, A., Darpel, K., & Mellor, P.S. (2008). Where does bluetongue virus sleep in the winter? *PLoS Biology*, 6, e210.

SECTION III

Ecological Interactions

CHAPTER 11

Mosquito—Virus Interactions

Christine M. Reitmayer, Michelle V. Evans, Kerri L. Miazgowicz, Philip M. Newberry, Nicole Solano, Blanka Tesla, and Courtney C. Murdock

11.1 Introduction

Mosquito-borne viruses are an emerging threat impacting human health and well-being. Epidemics of dengue, chikungunya, and Zika have spilled out of Africa to spread explosively throughout the world creating public health crises. Worldwide, an estimated 3.9 billion people living within 120 countries are at risk (Brady et al. 2012). In 2015–2016, Zika virus (ZIKV) spread throughout the Americas, including the continental U.S.A., resulting in over 360,000 suspected cases, with likely many more undetected (PAHO 2018). With the rise of neurological disorders and birth defects, such as Guillain-Barré and congenital Zika virus syndrome (Cao-Lormeau et al. 2016, Mlakar et al. 2016), ZIKV became widely feared and was declared a 'public health emergency of international concern' by the World Health Organization in 2016 (World Health Organization 2016). In spite of growing research efforts to develop new therapeutics, vaccines, and innovative mosquito control technologies, mitigating arbovirus disease spread still depends on conventional mosquito control methods, such as insecticide-based interventions, and public education on strategies to reduce mosquito breeding habitat and human to mosquito contacts. Yet, even these conventional approaches are now being threatened by evolution of insecticide resistance, and public education on source reduction has been met with mixed success.

When taking a blood meal, female mosquitoes can potentially ingest arboviruses present in the blood of infectious vertebrate hosts. In order to be a competent vector, mosquitoes have to acquire, sustain and transmit arboviruses. The virus in the blood meal enters the midgut; from here the virus has to infect midgut cells in order to utilize the host cell machinery to replicate. Once it has crossed the midgut epithelium, virus particles disseminate throughout the mosquito body and continue to replicate in various cell types. In order to be passed on to another vertebrate host, virus particles need to cross the salivary gland epithelium. Once the virus is present in mosquito saliva it can be transmitted—the mosquito is now infectious and can pass on the virus when taking up her next blood meal (Fig. 11.1). In addition to the likelihood of mosquitoes acquiring and transmitting an arbovirus infection (e.g. vector competence), and the speed at which the infection progresses through the mosquito (e.g. the extrinsic incubation period), transmission is also affected by mosquito traits that shape mosquito population densities (e.g. mosquito development rate, egg to adult survival, fecundity, and lifespan), as well as contact with vertebrate hosts (e.g. mosquito biting rates and feeding preferences).

We currently lack fundamental knowledge on many facets of transmission even in relatively well studied systems like dengue virus (DENV). Not only do we lack high quality data from a diversity of mosquito-arbovirus systems on the specific mosquito and viral traits that drive disease transmission (mentioned previously), the factors that contribute to variation in these traits and transmission remain largely unidentified. In this chapter, we outline and explore the following: 1. the specific mechanisms governing

Christine M. Reitmayer, Michelle V. Evans, Kerri L. Miazgowicz, Philip M. Newberry, Nicole Solano, Blanka Tesla, and Courtney C. Murdock, *Mosquito—Virus Interactions*
In: *Population Biology of Vector-Borne Diseases*. Edited by John M. Drake, Michael B. Bonsall, and Michael R. Strand: Oxford University Press (2021).
© Christine M. Reitmayer, Michelle V. Evans, Kerri L. Miazgowicz, Philip M. Newberry, Nicole Solano, Blanka Tesla, and Courtney C. Murdock.
DOI: 10.1093/oso/9780198853244.003.0011

Infection within the Mosquito

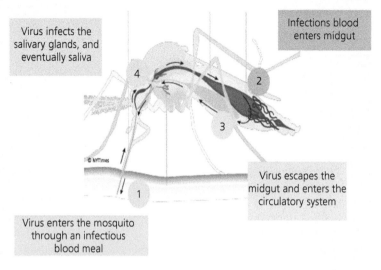

Figure 11.1 For the majority of mosquito species, female mosquitoes require vertebrate blood meal in order to produce a clutch of eggs and reproduce. Mosquitoes first become exposed to arboviruses after taking a blood meal from an infectious human host. The virus enters the midgut with the ingested blood meal, after which the virus needs to successfully invade and replicate within the midgut epithelium. After successfully replicating, the virus will then rupture out of midgut cells into the haemocoel (mosquito body cavity). Once in the haemocoel, the hemolymph (mosquito blood system) will circulate the virus to other tissues throughout the body. In order for the mosquito to become infectious to human hosts, the virus needs to invade, replicate within, and rupture out of the salivary gland epithelium. Then, when an infectious mosquito feeds on a susceptible human host, the virus is injected subcutaneously with the mosquito saliva while the mosquito is feeding on a human host.

the outcome of vector-virus interactions, such as tissue barriers to infection, mosquito immunity, and the strategies arboviruses employ to evade or suppress mosquito antiviral responses (Fig. 11.2), 2. how genetic variation across mosquito populations (i.e. in immune mechanisms) and viral strains, as well as environmental variation in abiotic (e.g. temperature, relative humidity) and biotic (e.g. coinfection) factors, shape the mosquito-virus interaction, and 3. the implications of these interactions for understanding and predicting arbovirus transmission, as well as for control of mosquitoes that transmit human pathogens. Throughout we highlight the current state of knowledge, critical knowledge gaps, and explore implications for our understanding of the transmission process as well as our ability to predict and mitigate disease transmission.

11.2 Specific Mechanisms Shaping Mosquito-Virus Interactions

Transmission of arboviruses among vertebrate hosts depends on a complicated interaction between

the mosquito vector's immune system and the arboviruses it transmits. Arboviruses have to cross four major physical tissue barriers within their mosquito host. In traversing these barriers, arboviruses will encounter specific cellular receptors that may or may not be compatible for viral entry, different physiological environments, and tissue specific immune responses. Mosquitoes elicit multiple immune mechanisms that are capable of eliminating or restricting infection. Arboviruses have in turn evolved a diversity of mechanisms to evade or suppress mosquito defenses or to alter tissue-specific cellular environments to enhance infection and persistence. In this section we outline the physical barriers to infection, the key antiviral immune responses mosquitoes mount in response to infection, and the strategies viruses employ to evade, suppress, or modify those responses to maximize viral spread.

11.2.1 Physical Barriers of Infection

Arboviruses have to overcome four major physical barriers within their mosquito host in order to

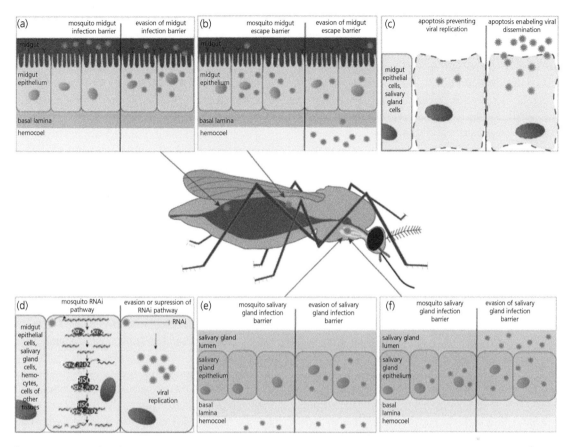

Figure 11.2 In order for infectious virus particles to be present in the saliva of a mosquito, virus particles that have been taken up via a blood meal need to cross the midgut infection barrier (A) and well as the midgut escape barrier (B) to escape the midgut and disseminate throughout the body. If the virus manages to cross the midgut infection barrier viral replication can take place inside midgut cells. In every cell infected by a virus, the RNAi pathway can block or slow down viral replication. If the virus manages to evade or suppress the RNAi pathway, viral replication can take place (D). Similarly, cell apoptosis can prevent viral replication as the virus needs a functioning cell replication machinery to replicate. However, during later stages of infection of a cell, apoptosis can actually facilitate viral dissemination by breaking down cell barriers (C). After dissemination throughout the body, the virus can ultimately reach the salivary glands. Similarly to the midgut, the virus needs to cross the salivary infection barrier (E) and can then further replicate within the salivary gland cells. The last barrier to cross before reaching the salivary gland lumen is the salivary escape barrier (F). Once this barrier has been crossed, the mosquito is likely to inject infectious saliva while blood feeding on a host.

transmit infection to the vertebrate host: the midgut infection barrier (MIB), the midgut escape barrier (MEB), the salivary gland infection barrier (SGIB), and the salivary gland escape barrier (SGEB). If a viral pathogen is unable to traverse any one of these barriers, transmission will be blocked. Thus, these barriers serve as major bottleneck that in turn limit the proportion of the mosquito population that successfully is infected, that go on to disseminate infection, and that ultimately become infectious, as well

as increase the average duration of time required for a mosquito to become infectious.

11.2.1.1 Infection of the Midgut

The first major physical barrier is the ability of the virus to infect the midgut epithelial cells, or the midgut infection barrier (MIB, Fig 11.2a). When a mosquito takes a blood meal from an infectious host, the blood meal, along with the viral pathogen travels to the mosquito midgut lumen. Upon ingestion of a blood meal, the peritrophic matrix (PM) is formed to encase the blood meal for digestion and is composed mainly of chitin, proteins, and proteoglycans secreted in response to blood feeding. Although the PM is semi-permeable, it is thought to form a barrier that can protect the midgut from pathogens (e.g. viruses Wang & Granados 2000), bacteria (Jin et al. 2019; Kuraishi et al. 2011), malaria (Rodgers et al. 2017, Abraham & Jacobs-Lorena 2004; Shahabuddin et al. 1993; Billingsley & Rudin 1992), and protozoa (Weiss et al. 2014) as well as other harmful substances present in the insect gut after a blood meal (Wang et al. 2004; Shibata et al., 2015). In general, arboviruses appear not to be as affected by PM formation as other pathogens (Kato et al. 2008) potentially because of their ability to invade midgut epithelial cells immediately after blood ingestion (Perrone & Spielman 1988; Han et al. 2000).

The degree of initial infection of the midgut epithelial lining varies across mosquito-pathogen pairings, but typically only a small portion of midgut cells are initially infected (Whitfield et al. 1973; Smith et al. 2008; Scholle et al. 2004). In general, several mechanisms have been invoked to explain the reason why a virus is unable to cross tissue barriers and they can be dose-dependent and/or-independent (Hardy et al. 1983). Dose-dependent mechanisms can be affected by innate antiviral responses (outlined below) specific to each tissue barrier, which has been shown to be the case for the midgut infection and escape barrier in *Aedes aegypti* after infection with Sindbis virus (SINV) (Khoo et al. 2010). Dose-independent barriers to arbovirus infection are highly dependent on the particular vector/virus pairing (such as virus/tissue receptor incompatibilities) (Forrester et al. 2014).

With respect to the MIB, the most supported hypothesis is the absence of receptors on the apical surface of epithelial cells, which prevent initial infection of midgut epithelial cells. Arboviruses typically enter the midgut epithelial cells through receptor-mediated endocytosis, yet alphaviruses have also been observed to enter through direct fusion between the virus and the microvillar membrane (Houk et al. 1985; Mrkic & Kempf 1996). The primary factor determining a virus' ability to infect midgut cells is the envelope (E) glycoprotein that interacts with cellular receptors (Pletnev et al. 2001;

Smith 2012; Pierro et al. 2003, Strauss et al. 1994). Numerous cellular receptors have been proposed to facilitate virus uptake into mosquito midgut epithelial cells, however, these interactions can be highly specific. Receptor identity, surface density, and receptor polymorphisms are likely major factors influencing the ability of viruses to infect different mosquito species. Key receptors identified to date include: DC-SIGN and L-SIGN (Klimstra et al. 2003; Liu et al. 2014), divalent metal ion transporter natural resistance-associated macrophage protein (NRAMP)(Rose et al. 2011), prohibitin (Kuadkitkan et al. 2010), enolase, beta-adrenergic receptor kinase (beta-ARK), translation elongation factor EF-1 alpha/Tu, and cadherin (Mercado-Curiel et al. 2008; Mercado-Curiel et al. 2006).

11.2.1.2 Spread of Midgut Epithelium Infection

The second major physical barrier to viral transmission is the midgut escape barrier (MEB, Fig 11.2b). For a virus to transmit to a new host, the virus needs to pass through the mosquito midgut epithelium layer to disseminate to secondary tissues within the mosquito. The MEB prevents the virus from escaping out of the midgut epithelial cell and to cross the basal lamina (BL). The BL is located on the basolateral side of the midgut epithelium and is comprised of a proteinaceous matrix which provides structure (Yurchenco & O'Rear, 1994). The BL matrix allows the flow of small molecules, but through size exclusion can block larger microorganisms. It was observed that larger *Ochlerotatus triseriatus* females, which have more layers of BL compared to smaller, malnourished females, experienced reduced dissemination of La Crosse virus (Grimstad & Walker, 1991, Thomas et al., 1993) suggesting that females with thicker BL were less competent vectors. Furthermore, the extensive BL restructuring, which occurs after a blood meal to accommodate the expansion of the midgut, was suggested to provide a transient opportunity for arboviruses to traverse the BL more readily (Dong et al., 2017). However, BL thickness failed to be a reliable determinant of dengue-1 virus infection in different strains of *Ae. albopictus* (Thomas et al., 1993).

Studies have observed a dose-dependency with the MEB, where dissemination is not observed unless viral titers reach a specific threshold within the midgut epithelium. Modulation of Western equine encephalitis virus (WEEV) replication in the midgut demonstrated an association between overall viral burden and midgut escape (Kramer et al., 1981). While, only a subset of midgut epithelial cells are required for the primary infection, replication and viral spread throughout the midgut epithelium can be propagated via cell-to-cell spread (Girard et al., 2004).

11.2.1.3 Dissemination to Secondary Tissues, Including the Salivary Glands

Once a virus escapes the midgut lining it can circulate through the hemocoel, providing opportunities to infect secondary tissues, including the salivary glands. Typically, a virus will spread from the posterior midgut, systemically infecting the fat body, followed by the anterior midgut epithelium and muscles (Girard et al. 2004). Abdominal and thoracic fat body cells serve as the most important amplifying tissues preceding salivary gland infection (Girard et al. 2004, Weaver et al. 1990).

In order for an arbovirus to transmit via biting, it must infect the salivary gland, and thus surpass the third major physical barrier, the salivary gland infection barrier (SGIB, Fig. 11.2e). The salivary gland is composed of two lateral lobes and one medial lobe with the lateral lobes having proximal and distal regions. Both dose-dependent and dose-independent mechanisms have been suggested to contribute to the salivary gland infection barrier (Kramer et al. 1981; Paulson et al.; 1989; Scott et al. 1990l; Turell et al., 2006). In support of a dose-dependent mechanism, increased viral loads in the haemolymph were found to be correlated with higher proportion of infectious mosquitoes (Hardy et al. 1983). However, it has also been postulated that the basal lamina surrounding the salivary glands, could represent a dose-independent physical barrier to infection, by physically blocking critical cell interactions required for invasion (Dourado et al. 2011; Romoser et al. 2005). In addition to dose-dependent and dose-independent mechanisms, certain mosquito-virus combinations exhibit distinct localization of infection among the salivary gland lobes. It is thought in these scenarios that there may be a receptor-mediated infection barrier to the salivary gland, analogous to the midgut (Barreau et al. 1999; Juhn et al. 2011).

11.2.1.4 Deposition of Virus into the Salivary Ducts

The last physical barrier to transmission of arboviruses is the salivary gland escape barrier (SGEB, Figure 11.2f). This barrier remains poorly characterized on a molecular level and has only been explicitly reported in a few instances. In these few studies, salivary gland infection was observed, however no viral transmission was achieved (Beaty et al. 1981, Grimstad et al. 1985; Jupp 1985; Paulson et al. 1992; Turell et al. 2013).

11.2.2 Mosquito Anti-viral Immune Mechanisms

In addition to physical tissue barriers, mosquitoes can also mount a series of innate immune responses that can actively suppress viral replication. There still is much less known in general about the insect antiviral immune response as compared to the antibacterial and antifungal response (Hoffmann 1995) or the immune response against pathogens such as *Plasmodium spp.* (Cirimotich et al. 2010). The main insect antiviral immune response restricting viral replication is mediated by RNA interference (RNAi, Figure 11.2d). Most of what we know about RNAi was originally discovered in *Drosophila melanogaster* (Wang et al. 2006, Kennerdell & Carthew, 1998) and *Caenorhabditis elegans* (Fire et al. 1998). RNA silencing utilizes RNAi to regulate gene expression and, in insects, RNAi also plays a crucial role in defense against RNA viruses. During viral replication, viral double-stranded RNA (dsRNA) accumulates in cells infected by a variety of RNA viruses. Long dsRNA is initially recognized by the mosquito immune system as a pathogen-associated molecular pattern (PAMP), which triggers the RNAi pathway. Once detected by the cell, dsRNA gets cleaved by ribonuclease III enzyme Dicer-2 (Dcr-2), resulting in viral small interfering RNA (siRNA) (Bernstein et al. 2001: Elbashir et al. 2001). Those siRNA fragments, in combination with Dcr-2 and the dsRNA-binding protein R2D2, get introduced into the Argonaute-2 (Ago2) containing, RNA-induced silencing complex (RISC) (Liu et al. 2003).

One of the siRNA strands is subsequently degraded and the remaining strand acts as a complementary strand to viral single-stranded RNA (ssRNA) present in the cell. Once a complementary ssRNA has been bound to the siRNA portion of the RISC complex, Ago2 mediates its degradation and thereby leading to an inhibition of viral replication (Okamura et al. 2004; Rand et al. 2004; Miyoshi et al. 2005).

Orthologs of involved immune genes, including dcr-2, r2d2 and ago2, were identified in the genome of several mosquito vectors, such as *Ae. aegypti* and *Anopheles gambiae* (Campbell et al. 2008a; Waterhouse et al. 2007). Functional studies in *Ae. aegypti* and *An. gambiae* confirmed the involvement of the RNAi pathway in the mosquito antiviral defense, showing that silencing of components of the RNAi pathway results in an increase in viral replication during different stages of the infection, a shortening of the extrinsic incubation period (Keene et al. 2004; Campbell et al. 2008b; Cirimotich et al. 2009b; Carissimo et al. 2015; Sánchez-Vargas et al. 2009) as well as an increased mortality rate of *Ae. aegypti* after infection with Sindbis virus (SINV) (Cirimotich et al. 2009b).

In addition to the siRNA pathway, another RNAi pathway has been under discussion to play a role in the mosquito antiviral response—the PIWI-interacting RNA (piRNA) pathway. Several studies suggest the potential involvement of piRNA mechanisms in mosquitoes infected with dengue virus (DENV) (Hess et al. 2011), Chikungunya virus (CHIKV) (Morazzani et al. 2012) and SINV (Vodovar et al. 2012). Furthermore, there is evidence of a complex interaction between both pathways as Varjak and colleagues showed that Piwi4 acts as a mediator for both siRNA and piRNA pathways in *Ae. aegypti* (Varjak et al. 2017).

Antimicrobial and antifungal properties of the Toll and Imd pathways have been well described, and there is also increasing evidence for their involvement in mosquito antiviral defense (Xi et al. 2008, Angleró-Rodríguez et al. 2017). Luplertlop and colleagues showed that infection with DENV leads to an upregulation of components of the Toll and Imd signaling pathway in salivary glands of *Ae. aegypti* (Luplertlop et al. 2011). Gene silencing of a Toll pathway component also led to an increase in DENV viral load in *Ae. aegypti* (Xi et al. 2008).

Another antiviral defense mechanism is apoptosis. Apoptosis is the highly regulated and controlled process of programmed cell death. It is necessary for development and homeostasis, but it may also serve in innate antiviral immunity in insects (Clarke & Clem 2003) (Figure 11.2C). Apoptosis was proposed as one of the limiting mechanisms of midgut infection and inhibition of dissemination in WNV-infected *Culex pipiens pipiens* (Vaidyanathan & Scott 2006) and reduced WNV transmission in *Cx. p. Cuinquefasciatus* (Girard et al. 2007). By orally infecting refractory and susceptible strains of *Ae. aegypti* with DENV, Liu et al. (2013) demonstrated that apoptosis is not simply one of the outcomes associated with viral infection but is also part of an early innate immune response. They observed rapid up-regulation of the pro-apoptotic gene only in refractory mosquitoes (Liu et al. 2013).

11.2.3 Viral Mechanisms to Enhance Infection

Arboviruses must replicate within the vector to a degree that is high enough for successful transmission to a vertebrate host, yet low enough to minimize effects on mosquito survival and reproduction (Hwang et al. 2016). Arboviral infection in mosquitoes is not inherently benign, yet it is a result of mosquito-virus adaptions that allow them to coexist long enough to transmit the virus (Forrester et al. 2014). As mosquitoes typically live a short time in the field (weeks to months) relative to the vertebrate hosts (years), arbovirus infections that persist across the lifespan of the mosquito increase the likelihood of successful transmission (Salas-Benito and De Nova-Ocampo 2015; Goic and Saleh 2012). Arboviruses have a variety of strategies that allow for more efficient invasion and replication of tissue barriers in the mosquito host, as well as active suppression or evasion of localized and systemic innate immune responses.

11.2.3.1 Overcoming Physical Barriers

Genetic diversity is an essential strategy for adaptation to a new environment and the ability to infect multiple hosts (Ciota et al. 2007). Many medically important arboviruses are RNA viruses (e.g. West

Nile virus (WNV), DENV, Zika virus (ZIKV), yellow fever virus (YFV), and CHIKV and have exceptionally high mutation rates due to a lack of proofreading by the RNA-dependent RNA polymerase (RdRP) (Brackney et al. 2011; Lauring and Andino 2010). As a result, RNA viruses exist as a heterogeneous population within the vertebrate and mosquito vector host, with mutations generating new viral variants arising during transmission events from the vertebrate host to the mosquito vector, and as the infection progresses throughout the mosquito (Brackney et al. 2011; Sim et al. 2015; Brackney et al. 2009). Mutations affecting the viral envelope are crucial because envelope proteins facilitate viral cell entry across various tissues, which in turn affects mosquito susceptibility and capacity to transmit the virus. For example, the 2005–2006 CHIKV epidemic in Reunion Island was caused by a strain of CHIKV that had a single amino acid change from alanine to valine at position 226 of the CHIKV envelope protein (E1 glycoprotein, E1-A226V). This mutation was directly responsible for increased efficiency of viral dissemination to secondary tissues and transmission to a vertebrate host in a novel mosquito vector, *Ae. albopictus*, relative to the established mosquito vector, *Ae. aegypti* (Tsetsarkin & Weaver 2011). Because of this mutation and the widespread global distribution of *Ae. albopictus*, there is concern that CHIKV could further extend its range into Europe and the Americas.

Arboviruses depend on and can manipulate the expression of several mosquito genes to facilitate mosquito infection, dissemination, and transmission (Hwang et al. 2016). For example, DENV induces transcription of cysteine-rich venom protein (CRVP379) that enhances virus entry in *Ae. aegypti*. Silencing of this gene was shown to reduce overall viral infection in vitro and *in vivo* (Hwang et al. 2016). Mosquito carbohydrate-binding C-type lectin (mosGCTL-1) is a secreted protein that is necessary for WNV dissemination in mosquitoes. WNV induces mosGCTL-1 and binds to secreted mosGCTL-1 in the hemolymph. The mosGCTL-1/WNV complex has the ability to interact with the membrane protein and facilitates cellular entry. The virus then rapidly replicates in the mosquito thorax, which induces additional mosGCTL-1 expression and more mosGCTL-1/WNV complexes. This, in

turn, enables WNV to invade different mosquito tissues enhancing viral dissemination throughout the body (Cheng et al. 2010).

Another mechanism to successfully achieve dissemination is by modulating insect miRNA expression. For example, DENV infection of *Ae. aegypti* mosquitoes (Campbell et al. 2014) and CHIKV infection of *Ae. albopictus* cells (Shrinet et al. 2014) affect miRNA expression of genes that are potentially important for virus replication and dissemination. Finally, some arboviruses may use apoptosis to facilitate dissemination. Inducing apoptosis, by silencing the anti-apoptotic *iap1* gene in *Ae. Aegypti*, increased midgut infection and dissemination of SINV, while inhibition of apoptosis, by silencing the expression of the initiating caspase Aedronc, had the opposite effect. The positive effect apoptosis had on SINV infection could be due to degradation of structural barriers or inhibition of innate immunity. SINV might induce low degree apoptosis that helps replication and escaping the midgut barrier by degrading basal laminae (Wang et al. 2012).

11.2.3.2 Suppression of Immune Mechanisms

Viruses employ a diversity of mechanisms to suppress the vertebrate immune response. Whether there are equivalent mechanisms in the mosquito host is still an open question (Sim & Dimopoulos 2010). In mosquito cell culture, Semliki Forest virus (SFV), CHIKV, and Japanese encephalitis virus have been shown to suppress JAK/STAT, Toll and Imd signaling pathways (Fragkoudis et al. 2008, McFarlane et al. 2014; Lin et al. 2004). For example, DENV downregulates multiple genes in an *Ae. aegypti* cell line (Aag2) resulting in inhibition of mosquito immune responses to infection such as antimicrobial peptide production (cecropin and defensin) in response to an *E. coli* challenge (Sim & Dimopoulos 2010). Immune suppression at early stages of an infection is likely to facilitate and establish midgut infection (Ramirez & Dimopoulos 2010). While there has been less work done in vivo, similar studies in mosquitoes suggest these viral mechanisms are important for virus establishment and infection. For example, SINV suppresses Toll signaling in the *Ae. aegypti* midgut (Sanders et al., 2005), and other flaviviruses such as WNV and YFV were also shown to actively suppress antiviral gene expression that limit viral replication in *Ae. aegypti* mosquitoes (Colpitts et al. 2011).

Many pathogenic insect viruses have viral suppressors of RNAi (VSRs) that bind and block crucial steps of the RNAi pathway, allowing for viruses to replicate and propagate (Ding & Voinnet 2007). VSRs have to date not been identified in pathogenic arboviruses (Fragkoudis et al. 2009). For the flavivirus DENV, there is evidence that the non-structural protein NS4B suppresses RNAi in mammalian and insect (Sf21, non-mosquito) cells (Kakumani et al., 2013). However, whether this is an important viral strategy to avoid mosquito innate immune responses is unclear. Mosquitoes infected with SINV and expressing flock house virus (an insect-only virus) VSRs show dramatically increased viral replication and dissemination. However, infected mosquitoes also experienced significantly reduced lifespans (Myles et al. 2008; Cirimotich et al. 2009a).

11.2.3.3 Evasion of RNAi

Another mechanism pathogens can utilize to promote infection and replication in an immune hostile environment is to evade host defenses. Currently, there is very little known on the existence of mechanisms viruses may use to evade both vertebrate host and mosquito vector immune defenses. However, if suppressing mosquito innate immune mechanisms are associated with steep fitness costs (as outlined above for VSRs), then arboviral strategies to evade host immunity (e.g. RNAi pathway) could be an evolutionarily favorable strategy (Fragkoudis et al. 2009) to support viral dissemination and transmission (Olson and Blair 2015). One suggested mechanism for RNAi evasion is that replication intermediates are sequestered in compartments inaccessible to the RNAi machinery (Kingsolver et al. 2013). It has been shown that DENV-2 evades antiviral RNAi response in-vivo in *Ae. aegypti* and in mosquito cells (Chotkowski et al. 2008). In *Culex pipiens quinquefasciatus*, regions of the WNV genome that were more intensely targeted by RNAi were more likely to have point mutations compared to regions that were weakly targeted. This suggests that positive selection of viral RNA within mosquitoes is stronger in regions that are highly targeted by the host RNAi response (Brackney et al. 2009).

11.2.3.4 Manipulation of Other Host Mechanisms

A midgut-specific miRNA, miR-281, was found to be upregulated after DENV-2 infection and expressed at a high level in the *Ae. albopictus* midgut. It enhances DENV-2 replication by targeting the viral transcript of host mRNA (Zhou et al. 2014).

Acidification of endosomes, lysosomes, and secretory vesicles is important for DENV replication because acidic pH in the late endosome triggers membrane fusion and the release of the viral genome into cells (Acosta et al. 2008; Clyde et al. 2006). DENV was shown to be capable of upregulating different vascular ATPase (vATPase) subunits in *Ae. aegypti* that are responsible for acidification of these organelles (Kang et al. 2014).

11.3 Effects of Genetic and Environmental Variation on the Vector-Virus Interaction

The within host mechanisms outlined above that govern the ability of mosquitoes to transmit virus are subject to factors that alter the environment viruses must traverse to establish infection, efficiently replicate, and transmit successfully to the next vertebrate host. Work in a range of invertebrate-pathogen systems, spanning fruit flies (Lazzaro et al. 2008), daphnia (Mitchell et al. 2005), and pea aphids (Bensadia et al. 2006) demonstrates that locale-specific conditions are key to understanding disease transmission. Almost always, both the underlying genetics of the specific host-pathogen combination and local environmental conditions shape transmission (Fig 11.3). There is growing evidence demonstrating that these factors also play a role in shaping vector competence across different mosquito-virus pairings for many recently emerging arboviruses including DENV, CHIKV, ZIKV (Gloria-Soria et al. 2017; Zouache et al. 2014; Lambrechts et al. 2009). Genetic variation in mosquito immunity, relative investment in immune defenses, and pathogen infection strategies likely play a role in shaping the observed variation in vector competence. Additionally, substantial evidence exists highlighting the effects of variation in abiotic (Tesla et al. 2018; Evans et al. 2018; Mordecai et al. 2017; Shocket et al. 2018) and biotic factors (Alto

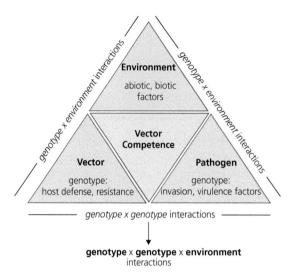

Figure 11.3 Vector competence (the proportion of mosquitoes that become infectious and capable of transmitting) will be subject to the underlying genetics of the specific host-pathogen combination, as well as local environmental conditions.

et al. 2008; Muturi et al. 2011; Minard et al. 2013; Weger-Lucarelli et al. 2018; Dennison et al. 2014) in the external environment on vector competence and infection outcomes. Environmental factors can have both direct effects on viral replication within the mosquito vector, or indirect effects that result from shifts in the rates and magnitude of key immune responses, the structure of physical barriers to infection, and resources available for virus replication due to changes in overall mosquito body condition. Finally, variation in the genetic background of the mosquito host and the infecting virus, as well as variation in the external environment will likely interact to ultimately shape the outcome of mosquito-virus interactions.

11.3.1 Virus Genotype (G)

In addition to host immune-physiology, vector-driven selection might play an important, and understudied, role in shaping the evolution of arboviruses selecting for strains that are more efficient at infecting the mosquito vector and that are potentially more virulent in the vertebrate host (Chevillon & Failloux 2003). For example, CHIKV has four major lineages, the East-Central-South Africa

(ECSA), West Africa, Asian, and the Indian Ocean lineage, all of which exhibit different transmission efficiencies across mosquito populations (Dupont-Rouzeyrol et al. 2012). As mentioned previously, during a CHIKV outbreak in the Indian Ocean region (2005-2006), CHIKV acquired a single adaptive amino acid substitution of the E1 envelope glycoprotein (Tsetsarkin et al. 2007). This mutation resulted in a selective advantage in locations where *Ae. albopictus* is more abundant than *Ae. aegypti* (the primary vector for DENV, CHIKV, and ZIKV (Tsetsarkin et al. 2007, Vazeille et al. 2007), primarily facilitating viral fusion, entry, and replication in the mosquito midgut (Arias-Goeta et al., 2013). Similarly, as WNV spread west throughout the United States after its introduction into the state of New York in 1999, a novel lineage was identified (WN02) (Moudy et al. 2007). This lineage is characterized by three amino acid changes that allow for more efficient establishment in the mosquito midgut and shorter extrinsic incubation periods relative to the colonizing lineage NY99 (Moudy et al. 2007). Consequently, WN02 experienced a competitive advantage, began to dominate circulation, and eventually displaced NY99 completely after 2004. Finally, there is evidence for similar phenomena occurring in the DENV and ZIKV systems, where different viral isolates have been demonstrated to vary in their ability to infect a certain mosquito strain (Enfissi et al. 2016; Lambrechts et al. 2009; Lambrechts et al. 2013; Chouin-Carneiro et al. 2016).

11.3.2 Vector Genotype (G)

Numerous studies have demonstrated that different geographic populations of *Ae. aegypti* and *Ae. albopictus* vary in vector competence for reference isolates of DENV (Lambrechts et al. 2009), CHIKV (Vega-Rua et al. 2015) and ZIKV (Chouin-Carneiro et al. 2016). Further, the variation in vector competence can be directly correlated to genetic differences across mosquito populations (e.g. DENV, Lambrechts et al. (2009)). What remains relatively unexplored are the underlying mechanisms (i.e. innate immune function, physical barriers, receptors on primary and secondary host tissues, etc.) that result in variation in vector competence across mosquito populations. Variation in arbovirus immunity across mosquito populations could potentially be shaped by differential exposure to parasites and pathogens in both larval and adult environments. There is some evidence in the dengue system that different levels of resistance to DENV infection in *Ae. aegypti* can be explained by genetic variation in the immune gene Dicer-2 (Lambrechts et al. 2013).

One major selective pressure leading to genetic variation across mosquito populations is insecticide pressure, one of the most important tools for controlling arbovirus transmission. The evolution of resistance to insecticides in mosquito populations is common and has been reported for every class of insecticide used (reviewed in Rivero et al. 2010). The genetic mechanisms of resistance are due to single point mutations (target site resistance, e.g. sodium channel *kdr* knockdowns, GABA receptor) and/or the overexpression of cuticular resistance (penetration resistance) or detoxification genes (metabolic resistance). However, significantly more data on resistance mechanisms in the field are needed. Due to variation in protective responses to different classes of insecticides, the physiological environment arboviruses experience is likely to be very different in susceptible mosquitoes and mosquitoes that are resistant to different classes of insecticides. This, in turn, could lead to changes in mosquito-virus interactions, vector competence, as well as mosquito life history traits that are relevant to transmission (Agnew et al. 2004; Boivin et al. 2001; Martins et al. 2012; Roberto & Omoto 2006; Yamamoto et al. 1995; Foster et al. 2007a; Foster et al. 2007b; Foster et al. 2003; Lee et al. 1999). A variety of mechanisms have been proposed to explain the effects of resistance on these traits ranging from resource based trade-offs (Hardstone et al. 2010), oxidative stress (Ortiz de Montellano and De Voss, 2005), overlapping substrate demands of enzymes/pathways involved in resistance and mosquito immunity (Vontas et al. 2005), and reduced neural sensitivity associated with target site resistance (Braks et al. 2001; Steib et al. 2001). However, the extent to which insecticide resistance alters mosquito-virus interactions, as well as the underlying physiological and cellular mechanisms, remain relatively unexplored (Rivero et al. 2010).

11.3.3 Environment (E)

The efficiency of virus infection in the insect vector is a dynamic phenotype, which is dependent upon both the specific mosquito-virus pairs (outlined previously) and variation in key environmental factors. Many aspects of the environment are likely important in shaping mosquito-virus interactions and disease transmission. Here we focus primarily on environmental temperature and mosquito interactions with other pathogens and microbiota, as they are key abiotic and biotic environmental drivers shaping mosquito-virus interactions.

There is substantial evidence demonstrating that temperature has profound effects on vector competence and arbovirus transmission (Evans et al. 2018; Mordecai et al. 2017; Shocket et al. 2018; Tesla et al. 2018). The relationship between temperature and vector competence is often unimodal, characterized by an intermediate temperature that maximizes mosquito infection and a low and hot temperature that minimizes mosquito infection. Yet, the constraints on infection at sub-optimal temperatures are likely regulated through different mechanisms (Tesla et al. 2018). Temperature variation in general could impact arbovirus infection and replication early in infection due to shifts in the midgut environment. The mosquito midgut environment is fairly complex, and arboviruses encounter microbiota (Dennison et al. 2014, see Chapter 13 this volume), oxidative and nitration stress associated with digestion of the blood meal (Graca-Souza et al. 2006), key immune factors (reviewed in Simoes et al. 2018), and potentially a wide range of blood derived nutrients that could be involved in pathogen defense as seen in the malaria system (Pakpour et al. 2014). The temperature effects on these diverse players need not be equivalent. Further, cellular damage (Nicolson 2004; Feder 1999) and chronic overexpression of heat shock proteins in response to hot temperatures have been implicated in reducing lifespan in other insect systems (Feder & Krebs 1998; Krebs & Feder 1997) and facilitating arbovirus infection in mosquitoes (Taguwa et al. 2015; Kuadkitkan et al. 2010). Yet, the extent to which the effects of temperature on the mosquito-virus interaction is driven by direct effects on pathogen biology (e.g. temperature effects on stability of viral envelope proteins, ability to fuse with host cell receptors, etc.) or indirect effects on mosquito immunity and physiology (Adelman et al. 2013, Murdock et al. 2013; Murdock et al. 2012a), remain largely unexplored. Thus, the net effect of temperature on mosquito-virus interactions is expected to depend on the relative thermal sensitivities of both mosquito and pathogen traits (Murdock et al. 2012b).

An important source of biotic environmental variation that could impact mosquito physiology, immunity, life history, and ability to transmit disease are interactions with other pathogens, as well as bacterial symbionts and commensals. Many of the mechanisms discussed previously concern a single infection, however mosquitoes in the field are often co-infected with other viruses (including non-pathogenic, insect only viruses), parasites, and microbiota (Hilgenboecker et al. 2008; Furuya-Kanamori et al. 2016; Dong et al. 2009). Co-infection in humans is common in endemic regions, particularly when anthropophilic species such as *Ae. aegypti* are present, with arbovirus co-infection prevalence in human populations as high as 36 percent (Furuya-Kanamori et al. 2016). While less is known about co-infection rates in wild mosquito populations, lab-based studies have found that co-infection of mosquitoes is possible (Goertz et al. 2017; Rückert et al. 2017). One major and underappreciated factor that likely contributes to variation in vector competence and transmission potential in mosquito populations could be coinfection with insect-specific viruses (Bolling et al. 2015b; Bolling et al. 2015a; Vasilakis and Tesh 2015; Halbach et al. 2017). For example, there have been many cases of co-infection of mosquito-specific viruses in *Aedes* and *Culex* cell lines resulting in reduced replication or blocking of important human arboviruses. Murray Valley encephalitis virus (Hobson-Peters et al., 2013), as well as ZIKV, DENV, La crosse virus, and Japanese encephalitis virus replication has been hindered in the presence of a co-infecting insect-specific viruses (Schultz et al. 2018; Kenney et al. 2014). However, whether reduced replication rates or complete inhibition occurs in mosquitoes naturally co-infected with mosquito-specific viruses has been less studied. There is some evidence in the WNV system that insect-specific viruses can both reduce or potentially enhance the vector competence of *Culex* mosquitoes (Bolling et al. 2012; Goenaga et al. 2015; Newman et al. 2011). Finally, infection with insect-specific viruses in the field could represent a strong evolutionary pressure that shapes the mosquito immune system and could contribute significantly to genetic variation in mosquito immunity across mosquito populations (Halbach et al. 2017).

Whether co-infection with another virus results in a reduction or enhancement in viral dissemination and infectiousness is a relatively unexplored question. Exactly how co-infection modulates mosquito immune function is likely system dependent. For example, flaviviruses predominantly tend to suppress RNAi, while alphaviruses evade RNAi (see RNAi pathway illustrated earlier) (Rückert et al. 2014). Whether viruses suppress or evade mosquito immunity (Potiwat et al. 2011), or result in changes in cell morphology (e.g. viral interference; Abrao & da Fonseca 2016) could have important consequences for the fitness of co-infecting viruses. Further, whether co-infection leads to viral competition or enhancement may depend on the stage of infection (Le Coupanec et al. 2017). Thus, mirroring species interactions in ecology, co-infection can result in a reduction, increase, or no change in the infection of secondary viruses.

There is growing evidence that the mosquito microbiome is an important factor influencing the mosquito's ability to disseminate and transmit virus (Dennison et al. 2014; Hegde et al. 2015; Jupatanakul et al., 2014b). Efforts to characterize the microbiota of the midgut, as well as intracellular bacterial populations, suggest that there is a high degree of variability and change in the mosquito microbiome depending on factors like species, life stage, and seasonality (Wang et al. 2011; Novakova et al. 2017; Valzania et al. 2018; Coon et al. 2016). The microbiota and viruses in the mosquito system have a complex set of interactions, which are possibly mediated by the mosquito immune system, such as activation of the Toll pathway and the expression of antimicrobial proteins (Xi et al. 2008). The presence of gut microbiota has been shown to affect DENV titers in infected *Ae. aegypti* (Ramirez et al. 2012), with mosquitoes treated with antibiotics prior to DENV infection experiencing two times higher virus titers compared to untreated, DENV infected mosquitoes. Further, the reintroduction of *Proteus sp.* and *Paenibacillus sp.* led to a reduction in DENV titers (Ramirez et al. 2012). Bacteria of the genus *Chromobacterium* have also been shown to produce bioactive factors with transmission blocking potential for DENV infections (Ramirez et al. 2014). Specifically, how natural microbiota regulate viral infection (e.g. resource competition or immune stimulation), or whether the overall abundance, diversity, or presence of specific bacterial species matter for viral infection, remain open and unexplored questions.

Perhaps the most well-studied interaction between co-infection with a symbiotic bacteria and human relevant arboviruses has been in the *Ae. aegypti—Wolbachia* system. Bacteria of the genus *Wolbachia* are maternally inherited, obligate intracellular symbionts that have been estimated to infect approximately 70 percent of all insects (Hilgenboecker et al. 2008). Over the past ten years, research has shown that stable transinfection of *Ae. aegypti* (naturally uninfected) with *Wolbachia* strains from naturally infected *Drosophila* or *Ae. albopictus* can inhibit infection with human relevant flaviviruses and alphaviruses like DENV, YFV, ZIKV, and CHIKV (Hilgenboecker et al. 2008). In some cases, however, *Wolbachia* transinfection—such as wAlbB to *Culex tarsalis*—has been shown to increase susceptibility of the mosquitoes to WNV (Dodson et al. 2014). The mechanisms contributing to reduced vector competence of *Wolbachia* infected *Ae. aegypti* still remain somewhat unclear but have been linked to upregulation of key immune pathways (Rances et al. 2012; Pan et al. 2012), modulating RNA interference (Mayoral et al. 2014; Osei-Amo et al. 2012; Zhang et al. 2013), increasing oxidative stress (Pan et al. 2012), and increasing competition for intracellular resources like lipids for example (Perera et al. 2012 Jupatanakul et al. 2014a; Heaton et al. 2010; Sinkins 2013; Moreira et al. 2009).

11.3.4 Combined (G x G x E)

Environmental variation will likely interact with genetic variation in mosquito cellular and physiological mechanisms that limit viral establishment and replication, as well as in the ability of viruses to fuse, enter, replicate, and escape mosquito tissues. Despite the plethora of research in other insect systems, e.g. (Bensadia et al. 200;, Lazzaro et al. 2008; Mitchell et al. 2005), very little work has been done in mosquito-borne disease systems. However, the limited existing work that has been done does suggests that variation in the environment could play a significant role in shaping interactions among genetically distinct viral lineages and mosquito populations. One study demonstrated significant interactions between CHIKV lineage (*Ae. aegpyti*-adapted CHIKV vs. *Ae. albopictus*-adapted CHIKV), mosquito population (tropical vs. temperate), and environmental temperature (cool vs. warm mean temperature) on the proportion of *Ae. albopictus* that become infectious (e.g. vector competence) (Zouache et al. 2014). Insights from this study suggests that the viral mutation conferring increased transmission of CHIKV in *Ae. albopictus* may not be successful under all environmental conditions. Alternatively, *Ae. albopictus* may become more permissive for *Ae. aegypti*-adapted CHIKV strains under certain environmental contexts. More recent work in the DENV-*Ae. aegypti* system reinforce this picture (Gloria-Soria et al., 2017), suggesting that these effects are likely important across a wider range of mosquito-transmitted viruses. However, the identification of the underpinning vector-pathogen mechanisms, as well as how environmental variation alters those mechanisms, remain unexplored.

11.4 Implications for Modeling

A recent review found that while 85 percent of mathematical models of mosquito-borne disease included infection dynamics within the host, only 62 percent included infection dynamics within the mosquito (Reiner et al. 2013). In those models that do include infection dynamics within the vector, the immune system is rarely explicitly modelled, often relying on a singular term for infection latency and vector competence. Yet, variation in host susceptibility, immunity, and infectiousness exists across vertebrate and mosquito populations as already outlined due to genetic and environmental variation. Further, models often do not account for relationships that may exist across organismal traits that influence transmission. Organismal fitness is often determined by balancing or distributing effort optimally to a diversity of life history demands (e.g. growth, survival, mating, and reproduction) in order to maximize fitness (Roff 2002; Schwenke et al. 2016; Koella & Boete 2002). As these life history demands are energetically costly, and energetic resources are limited, physiological trade-offs occur across life history demands (e.g. current reproduction and future survival). Thus, genetic and environmental variation that result in different levels of investment in immunity across mosquito populations could have indirect effects on disease transmission if physiological trade-offs occur across traits such as larval growth (Armbruster & Conn 2006; O'Donnell & Armbruster 2009), adult survival (Leisnham et al. 2008), reproduction and fecundity (Armbruster & Conn 2006; Leisnham et al. 2008), and overwintering survival (Hawley et al. 1989). This has been demonstrated in other invertebrate-pathogen systems (Schwenke et al. 2016; Cornet et al. 2009; Libert et al. 2006; Adamo & Parsons 2006), but has yet to be explored empirically in mosquito-pathogen interactions in depth and consequently has not been addressed in current modeling frameworks of vector-borne disease transmission. Finally, most arboviruses (e.g. YFV, WNV, DENV, CHIKV, and ZIKV) are zoonotic and exist in sylvatic transmission cycles or are transmitted by a diversity of vertebrate hosts and mosquito species (Weaver & Barrett 2004; Weaver 2005). This can have important implications for both the evolution of and constraints on viral infection strategies, mosquito resistance mechanisms, and virus persistence, as well as the ability of models to effectively predict transmission and the success of intervention efforts (Weaver 2013).

One area where these issues could be particularly important is with co-infection. Co-infection is a relatively new topic in the field of mosquito-borne disease, and, as such, the majority of models of arboviruses focus on one viral infection of a vector, although several models do include co-infection of multiple-strain pathogens within the host (Cummings et al. 2005; Ferguson et al. 1999). Given the ability of co-infection to suppress or enhance infection at the level of the individual mosquito and high rates of co-infection found in endemic areas,

integrating co-infection into models, and its effects on the mosquito immune system, is a promising step forward. When included in a model of malaria and lymphatic filariasis, competitive co-infection within the *Anopheles spp.* vector leads to a counterintuitive increase in malaria prevalence when lymphatic filariasis prevalence is reduced (Slater et al. 2013). Recent laboratory studies on arboviral co-infections (e.g. Goertz et al. 2017; Le Coupanec et al. 2017; Ruckert et al. 2017) provide realistic parameters with which to begin modeling co-infection within the vector. Further, co-infection with insect-specific viruses will also likely have important ramifications for our understanding of the transmission process as well as for predicting the efficacy of control strategies that manipulate mosquito immunity or rely on co-infections with other organisms, e.g. (Schnettler et al. 2016; Olson et al. 2002; Travanty et al. 2004). The issues outlined above are not unique to mosquito-borne disease, and the problem of integrating multi-scale dynamics into models is well-known in epidemiology (Matthews and Haydon 2007; Soubeyrand et al. 2007). However, given the importance of vector-virus interactions to transmission, it is clear that the within-host processes of immune functioning can scale up to impact disease at the level of the population (Lord et al. 2014). Explicitly modeling vector-virus interactions at the level of the individual, rather than with population-level parameters, allows for models to incorporate non-linear effects of vector-virus interactions, leading to phenomena such as bifurcations (Garba et al. 2008). Further, as with all multi-scale systems, individual-level stochasticity and heterogeneity lead to stochastic dynamics at the population level (Black & McKane 2012).

Modeling provides an important tool for scaling up the within-host vector-virus interactions between arboviruses and their mosquito hosts to the population. Our knowledge of the mosquito immune system is increasing, allowing for more accurate parameterization of these fine-scale parameters in models. However, a known weakness of past models is the tendency to focus on certain modeling themes and systems (i.e. mosquito vs. host ecology) (Reiner et al. 2013), limiting models to one spatial scale. Therefore, many models that include processes within the mosquito, such as immune response, omit processes at the level of the population, e.g. density dependence, and vice-versa. Interdisciplinary collaborations consisting of both empirical and theoretical biologists can help bridge this divide and incorporate individual-level immune functioning into population scale models, leading to more accurate predictions of disease risk.

11.5 Conclusions

Due to general lack of effective therapeutics and vaccines for many arboviral diseases, as well as rapidly evolving resistance to conventional insecticide-based approaches to control, there is a pressing need to develop novel mosquito control technologies. With growing knowledge of the mechanisms underpinning mosquito-virus interactions, there has been increased investment in developing genetics-based tools to limit pathogen transmission. With respect to arboviruses, targeting RNAi mechanisms has been explored as an approach to suppress viral replication in *Ae. aegypti* and therefore lower overall transmission risk. Transgenic animals expressing hairpin-RNA binding to DENV were shown to have a reduced transmission potential (Franz et al., 2006, Mathur et al. 2010).

However, these approaches have been met with some key challenges. Many mosquito-pathogen interactions are species-specific, suggesting that identification of mosquito and pathogen mechanisms cannot necessarily be extrapolated from one mosquito-pathogen combination to another. Further, evidence of G x G interactions in the field suggest that identifying within-host mechanisms for a given mosquito-virus combination might not extrapolate cleanly to all mosquito population–virus lineage pairings. Many of these mechanisms will also be influenced by variation in key abiotic and biotic environmental parameters, like temperature (Murdock et al. 2012a), suggesting that novel control tools designed around manipulating mosquito resistance to infection could vary in efficacy under different environmental contexts (Murdock et al. 2014, Ross et al. 2017). Finally, if manipulated phenotypes impose fitness costs either directly or indirectly (mediated through physiological trade-offs with other important life history traits), it will require a large number of modified mosquitoes to

be released, will decrease the likelihood of population replacement and trait persistence in the field, and will likely vary depending on the genetic background of local mosquito populations and infecting viral lineages. Incorporating mathematical modeling to better understand within-mosquito mechanisms will provide an explicitly defined framework for identifying the relative importance of different mechanisms in the infection process, understanding the nature of potential physiological trade-offs with other life history processes (e.g. development, digestion, egg production, etc.), and how genetic and environmental variation shapes these processes. Further, connecting within-mosquito models of the mosquito-pathogen interaction to population-scale models will provide insight into the mechanisms that have the largest impact on the transmission process, which will allow for streamlined identification of potential targets for novel mosquito control tools as well as the evaluation of their efficacy across variable field environments.

References

Abraham, E.G. & Jacobs-Lorena, M. (2004). Mosquito midgut barriers to malaria parasite development. *Insect Biochemistry and Molecular Biology*, 34, 667–71.

Abrao, E.P. and da Fonseca, B.A. 2016. Infection of mosquito cells (C6/36) by dengue-2 virus interferes with subsequent infection by yellow fever virus. *Vector Borne and Zoonotic Diseases*. 16, 124–30.

Acosta, E.G., Castill A, V. & Damonte, E.B. (2008). Functional entry of dengue virus into *Aedes albopictus* mosquito cells is dependent on clathrin-mediated endocytosis. *Journal of General Virology*, 89, 474–84.

Adamo, S.A. & Parsons, N.M. (2006). The emergency life-history stage and immunity in the cricket, *Gryllus texensis*. *Animal Behaviour*, 72, 235–44.

Adelman, Z.N, Anderson, M.A.E., Wiley, M.R., et al. (2013). Cooler temperatures destabilize RNA interference and increase susceptibility of disease vector mosquitoes to viral infection. *PLoS Neglected Tropical Diseases*, 7, e2239.

Agnew, P., Berticat, C., Bedhomme, S., Sidobre, C., & Michalakis, Y. (2004). Parasitism increases and decreases the costs of insecticide resistance in mosquitoes. *Evolution*, 58, 579–86.

Alto, B.W., Lounibos, L.P., Mores, C.N., & Reiskind, M.H. (2008). Larval competition alters susceptibility of adult Aedes mosquitoes to dengue infection. *Proceedings of the Royal Society B: Biological Sciences*, 275, 463–71.

Angleró-Rodríguez, Y.I., Macleod, H.J., Kang, S., Carlson, J.S., Jupatanakul, N. & Dimopoulos, G. (2017). *Aedes aegypti* molecular responses to Zika Virus: Modulation of infection by the toll and Jak/Stat immune pathways and virus host factors. *Frontiers in Microbiology*, 8, 2050.

Arias-Goeta, C., Mousson, L., Rougeon, F., & Failloux, A.-B. (2013). Dissemination and transmission of the E1-226V variant of chikungunya virus in *Aedes albopictus* are controlled at the midgut barrier level. *PLoS One*, 8, e57548.

Armbruster, P. & Conn, J.E. (2006). Geographic variation of larval growth in North American *Aedes albopictus* (Diptera: Culicidae). *Annals of the Entomological Society of America*, 99, 1234–43.

Barreau, C., Conrad, J., Fischer, E., Lujan, H.D., & Vernick, K.D. (1999). Identification of surface molecules on salivary glands of the mosquito, *Aedes aegypti*, by a panel of monoclonal antibodies. *Insect Biochemistry and Molecular Biology*, 29, 515–26.

Beaty, B.J., Holterman, M., Tabachnick, W., Shope, R.E., Rozhon, E.J., & Bishop, D.H. (1981). Molecular basis of bunyavirus transmission by mosquitoes: Role of the middle-sized RNA segment. *Science*, 211, 1433–5.

Bensadia, F., Boudreault, S., Guay, J.F., Michaud, D., & Cloutier, C. (2006). Aphid clonal resistance to a parasitoid fails under heat stress. *Journal of Insect Physiology*, 52, 146–57.

Bernstein, E., Caudy, A.A., Hammond, S.M., & Hannon, G.J. (2001). Role for a bidentate ribonuclease in the initiation step of RNA interference. *Nature*, 409, 363.

Billingsley, P.F. & Rudin, W. (1992). The role of the mosquito peritrophic membrane in bloodmeal digestion and infectivity of *Plasmodium* species. *Journal of Parasitology*, 78, 430–40.

Black, A.J. & Mckane, A.J. (2012). Stochastic formulation of ecological models and their applications. *Trends in Ecology & Evolution*, 27, 337–45.

Boivin, T., D'Hières, C.C., Bouvier, J.C., Beslay, D., & Sauphanor, B. (2001). Pleiotropy of insecticide resistance in the codling moth, *Cydia pomonella*. *Entomologia Experimentalis et Applicata*, 99, 381–6.

Bolling, B.G., Olea-Popelka, F.J., Eisen, L., Moore, C.G. & Blair, C.D. (2012). Transmission dynamics of an insect-specific flavivirus in a naturally infected *Culex pipiens* laboratory colony and effects of co-infection on vector competence for West Nile virus. *Virology*, 427, 90–7.

Bolling, B.G., Vasilakis, N., Guzman, H., et al. (2015a). Insect-specific viruses detected in laboratory mosquito colonies and their potential implications for experiments evaluating arbovirus vector competence. *American Journal of Tropical Medicine and Hygiene*, 92, 422–8.

Bolling, B.G., Weaver, S.C., Tesh, R.B., & Vasilakis, N. (2015b). Insect-specific virus discovery: Significance for the arbovirus community. *Viruses*, 7, 4911–28.

Brackney, D.E., Beane, J.E., & Ebel, G.D. (2009). RNAi targeting of West Nile virus in mosquito midguts promotes virus diversification. *PLoS Pathogens*, 5, e1000502.

Brackney, D.E., Pesko, K.N., Brown, I.K., Deardorff, E.R., Kawatachi, J., & Ebel, G.D. (2011). West Nile virus genetic diversity is maintained during transmission by *Culex pipiens* quinquefasciatus mosquitoes. *PLoS One*, 6, e24466.

Brady, O.J., Gething, P.W., Bhatt, S. et al. (2012). Refining the global spatial limits of dengue virus transmission by evidence-based consensus. *PLoS Neglected Tropical Diseases*, 6, e1760.

Braks, M.A.H., Meijerink, J., & Takken, W. (2001). The response of the malaria mosquito, *Anopheles gambiae*, to two components of human sweat, ammonia and l-lactic acid, in an olfactometer. *Physiological Entomology*, 26, 142–8.

Campbell, C.L., Black, W.C., Hess, A.M., & Foy, B.D. (2008a). Comparative genomics of small RNA regulatory pathway components in vector mosquitoes. *BMC Genomics*, 9, 425.

Campbell, C.L., Harrison, T., Hess, A.M., & Ebel, G.D. (2014). MicroRNA levels are modulated in *Aedes aegypti* after exposure to Dengue-2. *Insect Molecular Biology*, 23, 132–9.

Campbell, C.L., Keene, K.M., Brackney, D.E., Olson, K.E., Blair, C.D., Wilusz, J., & Foy, B.D. (2008b). *Aedes aegypti* uses RNA interference in defense against Sindbis virus infection. *BMC Microbiology*, 8, 47.

Cao-Lormeau, V.-M., Blake, A., Mons, S. et al. (2016). Guillain-Barre Syndrome outbreak associated with Zika virus infection in French Polynesia: A case-control study. *The Lancet*, 387, 1531–9.

Carissimo, G., Pondeville, E., Mcfarlane, M. et al. (2015). Antiviral immunity of *Anopheles gambiae* is highly compartmentalized, with distinct roles for RNA interference and gut microbiota. *Proceedings of the National Academy of Sciences of the United States of America*, 112, E176–85.

Cheng, G., Cox, J., Wang, P. (2010). A C-type lectin collaborates with a CD45 phosphatase homolog to facilitate West Nile virus infection of mosquitoes. *Cell*, 142, 714–25.

Chevillon, C. & Failloux, A.-B. (2003). Questions on viral population biology to complete dengue puzzle. *Trends in Microbiology*, 11, 415–21.

Chotkowski, H.L., Ciota, A.T., Jia, Y., et al. (2008). West Nile virus infection of *Drosophila melanogaster* induces a protective RNAi response. *Virology*, 377, 197–206.

Chouin-Carneiro, T., Vega-Rua, A., Vazeille, M., et al. (2016). Differential susceptibilities of *Aedes aegypti* and

Aedes albopictus from the Americas to Zika virus. *PLoS Neglected Tropical Diseases*, 10, e0004543.

Ciota, A.T., Ngo, K.A., Lovelace, A.O. (2007). Role of the mutant spectrum in adaptation and replication of West Nile virus. *Journal of General Virology*, 88, 865–74.

Cirimotich, C.M., Dong, Y., Garver, L.S., Sim, S., & Dimopoulos, G. (2010). Mosquito immune defenses against *Plasmodium* infection. *Developmental & Comparative Immunology*, 34, 387–95.

Cirimotich, C.M., Scott, J C., Phillips, A.T., Geiss, B.J., & Olson, K.E. (2009a). Suppression of RNA interference increases alphavirus replication and virus-associated mortality in *Aedes aegypti* mosquitoes. *BMC Microbiology*, 9, 49.

Cirimotich, C.M., Scott, J.C., Phillips, A.T., Geiss, B.J., & Olson, K.E. (2009b). Suppression of RNA interference increases alphavirus replication and virus-associated mortality in *Aedes aegypti* mosquitoes. *BMC Microbiology*, 9, 49.

Clarke, T.E. & Clem, R.J. (2003). Insect defenses against virus infection: the role of apoptosis. *International Review of Immunology*, 22, 401–24.

Clyde, K., Kyle, J.L., & Harris, E. (2006). Recent advances in deciphering viral and host determinants of dengue virus replication and pathogenesis. *Journal of Virology*, 80, 11418–31.

Colpitts, T.M., Cox, J., Vanlandingham, D.L., et al. (2011). Alterations in the *Aedes aegypti* transcriptome during infection with West Nile, dengue and yellow fever viruses. *PLoS Pathogens*, 7, e1002189.

Coon, K.L., Brown, M.R. & Strand, M.R. (2016). Mosquitoes host communities of bacteria that are essential for development but vary greatly between local habitats. *Molecular Ecology*, 25, 5806–26.

Cornet, S., Biard, C., & Moret, Y. (2009). Variation in immune defence among populations of *Gammerus pulex* (Crustacea: Amphipoda). *Oecologia*, 159, 257–69.

Cummings, D.A.T., Schwartz, I.B., Billings, L., Shaw, L.B., & Burke, D.S. (2005). Dynamic effects of antibody-dependent enhancement on the fitness of viruses. *Proceedings of the National Academy of Sciences of the United States of America*, 102, 15259–64.

Dennison, N.J., Jupatanakul, N., & Dimopoulos, G. (2014). The mosquito microbiota influences vector competence for human pathogens. *Current Opinion in Insect Science*, 3, 6–13.

Ding, S.W. & Voinnet, O. (2007). Antiviral immunity directed by small RNAs. *Cell*, 130, 413–26.

Dodson, B.L., Hughes, G.L., Paul, O., Matacchiero, A.C., Kramer, L.D., & Rasgon, J.L. (2014). *Wolbachia* enhances West Nile virus (WNV) infection in the mosquito *Culex tarsalis*. *PLoS Neglected Tropical Diseases*, 8, e2965.

Dong, S., Balaraman, V., Kantor, A.M., et al. (2017). Chikungunya virus dissemination from the midgut of *Aedes aegypti* is associated with temporal basal lamina degradation during bloodmeal digestion. *PLoS Neglected Tropical Diseases*, 11, e0005976.

Dong, Y., Manfredini, F., & Dimopoulos, G. (2009). Implication of the mosquito midgut microbiota in the defense against malaria parasites. *PLoS Pathogens*, 5, e1000423.

Dourado, L.A., Ribeiro, L.F., Brancalhao, R.M., Tavares, J., Borges, A.R., & Fernandez, M.A. (2011). Silkworm salivary glands are not susceptible to *Bombyx mori* nuclear polyhedrosis virus. *Genetics Molecular Research*, 10, 335–9.

Dupont-Rouzeyrol, M., Caro, V., Guillaumot, L. et al. (2012). Chikungunya virus and the mosquito vector *Aedes aegypti* in New Caledonia (South Pacific Region). *Vector Borne Zoonotic Diseases*, 12, 1036–41.

Feder, E.M. & Krebs, R. (1998). Natural and genetic engineering of the heat-shock protein Hsp70 in *Drosophila melanogaster*: Consequences for thermotolerance. *American Zoologist*, 38, 3, 503–517.

Elbashir, S.M., Lendeckel, W., & Tuschl, T. (2001). RNA interference is mediated by 21- and 22-nucleotide RNAs. *Genes and Development*, 15, 188–200.

Enfissi, A., Codrington, J., Roosblad, J., Kazanji, M., & Rousset, D. (2016). Zika virus genome from the Americas. *The Lancet*, 387, 227–8.

Evans, M.V., Shiau, J.C., Solano, N., Brindley, M.A., Drake, J.M., & Murdock, C.C. (2018). Carry-over effects of urban larval environments on the transmission potential of dengue-2 virus. *Parasites & Vectors*, 11, 426.

Feder, M.E. (1999). Engineering Candidate Genes in Studies of Adaptation: The Heat-Shock Protein Hsp70 in *Drosophila melanogaster*. *American Naturalist*, 154, S55–S66.

Ferguson, N., Anderson, R., & Gupta, S. (1999). The effect of antibody-dependent enhancement on the transmission dynamics and persistence of multiple-strain pathogens. *Proceedings of the National Academy of Sciences of the United States of America*, 96, 790–4.

Fire, A., Xu, S., Montgomery, M.K., Kostas, S.A., Driver, S.E., & Mello, C.C. (1998). Potent and specific genetic interference by double-stranded RNA in *Caenorhabditis elegans*. *Nature*, 391, 806–11.

Forrester, N.L., Coffey, L.L., & Weaver, S.C. (2014). Arboviral bottlenecks and challenges to maintaining diversity and fitness during mosquito transmission. *Viruses*, 6, 3991–4004.

Foster, S.P., Tomiczek, M., Thompson, R. et al. (2007a). Behavioural side-effects of insecticide resistance in aphids increase their vulnerability to parasitoid attack. *Animal Behaviour*, 74, 621–32.

Foster, S.P., Woodcock, C.M., Williamson, M.S., Devonshire, A.L., Denholm, I., & Thompson, R. (2007b). Reduced alarm response by peach–potato aphids,

Myzus persicae (Hemiptera: Aphididae), with knockdown resistance to insecticides (KDR) may impose a fitness cost through increased vulnerability to natural enemies. *Bulletin of Entomological Research*, 89, 133–8.

Foster, S.P., Young, S., Williamson, M.S., Duce, I., Denholm, I., & Devine, G.J. (2003). Analogous pleiotropic effects of insecticide resistance genotypes in peach-potato aphids and houseflies. *Heredity*, 91, 98–106.

Fragkoudis, R., Attarzadeh-Yazdi, G., Nash, A.A., Fazakerley, J.K., & Kohl, A. (2009). Advances in dissecting mosquito innate immune responses to arbovirus infection. *Journal of General Virology*, 90, 2061–72.

Fragkoudis, R., Chi, Y., Siu, R.W. et al. (2008). Semliki Forest virus strongly reduces mosquito host defence signaling. *Insect Molecular Biology*, 17, 647–56.

Franz, A.W., Sanchez-Vargas, I., Adelman, Z.N. et al. (2006). Engineering RNA interference-based resistance to dengue virus type 2 in genetically modified *Aedes aegypti*. *Proceedings of the National Academy of Sciences of the United States of America*, 103, 4198–203.

Furuya-Kanamori, L., Liang, S., Milinovich, G. et al. (2016). Co-distribution and co-infection of chikungunya and dengue viruses. *BMC Infectious Diseases*, 16, 84.

Garba, S.M., Gumel, A.B., & Abu Bakar, M.R. (2008). Backward bifurcations in dengue transmission dynamics. *Mathematical Biosciences*, 215, 11–25.

Girard, Y.A., Klingler, K.A., & Higgs, S. (2004). West Nile virus dissemination and tissue tropisms in orally infected *Culex pipiens quinquefasciatus*. *Vector Borne Zoonotic Diseases*, 4, 109–22.

Girard, Y.A., Schneider, B.S., Mcgee, C.E. et al. (2007). Salivary gland morphology and virus transmission during long-term cytopathologic West Nile virus infection in Culex mosquitoes. *American Journal of Tropical Medicine and Hygiene*, 76, 118–28.

Gloria-Soria, A., Armstrong, P.M., Powell, J.R., & Turner, P.E. (2017). Infection rate of *Aedes aegypti* mosquitoes with dengue virus depends on the interaction between temperature and mosquito genotype. *Proceedings of the Royal Society B: Biological Sciences*, 284.

Goenaga, S., Kenney, J.L., Duggal, N.K. et al. (2015). Potential for co-infection of a mosquito-specific flavivirus, nhumirim virus, to block West Nile virus transmission in mosquitoes. *Viruses*, 7, 5801–12.

Goertz, G.P., Vogels, C.B.F., Geertsema, C., Koenraadt, C.J.M., & Pijlman, G.P. (2017). Mosquito co-infection with Zika and chikungunya virus allows simultaneous transmission without affecting vector competence of Aedes aegypti. *PLoS Neglected Tropical Diseases*, 11, e0005654.

Goic, B. & Saleh, M.C. (2012). Living with the enemy: Viral persistent infections from a friendly viewpoint. *Current Opinion in Microbiology*, 15, 531–7.

Graca-Souza, A.V., Maya-Monteiro, C., Paiva-Silva, G.O. et al. (2006). Adaptations against heme toxicity in blood-feeding arthropods. *Insect Biochemistry and Molecular Biology*, 36, 322–35.

Grimstad, P.R., Paulson, S.L. & Craig, G.B., Jr. (1985). Vector competence of *Aedes hendersoni* (Diptera: Culicidae) for La Crosse virus and evidence of a salivary-gland escape barrier. *Journal of Medical Entomology*, 22, 447–53.

Grimstad, P.R. & Walker, E.D. (1991). *Aedes triseriatus* (Diptera: Culicidae) and La Crosse virus. IV. Nutritional deprivation of larvae affects the adult barriers to infection and transmission. *Journal of Medical Entomology*, 28, 378–86.

Halbach, R., Junglen, S., & Van Rij, R.P. (2017). Mosquito-specific and mosquito-borne viruses: evolution, infection, and host defense. *Current Opinion in Insect Science*, 22, 16–27.

Han, Y.S., Thompson, J., Kafatos, F.C., & Barillas-Mury, C. (2000). Molecular interactions between *Anopheles stephensi* midgut cells and *Plasmodium berghei*: The time bomb theory of ookinete invasion of mosquitoes. *EMBO Journal*, 19, 6030–40.

Hardstone, M.C., Huang, X., Harrington, L.C., & Scott, J.G. (2010). Differences in development, glycogen, and lipid content associated with cytochrome P.450-mediated permethrin resistance in *Culex pipiens quinquefasciatus* (Diptera: Culicidae). *Journal of Medical Entomology*, 47, 188–98.

Hardy, J.L., Houk, E.J., Kramer, L.D., & Reeves, W.C. (1983). Intrinsic factors affecting vector competence of mosquitoes for arboviruses. *Annual Review of Entomology*, 28, 229–62.

Hawley, W.A., Pumpuni, C.B., Brady, R.H., & Craig, G.B. Jr. (1989). Overwintering survival of *Aedes albopictus* (Diptera: Culicidae) eggs in Indiana. *Journal of Medical Entomology*, 26, 122–9.

Heaton, N.S., Perera, R., Berger, K.L., Khadka, S., Lacount, D.J., Kuhn, R.J., & Randall, G. (2010). Dengue virus nonstructural protein 3 redistributes fatty acid synthase to sites of viral replication and increases cellular fatty acid synthesis. *Proceedings of the National Academy of Sciences of the United States of America*, 107, 17345–50.

Hegde, S., Rasgon, J.L., & Hughes, G.L. (2015). The microbiome modulates arbovirus transmission in mosquitoes. *Current Opinion in Virology*, 15, 97–102.

Hess, A.M., Prasad, A.N., Ptitsyn, A. et al. (2011). Small RNA profiling of Dengue virus-mosquito interactions implicates the PIWI RNA pathway in anti-viral defense. *BMC Microbiology*, 11, 45.

Hilgenboecker, K., Hammerstein, P., Schlattmann, P., Telschow, A., & Werren, J.H. (2008). How many species are infected with *Wolbachia*–a statistical analysis of current data. *FEMS Microbiology Letters*, 281, 215–20.

Hobson-Peters, J., Yam, A.W.Y., Lu, J.W.F. et al. (2013). A new insect-specific flavivirus from Northern Australia suppresses replication of West Nile virus and Murray Valley encephalitis virus in co-infected mosquito cells. *PLoS One*, 8, e56534.

Hoffmann, J.A. (1995). Innate immunity of insects. *Current Opinion in Immunology*, 7, 4–10.

Houk, E.J., Kramer, L.D., Hardy, J.L., & Chiles, R.E. (1985). Western equine encephalomyelitis virus: In vivo infection and morphogenesis in mosquito mesenteronal epithelial cells. *Virus Research*, 2, 123–38.

Hwang, J., Jurado, K.A,. & Fikrig, E. (2016). Genetics of War and Truce between Mosquitos and Emerging Viruses. *Cell Host and Microbe*, 19, 583–7.

Jin, M., Liao, C., Chakrabarty, S., Wu, K., & Xiao, Y. (2019). Comparative proteomics of peritrophic matrix provides an insight into its role in Cry1Ac resistance of cotton bollworm helicoverpa armigera. *Toxins* 11, 92.

Juhn, J., Naeem-Ullah, U., Maciel Guedes, B.A. et al. (2011). Spatial mapping of gene expression in the salivary glands of the dengue vector mosquito, Aedes aegypti. *Parasites & Vectors*, 4, 1.

Jupatanakul, N., Sim, S., & Dimopoulos, G. (2014). Aedes aegypti ML and Niemann-Pick type C family members are agonists of dengue virus infection. *Developmental and Comparative Immunology*, 43, 1–9.

Jupatanakul, N., Sim, S., & Dimopoulos, G. (2014b). The insect microbiome modulates vector competence for arboviruses. *Viruses*, 6, 4294–313.

Jupp, P.G. (1985). *Culex theileri* and Sindbis virus; salivary glands infection in relation to transmission. *Journal of the American Mosquito Control Association*, 1, 374–6.

Kakumani, P.K., Ponia, S.S., Sood, S.R.K., et al. (2013). Role of RNA interference (RNAi) in dengue virus replication and identification of NS4B as an RNAi suppressor. *Journal of Virology*, 87, 8870–83.

Kang, S., Shields, A.R., Jupatanakul, N., & Dimopoulos, G. (2014). Suppressing dengue-2 infection by chemical inhibition of *Aedes aegypti* host factors. *PLoS Neglected Tropical Disease*, 8, e3084.

Kato, N., Mueller, C.R., Fuchs, J.F. et al. (2008). Evaluation of the function of a type i peritrophic matrix as a physical barrier for midgut epithelium invasion by mosquito-borne pathogens in *Aedes aegypti*. *Vector Borne and Zoonotic Diseases*, 8, 701–12.

Keene, K.M., Foy, B.D., Sanchez-Vargas, I., Beaty, B.J., Blair, C.D., & Olson, K.E. (2004). RNA interference acts as a natural antiviral response to O'nyong-nyong virus (*Alphavirus*; Togaviridae) infection of *Anopheles gambiae*. *Proceedings of the National Academy of Sciences of the United States of America*, 101, 17240–5.

Kennerdell, J.R. & Carthew, R.W. (1998). Use of dsRNA-mediated genetic interference to demonstrate that friz-

zled and frizzled 2 act in the wingless pathway. *Cell*, 95, 1017–26.

Kenney, J.L., Solberg, O.D., Langevin, S.A., & Brault, A.C. (2014). Characterization of a novel insect-specific flavivirus from Brazil: Potential for inhibition of infection of arthropod cells with medically important flaviviruses. *Journal of General Virology*, 95, 2796–808.

Khoo, C.C., Piper, J., Sanchez-Vargas, I., Olson, K.E., & Franz, A.W. (2010). The RNA interference pathway affects midgut infection- and escape barriers for Sindbis virus in *Aedes aegypti*. *BMC Microbiology*, 10, 130.

Kingsolver, M.B., Huang, Z,. & Hardy, R.W. (2013). Insect antiviral innate immunity: Pathways, effectors, and connections. *Journal of Molecular Biology*, 425, 4921–36.

Klimstra, W.B., Nangle, E.M., Smith, M.S., Yurochko, A.D., & Ryman, K.D. (2003). DC-SIGN and L-SIGN can act as attachment receptors for alphaviruses and distinguish between mosquito cell- and mammalian cell-derived viruses. *Journal of Virology*, 77, 12022–32.

Koella, J.C. & Boete, C. (2002). A genetic correlation between age at pupation and melanization immune response of the yellow fever mosquito *Aedes aegypti*. *Evolution*, 56, 1074–9.

Kramer, L.D., Hardy, J.L., Presser, S.B., & Houk, E.J. (1981). Dissemination barriers for western equine encephalomyelitis virus in *Culex tarsalis* infected after ingestion of low viral doses. *American Journal of Tropical Medicine and Hygiene*, 30, 190–7.

Krebs, R.A. & Feder, M.E. (1997). Natural variation in the expression of the heat-shock protein HSP70 in a population of *Drosophila melanogaster* and its correlation with tolerance of ecologically relevant thermal stress. *Evolution*, 51, 173–9.

Kuadkitkan, A., Wikan, N., Fongsaran, C., & Smith, D.R. (2010). Identification and characterization of prohibitin as a receptor protein mediating DENV-2 entry into insect cells. *Virology*, 406, 149–61.

Kuraishi, T., Binggeli, O., Opota, O., Buchon, N., & Lemaitre, B. (2011). Genetic evidence for a protective role of the peritrophic matrix against intestinal bacterial infection in *Drosophila melanogaster*. *Proceedings of the National Academy of Sciences of the United States of America* 108, 15966–71.

Lambrechts, L., Chevillon, C., Albright, R.G. et al. (2009). Genetic specificity and potential for local adaptation between dengue viruses and mosquito vectors. *BMC Evolutionary Biology*, 9, 160.

Lambrechts, L., Quillery, E., Noel, V. et al. (2013). Specificity of resistance to dengue virus isolates is associated with genotypes of the mosquito antiviral gene Dicer-2. *Proceedings of the Royal Society B: Biological Sciences*, 280, 20122437.

Lauring, A.S. & Andino, R. (2010). Quasispecies theory and the behavior of RNA viruses. *PLoS Pathogens*, 6, e1001005.

Lazzaro, B.P., Flores, H.A., Lorigan, J.G., & Yourth, C.P. (2008). Genotype-by-environment interactions and adaptation to local temperature affect immunity and fecundity in *Drosophila melanogaster*. *PLoS Pathogens*, 4, e1000025.

Le Coupanec, A., Tchankouo-Nguetcheu, S., Roux, P. et al. (2017). Co-infection of mosquitoes with chikungunya and dengue viruses reveals modulation of the replication of both viruses in midguts and salivary glands of *Aedes aegypti* mosquitoes. *International Journal of Molecular Sciences*, 18, 1708.

Lee, S.H., Smith, T.J., Knipple, D.C., & Soderlund, D.M. (1999). Mutations in the house fly Vssc1 sodium channel gene associated with super-kdr resistance abolish the pyrethroid sensitivity of Vssc1/tipE sodium channels expressed in Xenopus oocytes. *Insect Biochemistry and Molecular Biology*, 29, 185–94.

Leisnham, P.T., Sala, L.M., & Juliano, S.A. (2008). Geographic variation in adult survival and reproductive tactics of the mosquito *Aedes albopictus*. *Journal of Medical Entomology*, 45, 210–21.

Libert, S., Chao, Y., Chu, X., & Pletcher, S.D. (2006). Trade-offs between longevity and pathogen resistance in *Drosophila melanogaster* are mediated by NFkappaB signaling. *Aging Cell*, 5, 533–43.

Lin, C.C., Chou, C.M., Hsu, Y.L. et al. (2004). Characterization of two mosquito STATs, AaSTAT and CtSTAT. Differential regulation of tyrosine phosphorylation and DNA binding activity by lipopolysaccharide treatment and by Japanese encephalitis virus infection. *Journal of Biological Chemistry*, 279, 3308–17.

Liu, B., Behura, S.K., Clem, R.J. et al. (2013). P.53-mediated rapid induction of apoptosis conveys resistance to viral infection in *Drosophila melanogaster*. *PLoS Pathogens*, 9, e1003137.

Liu, Q., Rand, T.A., Kalidas, S. et al. (2003). R2D2, a bridge between the initiation and effector steps of the *Drosophila* RNAi pathway. *Science*, 301, 1921–5.

Liu, Y., Zhang, F., Liu, J. (2014). Transmission-blocking antibodies against mosquito C-type lectins for dengue prevention. *PLoS Pathogens*, 10, e1003931.

Lord, C.C., Alto, B.W., Anderson, S.L. et al. (2014). Can Horton hear the whos? The importance of scale in mosquito-borne disease. *Journal of Medical Entomology*, 51, 297–313.

Luplertlop, N., Surasombatpattana, P., Patramool, S. et al. (2011). Induction of a peptide with activity against a broad spectrum of pathogens in the *Aedes aegypti* salivary gland, following infection with dengue virus. *PLoS Pathogens*, 7, e1001252.

Martins, A.J., Ribeiro, C.D.E.M., Bellinato, D.F., Peixoto, A.A., Valle, D., & Lima, J.B.P. (2012). Effect of insecticide resistance on development, longevity and reproduction of field or laboratory selected *Aedes aegypti* Populations. *PLoS One*, 7, e31889.

Mathur, G., Sanchez-Vargas, I., Alvarez, D., Olson, K.E., Marinotti, O., & James, A.A. (2010). Transgene-mediated suppression of dengue viruses in the salivary glands of the yellow fever mosquito, *Aedes aegypti*. *Insect Molecular Biology*, 19, 753–63.

Matthews, L. & Haydon, D. (2007). Introduction. Cross-scale influences on epidemiological dynamics: from genes to ecosystems. *Journal of The Royal Society Interface*, 4, 763–5.

Mayoral, J.G., Etebari, K., Hussain, M., Khromykh, A.A., & Asgari, S. (2014). *Wolbachia* infection modifies the profile, shuttling and structure of microRNAs in a mosquito cell line. *PLoS One*, 9, e96107.

McFarlane, M., Arias-Goeta, C., Martin, E. et al. (2014). Characterization of *Aedes aegypti* innate-immune pathways that limit Chikungunya virus replication. *PLoS Neglected Tropical Diseases*, 8, e2994.

Mercado-Curiel, R.F., Black, W.C.T., & Munoz Mde, L. (2008). A dengue receptor as possible genetic marker of vector competence in *Aedes aegypti*. *BMC Microbiology*, 8, 118.

Mercado-Curiel, R.F., Esquinca-Aviles, H.A., Tovar, R., Diaz-Badillo, A., Camacho-NUEZ, M., & Munoz Mde, L. (2006). The four serotypes of dengue recognize the same putative receptors in *Aedes aegypti* midgut and Ae. albopictus cells. *BMC Microbiology*, 6, 85.

Minard, G., Mavingui, P., & Moro, C.V. (2013). Diversity and function of bacterial microbiota in the mosquito holobiont. *Parasites & Vectors*, 6, 146.

Mitchell, S.E., Rogers, E.S., Little, T.J., & Read, A.F. (2005). Host-parasite and genotype-by-environment interactions: Temperature modifies potential for selection by a sterilizing pathogen. *Evolution*, 59, 70–80.

Miyoshi, K., Tsukumo, H., Nagami, T., Siomi, H., & Siomi, M.C. (2005). Slicer function of *Drosophila* Argonautes and its involvement in RISC formation. *Genes & Development*, 19, 2837–48.

Mlakar, J., Korva, M., Tul, N. et al. (2016). Zika virus associated with microcephaly. *New England Journal of Medicine*, 374, 951–8.

Morazzani, E.M., Wiley, M.R., Murreddu, M.G., Adelman, Z.N., & Myles, K.M. (2012). Production of virus-derived ping-pong-dependent piRNA-like small RNAs in the mosquito soma. *PLoS Pathogens*, 8, e1002470.

Mordecai, E.A., Cohen, J.M., Evans, M.V. et al. (2017). Detecting the impact of temperature on transmission of Zika, dengue, and chikungunya using mechanistic models. *PLoS Neglected Tropical Diseases*, 11, e0005568.

Moreira, L.A., Iturbe-Ormaetxe, I., Jeffery, J.A. et al. (2009). A *Wolbachia* symbiont in *Aedes aegypti* limits infection with dengue, chikungunya, and plasmodium. *Cell*, 139, 1268–78.

Moudy, R.M., Meola, M.A., Morin, L.-L.L., Ebel, G.D., & Kramer, L.D. 2007. A newly emergent genotype of West Nile virus is transmitted earlier and more efficiently by Culex mosquitoes. *American Journal of Tropical Medicine and Hygeine*, 77, 365–70.

Mrkic, B. & Kempf, C. (1996). The fragmentation of incoming Semliki Forest virus nucleocapsids in mosquito (*Aedes albopictus*) cells might be coupled to virion uncoating. *Archives of Virology*, 141, 1805–21.

Murdock, C.C., Blanford, S., Hughes, G.L., Rasgon, J.L., & Thomas, M.B. (2014). Temperature alters *Plasmoidum* blocking by *Wolbachia Scientific reports*, 4, 3932–2.

Murdock, C.C., Moller-Jacobs, L.L., & Thomas, M.B. (2013). Complex environmental drivers of immunity and resistance in malaria mosquitoes. *Proceedings of the Royal Society B: Biological Sciences*, 280, 20132030.

Murdock, C.C., Paaijmans, K.P., Bell, A.S., et al. (2012a). Complex effects of temperature on mosquito immune function. *Proceedings of the Royal Society B. Biological Sciences*, 279, 3357–66.

Murdock, C.C., Paaijmans, K.P., Cox-Foster, D., Read, A.F., & Thomas, M.B. (2012b). Rethinking vector immunology: The role of environmental temperature in shaping resistance. *Nature Reviews Microbiology*, 10, 869–76.

Muturi, E.J., Kim, C.H., Alto, B.W., Berenbaum, M.R., & Schuler, M.A. (2011). Larval environmental stress alters *Aedes aegypti* competence for Sindbis virus. *Tropical Medicine International Health*, 16, 955–64.

Myles, K.M., Wiley, M.R., Morazzani, E.M., & Adelman, Z.N. (2008). Alphavirus-derived small RNAs modulate pathogenesis in disease vector mosquitoes. *Proceedings of the National Academy of Sciences of the United States of America*, 105, 19938–43.

Newman, C.M., Cerutti, F., Anderson, T.K. et al. (2011). *Culex* flavivirus and West Nile virus mosquito coinfection and positive ecological association in Chicago, United States. *Vector Borne Zoonotic Diseases*, 11, 1099–105.

Nicolson, C. (2004). *Insect Physiological Ecology: Mechanisms and Patterns*. Oxford University Press, New York.

Novakova, E., Woodhams, D.C., Rodríguez-Ruano, S.M. et al. (2017). Mosquito microbiome dynamics, a background for prevalence and seasonality of West Nile virus. *Frontiers in Microbiology*, 8, 26.

O'Donnell, D. & Armbruster, P. (2009). evolutionary differentiation of fitness traits across multiple geographic scales in *Aedes albopictus* (Diptera: Culicidae). *Annals of the Entomological Society of America*, 102, 1135–44.

Okamura, K., Ishizuka, A., Siomi, H., & Siomi, M.C. (2004). Distinct roles for Argonaute proteins in small RNA-directed RNA cleavage pathways. *Genes & Developmen*, 18, 1655–66.

Olson, K.E., Adelman, Z.N., Travanty, E.A., Sanchez-Vargas, I., Beaty, B.J., & Blair, C.D. (2002). Developing arbovirus resistance in mosquitoes. *Insect Biochemistry and Molecular Biology*, 32, 1333–43.

Olson, K.E. & Blair, C.D. (2015). Arbovirus-mosquito interactions: RNAi pathway. *Current Opinion in Virology*, 15, 119–26.

Osei-Amo, S., Hussain, M., O'Neill, S.L., & Asgari, S. (2012). *Wolbachia*-induced aae-miR-12 miRNA negatively regulates the expression of MCT1 and MCM6 genes in *Wolbachia*-infected mosquito cell line. *PLoS One*, 7, e50049.

PAHO. 2018. *Pan American Health Organisation* [Online]. Available: http://www.paho.org/ [Accessed January 15 2018].

Pakpour, N., Riehle, M.A., & Luckhart, S. (2014). Effects of ingested vertebrate-derived factors on insect immune responses. *Current Opinion in Insect Science*, 3, 1–5.

Pan, X., Zhou, G., Wu, J. et al. (2012). *Wolbachia* induces reactive oxygen species (ROS)-dependent activation of the toll pathway to control dengue virus in the mosquito *Aedes aegypti*. *Proceedings of the National Academy of Sciences of the United States of America*, 109, E23–31.

Paulson, S.L., Grimstad, P.R. & Craig, G.B., Jr. (1989). Midgut and salivary gland barriers to La Crosse virus dissemination in mosquitoes of the *Aedes triseriatus* group. *Medical and Veterinary Entomology*, 3, 113–23.

Paulson, S.L., Poirier, S.J., Grimstad, P.R., & Craig, G.B., Jr. (1992). Vector competence of *Aedes hendersoni* (Diptera: Culicidae) for La Crosse virus: Lack of impaired function in virus-infected salivary glands and enhanced virus transmission by sporozoite-infected mosquitoes. *Journal of Medical Entomology*, 29, 483–8.

Perera, R., Riley, C., Isaac, G. et al. (2012). Dengue virus infection perturbs lipid homeostasis in infected mosquito cells. *PLoS Pathogens*, 8, e1002584.

Perrone, J.B. & Spielman, A. (1988). Time and site of assembly of the peritrophic membrane of the mosquito *Aedes aegypti*. *Cell and Tissue Researches*, 252, 473–8.

Pierro, D.J., Myles, K.M., Foy, B.D., Beaty, B.J., & Olson, K.E. (2003). Development of an orally infectious Sindbis virus transducing system that efficiently disseminates and expresses green fluorescent protein in *Aedes aegypti*. *Insect Molecular Biology*, 12, 107–16.

Pletnev, S.V., Zhang, W., Mukhopadhyay, S. et al. (2001). Locations of carbohydrate sites on alphavirus glycoproteins show that E1 forms an icosahedral scaffold. *Cell*, 105, 127–36.

Potiwat, R., Komalamisra, N., Thavara, U., Tawatsin, A., & Siriyasatien, P. (2011). Competitive suppression between chikungunya and dengue virus in *Aedes albopictus* c6/36 cell line. *Southeast Asian Journal of Tropical Medicine and Public Health*, 42, 1388–94.

Ramirez, J.L. & Dimopoulos, G. (2010). The Toll immune signaling pathway control conserved anti-dengue defenses across diverse Ae. aegypti strains and against multiple dengue virus serotypes. *Developmental and Comparative Immunology*, 34, 625–9.

Ramirez, J.L., Short, S.M., Bahia, A.C. et al. (2014). Chromobacterium Csp_P reduces malaria and dengue infection in vector mosquitoes and has entomopathogenic and in vitro anti-pathogen activities. *PLoS Pathogens*, 10, e1004398.

Ramirez, J.L., Souza-Neto, J., Torres Cosme, R. et al. (2012). Reciprocal tripartite interactions between the *Aedes aegypti* midgut microbiota, innate immune system and dengue virus influences vector competence. *PLoS Neglected Tropical Diseases*, 6, e1561.

Rances, E., Ye, Y.H., Woolfit, M., McGraw, E.A., & O'Neill, S.L. (2012). The relative importance of innate immune priming in *Wolbachia*-mediated dengue interference. *PLoS Pathogens*, 8, e1002548.

Rand, T.A., Ginalski, K., Grishin, N.V., & Wang, X. (2004). Biochemical identification of Argonaute 2 as the sole protein required for RNA-induced silencing complex activity. *Proceedings of the National Academy of Sciences of the United States of America*, 101, 14385–9.

Reiner, R.C., Perkins, T.A., Barker, C.M. et al. (2013). A systematic review of mathematical models of mosquito-borne pathogen transmission: 1970–2010. *Journal of The Royal Society Interface*, 10, 20120921.

Rivero, A., Vézilier, J., Weill, M., Read, A.F., & Gandon, S. (2010). Insecticide control of vector-borne diseases: When is insecticide resistance a problem? *PLoS Pathogens*, 6, e1001000.

Roberto, H.K. & Omoto, C. (2006). Fitness cost associated with carbosulfan resistance in *Aphis gossypii* Glover (Hemiptera: Aphididae). *Neotropical Entomology*, 35, 246–50.

Rodgers, F.H., Gendrin, M., Wyer, C.A.S., & Christophides, G.K. (2017). Microbiota-induced peritrophic matrix regulates midgut homeostasis and prevents systemic infection of malaria vector mosquitoes. *PLoS Pathogens*, 13, e1006391.

Roff, D.A. (2002). Life History Evolution, *Encyclopedia of Biodiversity*, 2, 631–41.

Romoser, W.S., Turell, M.J., Lerdthusnee, K. et al. (2005). Pathogenesis of Rift Valley fever virus in mosquitoes—tracheal conduits & the basal lamina as an extra-cellular barrier. *Archives of Virology Supplement*, 89–100.

Rose, P.P., Hanna, S.L., SpiridigliozzI, A. et al. (2011). Natural resistance-associated macrophage protein is a cellular receptor for sindbis virus in both insect and mammalian hosts. *Cell Host and Microbe*, 10, 97–104.

Ross, P.A., Wiwatanaratanabutr, I., Axford, J.K., White, V.L., Endersby-Harshman, N.M., & Hoffmann, A.A. (2017). *Wolbachia* infections in *Aedes aegypti* differ markedly in their response to cyclical heat stress. *PLoS Pathogens*, 13, 6–e1006006.

Rückert, C., Bell-Sakyi, L., Fazakerley, J.K., & Fragkoudis, R. (2014). Antiviral responses of arthropod vectors: An update on recent advances. *VirusDisease*, 25, 249–60.

Rückert, C., Weger-Lucarelli, J., Garcia-Luna, S.M. et al. (2017). Impact of simultaneous exposure to arboviruses on infection and transmission by *Aedes aegypti* mosquitoes. *Nature Communications*, 8, 15412.

Salas-Benito, J.S. & de Nova-Ocampo, M. (2015). Viral Interference and Persistence in Mosquito-Borne Flaviviruses. *Journal of Immunological Research*, 873404.

Sánchez-Vargas, I., Scott, J.C., Poole-Smith, B.K. et al. (2009). Dengue virus type 2 infections of *Aedes aegypti* are modulated by the mosquito's RNA interference pathway. *PLoS Pathogens*, 5, e1000299.

Sanders, H.R., Foy, B.D., Evans, A.M. et al. (2005). Sindbis virus induces transport processes and alters expression of innate immunity pathway genes in the midgut of the disease vector, *Aedes aegypti*. *Insect Biochemistry and Molecular Biology*, 35, 1293–307.

Schnettler, E., Sreenu, V.B., Mottram, T., & McFarlane, M. (2016). *Wolbachia* restricts insect-specific flavivirus infection in *Aedes aegypti* cells. *Journal of General Virology*, 97, 3024–9.

Scholle, F., Girard, Y.A., Zhao, Q., Higgs, S., & Mason, P.W. (2004). Trans-packaged West Nile virus-like particles: Infectious properties in vitro and in infected mosquito vectors. *Journal of Virology*, 78, 11605–14.

Schultz, M.J., Frydman, H.M., & Connor, J.H. (2018). Dual insect specific virus infection limits Arbovirus replication in *Aedes* mosquito cells. *Virology*, 518, 406–13.

Schwenke, R.A., Lazzaro, B P. & Wolfner, M. F. (2016). Reproduction-immunity trade-offs in insects. *Annual Review of Entomology* 61, 239–56.

Scott, T.W., Lorenz, L.H., & Weaver, S.C. (1990). Susceptibility of *Aedes albopictus* to infection with eastern equine encephalomyelitis virus. *Journal of the American Mosquito Control Association*, 6, 274–8.

Shahabuddin, M., Toyoshima, T., Aikawa, M., & Kaslow, D.C. (1993). Transmission-blocking activity of a chitinase inhibitor and activation of malarial parasite chitinase by mosquito protease. *Proceedings of the National Academy of Sciences of the United States of America*, 90, 4266–70.

Shibata, T., Maki, K., Hadano, J., Fujikawa, T., Kitazaki, K., Koshiba, T., & Kawabata, S. (2015). Crosslinking of a peritrophic matrix protein protects gut epithelia from bacterial exotoxins. *PLoS Pathogens*, 11, e1005244.

et al. (2018). Temperature explains broad patterns of Ross river virus transmission. *eLife*, 7, e37762.

Shrinet, J., Jain, S., Jain, J., Bhatnagar, R.K., & Sunil, S. (2014). Next generation sequencing reveals regulation of distinct *Aedes* microRNAs during chikungunya virus development. *PLoS Neglected Tropical Diseases*, 8, e2616.

Sim, S., Aw, P.P., Wilm, A. et al. (2015). Tracking dengue virus intra-host genetic diversity during human-to-mosquito transmission. *PLoS Neglected Tropical Diseases*, 9, e0004052.

Sim, S. & Dimopoulos, G. (2010). Dengue virus inhibits immune responses in *Aedes aegypti* cells. *PLoS One*, 5, e10678.

Simoes, M.L., Caragata, E. P., and Dimopoulos, G. (2018). Diverse host and restriction factors regulate mosquito-pathogen interactions. *Trends in Parasitology*, 34 (7), 603–616.

Sinkins, S.P. (2013). *Wolbachia* and arbovirus inhibition in mosquitoes. *Future Microbiology*, 8, 1249–56.

Slater, H.C., Gambhir, M., Parham, P.E., & Michael, E. (2013). Modelling co-infection with malaria and lymphatic filariasis. *PLoS Computational Biology*, 9, e1003096.

Smith, D.R. (2012). An update on mosquito cell expressed dengue virus receptor proteins. *Insect Molecular Biology*, 21, 1–7.

Smith, D.R., Adams, A.P., Kenney, J.L., Wang, E., & Weaver, S.C. (2008). Venezuelan equine encephalitis virus in the mosquito vector *Aedes taeniorhynchus*: Infection initiated by a small number of susceptible epithelial cells and a population bottleneck. *Virology*, 372, 176–86.

Soubeyrand, S., Thébaud, G., & Chadœuf, J. (2007). Accounting for biological variability and sampling scale: A multi-scale approach to building epidemic models. *Journal of The Royal Society Interface*, 4, 985–97.

Steib, B.M., Geier, M., & Boeckh, J. (2001). The effect of lactic acid on odour-related host preference of yellow fever mosquitoes. *Chemical Senses*, 26, 523–8.

Strauss, J.H., Wang, K.S., Schmaljohn, A.L., Kuhn, R.J., & Strauss, E.G. (1994). Host-cell receptors for Sindbis virus. *Archives of Virology Supplement*, 9, 473–84.

Taguwa, S., Maringer, K., Li, X. et al. (2015). Defining HSP70 subnetworks in dengue virus replication reveals key vulnerability in flavivirus infection. *Cell*, 163, 1108–23.

Tesla, B., Demakovsky, L.R., Mordecai, E.A. et al. (2018). Temperature drives Zika virus transmission: evidence from empirical and mathematical models. *Proceedings of the Royal Society B: Biological Sciences*, 285, 1884.

Thomas, R.E., Wu, W.K., Verleye, D., & Rai, K.S. (1993). Midgut basal lamina thickness and dengue-1 virus dissemination rates in laboratory strains of *Aedes albopictus*

(Diptera: Culicidae). *Journal of Medical Entomology*, 30, 326–31.

Travanty, E.A., Adelman, Z.N., Franz, A.W. et al. (2004). Using RNA interference to develop dengue virus resistance in genetically modified *Aedes aegypti*. *Insect Biochemistry and Molecular Biology*, 34, 607–13.

Tsetsarkin, K.A., Vanlandingham, D.L., McGee, C.E., & Higgs, S. (2007). A single mutation in chikungunya virus affects vector specificity and epidemic potential. *PLoS Pathogens*, 3, e201.

Tsetsarkin, K.A. & Weaver, S.C. (2011). Sequential adaptive mutations enhance efficient vector switching by Chikungunya virus and its epidemic emergence. *PLoS Pathogens* 7, e1002412.

Turell, M.J., Britch, S.C., Aldridge, R.L., Kline, D.L., Boohene, C., & Linthicum, K.J. (2013). Potential for mosquitoes (Diptera: Culicidae) from Florida to transmit Rift Valley fever virus. *Journal of Medical Entomology*, 50, 1111–17.

Turell, M.J., Mores, C.N., Dohm, D.J., Lee, W.J., Kim, H.C., & Klein, T.A. (2006). Laboratory transmission of Japanese encephalitis, West Nile, and Getah viruses by mosquitoes (Diptera: Culicidae) collected near Camp Greaves, Gyeonggi Province, Republic of Korea 2003. *Journal of Medical Entomology*, 43, 1076–81.

Vaidyanathan, R. & Scott, T.W. (2006). Apoptosis in mosquito midgut epithelia associated with West Nile virus infection. *Apoptosis*, 11, 1643–51.

Valzania, L., Coon, K.L., Vogel, K.J., Brown, M.R., & Strand, M.R. (2018). Hypoxia-induced transcription factor signaling is essential for larval growth of the mosquito *Aedes aegypti*. *Proceedings of the National Academy of Sciences of the United States of America* 115, 457–65.

Varjak, M., Maringer, K., Watson, M. et al. (2017). *Aedes aegypti* Piwi4 is a noncanonical PIWI protein involved in antiviral responses. *mSphere*, 2.

Vasilakis, N. & Tesh, R.B. 2015. Insect-specific viruses and their potential impact on arbovirus transmission. *Current Opinion in Virology*, 15, 69–74.

Vazeille, M., Moutailler, S., Coudrier, D. et al. (2007). Two chikungunya isolates from the outbreak of La Reunion (Indian Ocean) exhibit different patterns of infection in the mosquito, *Aedes albopictus*. *PLoS One* 2, e1168.

Vega-Rua, A., Lourenco-de-Oliveira, R., Mousson, L. et al. Nougairede, A., De Lamballerie, X., & Failloux, A.B. (2015). Chikungunya virus transmission potential by local *Aedes* mosquitoes in the Americas and Europe. *PLoS Neglected Tropical Diseases* 9, e0003780.

Vodovar, N., Bronkhorst, A.W., Van Cleef, K.W.R. et al. (2012). Arbovirus-Derived piRNAs Exhibit a Ping-Pong Signature in Mosquito Cells. *PLoS One*, 7, e30861.

Vontas, J., Blass, C., Koutsos, A.C. et al. (2005). Gene expression in insecticide resistant and susceptible *Anopheles gambiae* strains constitutively or after insecticide exposure. *Insect Molecular Biology*, 14, 509–21.

Wang, H., Gort, T., Boyle, D.L., & Clem, R.J. (2012). Effects of manipulating apoptosis on Sindbis virus infection of *Aedes aegypti* mosquitoes. *Journal of Virology*, 86, 6546–54.

Wang, P. & Granados, R.R. (2000). Calcofluor disrupts the midgut defense system in insects. *Insect Biochemistry and Molecular Biology* 30, 135–43.

Wang, P., Li, G., & Granados, R.R. (2004). Identification of two new peritrophic membrane proteins from larval *Trichoplusia ni*: Structural characteristics and their functions in the protease rich insect gut. *Insect Biochemistry and Molecular Biology* 34, 215–27.

Wang, X.H., Aliyari, R., Li, W.X. et al. (2006). RNA interference directs innate immunity against viruses in adult *Drosophila*. *Science*, 312, 452–4.

Wang, Y., Gilbreath, T.M., III, Kukutla, P., Yan, G., & Xu, J. (2011). Dynamic Gut Microbiome across Life History of the Malaria Mosquito *Anopheles gambaie* in Kenya. *PLoS One*, 6, e24767.

Waterhouse, R.M., Kriventseva, E.V., Meister, S. et al. (2007). Evolutionary Dynamics of Immune-Related Genes and Pathways in Disease-Vector Mosquitoes. *Science*, 316, 1738–43.

Weaver, S.C. (2005. Host range, amplification and arboviral disease emergence. *Archives of Virology Supplement*, 33–44.

Weaver, S.C. (2013). Urbanization and geographic expansion of zoonotic arboviral diseases: Mechanisms and potential strategies for prevention. *Trends in Microbiology*, 21, 360–3.

Weaver, S.C. & Barrett, A.D. (2004). Transmission cycles, host range, evolution and emergence of arboviral disease. *Nature Reviews Microbiology*, 2, 789–801.

Weaver, S.C., Scott, T.W., & Lorenz, L.H. (1990). Patterns of eastern equine encephalomyelitis virus infection in *Culiseta melanura* (Diptera: Culicidae). *Journal of Medical Entomology*, 27, 878–91.

Weger-Lucarelli, J., Auerswald, H., Vignuzzi, M., Dussart, P., & Karlsson, E.A. (2018). Taking a bite out of nutrition and arbovirus infection. *PLoS Neglected Tropical Diseases*, 12, e0006247.

Weiss, B.L., Savage, A.F., Griffith, B.C., Wu, Y., & Aksoy, S. (2014). The peritrophic matrix mediates differential infection outcomes in the tsetse fly gut following challenge with commensal, pathogenic, and parasitic microbes. *Journal of Immunology*, 193, 773–82.

Whitfield, S.G., Murphy, F.A., & Sudia, W.D. (1973). St. Louis encephalitis virus: an ultrastructural study of infection in a mosquito vector. *Virology*, 56, 70–87.

World Health Organization. 2016. *WHO statement on the first meeting of the International Health Regulations (2005) (IHR 2005) Emergency Committee on Zika virus and observed increase in neurological disorders and neonatal malformations* [Online]. Available: http://www.who.int/mediacentre/news/statements/2016/1st-emergency-committee-zika/en/ [Accessed July 19 2017].

Xi, Z., Ramirez, J.L., & Dimopoulos, G. (2008). The *Aedes aegypti* toll pathway controls dengue virus infection. *PLoS Pathogens*, 4, e1000098.

Yamamoto, A., Yoneda, H., Hatano, R., & Asada, M. (1995). Influence of hexythiazox resistance on life history parameters in the citrus red mite, *Panonychus citri* (McGregor) studies on hexythiazox resistance in phytophagous mites (Part 4). *Journal of Pesticide Science*, 20, 521–7.

Yurchenco, P.D. & O'Rear, J.J. (1994). Basal lamina assembly. *Current Opinion in Cell Biology* 6, 674–81.

Zhang, G., Hussain, M., O'Neill, S.L., & Asgari, S. (2013). *Wolbachia* uses a host microRNA to regulate transcripts of a methyltransferase, contributing to dengue virus inhibition in *Aedes aegypti*. *Proceedings of the National Academy of Sciences of the United States of America*, 110, 10276–81.

Zhou, Y., Liu, Y., Yan, H., Li, Y., Zhang, H., Xu, J., Puthiyakunnon, S., & Chen, X. (2014). miR-281, an abundant midgut-specific miRNA of the vector mosquito *Aedes albopictus* enhances dengue virus replication. *Parasites & Vectors*, 7, 488.

Zouache, K., Fontaine, A., Vega-Rua, A. et al. (2014). Three-way interactions between mosquito population, viral strain and temperature underlying chikungunya virus transmission potential. *Proceedings of the Royal Society B: Biological Sciences*, 281.

Kindling, Logs, and Coals: The Dynamics of *Trypanosoma cruzi*, the Etiological Agent of Chagas Disease in Arequipa, Peru

Michael Z. Levy

Chagas disease affects millions of people in the Americas, and, through migration, thousands more on other continents. The agent of the disease, *Trypanosoma cruzi*, is a slender, highly-motile, unicellular parasite. *T. cruzi* does not migrate to the salivary glands of its insect vector–the blood-sucking triatomine insects–as many other vector-borne parasites do. Instead it makes its home in the gut, and replicates there before passing, along with the insect's feces, onto the skin of a host. Humans are infected, usually, through contact with infectious feces. Epidemics of vector-borne *T. cruzi* transmission are remarkably slow, and, given the longer time scale, the forces that lead to the emergence of *T. cruzi* are not often those that maintain its transmission, and these again are not those that sustain it over the course of decades. Consider a fire. Kindling, logs, and coals all burn, but each contributes to the fire in a very different manner. Whether or not the fire is sustained depends on the precise mixture of each fuel, the contact between them, some chance, and, often, a good deal of human carelessness. If we are to understand the fire we must understand each kind of fuel: its nature and propensity to burn as well as its conformation—how it is piled up and otherwise conditioned to alight.

12.1 Discovery

Carlos Chagas, a Brazilian physician and scientist, discovered *T. cruzi* in its insect vector before he saw a single human case of the disease that now bears his name (Chagas 1909). It was 1909, the height of the rubber boom in the Amazon, and Chagas had been sent by his mentor, Oswaldo Cruz, to investigate a malaria outbreak among railway workers in the town of Lassance, Belo Horizonte. While in Lassance, Chagas became familiar with the local triatomine–*Panstrongylus megistus* (Lewinsohn 2003). Even working in a makeshift lab in an abandoned railway car it was easy for him to observe the parasite in the feces of its vector. He would have seen the frenetic movement of the epimastigotes, the form that reproduces in the insect, broken occasionally by the directional undulating swim of the infectious form—the trypomastigotes (Levy 2007).

Since Chagas' discovery, the triatomine species that vector *T. cruzi* have played a large role in the study and control of the disease, perhaps because they are larger than most other vector arthropods. Adult triatomines of the larger species are similar in size to cockroaches and can imbibe nearly 1ml of blood in one sitting (Rabinovich 1972). Over eighty species of triatomines are known to transmit *T. cruzi*. Species are classified as, or sometimes

Michael Z. Levy, *Kindling, Logs, and Coals: The Dynamics of* Trypanosoma cruzi, *the Etiological Agent of Chagas Disease, in Arequipa, Peru* In: *Population Biology of Vector-Borne Diseases*. Edited by: John M. Drake, Michael B. Bonsall, and Michael R. Strand: Oxford University Press (2021). © Michael Z. Levy.
DOI: 10.1093/oso/9780198853244.003.0012

placed along a spectrum of, domestic and sylvatic. The most domestic species tend to be the most dangerous to humans. *Triatoma infestans*, *Rhodnius prolixius* and *Triatoma dimidiata*, historically, have been responsible for the most infections (Rassi Jr, Rassi Marin-Neto 2010). Sylvatic populations occur in a wide variety of habitats. Many reside or pass through palm trees, where the base of the palm crown provides them with shelter and protection. Nestled in these micro-habitats they feed on, and are fed upon by, the great diversity of species that also inhabit the trees. Others are commonly found in armadillo or rodent burrows, or bird nests (Zeledon & Rabinovich 1981).

Chagas sent the triatomines in which he observed the peculiar parasite to his mentor, Oswaldo Cruz. Cruz, working in Rio de Janeiro, exposed various animals he had in his lab, and, after some effort, succeeded in observing the parasites in the blood of a marmoset (Miles 2004). *Trypanosoma cruzi* has one of the largest host ranges of any parasite, it seems able to infect, and replicate in, every mammal with which it has come into contact. The ease at which the parasite can thrive in such a variety of hosts is balanced by the difficulties of its transmission. Stercocarian (feces mediated) transmission, which is considered the principal means by which the parasite infects mammals, is incredibly inefficient. Thousands of contacts with an infectious vector are necessary for a host to become infected, on average (Nouvellet 2015). Cruz was lucky to infect the marmoset—though it is possible that his luck was aided by his choice of experimental animal. Marmosets are insectivores and it is not unlikely that his subjects ingested infectious insects. Transmission by the oral route, in which the parasite enters the body through the mucosa, is much more efficient (Hoft 1996).

12.2 Burden

Approximately six million individuals are infected with *T. cruzi*, making it the parasitic disease with the highest burden in the Americas, but very few infected people are aware of their infection (Pérez-Molina & Molina 2018). Chagas disease is divided into two phases, acute and chronic. The acute phase begins one-to-two weeks following infection and generally consists of mild and nonspecific symptoms (fever, malaise among others). Textbooks often show images of the Romaña sign, a prolonged painless swelling of an eyelid, or of a Chagoma, a skin nodule which may be present, presumably, at the site of inoculation. In fact these dramatic and visual signs of acute disease occur in only a small percentage of infections (Bern 2015). The acute phase lasts four-to-eight weeks, and patients then enter the long (often lifelong) chronic phase. The chronic phase is subdivided into two forms of the disease. The indeterminate form has no signs or symptoms; most people remain in this form, infected but asymptomatic, for life. Some 20–30 percent of individuals progress to the determinate form, most commonly characterized by cardiomyopathy; digestive disease, including megacolon, also can occur (Bern 2015). Two drugs are currently available for treatment of Chagas disease: Benznidazole and Nifurtimox. Both can eliminate the parasite in the acute phase of infection (Coura & De Castro 2002). However, the efficacy of the drugs in the chronic phase is a matter of contention, in part due to the dearth of appropriately powered controlled trials (Maguire 2015).

12.3 Control

Control of Chagas disease through indoor residual spraying of insecticides has been one of the greatest public health successes of recent decades. A concerted effort by Brazil, Argentina, Paraguay, Uruguay, Bolivia, and occasionally Peru, known as the Southern Cone Initiative, was especially successful in controlling *Triatoma infestans*, one of the most dangerous vectors of *T. cruzi* (Schofield & Dias Dias 1999). A Central American Initiative (IPCA, by its Spanish acronym) may have regionally eliminated another dangerous insect, *Rhodnius prolixius* (Hashimoto & Schofield 2012). The success of these campaigns is surprising, perhaps, given the immense challenges facing them. Chagas disease is especially prevalent in some of the poorest communities, and many of these are remote, making insecticide application, and especially surveillance following insecticide application, difficult (Gürtler 2009). In recent years *T. infestans* has also invaded major urban centers (Levy et al. 2004). While control of the vector in cities is on one hand easier, as there are generally more resources available and better infrastructure, eliminating an insect from a dense and sprawling urban environment is a Sisyphean task, as areas under control are at constant risk of reinfestation from areas still infested, some of which are only a stone's throw away (Levy et al. 2010).

The success against *T. infestans* and *R. prolixus* seems miraculous, though there are some clues in the population biology of the vectors that perhaps might explain the miracle. Both species are essentially invasive across most of the range of the most successful control campaigns. *R. prolixus*, native to northern South America, was most likely released in Central America accidentally (Zeledón-Araya 2004). *Triatoma infestans* originated in the Chaco region of South America, expanding its range with humans, but is limited in most areas to domestic and peridomestic spaces (Schofield & Dias 1999). The triatomines are also relatively long-lived and reproduce slowly. Insecticide resistance has emerged in populations of triatomines, but the degree and extent of resistance is very much constrained (Fronza et al. 2016). To date, resistant strains of insects have been susceptible to second-line insecticides (Fronza et al. 2016), though counting on such a pattern to continue would be unwise.

12.4 *T. cruzi* Emergence and Control in Arequipa, Peru

Nouvellet et. al. (2015) recently produced a review and description of mathematical models of multiple facets of Chagas disease epidemiology and control. Here I focus on a single focus of transmission, the city of Arequipa, a metropolis of nearly 1 million people (*de Estadı́stica e Informática Perú*) in southern Peru. Chagas disease is endemic in the fertile valleys that channel water from the Andes to the Pacific (Náquira 2014 Cornejo & Lumbreras 1950). *Trypanosoma cruzi* has been detected in mummies from the area (Aufderheide et al. 2004). But the emergence in Arequipa is recent. Spatial force of infection models suggest that the epidemic likely initiated in the 1980's (Levy et al. 2011), but it only came to light in 2001, following the death of an infant from acute Chagas disease (Fraser 2008). By 2004, 5 percent of children were infected in poorer peri-urban districts of the

city (Bowman et al. 2008). More urban areas were also hit, though at relatively lower prevalences [2 percent Socabaya, Paucarpata] (Hunter et al. 2012) [1 percent Mariano Melgar] (Levy et al. 2014). A large vector control campaign running from 2003 to 2018 doused 80,000 houses in affected districts of the city with deltamethrin insecticide and effectively put out the fire. But the parasite is not gone, it smoulders in the bodies of long-lived hosts, the insect is not entirely gone, sites of vector recurrence are reported or uncovered regularly (Barbu et al. 2014). The coals are still hot.

12.4.1 Kindling

The kindling for *T. cruzi* transmission in southern Peru is the guinea pig. Guinea pigs are commonly raised throughout the Andes for human consumption; the vector *T. infestans* also feeds upon them. From the point of view of *T. infestans*, domesticated guinea pigs must seem the perfect host. They have no means to protect themselves from triatomine bites. Their legs are short and entirely incapable of swatting, and, unlike mice and many other rodents, domesticated guinea pigs show neither insectivorous interest nor ability.[1] Their tendency to huddle against the walls of their enclosures, perhaps for warmth, brings them in very close contact with triatomines hiding in these walls.

Trypanosoma cruzi also finds the domesticated guinea pig to be an excellent host. Guinea pigs, like most mammals, become parasitemic following infection, and, when triatomines are fed on them during the acute phase, nearly all of the insects pick up the parasite (Castillo-Neyra et al. 2016). In addition, some guinea pigs remain infectious for a much longer period, infecting bugs at a high rate for as long as a year (Castillo-Neyra et al. 2016). The biology of the domestic guinea pig makes it good kindling, but it takes more than the composition of kindling to start a fire. Mulitiple observations of triatomines from guinea pig pens yielded an extremely high index of parasitic infection, with over 85 percent of triatomines collected from a pen carrying the parasite (Levy et al. 2015, see Fig. 12.1).

---

[1] After months of developing protocols and obtaining IACUC approvals we tried to feed live and dead triatomines to guinea pigs with no success.

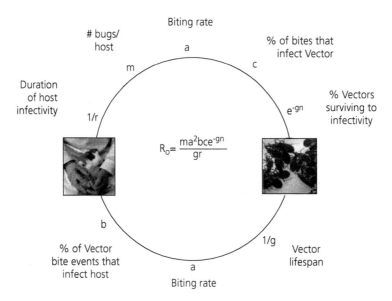

Figure 12.1 A Ross-MacDonald Model of *T. cruzi* transmission between guinea pigs and *Triatoma infestans*.

Such fires are not readily explained from the biology of the animals alone. To illustrate the point, consider a version of the Ross–MacDonald equation that estimates the expected number of secondary guinea pig infections that might occur when a single, average infectious guinea pig were introduced into an otherwise susceptible pen:

$$R_0 = \frac{ma^2bce^{-gn}}{gr}$$

where m is the ratio of vectors to hosts, a the vector's biting rate, b the transmission efficiency from vectors to hosts, $1/r$ is the average duration of the infectious period in hosts, c the transmission efficiency from hosts to vectors, g the rate of vector mortality, and n the average length of the parasite incubation period in the vector. A little more math borrowed from Smith et al. (2007) yields the following expectations at equilibrium:

$$\bar{X} = \text{proportion of infected hosts}$$

$$\bar{Y} = \text{proportion of infected vectors}$$

$$\bar{X} = \frac{R_0 - 1}{R_0 + ca/g}$$

$$\bar{Y} = \frac{ac\bar{X}}{g + ac\bar{X}}$$

The parameters, some of which are only poorly estimated, are described in Levy et al. (2015). The expected prevalence, based on these best guesses, is much less than the 85 percent observed in the field. The model thereby serves to demonstrate a gap in understanding of the system, and Levy et al. (2015) recounts a number of experiments to try to fill that gap. A possibility, mentioned above, is that the presence of guinea pigs with extended parasitemia might account for the extremely high prevalence of parasites in the vectors. This hypothesis fails to account for the observation—because guinea pigs, in general, do not live all that long. In fact it may be the idiosyncrasies of guinea pig husbandry that confers upon them their role as kindling in the urban epidemic.

The manner by which guinea pigs are raised in southern Peru is shaped in nearly equal parts by centuries of tradition and very recent economic phenomena. Guinea pigs grow quickly, feed primarily on alfalfa, and, as natural borrowers of burrows, are easily raised in small enclosures. The last characteristic especially pre-adapts guinea pig husbandry for urban areas. Many migrants came to Arequipa from farmlands, and adapted what husbandry practices that they could to the small footprint of urban houses. Procuring alfalfa poses a challenge in a city. Alfalfa is a water dependent crop, growing most readily in the rainy season (January–March) and its price fluctuates naturally with the seasons. The demands on alfalfa from the large dairy industry centered in Arequipa can further aggravate price fluctuations so much so that the price in the midst of the dry season (July and August) reaches three times that of the rainy one (Levy et al. 2015). Households raising guinea pigs often sell or eat the animals when the price of alfalfa rises steeply; occasionally maintaining a small number of guinea pigs that are perhaps of greater value for breeding purposes. The resultant bottleneck in guinea pig populations, is, by the fire analogy, grouping together a small pile of kindling. If one of the pieces is alight the others quickly catch— because, while the host may have undergone a bottleneck, the vector population, which is often quite large and impervious to periods of starvation, does not, immediately, follow. The resultant mismatch between host and vector population sizes is sufficient to drive the prevalence of *T. cruzi* to dizzying heights among the vector populations (Fig. 12.2).

Kindling burns fast. The guinea pigs that are not kept through bottleneck periods can only expect to live about four months before they are sold or eaten. The conditions needed to light the kindling–the necessity of an infected guinea pig being among those retained during a population bottleneck–cannot be expected to repeat over longer time scales. How does the parasite persist? One possibility is that the system is stabilized spatially. If the movement of infected insects and/or infected guinea pigs is sufficient to seed a new pen, such that the rate of seeding exceeds the rate of extinction, a stable metapopulation may ensue. Stabilization through metapopulation dynamics is common in infectious disease dynamics (Grenfell & Harwood 1997). For the system in Arequipa, at least, the possibility has not been explored fully. If it were stable, the parasite might be maintained through guinea pigs alone, as has been suggested by Gürtler &

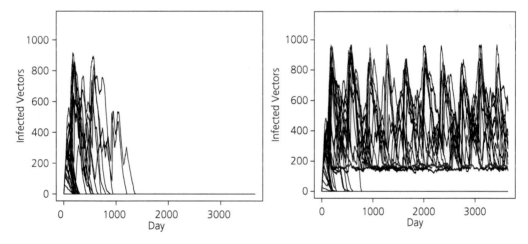

Figure 12.2 Thirty simulations of *T. cruzi* transmission through a population of 1000 vectors in a guinea pig enclosure. Initially ten animals occupy the pen; this number is reduced to two each year for 90 days to mimic the dry season when alfalfa is at a premium. Each bug is assigned to feed every 14 days, though the particular set of days is different for each insect. On a feeding day the bug is matched, at random, with one of the hosts in the pen. At each feeding event the parasite can, with some probability, be transmitted from an infectious insect to a susceptible host or from an infectious host to an uninfected insect. The right figure includes, in addition, one susceptible dog, near the pen. The dog lives longer (five years), but is otherwise treated like the other hosts, and fed upon by the insects at an equal probability as the guinea pigs. Note that the host bottlenecks can drive the prevalence of the parasite in the vector colony to very high prevalence, but that transmission ultimately fizzles out over a short time period (Left), while the addition of a dog (Right) is sufficient to stabilize transmission of the parasite over time.

Cardinal (2015). A second possibility is that the system is stabilized temporally—with the addition of a longer-lived host.

12.4.2 Logs

The logs of the epidemic in Arequipa and elsewhere–responsible for keeping *T. cruzi* burning–are dogs. Dogs live much longer than guinea pigs. They are infectious to insects, though, for a variety of reasons, may not support as large an insect population (Levy et al. 2006). Dogs also do not infect insects as consistently as guinea pigs—their longer lifespan means that dogs spend much more of their life in the chronic phase of the disease during which their infectiousness to bugs undergoes a time-dependent decline (Gürtler & Cardinal 2015). Fig. 12.2 (Right panel) demonstrates the effect of adding a single dog to the simulation described above. The dog buffers against the stochasticity of the guinea pig bottlenecks, stabilizing the transmission dynamics so that, in the majority of cases, the parasite does not go extinct, see Table 12.1.

12.4.3 Coals

The massive insecticide control campaign initiated in 2003 in Arequipa was not the first in the region. A campaign was undertaken in 1970, (Náquira & Cabrera 2009) house codes visible on some houses suggest spraying occurred in the 1980s and at least one district was treated in 1995 (Delgado et al. 2011). Even imperfect campaigns can be devastating to the vector population. How does the parasite survive extended periods in the absence of vectors? In the bodies of its host. In cases of exceptionally poor control campaigns, with rapid reinfestation, dogs may suffice as the temporal refuge for the parasite, and re-seed it into the returning vector population. In the case of reasonably sustained control campaigns, in which the vectors are absent, or at very low levels, for decades, dogs are insufficient to bridge the temporal gap between infestations. Humans are the coals with the potential to rekindle the fire.

How does the parasite survive in the bodies of its host in the absence of vectors? There are a couple of possible mechanisms. First, the parasite can be transmitted from mother to child, raising the possibility that vertical transmission chains might allow it to weather the drought of vectors. Vertical transmission, also known as congenital transmission, is well-studied in humans and also has been observed in animals (Gürtler et al. 2015). However, the probability of infection by this route is very low, such that sustaining a chain of vertical transmission for twenty years, say, is exceedingly difficult. The parasite,

Table 12.1 Parameter values for simulations of *T. cruzi* transmission through a population of 1000 vectors in a guinea pig enclosure. Simulation results are presented in Figure 12.2.

parameter	description	simulation model value	sources
a	number of days between bloodmeals	1/14	estimated from laboratory feeding (Levy, unpublished data)
n	length of incubation period of parasite in vector (days)	45[2]	(Levy et al. 2015)
g	lifespan of the vector (days)	180	(Cruz-Pancheco et al. 2012)
b	infected vector to susceptible host infection probability	0.00068	(Catala et al. 1992)
c	infectious host to susceptible vector infection probability	1	(Gürtler et al. 1996)
$1/r$	duration of parasitemia in host (days)	56	(Castro-Sesquen et al. 2011)

[2] Parasites were observed to reach equilibrium densities in vectors approximately 45 days after vectors ingested an infectious blood meal. Vector infectiousness (b) was therefore modeled as a logistic function, $b/1+20e^{-0.20t}$ with infectiousness reaching a peak of 0.00068 in approximately 45 days.

most likely and most often, survives within the bodies of a single host.

How does *T. cruzi* manage to evade the immune system for so long? The motile trypomastigotes quickly enter cells upon infecting a mammal. Within the cell the trypomastigote transforms into circular amastigotes, which replicate, burst the cell and transform into trypomastigotes which then seek and infect additional cells. Recently, work by Sánchez-Valdéz et al. (2018) demonstrated that a sub-population of these amastigotes goes dormant—remaining in the cell, without replicating, for extended periods of time, but, regaining the ability to transform into trypomastigotes upon exiting dormancy. The dormancy mechanism, along with multiple, more widely recognized mechanisms for evading the immune system (Cardoso et al. 2016), may permit parasite populations to persist in humans for decades.

The parasite does not just avoid the immune system, it also avoids treatment. There are two drugs available to treat *T. cruzi* infected individuals. A recent clinical trial of the more common of the two, Benznidazole, failed to show efficacy in the treatment of symptomatic chronic Chagas disease. Whether the drug is efficacious in treating the much larger population of asymptomatic individuals with chronic Chagas disease is an open question (Maguire 2015). What is clear from clinical trials (most recently the BENEFIT trial (Morillo et al. 2015) is that Benznidazole shows strong trypanosidal activity. However, in these trials, as well as in animal studies (Bustamante et al. 2008), the parasite can rebound following cessation of treatment. The same dormancy mechanism that likely evolved to facilitate immune evasion may also protect the parasite from the drug's action, which targets dividing parasites (Sánchez-Valdéz et al. 2018). Drug evasion is one impediment to extinguishing the coals of Chagas disease. Only a tiny fraction of the population at risk for Chagas disease is tested and diagnosed, and only a small fraction of those diagnosed receive treatment. These are the causes and consequences of neglect, and they, more than any trick of trypanosome biology, allow the coals to burn on. If things continue in the manner they are now, the expectation must be that Chagas disease will

re-emerge following the recent successes of vector control in much the same way that malaria rebounded following the massive eradication efforts in the 1950s (Shah 2010).

12.5 Other Fires

The combined socio-ecological phenomena that permitted the emergence and maintenance of *T. cruzi* in Southern Peru, and may permit re-emergence in the future, do not occur widely, and certainly not nearly as widely as Chagas disease. But other similar dynamics, with different species of hosts and insects involved, occur across the range of *T. cruzi* in the Americas.

12.5.1 Wildfires

While insecticide application campaigns have been extremely successful against domestic insects, most of which were essentially invasive species across most of their ranges, control of sylvatic vectors is only in its infancy. Sylvatic vectors, of which there are dozens of species, live in a variety of habitats, many associated with palm trees, and enter human dwellings only occasionally. Often they are infected with *T. cruzi*; the infection occurs outside of the residence, and is probably related to the dynamics of hosts and vectors there. These dynamics are only very poorly understood (Abad-Franch et al. 2015). The fire, and the conditions that light it, are external. The problem, to human health, is the encroachment of the parasite into human dwellings, or, more often, the encroachment of human dwellings into the vector and parasite's habitat. Sparks—infectious vectors—land in human dwellings unprotected by window screens (or even walls), in search of a bloodmeal. These sparks generally die out, but with enough sparks the prevalence among communities can climb. Efforts to prevent or extinguish sylvatic transmission cycles seem far-fetched, most control programs focus on protecting human dwellings.

12.5.2 Chickens as Kindling

The Chaco region of Northern Argentina is one of the best studied foci of Chagas disease transmission. As in Arequipa, the main vector is *T. infestans* and it is predominantly domestic. Prevalence among vectors can climb quite high (Gürtler et al. 1993), though guinea pigs are not commonly raised for food in the area. A model (Cohen & Gürtler 2001) addresses the role of chickens in the transmission of the parasite. Chickens are not suitable hosts for *T. cruzi*, they are, however, good hosts for the vector. The model addresses a perhaps antiquated idea of zooprophylaxis—whether chickens in households might attract potentially infectious bites that otherwise would land on people or dogs. The model's answer is that, under most conditions, the increase in vector populations due to the presence of chickens overcomes any potential protective zooprophylatic effect they might have (Cohen & Gürtler 2001). But there is an important detail in the model that closely mimics the situation in Arequipa—the chickens remain inside the house for part of the year, and then leave to roost elsewhere. During their time in the house the chickens support the growth of a large vector population, which, after their departure, concentrates on the dogs and people in the house. The chickens thereby serve a partially analogous role as the guinea pig population bottlenecks—piling up kindling that creates the conditions for the parasite to spread through the vector population, even though the chickens are not infected themselves. Dogs again serve the role of logs. Recent insecticide control campaigns in the Chaco have been successful (Gaspe et al. 2018), yet the high prevalence of infection in some human populations (Samuels et al. 2013) makes the re-emergence of the parasite extremely likely if the vector is allowed to re-disperse.

12.5.3 Will Bed Bugs Catch?

Just a few years following Carlos Chagas' description of *T. cruzi* in a triatomine, a French scientist, Emilie Brumpt, reported that the common bed bug was also a competent vector of the parasite (Brumpt 1912). His work has been repeated and expanded over the years (Salazar et al. 2015), leaving little doubt that bed bugs can transmit *T. cruzi*. Whether they are or are likely to become important in the transmission under field conditions (meaning

bedrooms) is unclear and unstudied. Based on the basic Ross–MacDonald model shown in Fig. 12.1, there are many reasons to worry about bed bugs. The number of bed bugs per host (m) has not been measured, but is likely an order of magnitude higher than that of triatomines. The biting rate (a) is also around twice as frequent as that in triatomines (Peterson et al. 2018). In the Ross–MacDonald model this term is squared, meaning it could have a four-fold increase in the R0. Peterson et al. (2018) found no evidence that the death rate ($1/g$) of bed bugs might be increased when these are infected with *T. cruzi*, which could significantly affect their vectorial capacity. What remains unknown, and what is critical to assessing the potential of emergence and maintenance over the short and long term, are the feeding patterns of bed bugs. In laboratory conditions bed bugs will feed on pretty much any warm-blooded animal, though there is some evidence that there survival is greater when fed on human blood (Barbarin et al. 2013). To date no study has identified the bloodmeals of bed bugs caught in the field. Such a study is non-trivial, as it would require careful spatial and temporal sampling. Bed bugs, for reasons that are not clear, seem to bother people much less in the winter months (Mabud et al. 2014; Sentana-Lledo et al. 2015), and it is possible that they are feeding on mice or other mammalian hosts during these periods.

12.6 Final Thoughts

The fate of Chagas disease is embroiled in social, political and economic factors. The key epidemiological phenomena in the system often occur on a timescale longer than the course of world events, and the questions presently surrounding the disease—including whether urban transmission by *T. infestans* can be eliminated—cannot be answered without reference to these. Framing the epidemiology of Chagas disease as a fire, in which discrete and different processes work over different timescales, is useful, first and foremost, because it breaks away from modeling frameworks successful for quicker disease systems. The framework also may help in understanding the nature of emergence, maintenance, and re-emergence of other diseases that occur over the course of generations.

12.7 Acknowledgements

I would like to thank Corentin Barbu, who originally formulated the kindling, logs, and coals analogy and also Aaron Tustin who wrote some of the simulation codes. I also thank my collaborators, especially Ricardo Castillo-Neyra, Justin Sheen, Victor Quispe-Machaca, Jenny Ancca-Juarez, Katty Borrini-Mayori, Cesar Naquira-Velarde, Eleazar Cordova-Benzaquen, and all in the Zoonotic Disease Research Lab of the Universidad Peruana Cayetano Heredia, Arequipa, Peru. I acknowledge the efforts of the Laboratorio de Ministerio de Salud del Perú (MINSA), the Dirección General de Salud de las Personas (DGSP), the Estrategia Sanitaria Nacional de Prevención y Control de Enfermedades Metaxenicas y Otras Transmitidas por Vectores (ESNPCEMOTVS), the Dirección General de Salud Ambiental (DIGESA), the Gobierno Regional de Arequipa, the Gerencia Regional de Salud de Arequipa (GRSA) without whom none of these studies would have been possible. This work received financial support from National Institutes of Health NIAID P50 AI074285 and 5R01 AI101229.

References

Abad-Franch, F., Lima, M.M. Sarquis, O. et al. (2015). On palms, bugs, and Chagas disease in the Americas. *Acta Tropica*, 151, 126–41.

Aufderheide, A.C., Salo, W., Madden, M. et al. (2004). A 9,000-year record of Chagas' disease. *Proceedings of the National Academy of Sciences of the United States of America*, 101, 2034–9.

Barbarin, A., Gebhardtsbauer, R., & Rajotte, E. (2013). Evaluation of blood regimen on the survival of *Cimex lectularius* L. using life table parameters. *Insects*, 4, 273–86.

Barbu, C.M., Butteenheim, A. M., Pumahuanca, M.L. et al. (2014). Residual infestation and recolonization during urban *Triatome infestans* bug control campaign, Peru. *Emerging Infectious Diseases*, 20, 2055.

Bern, C. (2015). Chagas' disease. *New England Journal of Medicine*, 373, 456–66.

Bowman, N.M., Kawai, V., Levy, M.Z. et al. (2008). Chagas disease transmission in periurban communities of Arequipa, Peru. *Clinical Infectious Diseases*, 461, 1822–8.

Brumpt, E. (1912). Le Trypanosoma cruzi évolue chez *Conorhinus megistus, Cimex lectularius* et *Ornithodorus moubata*. Cycle evolutif de ce parasite. *Bulletin de la Societe de Pathologie Exotique Filiales*, 5, 360–7.

Bustamante, J.M., Bixby, L.M., & Tarleton, R.L. (2008). Drug-induced cure drives conversion to a stable and protective CD8+ T central memory response in chronic Chagas disease. *Nature Medicine*, 14, 542.

Cardoso, M.S., Reis-Cunha, J.L., & Bartholomeu, D.C. (2016). Evasion of the immune response by *Trypanosoma cruzi* during acute infection. *Frontiers in Immunology*, 6, 659.

Castillo-Neyra, R., Borrini Mayori, K., Salazar Sanchez, R. et al. (2016). Heterogeneous infectiousness in guinea pigs experimentally infected with *Trypanosoma cruzi*. *Parasitology International*, 65, 50–54.

Castro-Sesquen, Y.E., Giman, R.H., Yauri, V. et al. (2011). *Cavia porcellus* as a model for experimental infection by *Trypanosoma cruzi*. *American Journal of Pathology*, 179, 281–8.

Catala, S.S., Gorla, D.E., & Basombrio, M.A. (1992). Vectorial transmission of *Trypanosoma cruzi*: an experimental field study with susceptible and immunized hosts. *American Journal of Tropical Medicine and Hygiene*, 47, 20–6.

Chagas, C. (1909). Nova tripanozomiaze humana: estudos sobre a morfolojia e o ciclo evolutivo do *Schizotrypanum cruzi* n. gen., n. sp., ajente etiolojico de nova entidade morbida do homem. *Memórias do Instituto Oswaldo Cruz*, 1, 159–218.

Cohen, J.E., & Gürtler, R.E. (2001). Modeling household transmission of American trypanosomiasis. *Science*, 293, 694–8.

Cornejo, A., & Lumbreras, H. (1950). Estudios preliminares sobre epidemiologi'a de la enfermedad de Chagas en el valle de Majes. *Archivio di Patologia e Clinica*, 4, 121–30.

Coura, J.R, & De Castro, S.L. (2002). A critical review on Chagas disease chemotherapy. *Memórias do Instituto Oswaldo Cruz*, 97, 3–24.

Cruz-Pancheco, G., Esteva, L., & Vargas, E. (2012). Control measures for Chagas disease. *Mathematical Biosciences*, 237, 49–60.

Instituto Nacional de Estadística e Informática Perú, Instituto (2008). *Población y Vivienda. Estimaciones y Proyecciones de Población. Población total al 30 de junio del 2013, por grupos quinquenales de edad, según departamento, provincia y distrito*. inei.gob.pe accessed July 2019.

Delgado, S., Castillo Neyra, R., Quispe Machaca, V.R. et al. (2011). A history of chagas disease transmission, control, and re-emergence in peri-rural La Joya, Peru. *PLoS Neglected Tropical Diseases*, 5, e970.

Dias, J.C.P. (2007). Southern Cone Initiative for the elimination of domestic populations of *Triatome infestans* and the interruption of transfusion Chagas disease: Historical aspects, present situation, and perspectives. *Memórias do Instituto Oswaldo Cruz*, 102, 11–18.

Fraser, B. (2008). Controlling Chagas' disease in urban Peru. *The Lancet*, 372, 16–17.

Fronza, G., Toloza, A.C., Picollo, M.I., Spillmann, C., & Mougabure-Cueto, G.A. (2016). Geographical variation of deltamethrin susceptibility of *Triatome infestans* (Hemiptera: Reduviidae) in Argentina with emphasis on a resistant focus in the Gran Chaco. *Journal of Medical Entomology*, 53, 880–7.

Gaspe, M. S., Provecho, Y.M., Fernandez, M.P., Vassena, C.V., Santo Orihuela, P.L, Gurtler, R.E. (2018). Beating the odds: Sustained Chagas disease vector control in remote indigenous communities of the Argentine Chaco over a seven-year period. *PLoS Neglected Tropical Diseases*, 12, e0006804.

Grenfell, B. & Harwood, J. (1997). (Meta) population dynamics of infectious diseases. *Trends in Ecology & Evolution*, 12, 395–9.

Gürtler, R.E. Cecere, M.C., Lauricella, M.A., Cardinal, M.V., Kitron, U., & Cohen, J.E. (2007). Domestic dogs and cats as sources of *Trypanosoma cruzi* infection in rural northwestern Argentina. *Parasitology*, 134, 69–82.

Gürtler, R.E. (2009). Sustainability of vector control strategies in the Gran Chaco Region: Current challenges and possible approaches. *Memórias do Instituto Oswaldo Cruz*, 104, 52–9.

Gürtler, R.E., & Cardinal, M.V. (2015). Reservoir host competence and the role of domestic and commensal hosts in the transmission of *Trypanosoma cruzi*. *Acta Tropica*, 151, 32–50.

Gürtler, R.E., Cecere, M.C., Petersen, R.M., Rubel, D.N., & Schweigmann, N.J. (1993). Chagas disease in north-west Argentina: association between *Trypanosoma cruzi* parasitaemia in dogs and cats and infection rates in domestic *Triatoma infestans*. *Transactions of The Royal Society of Tropical Medicine and Hygiene*, 87, 12–15.

Gürtler, R.E., Cecere, M.C., Castanera, M.B. et al. (1996). Probability of infection with *Trypanosoma cruzi* of the vector *Triatoma infestans* fed on infected humans and dogs in northwest Argentina. *American Journal of Tropical Medicine and Hygiene*, 55, 24–31.

Hashimoto, K., & Schofield, C.J. (2012). Elimination of *Rhodnius prolixus* in central America. *Parasites & Vectors*, 5, 45.

Hoft, D.F. (1996). Differential mucosal infectivity of different life stages of *Trypanosoma cruzi*. *American Journal of Tropical Medicine and Hygiene* 55, 360–4.

Hunter, G.C., Borrini-Mayori, K., Juarez, J.A. et al. (2012). A field trial of alternative targeted screening strategies for Chagas disease in Arequipa, Peru. *PLoS Neglected Tropical Diseases*, 6, e1468.

Levy, M.Z. (2007). *Control of vector-borne Chagas disease in the urban environment*, Arequipa, Peru: Thesis to Emory University.

Levy, M.Z., Bowman, N.M., Kawai, V. et al. (2006). Periurban Trypanosoma cruzi–infected *Triatoma infestans*, Arequipa, Peru. *Emerging Infectious Diseases*, 12, 1345.

Levy, M.Z., Malaga Chavez, F.S., Cornejo del Carpio, J.G., (2010). Rational spatio-temporal strategies for controlling a Chagas disease vector in urban environments. *Journal of the Royal Society Interface*, 7, 1061–1070.

Levy, M.Z., Small, D.S., Vilhena, D.A. et al. (2011). Retracing micro-epidemics of Chagas disease using epicenter regression. *PLoS Computational Biology*, 7, e1002146.

Levy, M.Z., Barbu, C.M., Castillo-Neyra, R. et al. (2014). Urbanization, land tenure security and vector-borne Chagas disease. *Proceedings of the Royal Society B: Biological Sciences* 281, 20141003.

Levy, M.Z., Tustin, A., Castillo-Neyra, R. et al. (2015). Bottlenecks in domestic animal populations can facilitate the emergence of *Trypanosoma cruzi*, the aetiological agent of Chagas disease. *Proceedings of the Royal Society B: Biological Sciences*, 282, 20142807.

Lewinsohn, R. (2003). Prophet in his own country: Carlos Chagas and the Nobel Prize. *Perspectives in Biology and Medicine*, 46, 532–49.

Mabud, M.S., Barbarin, A.M., Barbu, C.M., Levy, K.H., Edinger, J., & Levy, M.Z. (2014). Spatial and temporal patterns in *Cimex lectularius* (Hemiptera: Cimicidae) reporting in Philadelphia, PA. *Journal of Medical Entomology*, 51(1), 50–4.

Maguire, J.H. (2015). Treatment of chagas' disease–time is running out *New England Journal of Medicine*. 373, 1369–1370.

Miles, M.A. (2004). The discovery of Chagas disease: Progress and prejudice. *Infectious Disease Clinics*, 18, 247–60.

Morillo, C.A., Marin-Neto, J.A., Avezum, A. et al. (2015). Randomized trial of benznidazole for chronic Chagas' cardiomyopathy. *New England Journal of Medicine*, 373, 1295–306.

Náquira, C. (2014). Urbanización de la enfermedad de chagas en el Perú: Experiencias en su prevención y control. *Revista Peruana de Medicina Experimental y Salud Pública*, 31, 343–7.

Náquira, C. & Cabrera, R. (2009). Breve reseña histórica de la enfermedad de Chagas, a cien años de su descubrimiento y situación actual en el Perú. *Revista Peruana de Medicina Experimental y Salud Pública*, 26, 494–504.

Nouvellet, P., Cucunubá, Z.M., & Gourbière, S. (2015). Ecology, evolution and control of Chagas disease: A century of neglected modelling and a promising future. *Advances in Parasitology*, 87, 135–91.

Pérez-Molina, José A. & Molina, I. (2018). Chagas disease. *Lancet*, 391: 82–94.

Peterson J.K., Salazar R., Castillo-Neyra R., Borrini-Mayori K. et al. (2018). *Trypanosoma cruzi* infection does not decrease survival or reproduction of the common bed bug, cimex lectularius. *American Journal of Tropical Medicine and Hygiene*.

Rabinovich, J.E. (1972). Vital statistics of Triatominae (Hemiptera: Reduviidae) under laboratory conditions. I. *Triatoma infestans* Klug. *Journal of Medical Entomology*, 9, 351–70.

Rassi A. Jr., Rassi, A., Marin-Neto, J.A. (2010). Chagas disease. *Lancet*, 375, 1388–402.

Salazar, R., Castillo-Neyra, R., Tustin, A.W., Borrini-Mayori, K., Naquira, C., & Levy, M.Z. (2015). Bed bugs (*Cimex lectularius*) as vectors of *Trypanosoma cruzi*. *The American Journal of Tropical Medicine and Hygiene*, 92, 331–5.

Samuels, A.M., Clark, E.H., Galdos-Cardenas, G. et al. (2013). Epidemiology of and impact of insecticide spraying on Chagas disease in communities in the Bolivian Chaco. *PLoS Neglected Tropical Diseases*, 7, e2358.

Sánchez-Valdéz, F.J., Padilla, A., Wang, W., Orr, D., & Tarleton, R.L. (2018). Spontaneous dormancy protects *Trypanosoma cruzi* during extended drug exposure. *Elife*, 7, 34039.

Schofield, C.J. & Pinto Dias, J.C. (1999). The southern cone initiative against Chagas disease. *Advances in Parasitology*, 42, 1–27.

Sentana-Lledo, D., Barbu, C.M., Ngo, M.N., Wu, Y. Sethuraman, K., & Levy, M.Z. (2015). Seasons, searches, and intentions: What the internet can tell us about the bed bug (Hemiptera: Cimicidae) epidemic. *Journal of Medical Entomology* 53, 116–21.

Shah, S. (2010). *The Fever: How Malaria has Ruled Humankind for 500,000 Years*. Sarah Crichton Books, New York.

Smith, D.L., McKenzie, F.E., Snow, R.W., Hay, S. (2007). Revisiting the basic reproductive number for malaria and its implications for malaria control. *PLoS Biology*, 5(3), e42.

Zeledon, R. & Rabinovich, R.E. (1981). Chagas disease: An ecological appraisal with special emphasis on its insect vectors. *Annual Review of Entomology*, 26, 101–133.

Zeledón-Araya, R. (2004). *Algunos hechos históricos y datos recientes relacionados con la presencia de Rhodnius prolixus (Stål, 1859) (Hemiptera: Reduviidae) en América Central. Entomologi'a y Vectores.* [Some historical facts and recent issues related to the presence of Rhodnius prolixus 1859) (Hemiptera: Reduviidae) in Central America.] 11, 233–46.

CHAPTER 13

Gut Microbiome Assembly and Function in Mosquitoes

Kerri L. Coon and Michael R. Strand

13.1 Introduction

The most important insect vectors of vertebrate pathogens are mosquitoes, which transmit the causative agents of many important human diseases including dengue, malaria, chikungunya, yellow fever, and Zika. All mosquitoes belong to the family Culicidae (Order: Diptera), which to date contains about 3,500 recognized species and subspecies (Reidenbach et al. 2009; Wilkerson et al. 2015). Species within the Culicidae can be found in almost every region of the world, where immature stages (larvae) develop in practically any environment in which water occurs. Larvae in aquatic environments undergo complete metamorphosis to produce adults that are terrestrial and feed on sugar from plant nectar or sap. Adult females of most species also feed on blood from a vertebrate host to produce eggs, which is how mosquitoes acquire and transmit disease-causing organisms.

As in other animals, the digestive tract of both larval and adult stage mosquitoes is inhabited by a community of microbes that is collectively referred to as the 'gut microbiota' (Ley et al. 2008; Nicholson et al. 2012; Sommer & Backhed 2013; Yun et al. 2014). Recent studies reveal important contributions of the gut microbiota to the physiology of mosquitoes, which has stimulated broad interest in understanding the diversity and function of these communities. In this chapter, we provide a synthesis of current knowledge on mosquito gut microbiota,

with an emphasis on recent efforts to understand the factors that shape mosquito gut microbial communities and their effects on the transmission of mosquito-borne pathogens.

13.2 The Mosquito Gut Ecosystem

It has long been known that the digestive tract of larval and adult stage mosquitoes contains microbes (Hinman 1930). Much of our early knowledge on microbes associated with mosquitoes originated from traditional culture-based studies, in which gut homogenates from individual mosquitoes were cultured under laboratory conditions and the colony-forming microbes were identified based on morphological, metabolic, and other characteristics (Hinman 1930; Rozeboom 1935; Chao & Wistreich 1959; Jones & DeLong 1961; Ferguson & Micks 1961). However, a limitation of such studies is that the characterization of gut microbes is inherently biased towards those that are readily culturable in the laboratory, while microbes for which suitable laboratory growth conditions are not known remain undetected. Recent studies have therefore primarily relied on high-throughput sequencing approaches that more accurately estimate community richness (the total number of species) and diversity (the relative abundance of individual species). While numerous types of microorganisms (including bacteria, algae, protists, fungi, and rotifers) have been identified from the mosquito gut (Hinman 1930;

Kerri L. Coon and Michael R. Strand, *Gut Microbiome Assembly and Function in Mosquitoes* In: *Population Biology of Vector-Borne Diseases.*
Edited by: John M. Drake, Michael B. Bonsall, and Michael R. Strand: Oxford University Press (2021). © Kerri L. Coon and Michael R. Strand.
DOI: 10.1093/oso/9780198853244.003.0013

Walker et al. 1988; Merritt et al. 1990; DeMaio et al. 1996), most studies have focused on the characterization of bacteria using the 16S rRNA gene (Minard et al. 2013a). These surveys have significantly advanced understanding of the composition and structure of mosquito microbiomes and the factors that shape them.

13.2.1 Routes of Acquisition of Mosquito Gut Microbiota

As larvae, mosquitoes inhabit a wide variety of aquatic habitats where most species filter feed on detritus and other organic matter, including bacteria and eukaryotic microorganisms present in the water column (Clements 1992; Merritt et al. 1992) (see Fig. 13.1). Deep sequencing of bacterial 16S rRNA gene amplicons shows that the number of bacterial species in larvae is relatively low (<200) compared to humans and other mammals (>1000) (Minard et al. 2013a). The bacteria present in larvae also near fully overlap with the bacteria present in their aquatic environment, although community diversity is lower in larvae and the abundance of specific community members differs (Coon et al. 2014; Gimonneau et al. 2014; Coon et al. 2016a; Bascuñán et al. 2018; Wang et al. 2018) (see Fig. 13.1). Controlled experiments further indicate that mosquito larvae hatch from eggs with no microbes in their digestive tract (Coon et al. 2014). Altogether, these data strongly support the conclusion that mosquito larvae acquire their gut microbiota from the environment in which they feed.

Mosquito pupae are mobile in the aquatic habitat but do not feed, which results in no new microorganisms being introduced into the gut during

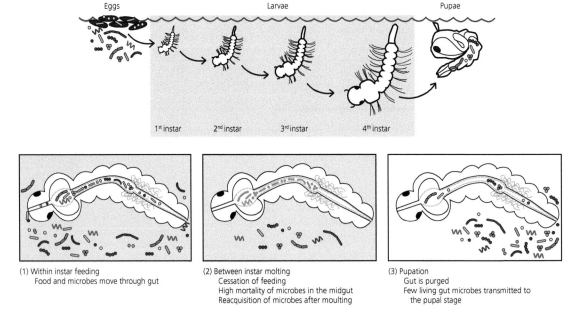

Figure 13.1 Schematic illustrating acquisition and turnover of the gut microbiota in larval and pupal stage mosquitoes. Mosquito larvae hatch from eggs in aquatic habitats, where they develop through four instars. Each instar is punctuated by a molt, with the fourth instar molting to a pupa. During active feeding periods, larvae ingest bacteria and other microorganisms present in their aquatic habitat, some of which increase in abundance in the gut to form a gut microbiota (1). Larvae also secrete a peritrophic matrix (dotted line) that lines the gut to form a protective barrier between ingested materials and the gut epithelium. Within each instar, the gut microbiota and surrounding peritrophic matrix are continuously shed and replaced as food passes through the midgut, where digestion occurs under strongly alkaline (pH 11) conditions. Larvae stop feeding shortly before molting, which results in retention of the gut microbiota and pH-mediated mortality (2). The fore- and hindgut cuticle plus the peritrophic matrix are then shed and replaced with each molt and the gut microbiota is subsequently acquired anew from the environment during the next active feeding period. Prior to pupation, fourth instars purge gut contents and only a subset of the larval gut microbiota persists to the pupal stage (3).

this stage of the life cycle (see Fig. 13.1). As previously noted, adults emerge from the pupal stage and persist in terrestrial habitats where both sexes feed on plant nectar and other sugar sources (Foster 1995), while females usually must also blood feed to produce eggs (Briegel 2003). Newly emerged adults contain very few bacteria (Moll et al. 2001; Lindh et al. 2008a; Terenius et al. 2012). This is because larvae void the contents of their gut prior to pupation, and any remaining bacteria are sequestered with the degenerating larval gut during metamorphosis to form a meconium that is egested by adults immediately after emergence (Moll et al. 2001; Moncayo et al. 2005). Culture-based studies support a drastic reduction in bacterial counts just prior to and following metamorphosis (Moll et al. 2001; Coon et al. 2016b). However, high throughput sequencing studies also indicate that a subset of the bacteria present in larvae persist to the adult stage (termed transstadial transmission) (Coon et al. 2014). This means that the adult microbiota is initially seeded by bacteria from larvae. Thereafter, however, the adult gut microbiota may change in response to consumption of water from breeding sites, nectar or other food sources including a blood meal, although bacterial diversity in adults is consistently much lower than in larvae (Pumpuni et al. 1996; Lindh et al. 2008a; Wang et al. 2011; Alvarez-Perez et al. 2012; Coon et al. 2014; Gimonneau et al. 2014; Duguma et al. 2015). Some bacteria can also be passed from males to females during mating or transmitted directly from females to progeny via deposition on the surface of eggs masses and consumption by larvae after hatching (Strand 2017). These include species of *Asaia* and *Serratia*, which have been detected in the gut, salivary glands, and/or reproductive organs of some adult mosquitoes (Favia et al. 2007; Damiani et al. 2008; Crotti et al. 2009; Wang et al. 2017). The mechanisms underlying how certain gut bacteria colonize different tissues in their mosquito hosts are not known. Whether the bacteria females deposit on the surface of eggs are primarily derived from the gut or reproductive tissues is also unclear, although experimental evidence suggests that both *Asaia* and *Serratia* colonize the ovaries and adhere to eggs prior to egg laying (Damiani et al. 2010; Wang et al. 2017).

13.2.2 The Gut as a Selective Habitat For Microbes

That only a subset of bacteria present in aquatic habitats persist in larvae and adults points to a selective role for the gut environment in assembly of the mosquito microbiota. The insect gut provides a number of barriers to microbial colonization and persistence, including potentially unfavorable physiochemical conditions (e.g. pH, redox potential), the presence of lytic enzymes and other immune-related compounds, and physical disruption caused by peristalsis of gut contents and loss of habitat during insect molting and metamorphosis.

As in other insects, the mosquito gut consists of epithelial cells that form three regions: the foregut, the midgut, and the hindgut (Chapman et al. 2013). The foregut and hindgut are lined by a thin layer of cuticle. In contrast, the midgut is lined by a semipermeable membrane called the peritrophic matrix, which surrounds the food bolus, assists in digestive processes and nutrient absorption, and protects the underlying epithelium from physical and chemical damage (Shao et al. 2001). Extensions of the anterior hindgut called the Malpighian tubules serve as the main osmoregulatory and excretory organs and deliver nitrogenous waste to the hindgut for excretion along with food waste (Piermarini 2016). The hindgut also serves important functions in the regulation of water and ion homeostasis. Little is known about the spatial distribution of bacteria in different regions of the gut, although studies in both larval and adult stage mosquitoes suggest most bacteria reside in the endoperitrophic space of the midgut (Walker et al. 1988; Merritt et al. 1990; Gusmão et al. 2010; Vogel et al. 2017).

During the larval stage, ingested microorganisms experience strongly alkaline (pH 11) conditions in the midgut, which aid in digestion of the tannin-rich plant detritus that larvae feed on but may also inhibit growth of most bacteria (Boudko et al. 2001) (see Fig. 13.1). Both the cuticle of the fore- and hindgut as well as the peritrophic matrix of the midgut are also shed and replaced at each larval molt, and most of the bacteria within the midgut are excreted in the feces larvae expel before pupating (Moll et al. 2001) (see Fig. 13.1). The physical turnover of bacteria through mechanisms that both displace and/or induce cell death in bacterial populations suggests that most if not all of the larval gut microbiota of mosquitoes is represented by transient microbes that do not persist in the host gut environment (see Fig. 13.1).

Gut bacteria transmitted to the pupal stage also experience significant population bottlenecks during metamorphosis, which includes histolysis of the larval gut and subsequent remodeling to produce the adult gut. As noted previously, a large proportion of the bacteria present at this stage is expelled in the meconium shortly after adult emergence (Moll et al. 2001), yet some bacteria also persist and are present in newly emerged adults (Pumpuni et al. 1996; Lindh et al. 2008a; Coon et al. 2014; Gimonneau et al. 2014; Duguma et al. 2015). Precisely how these bacteria survive is unclear, although some bacteria have been detected in the Malpighian tubules which could facilitate transstadial transmission since these organs are not remodeled during metamorphosis (Chavshin et al. 2015).

The newly formed adult gut differs from the larval gut in several ways, including the differentiation of the foregut to form a crop, which stores ingested sugar to provide energy for mating, host seeking, and egg laying (Calkins et al. 2017). Ingested blood, in contrast, bypasses the crop and is directed into the midgut, where it is digested and absorbed. Dietary shifts from being carbohydrate-rich in sugar-fed to protein-rich in blood-fed adult females can have marked impacts on resident microbial communities. Consumption of both a sugar and/or blood meal is known to reduce overall community diversity in the gut (Wang et al. 2011; Terenius et al. 2012), while consumption of a blood meal enables proliferation of some bacterial species that can utilize the blood and at the same time survive the oxidative stress that follows blood meal digestion (Gusmão et al. 2010; Gaio et al. 2011; Oliveira et al. 2011; Wang et al. 2011; Coon et al. 2014). Differences between the larval and adult gut also include formation of the peritrophic matrix, which in larvae occurs continuously but in adults is only formed in females in response to blood feeding (Shao et al. 2001). This, combined with the fact that adults do not molt, likely makes the adult gut a much more stable environment for microbes, allowing for the formation of resident microbial communities that persist in the gut over time.

13.2.3 Gut Community Composition and Evidence for a Core Gut Microbiota

To date, studies have failed to reach a consensus on the identity of a distinct core gut microbiota in mosquitoes, although several broad-scale patterns in community composition are emerging. First, most bacteria identified from larval and adult stage mosquitoes are Gram-negative aerobes or facultative anaerobes belonging to one of four phyla: Actinobacteria, Bacteroidetes, Firmicutes, and Proteobacteria (Strand 2017). Within these phyla, several bacterial families are also reliably detected across mosquito species, including members of the Enterobacteriaceae, Flavobacteriaceae, and Acetobacteriaceae (Gendrin & Christophides 2013). Second, while bacterial communities are very predictable at high taxonomic levels, they are far less so at lower ones. Mosquito gut bacterial communities tend to be dominated by only one or several genera or species that vary irregularly both within and among individuals or populations irrespective of host species or life stage (Wang et al. 2011; Zouache et al. 2011; Boissière et al. 2012; Osei-Poku et al. 2012; Coon et al. 2014; Gimonneau et al. 2014; Duguma et al. 2015; Muturi et al. 2016; Muturi et al. 2017; Thongsripong et al. 2017; Wang et al. 2018). In fact, even commonly detected taxa—from species to entire phyla—can vary in their reported relative abundance by more than an order of magnitude between different studies (Wang et al. 2011; Zouache et al. 2011; Boissière et al. 2012; Osei-Poku et al. 2012; Coon et al. 2014). Finally, numerous studies show that larval and adult stage mosquitoes of the same species sampled from different geographic locations harbor different gut microbial communities, even at very local scales (Zouache et al. 2011; Boissière et al. 2012; Buck et al. 2016; Coon et al. 2016a; Tchioffo et al. 2016a; Dickson et al. 2017; Bascuñán et al. 2018; Muturi et al. 2018). Studies also report that different mosquito species or strains sampled from the same site or reared under the same environmental conditions in the laboratory harbor similar gut microbial communities (Coon et al. 2016a; Dickson et al. 2018). These results strongly suggest that environment is the dominant factor shaping variation in mosquito gut microbiota while also indicating that mosquitoes do not harbor a 'core microbiome' consisting of specific microbes.

Other factors may also play a role in shaping mosquito gut microbiota. Short et al. (2017) recently reported that silencing of genes involved in amino acid metabolism can eliminate differences in gut bacterial load between *Aedes aegypti* strains reared in the same facility. These results build on a prior study, which reported up to a 100-fold difference in bacterial abundance among different *Ae. aegypti* strains (Charan et al. 2013). Thus, while community membership may be largely driven by the environment, community features such as total and taxon-specific abundances may be shaped by host genetics or physiochemical variables such as gut pH or oxygen tension. The random dispersal of microorganisms in the environment, drift associated with population bottlenecking during molting and metamorphosis, and intra- and interspecific interactions between microbial populations within the gut could also influence the composition, persistence, and abundance of gut microbial communities. Hegde et al. (2018) detected both positive and negative associations between bacterial species that were consistent between the gut microbiota sampled from multiple mosquito species in the laboratory and field. Experimental studies have also demonstrated that resident microbiota can both inhibit and/or facilitate colonization by other bacterial taxa (Terenius et al. 2012; Bahia et al. 2014; Coon et al. 2014; Hegde et al. 2018).

13.3 Impacts of Gut Microbiota on Mosquito Vector Competence and Pathogen Transmission

Mosquitoes, especially species within the genera *Aedes, Anopheles,* and *Culex,* are responsible for the transmission of a number of prominent pathogens relevant to human health. Despite centuries of control efforts, hundreds of millions of humans continue to become infected with mosquito-borne pathogens every year. Female mosquitoes ingest human or other vertebrate pathogens, which include protozoa, helminths, and viruses, when they take a blood meal from an infected host (see Fig.13.2). Ingested pathogens must thereafter cross the midgut epithelium, replicate, and travel through the hemocoel (insect body cavity) to the salivary glands, from where they will be delivered to a new, susceptible host when the mosquito takes her next blood meal (Hardy et al. 1983; Abraham & Jacobs-Lorena 2004). Each step in this process presents barriers to invading pathogens. In the midgut, these include the peritrophic matrix, proteolytic enzymes and toxic products associated with blood meal digestion, and immune defence molecules. Mosquitoes in which a particular pathogen is able to establish a transmissible infection are referred to as 'vector competent'.

Endogenous microbes present in the digestive tracts of animals are also known to modulate host interactions with invading pathogens, through direct and indirect mechanisms. Early evidence that native gut microbiota can impact vector competence in mosquitoes came from studies in *Anopheles* mosquitoes, where depletion of midgut bacteria using antibiotics enhanced susceptibility to infection by different species of the protozoan parasite *Plasmodium*, the causative agent of malaria (Pumpuni et al. 1993, 1996; Gonzalez-Ceron et al. 2003; Gendrin et al. 2015). Antibiotic clearance of gut bacteria also increased permissiveness of *Ae. aegypti* mosquitoes to dengue virus (DENV), while reintroduction of certain Gram-negative bacterial species increased resistance to pathogen infection in both *Anopheles gambiae* and *Ae. aegypti* (Gonzalez-Ceron et al. 2003; Xi et al. 2008; Dong et al. 2009; Cirimotich et al. 2011; Bahia et al. 2014; Ramirez et al. 2014). Owing to these results, most functional studies in mosquitoes have since focused on understanding the mechanisms by which gut microbiota influence mosquito vector competence to human pathogens. These studies have been extensively reviewed elsewhere (Minard et al. 2013; Dennison et al. 2014; Rodgers et al. 2015; Hegde et al. 2015; van Tol and Dimopoulos 2016).

Overall, current results support three mechanisms for why reducing gut bacterial populations increases *Plasmodium* or DENV infection in mosquitoes. The first is that the gut microbiota induces proper formation of the peritrophic matrix, which restricts access of invading pathogens to the midgut epithelium (Rodgers et al. 2017; Song et al. 2018) (see Fig. 3.2). Second, proliferation of bacteria in the gut following a blood meal stimulates the innate immune system of the mosquito to produce factors that have broad antimicrobial activity (Xi et al. 2008; Meister et al. 2009; Rodrigues et al. 2010; Dong et al. 2009, 2011; Eappen et al. 2013; Wang et al. 2013a; Wang et al. 2013b; Bahia et al. 2014; Ramirez et al. 2012, 2014; Stathopoulos et al. 2014; Dennison et al. 2015; Gendrin et al. 2015, 2017) (see Fig. 13.2). Third, gut bacteria directly inhibit pathogen development through the production of anti-pathogen molecules

(Joyce et al. 2011; Cirimotich et al. 2011; Bahia et al. 2014; Ramirez et al. 2014; Bongio & Lampe 2015) (see Fig.13.2).

Despite compelling evidence for the ability of mosquito gut microbiota to limit infection by human pathogens, the complex variation in mosquito gut microbial communities has made it difficult to assign antagonistic functions to specific taxa. Most functional studies have focused on members of the Enterobacteriaceae, which have previously

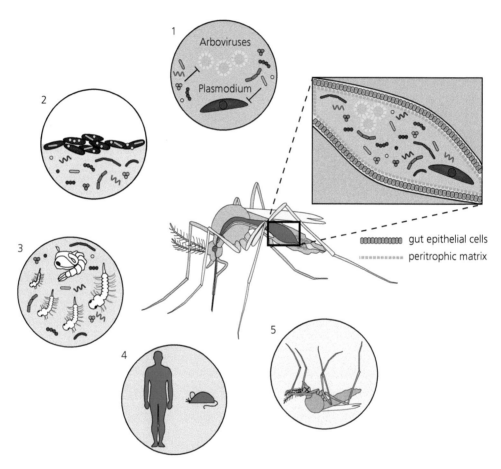

Figure 13.2 Impacts of the mosquito gut microbiota on vectorial capacity traits. The center of the schematic shows an adult female after consuming a blood meal. To the right is a schematic of the midgut. Female mosquitoes acquire pathogens when they take a blood meal from an infected host. Blood feeding results in proliferation of the microbiota within the midgut lumen. The gut microbiota contributes to normal formation of the peritrophic matrix, which physically separates the blood meal from the midgut epithelium and serves as the first barrier to pathogen infection. Certain members of the gut microbiota also negatively affect pathogen infection by producing anti-pathogen molecules or priming the mosquito immune system (1). Gut microbiota affect female egg production by aiding in blood meal digestion and/or nutrient acquisition and absorption (2). Bacteria on the surface of eggs or in the water in which they are laid also affect oviposition and egg hatching (2). Multiple mosquito species require living microbes in their gut in order to develop into adults (3). Composition of skin microbiota affects attractiveness of vertebrate hosts to adult female mosquitoes (4). The gut microbiota can also modulate traits associated with adult longevity, including insecticide resistance and susceptibility to entomopathogenic bacteria and fungi (5).

been documented to increase in abundance in the midgut following a blood meal (Gusmão et al. 2010; Wang et al. 2011). The abundance of Enterobacteriaceae in the midgut has also been shown to correlate with *Plasmodium* and chikungunya virus (CHIKV) infection status in *An. gambiae* and *Aedes albopictus* mosquitoes collected from the field (Boissière et al. 2012; Zouache et al. 2012). However, while certain genera within the Enterobacteriaceae (*Enterobacter, Pantoea, Proteus, Serratia*) inhibit pathogen development in vivo, intra-specific diversity between species or strains within these genera can have marked impacts on the level of inhibition (Cirimotich et al. 2011; Ramirez et al. 2012; Bando et al. 2013; Tchioffo et al. 2013, 2016b; Bahia et al. 2014; Dennison et al. 2016; Dickson et al. 2017). In some cases, certain bacterial species can also inhibit infection by some human pathogens while promoting infection by others. For example, certain *Serratia* species inhibit *Plasmodium* development in *Anopheles* mosquitoes, while others increase replication of DENV and CHKV in *Ae. aegypti* (Apte-Deshpande et al. 2012, 2014; Bando et al. 2013; Bahia et al. 2014; Wu et al. 2019). Some pathogens may even require the presence of gut microbiota for full midgut infectivity (Carissimo et al. 2015).

13.4 Microbial Functions Across Mosquito Life History

While a particular mosquito species may serve as a competent vector for a given pathogen, pathogen transmission dynamics at the population level may be influenced by a number of intrinsic and extrinsic factors. The potential for a mosquito population to transmit a pathogen within a susceptible host population is referred to as 'vectorial capacity' (Brady et al. 2016). Multiple elements of mosquito biology govern vectorial capacity, including population density, average lifespan, host feeding preferences, and vector competence. In this way, factors that influence basic physiological processes of mosquitoes, such as their survival, reproduction, growth, and development, have the potential to impact the spread of the pathogens they transmit.

13.4.1 Microbial Regulation of Larval Growth and Molting

Studies in the last few years have used axenic and gnotobiotic mosquitoes to demonstrate that the gut microbiota is required for the development of mosquito larvae into adults. The term 'axenic' (or germ-free) refers to an animal that is demonstrably free of associated forms of life, including bacteria, fungi, and other microbes. Axenic animals that are selectively colonized with one or more known microbial species are referred to as 'gnotobiotic', a term derived from the Greek 'gnotos', meaning known, and 'bios', meaning life. These studies were based on earlier observations that reducing bacterial abundance in the aquatic environment delays larval growth and elevates mortality, while adding certain bacterial species promotes growth and survival to the adult stage (Hinman 1930; Rozeboom 1935; Ferguson and Micks 1961; Chao et al. 1963; Chouaia et al. 2012; Mitraka et al. 2013). However, a major disadvantage of these earlier studies was that most relied on antibiotics or other techniques that lower bacterial abundance and diversity but do not eliminate all bacteria or other microbes from the mosquito gut. In contrast, Coon et al. (2014) produced axenic *Ae. aegypti* larvae by sterilizing the surface of egg masses and hatching first instars in sterile water. Surprisingly, axenic *Ae. aegypti* larvae reared under a standard photoperiod and provided a sterile diet failed to develop past the first larval instar and died without molting but were rescued if inoculated with the community of microbes present in conventional (non-sterile) larval cultures (see Fig. 13.2). Inoculation of axenic larvae with live cultures of several bacterial species from this community and *Escherichia coli* produced gnotobiotic larvae that developed normally into adults, while axenic larvae inoculated with dead bacteria never molted (Coon et al. 2014). Field-collected *Ae. aegypti* and several other mosquito species also exhibited the same defect in the absence of living bacteria, which suggests that the requirement of gut bacteria for development may be general among mosquitoes (Coon et al. 2014; Coon et al. 2016a,b) (see Fig. 13.2).

Subsequent studies took advantage of *E. coli* as a model bacterium to gain insights into how living bacteria support mosquito development. Using an *E. coli* single-gene knockout library, Coon et al. (2017) identified cytochrome *bd* oxidase, an enzyme with roles in aerobic respiration, as a bacterial product involved in mosquito development. Bioassays also demonstrated that non-sterile *Ae. aegypti* larvae and gnotobiotic larvae colonized by wild-type *E. coli* exhibit much lower gut oxygen levels (gut hypoxia) than axenic larvae or gnotobiotic larvae colonized by an *E. coli* mutant defective for cytochrome *bd* oxidase. Axenic larvae and gnotobiotic larvae colonized by the cytochrome *bd* oxidase mutant also exhibited defects in the release of the steroid molting hormone ecdysone, which resulted in failure of these larvae to molt past the first instar. This led Coon et al. (2017) to hypothesize that microbe-mediated gut hypoxia acts as a signal for ecdysone-induced molting.

In other animals, hypoxia has been shown to induce the stabilization of conserved hypoxia-inducible transcription factors (HIFs) that activate downstream target genes in diverse pathways (Semenza 2007; Xia & Kung 2009; Dekanty et al. 2010). Valzania et al. (2018a) demonstrated that HIFs are stabilized in response to gut hypoxia in *Ae. aegypti* larvae, and that HIF signaling activates downstream pathways with essential functions in growth and metabolism. These results are also consistent with axenic larvae displaying defects in nutrient acquisition and assimilation that resulted in larvae never achieving the critical size associated with molting (Vogel et al. 2017).

The above studies collectively indicate that living bacteria rescue development of *Ae. aegypti* by inducing gut hypoxia in larvae, which initiates a hypoxia-inducible transcriptional program that governs the endocrine and nutritional events leading up to molting. However, other living microorganisms also induce a gut hypoxia signal that rescues molting. Valzania et al. (2018b) recently reported that living yeast and algae induce gut hypoxia, HIF signaling, and molting in axenic *Ae. aegypti* and *An. gambiae* larvae, while the same organisms fail to support molting if heat-killed. Valzania et al. (2018b) also considered whether different diets that vary in nutrient composition may obviate or limit the requirement for living microbes.

No axenic larvae molted past the first instar when provided any of these diets in the absence of living microbes (Valzania et al. 2018b). However, recent results with *A. aegypti* report that yeast extract and heat-killed bacteria can also promote growth but only when provided at very high doses with other nutrient-rich dietary components in darkness (Correa et al. 2018). This finding suggests that yeast, bacteria and potentially other microbes produce photolabile nutrients or other factors that affect larval growth. It also suggests viable microbes at the lower densities that are present in aquatic habitats and the larval gut are able to produce sufficient amounts of this factor under normal photoperiodic conditions, whereas non-living microbes cannot.

13.4.2 Microbes as a Source of Nutrition for Larvae

It has long been stated that bacteria and other microorganisms serve as an important source of nutrition for developing mosquito larvae (Merritt et al. 1992). This is because most mosquito larvae are detritivores and rely on microbial conditioning to improve the nutritional quality of the dead plant material and other organic matter they feed on (Merritt et al. 1992). However, unlike other detritivoros insects that harbor specialized gut microbiota for detritus degradation (Kaufman et al. 2000; Engel & Moran 2013), several lines of evidence suggest that mosquito larvae primarily rely on environmental microorganisms to serve this function rather than members of their gut microbiota. First, multiple observations support that mosquito larvae lack a resident gut microbiota. In vertebrates, resident (or autochthonous) microbial communities are characterized by their ability to stably colonize the gut epithelium, while transient (or allochthonous) microorganisms are acquired from the environment with food and digested or passed as waste (Nava & Stappenbeck 2011). As previously mentioned, bacteria in mosquito larvae are confined to the endoperitrophic space in the midgut lumen, and therefore do not come into direct contact with the gut epithelium (Walker et al. 1988; Merritt et al. 1990; Gusmao et al. 2010; Vogel et al. 2017). Studies in mosquitoes also report rapid transit times for bacteria through the larval gut during active feeding periods, with many bacteria surviving as they are excreted with undigested food and waste (Coon et al. 2017) (see Fig. 13.1/). As larvae approach molting and cease feeding, however, ingested bacteria rapidly die due to protracted exposure to the highly alkaline pH of the midgut (Coon et al. 2017) (see Fig.13.1). Second, a portion of the microbes larvae consume may be digested as food. Microbial abundance is lower in aquatic habitats where larvae are present (Walker et al. 1991; Kaufman et al. 1999), and there are dramatic shifts in the composition and structure of microbial communities in the water where larvae feed (Kaufman et al 1999). However, the extent to which these trends are caused by digestion of certain microbial species in the gut versus the exposure of microbes to high midgut pH is unknown. Finally, results from several studies indicate that growth and survival rates can vary substantially between larvae reared in the presence of different bacteria and other microorganisms (Coon et al. 2016b; Dickson et al. 2017; Valzania et al. 2018). The composition of the microbiota present in the environment in which larvae develop also strongly determines adult fitness traits such as individual body size, teneral reserves, fecundity, longevity, and vector competence, which are all factors influencing vectorial capacity (Coon et al. 2016b; Dickson et al. 2017). Thus, while the presence of living microbes in the gut is sufficient to induce gut hypoxia and molting, certain microbial species and assemblages also appear to provide nutrients or perform metabolic functions that improve or reduce mosquito fitness (Valzania et al. 2018b).

13.4.3 Functional Roles of Gut Microbiota in Adults

Although numerous studies have examined the impact of the gut microbiota on vector competence, effects on other biological features of adult mosquitoes largely remain unclear. Several other hematophagous insects harbor obligate intracellular bacterial symbionts that are localized to a specialized organ-like structure called the bacteriome. These bacteria play a critical role in host reproduction by providing essential B vitamins that are defi-

cient in the vertebrate blood females consume (Rio et al. 2017). No mosquitoes are known to harbor obligate symbionts, however some evidence suggests that the mosquito gut microbiota is involved in certain nutritional processes.

First, members of the Enterobacteriaceae (e.g. *Serratia* and *Enterobacter* spp.) produce hemolytic enzymes, and these bacteria have consistently been shown to increase in abundance in the midgut of mosquitoes following a blood meal (Gusmão et al. 2010; Gaio et al. 2011; Wang et al. 2011). Depletion of these bacteria using antibiotics also reduces red blood cell lysis in *Ae. aegypti*, which correlates with delays in blood protein digestion and a concomitant reduction in fecundity (Gaio et al. 2011) (see Fig. 13.2). While proteins released from lysed red blood cells are known to be digested by trypsin-like proteases secreted by mosquito midgut cells (Isoe et al. 2009), bacteria may also contribute to blood meal digestion by metabolizing certain blood meal components or interfering with host trypsin activity (Minard et al. 2013b).

Some evidence also supports roles for gut microbiota in nutrient acquisition by autogenous mosquitoes, which produce eggs without blood feeding. The rockpool mosquito, *Aedes atropalpus*, is closely related to *Ae. aegypti* but is facultatively autogenous (Wilkerson et al. 2015). *Aedes aegypti* females must blood feed to produce eggs, while *Ae. atropalpus* females produce a first clutch of eggs without blood feeding (Clements 1992). The ability to produce eggs without blood feeding is due in part to enhanced nutrient acquisition by larvae, which provides resources for the first gonadotrophic cycle after emergence (Chambers & Klowden 1994; Su & Mulla 1997; Telang et al. 2006). In a recent study, Coon et al. (2016b) assessed the effects of specific members of the gut microbiota in *Ae. aegypti* and *Ae. atropalpus* larvae on female fitness traits including egg production. Several bacterial isolates supported growth and egg production by *Ae. aegypti* mosquitoes to levels consistent with conventionally reared individuals colonized by a mixed community of bacteria. However, only one isolate (a *Comamonas* sp.) supported growth and egg production by *Ae. atropalpus* to normal levels (Coon et al. 2016b). Interestingly, gnotobiotic *Ae. atropalpus* females colonized by this *Comamonas* sp. also emerged with higher glycogen and protein stores than gnotobiotic females colonized by other species of bacteria (Coon et al. 2016b). While the authors did not measure nutrient levels in gnotobiotic *Ae. aegypti* females, these results strongly suggest that egg production by autogenous mosquitoes like *Ae. atropalpus* depends on the composition of the gut microbiota and the presence of certain community members. In contrast, the added nutrients obtained through blood feeding obviate such dependence in anautogenous mosquitoes like *Ae. aegypti*.

Gut microbiota may also modulate processes with the potential to enhance or attenuate the lifespan of mosquito hosts (see Fig.13.2). For example, a recent study by Dada et al. (2018) used metagenomic sequencing to show that genes involved in the degradation of xenobiotic compounds were enriched in the gut microbiome of insecticide-resistant but not susceptible mosquito populations. Several bacterial species isolated from resistant mosquitoes also metabolized the organophosphate insecticide fenitrothion in vitro (Dada et al. 2018). Gut microbiota may also influence the efficacy of other mosquito control agents such as entomopathogenic bacteria that gain entry via the midgut, through both synergistic and antagonistic interactions. For example, the presence of an intact gut microbiota has been shown to inhibit colonization of the *An. gambiae* and *Ae. aegypti* midgut by the pathogenic bacterium *Chromobacterium Csp_P*, while disruption of the microbiota with antibiotics results in rapid colonization and host death (Ramirez et al. 2014). Interestingly, the gut microbiota has also been shown to accelerate infection by other pathogens like the fungus *Beauveria bassiana*, which gains entry through the external cuticle (Wei et al. 2017). While infection with *B. bassiana* induces a strong systemic immune response, immune responses in the mosquito's midgut are significantly down-regulated, causing overgrowth of the gut microbiota and dissemination of bacteria from the gut into the hemocoel (Wei et al. 2017).

13.4.4 Impacts of Bacteria in the Environment

Naturally occurring bacteria present in environments where mosquitoes persist also have the capacity to impact their behavior (see Fig. 13.5). Several studies have demonstrated that odors produced by skin microbiota are attractive to mosquitoes and that the composition of the skin microbiota affects the degree of attractiveness of humans to different mosquito species (Verhulst et al. 2009; Verhulst et al. 2010; Verhulst et al. 2011) (see Fig.2). Bacteria or water-soluble compounds secreted by bacteria in aquatic habitats can also act as cues for the selection of oviposition (egg-laying) sites by females, or as hatching stimuli for the eggs females lay (Trexler et al. 2003; Lindh et al. 2008b; Ponnusamy et al. 2010) (see Fig. 13.2). This is particularly interesting from both an evolutionary and life history perspective, given the role microbes play in larval development and nutrition. In this regard, the mere presence of any living microbe could serve as an indicator for oviposition sites that would support egg hatching, larval growth, and moulting. Certain metabolites and other microbial products may also serve as signals for microbial species or assemblages that promote larval nutrition and adult fitness.

13.5 Future Perspectives

By altering vector competence and other important biological traits, it is now well established that the mosquito gut microbiota has the potential to affect the vectorial capacity of mosquitoes to transmit human pathogens. As such, there is a growing interest in developing strategies for manipulating mosquito microbiota for disease control. For example, bacteria that naturally colonize the mosquito gut could be genetically modified to produce effector molecules that alter the mosquito's ability to become infected with and transmit pathogens, or that reduce mosquito fecundity or lifespan (i.e. paratransgenesis). Such paratransgenic approaches have shown promise in *Anopheles* mosquitoes for the control of malaria (Wang & Jacobs-Lorena 2013). Alternatively, unmodified gut bacteria that naturally inhibit pathogen colonization or mosquito fitness could be disseminated to mosquito populations.

The successful implementation of microbe-based control strategies, whether through the use of microbes with natural or engineered anti-pathogen or anti-mosquito properties, poses several significant challenges. First, the candidate microbe must be able to establish a stable association with the mosquito host and transformation should not compromise the ability of genetically modified microbes to colonize new hosts. Second, the candidate microbe must be able to effectively interact with its intended target in the mosquito host without undesired effects on the mosquito's fitness. Finally, a method must exist for dissemination of the microbe into mosquito populations in the field. Microbes have been successfully introduced into adult populations using attractive sugar baits (Mancini et al. 2016; Bilgo et al. 2018), which could be an effective method for disseminating microbes that are sexually and/or vertically transmitted. Microbes could also be disseminated via introduction into larval habitats. However, this would rely on the ability of the microbes to be transstadialy transmitted and/or persist in the aquatic environment long enough to be imbibed by newly emerged adults.

The suitability of different microbial candidates as control agents is also likely affected by host genetics or environmental factors like temperature, which can independently impact mosquito susceptibility to human pathogens and vary substantially over time and space (Murdock et al. 2012). Variation in the composition of the native gut microbiota between individuals and populations of mosquitoes could also have unpredictable effects on introduced microbes due to competition or other interactions. Thus, results from functional studies in laboratory-reared mosquitoes, which harbor gut microbial communities that are distinct from field-collected mosquitoes, may not be representative of naturally occurring interactions relevant for the spread of mosquito-borne pathogens. Additional information is also needed about the diversity and function of non-bacterial members of the mosquito gut microbiota, including yeast and fungi that may produce anti-pathogen molecules or alter host immunity or fitness (Tawidian et al. 2019). A deeper understanding of the factors that influence the acquisition, maintenance, and transmission of mosquito gut microbiota and the mechanisms that underlie how individual microbial species and assemblages impact mosquito

biology and vector competence will be essential for identifying suitable microbial candidates for pathogen or vector control. The development of methods to generate laboratory strains of mosquitoes colonized by standardized microbial communities, including representative communities from natural populations of interest, will also be essential for predicting the success of individual microbial candidates in different host backgrounds and under variable environmental conditions in the field.

That the requirement for a living gut microbiota for development under conditions typically encountered in the field is a general feature of mosquito biology raises the intriguing possibility that growth-related signaling pathways microbes activate in larvae may serve as conserved targets for mosquito control. The key challenges here would be to develop a compound that would disrupt microbe recognition and/or pathway activation in mosquitoes without affecting non-target animal species. Such a compound must also be able to be delivered to its site of action in the mosquito without being deactivated, which can be impacted by both environmental factors in the larval habitat and operational factors like application timing, frequency, and coverage. Compounds with multiple target sites or administration of compounds with other growth inhibitors or larvicides could also help prevent the evolution and spread of resistance in mosquito populations. Future studies to characterize the complete repertoire of genes that are activated in response to a living gut microbiota, as well as their expression in different vector mosquito species in the presence or absence of living microbes, will be necessary to identify potential targets for compounds that disrupt mosquito development and therefore reduce adult populations in areas of human disease transmission.

Writing of this chapter and some of the results reported herein were supported by funds from the University of Wisconsin-Madison (to K.L.C.) and grants from the National Institute of Food and Agriculture (2018-67,012-28,009 to K.L.C.), National Institutes of Health (R01AI106892 to M.R.S.), and National Science Foundation (IOS 1656236 to M.R.S.). We thank J.A. Johnson for assistance with figure preparation.

References

Abraham, E.G. & Jacobs-Lorena, M. (2004). Mosquito midgut barriers to malaria parasite development. *Insect Biochemistry and Molecular Biology*, 34, 667–71.

Alvarez-Perez, S., Herrera, C., & Vega, C. (2012). Zooming-in on floral nectar: A first exploration of nectar-associated bacteria in wild plant communities. *FEMS Microbiology Ecology*, 80, 591–602.

Apte-Deshpande, A., Paingankar, M., Gokhale, M.D., & Deobagkar, D.N. (2012). *Serratia odorifera* a midgut inhabitant of *Aedes aegypti* mosquito enhances its susceptibility to dengue-2 virus. *PLoS One*, 7, e40401.

Apte-Deshpande, A.D., Paingankar, M.S., Gkhae, M.D., & Deobagkar, D.N. (2014). *Serratia odorifera* mediated enhancement in susceptibility of *Aedes aegypti* for chikungunya virus. *Indian Journal of Medical Research*, 139, 762–8.

Bahia, A.C., Dong, Y., Blumberg, B.J. et al. (2014). Exploring *Anopheles* gut bacteria for *Plasmodium* blocking activity. *Environmental Microbiology*, 16, 2980–94.

Bando, H., Okado, K., Guelbeogo, W. et al. (2013). Intra-specific diversity of *Serratia marcescens* in *Anopheles* mosquito midgut defines *Plasmodium* transmission capacity. *Scientific Reports*, 3, 1641.

Bascuñán, P., Niño-Garcia, J.P., Galeano-Castañeda Y., Serre, D., & Correa, M.M. (2018). Factors shaping the gut bacterial community assembly in two main Colombian malaria vectors. *Microbiome*, 6, 148.

Bilgo, E., Vantaux, A., Sanon, A., Ilboudo, S. et al. (2018). Field assessment of potential sugar feeding stations for disseminating bacteria in a paratransgenic approach to control malaria. *Malaria Journal*, 17, 367.

Boissière, A., Tchioffo, M.T., Bachar, D. et al. (2012). Midgut microbiota of the malaria mosquito vector *Anopheles gambiae* and interactions with *Plasmodium falciparum* infection. *PLoS Pathogens*, 8, e1002742.

Bongio, N.J. & Lampe, D.J. (2015). Inhibition of *Plasmodium berghei* development in mosquitoes by effector proteins secreted from *Asaia* sp. bacteria using a novel native secretion signal. *PLoS One*, 10, e0143541.

Boudko, D.Y., Moroz, L.L., Harvey, W.R., & Linser, P.J. (2001). Alkalinization by chloride/bicarbonate pathway in larval mosquito midgut. *Proceedings of the National Academy of Sciences of the United States of America*, 98, 15,354–9.

Brady, O.J., Godfray, H.C.J., Tatem, A.J. et al. (2016). Vectorial capacity and vector control: reconsidering sensitivity to parameters for malaria elimination. *Transactions of the Royal Society of Tropical Medicine and Hygiene*, 110, 107–17.

Briegel, H. (2003). Physiological bases of mosquito ecology. *Journal of Vector Ecology*, 28, 1–11.

Buck, M., Nilsson, L.K., Brunius, C., Dabire, R.K., Hopkins, R., & Terenius, O. (2016). Bacterial associations reveal spatial population dynamics in *Anopheles gambiae* mosquitoes. *Scientific Reports*, 10, 22,806.

Calkins, T.L., DeLaat, A., & Piermarini, P.M. (2017). Physiological characterization and regulation of the contractile properties of the mosquito ventral diverticulum (crop). *Journal of Insect Physiology*, 103, 98–106.

Carissimo, G., Pondeville, E., McFarlane, et al. (2015). Antiviral immunity of *Anopheles gambiae* is highly compartmentalized, with distinct roles for RNA interference and gut bacteria. *Proceedings of the National Academy of Sciences of the United States of America*, 112, E176–85.

Chambers, G.M. & Klowden, M.J. (1994). Nutritional reserves of autogenous and anautogenous selected strains of *Aedes albopictus* (Diptera: Culicidae). *Journal of Medical Entomology*, 31, 554–60.

Chao, J., Wistreich, G.A., & Moore, J. (1963). Failure to isolate microorganisms from within mosquito eggs. *Annals of the Entomological Society of America*, 56, 559–61.

Chao, J., Wistreich, G.A. (1959). Microbial isolations from the midgut of *Culex tarsalis* Coquillet. *Journal of Insect Pathology*, 1, 311–18.

Chapman, R.F., Simpson, S.J., & Douglas, A.E. (2013). *The Insects: Structure and Function*. 5th edn. Cambridge: Cambridge University Press.

Charan, S.S., Pawar, K.D., Severson, D.W., Patole, M.S., & Shouche, Y.S. (2013). Comparative analysis of midgut bacterial communities of *Aedes aegypti* mosquito strains varying in vector competence to dengue virus. *Parasitology Research*, 112, 2627–37.

Chavshin, A.R., Oshaghi, M.A., Vatandoost, H., Yakhchali, B., Zarenejad, F., & Terenius, O. (2015). Malpighian tubules are important determinants of *Pseudomonas transstadial* transmission and longtime persistence in *Anopheles stephensi*. *Parasites & Vectors*, 21, 36.

Chouaia, B., Rossi, P., Epis, S. et al. (2012). Delayed larval development in *Anopheles* mosquitoes deprived of *Asaia* bacterial symbionts. *BMC Microbiology*, 12, S2.

Cirimotich, C.M., Dong, Y.M., Clayton, A.M. et al. (2011). Natural microbe-mediated refractoriness to *Plasmodium* infection in *Anopheles gambiae*. *Science*, 332, 855–8.

Clements, A.N. (1992). *The Biology of Mosquitoes, Volume 1: Development, Nutrition, and Reproduction*. New York: Chapman & Hall.

Coon, K.L., Vogel, K.J., Brown, M.R., & Strand, M.R. (2014). Mosquitoes rely on their gut microbiota for development. *Molecular Ecology*, 23, 2727–39.

Coon, K.L., Brown, M.R., & Strand, M.R. (2016a). Mosquitoes host communities of bacteria that are essential for development but vary greatly between local habitats. *Molecular Ecology*, 22, 5806–26.

Coon, K.L., Brown, M.R., & Strand, M.R. (2016b). Gut bacteria differentially affect egg production in the anautogenous mosquito *Aedes aegypti* and autogenous mosquito *Aedes atropalpus* (Diptera: Culicidae). *Parasites & Vectors*, 9, 375.

Coon, K.L., Valzania, L., McKinney, D.A., Vogel, K.J., Brown, M.R., & Strand, M.R. (2017). Bacteria-mediated hypoxia functions as a signal for mosquito development. *Proceedings of the National Academy of Sciences of the United States of America*, 114, E5362–9.

Correa, M.A., Matusovsky, B., Brackney, D.E., & Steven, B. (2018). Generation of axenic *Aedes aegypti* demonstrate live bacteria are not required for mosquito development. *Nature Communications*, 9, 4464.

Crotti, E., Damiani, C., Pajoro, M. et al. (2009). *Asaia*, a versatile acetic acid bacterial symbiont, capable of cross-colonizing insects of phylogenetically-distant genera and orders. *Environmental Microbiology*, 11, 3252–64.

Dada, N., Sheth, M., Liebman, K., Pinto, J., & Lenhart, A. (2018). Whole metagenome sequencing reveals links between mosquito microbiota and insecticide resistance in malaria vectors. *Scientific Reports*, 8, 2084.

Damiani, C., Ricci, I., Crotti, E. et al. (2008). Paternal transmission of symbiotic bacteria in malaria vectors. *Current Biology*, 18, R1087–8.

Damiani, C., Ricci, I., Crotti, E. et al. (2010). Mosquito-bacteria symbiosis: The case of *Anopheles gambiae* and *Asaia*. *Microbial Ecology*, 60, 644–54.

Dekanty, A., Romero, N.M., Bertolin, A.P. (2010). *Drosophila* genome-wide RNAi screen identifies multiple regulators of HIF-dependent transcription in hypoxia. *PLoS Genetics*, 6, e1000994.

DeMaio, J., Pumpuni, C.B., Kent, M., & Beier, J.C. (1996). The midgut bacterial flora of wild *Aedes triseriatus*, *Culex pipiens*, and *Psorophora columbiae* mosquitoes. *American Journal of Tropical Medicine and Hygiene*, 54, 219–23.

Dennison, N.J., Jupatanakul, N., Dimopoulos, G. (2014). The mosquito microbiota influences vector competence for human pathogens. *Current Opinion in Insect Science*, 1, 6–13.

Dennison, N., BenMarzouk-Hidalgo, O., & Dimopoulos, G. (2015). MicroRNA-regulation of *Anopheles gambiae* immunity to *Plasmodium falciparum* infection and midgut microbiota. *Developmental & Comparative Immunology*, 49, 170–8.

Dennison, N.J., Saraiva, R.G., Cirimotich, C.M., Mlambo, G., Mongodin, E.F., & Dimopoulos, G. (2016). Functional genomic analyses of *Enterobacter*, *Anopheles* and *Plasmodium* reciprocal interactions that impact vector competence. *Malaria Journal*, 15, 425.

Dickson, L.B., Jiolle, D., Minard, G. et al. (2017). Carryover effects of larval exposure to different environmental bacteria drive adult trait variation in a mosquito vector. *Scientific Advances*, 3, e1700585.

Dickson, L.B., Ghozlane, A., Volant, S. et al. (2018). Diverse laboratory colonies of *Aedes aegypti* harbor the same adult midgut bacterial microbiome. *Parasites & Vectors*, 11, 207.

Dong, Y.M., Manfredini, F., & Dimopoulos, G. (2009). Implication of the mosquito midgut microbiota in defense against malaria parasites. *PLoS Pathogens*, 5, 1,000,423.

Dong, Y., Das, S., Cirimotich, C., Souza-Neto, J.A., McLean, K.J., & Dimopoulos, G. (2011). Engineered *Anopheles* immunity to *Plasmodium* infection. *PLoS Pathogens*, 7, e1002458.

Duguma, D., Hall, M.W., Rugman-Jones, P. et al., (2015). Developmental succession of the microbiome of *Culex* mosquitoes. *BMC Microbiology*, 15, 140.

Eappen, A.G., Smith, R.C., & Jacobs-Lorena, M. (2013). *Enterobacter*-activated mosquito immune responses to *Plasmodium* involve activation of SRPN6 in *Anopheles stephensi*. *PLoS One*, 8, e62937.

Engel, P. & Moran, N.A. (2013). The gut microbiota of insects- diversity in structure and function. *FEMS Microbiology Reviews*, 37, 699–735.

Favia, G., Ricci, I., Damiani, C. et al. (2007). Bacteria of the genus *Asaia* stably associate with *Anopheles stephensi*, an Asian malarial mosquito vector. *Proceedings of the National Academy of Sciences of the United States of America*, 104, 9047–51.

Ferguson, M.J. & Micks, D.W. (1961). Microorganisms associated with mosquitoes: I. Bacteria isolated from the midgut of adult *Culex fatigans* Wiedemann. *Journal of Insect Pathology*, 3, 112–19.

Foster, W.A. (1995). Mosquito sugar feeding and reproductive energetics. *Annual Review of Entomology*, 40, 443–74.

Gaio, AD, Gusmão, D.S., Santos, A.V., Berbert-Molina, M.A., Pimenta, P.F.P., & Lemos, F.J.A. (2011). Contribution of midgut bacteria to blood digestion and egg production in *Aedes aegypti* (Diptera: Culicidae) (L.). *Parasites & Vectors*, 4, 105.

Gendrin, M. & Christophides, G.K. (2013). The *Anopheles* mosquito microbiota and their impact on pathogen transmission. In: S. Manguin, ed. *Anopheles mosquitoes—new insights into malaria vectors*, London: InTechOpen, pp. 525–48.

Gendrin, M., Rodgers, F.H., & Yerbanga, R.S. et al. (2015). Antibiotics in ingested human blood affect the mosquito microbiota and capacity to transmit malaria. *Nature Communications*, 6, 5921.

Gendrin, M., Turlure, F., Rodgers, F.H., Cohuet, A., Morlais, I., & Christophides, G.K. (2017). The peptidoglycan recognition proteins PGRPLA and PGRPLB regulate *Anopheles* immunity to bacteria and affect infection by *Plasmodium*. *Journal of Innate Immunity*, 9, 333–42.

Gimonneau, G., Tchioffo, M.T., Abate, L., et al. (2014). Composition of *Anopheles coluzzii* and *Anopheles gambiae* microbiota from larval to adult stages. *Infection, Genetics and Evolution*, 28, 715–24.

Gonzalez-Ceron, L., Santilian, F., Rodriguez, M.H., Mendez, D., & Hernandez-Avila, J.E. (2003). Bacteria in midguts of field-collected *Anopheles albimanus* block *Plasmodium vivax* sporogonic development. *Journal of Medical Entomology*, 40, 371–4.

Gusmão, D., Santos, A., Marini, D., Bacci, M., Berbert-Molina, M., & Lemos, F. (2010). Culture-dependent and culture-independent characterization of microorganisms associated with *Aedes aegypti* (Diptera: Culicidae) (L.) and dynamics of bacterial colonization in the midgut. *Acta Tropica*, 115, 275–81.

Hardy, J.L., Houk, E.J., Kramer, L.D., & Reeves, W.C. (1983). Intrinsic factors affecting vector competence of mosquitoes for arboviruses. *Annual Review of Entomology*, 28, 229–62.

Hegde, S., Rasgon, J.L., & Hughes, G.L. (2015). The microbiome modulates arbovirus transmission in mosquitoes. *Current Opinion in Virology*, 15, 97–102.

Hegde S., Khanipov K., & Albayrak L., et al. (2018). Microbiome interaction networks and community structure from laboratory-reared and field-collected *Aedes aegypti*, *Aedes albopictus*, and *Culex quinquefasciatus* mosquito vectors. *Frontiers in Microbiology*, 9, 2160.

Hinman, E.H. (1930). A study of the food of mosquito larvae. *American Journal of Hygiene*, 12, 238–70.

Isoe, J., Rascon Jr., A. A., Kunz, S., & Miesfeld, R.L. (2009). Molecular genetic analysis of midgut serine proteases in *Aedes aegypti*. *Insect Biochemistry and Molecular Biology*, 39, 903–12.

Jones, W.L. & DeLong, D.M. (1961). A simplified technique for sterilizing the surface of *Aedes aegypti* eggs. *Journal of Economic Entomology*, 54, 813–14.

Joyce, J.D., Nogueira, J.R., Bales, A.A., Pittman, K.E., & Anderson, J.R. (2011). Interactions between La Crosse virus and bacteria isolated from the digestive tract of *Aedes albopictus* (Diptera: Culicidae). *Journal of Medical Entomology*, 48, 389–94.

Kaufman, M.G., Walker, E.D., Smith, T.W., Merritt, R.W., & Klug, M.J. (1999). The effects of larval mosquitoes (*Aedes triseriatus*) and stemflow on microbial community dynamics in container habitats. *Applied Environmental Microbiology*, 65, 2661–73.

Kaufman, M.G., Walker, E.D., Odelson, D.A., & Klug, M.J. (2000). Microbial community ecology & insect nutrition, *American Entomologist*, 46, 173–85.

Ley, R.E., Hamady, M., Lozupone, C. et al. (2008). Evolution of mammals and their gut microbes. *Science*, 320, 1647–51.

Lindh, J.M., Borg-Karlson, A.K., & Faye, I., (2008a). Transstadial and horizontal transfer of bacteria within a colony of *Anopheles gambiae* (Diptera: Culicidae) and oviposition response to bacteria-containing water. *Acta Tropica*, 107, 242–50.

Lindh, J.M., Kännaste, A., Knols, B.G.J., Faye, I., & Borg-Karlson, A.K. (2008b). Oviposition responses of *Anopheles gambiae* s.s. (Diptera: Culicidae) and identification of volatiles from bacteria-containing solutions. *Journal of Medical Entomology*, 45, 1039–49.

Mancini, M.V., Spaccapelo, R., Damiani, C. et al. (2016). Paratransgenesis to control malaria vectors: A semi-field pilot study. *Parasites & Vectors*, 9, 140.

Meister, S., Agianian, B., Turlure, F. et al. (2009). *Anopheles gambiae* PGRPLC- mediated defense against bacteria modulates infections with malaria parasites. *PLoS Pathogens*, 5, e1000542.

Merritt, R.W., Olds, E.J., & Walker, E.D. (1990). Natural food and feeding ecology of larval *Coquillettida perturbans*. *Journal of the American Mosquito Control Association*, 6, 35–42.

Merritt, R.W., Dadd, R.H., Walker, E.D. (1992). Feeding behavior, natural food, and nutritional relationships of larval mosquitoes. *Annual Review of Entomology*, 37, 349–76.

Minard, G., Mavingui, P., & Moro, C.V. (2013a). Diversity and function of bacterial microbiota in the mosquito holobiont. *Parasites & Vectors*, 6, 146.

Minard, G., Tran, F.H., Raharimalala, F.N. et al. (2013b). Prevalence, genomic and metabolic profiles of *Acinetobacter* and *Asaia* associated with field-caught *Aedes albopictus* from Madagascar. *FEMS Microbiology Ecology*, 83, 63–73.

Mitraka, E., Stathopoulos, S., Siden-Kiamos, I., Christophides, G.K., & Louis, C. (2013). *Asaia* accelerates larval development of *Anopheles gambiae*. *Pathogens and Global Health*, 107, 305–11.

Moll, R.M., Romoser, W.S., Modrzakowski, M.C., Moncayo, A.C., & Lerdthusnee, K. (2001). Meconial peritrophic membranes and the fate of midgut bacteria during mosquito (Diptera: Culicidae) metamorphosis. *Journal of Medical Entomology*, 38, 29–32.

Moncayo, A., Lerdthusnee, K., Leon, R., Robich, R., & Romoser, W. (2005). Meconial peritrophic matrix structure, formation, and meconial degeneration in mosquito pupae/pharate adults: Histological and ultrastructural aspects. *Journal of Medical Entomology*, 42, 939–44.

Murdock, C.C., Paaijmans, K.P., Cox-Foster, D., Read, A.F., & Thomas, M.B. (2012). Rethinking vector immunology: The role of environmental temperature in shaping resistance. *Nature Reviews Microbiology*, 10, 869–76.

Muturi, E.J., Kim, C., Bara, J., Bach, E.M., & Siddappaji, M.H. (2016). *Culex pipiens* and *Culex restuans* mosquitoes harbor distinct microbiota dominated by few bacterial taxa. *Parasites & Vectors*, 9, 18.

Muturi, E.J., Ramirez, J.L., Rooney, A.P., & Kim, C.H. (2017). Comparative analysis of gut microbiota of mosquito communities in central Illinois. *PLoS Neglected Tropical Diseases*, 11, e0005377.

Muturi, E.J., Lagos-Kutz, D., Dunlap, C. et al. (2018). Mosquito microbiota cluster by host sampling location. *Parasites & Vectors*, 11, 468.

Nava, G.M. & Stappenbeck, T.S. (2011). Diversity of the autochthonous colonic microbiota. *Gut Microbes*, 2, 99–104.

Nicholson, J.K., Holmes, E., Kinross, J., Burcelin, R., Gibson, G., Jia, W., & Pattersson, S. (2012). Host-gut microbiota metabolic interactions. *Science*, 336, 1262–7.

Oliveira, J.H.M., Goncalves, R.L.S., Lara, F.A. et al. (2011). Blood meal-derived heme decreases ROS levels in the midgut of *Aedes aegypti* and allows proliferation of intestinal microbiota. *PLoS Pathogens*, 7, 1,001,320.

Osei-Poku, J., Mbogo, C.M., Palmer, W.J., & Jiggins, F.M. (2012). Deep sequencing reveals extensive variation in the gut microbiota of wild mosquitoes from Kenya. *Molecular Ecology*, 21, 5138–50.

Piermarini, P.M. (2016). Renal excretory processes in mosquitoes. In A.S. Raikhel, ed. *Advances in Insect Physiology*, Oxford: Academic Press, pp. 393–422.

Ponnusamy, L., Böröczky, K., Wesson, D.M., Schal, C., & Apperson, C.S. (2010). Bacteria stimulate hatching of yellow fever mosquito eggs. *PLoS One*, 6, e24409.

Pumpuni, C.B., Beier, M.S., Nataro, J.P., Guers, L.D., & Davis, J.R. (1993). *Plasmodium falciparum*: Inhibition of sporogonic development in *Anopheles stephensi* by Gram-negative bacteria. *Experimental Parasitology*, 77, 195–9.

Pumpuni, C., Demaio, J., Kent, M., Davis, J., & Beier, J. (1996). Bacterial population dynamics in three anopheline species: The impact on *Plasmodium* sporogonic development. *American Journal of Tropical Medicine and Hygiene*, 54, 214–18.

Ramirez, J., Souza-Neto, J., Torres, R. et al. (2012). Reciprocal tripartite interactions between the *Aedes aegypti* midgut microbiota, innate immune system and dengue virus influences vector competence. *PLoS Neglected Tropical Diseases*, 6, e1561.

Ramirez, J.L., Short, S.M., Bahia, A.C. et al. (2014). *Chromobacterium* Csp_P reduces malaria and dengue infection in vector mosquitoes and has entomopatho-genic and in vitro anti-pathogen activities. *PLoS Pathogens*, 23, e1004398.

Reidenbach, K.R., Cook, S., Bertone, M.A., Harbach, R.E., Wiegmann, B.M., & Besansky, N.J. (2009). Phylogenetic analysis and temporal diversification of mosquitoes (Diptera: Culicidae) based on nuclear genes and morphology. *BMC Evolutionary Biology*, 9, 298.

Rio, R.V.M., Attardo, G.M., & Weiss, B.L. (2017). Grandeur alliances: Symbiont metabolic integration and obligate arthropod hematophagy. *Trends in Parasitology*, 32, 739–49.

Rodgers, F.H., Gendrin, M., & Christophides, G.K. (2015). The mosquito immune system and its interactions with the microbiota: Implications for disease transmission. In: S.K. Wikel, S. Aksoy, & G. Dimopoulos, eds. *Arthropod Vector: Controller of Disease Transmission*, Academic Press, London, pp. 185–99.

Rodgers, F.H., Gendrin, M., Wyer, C.A.S., & Christophides, G.K. (2017). Microbiota-induced peritrophic matrix regulates midgut homeostasis and prevents systemic infection of malaria vector mosquitoes. *PLoS Pathogens*, 13, e1006391.

Rodrigues, J., Brayner, F., Alves, L., Dixit, R., & Barillas-Mury, C. (2010). Hemocyte differentiation mediates innate immune memory in *Anopheles gambiae* mosquitoes. *Science*, 329, 1353–5.

Rozeboom, L.E. (1935). The relation of bacteria and bacterial filtrates to the development of mosquito larvae. *American Journal of Hygiene*, 21, 167–79.

Semenza, G.L. (2007). Hypoxia-inducible factor 1 (HIF-1) pathway. Science's STKE, 2007, cm8.

Shao, L., Devenport, M., & Jacobs-Lorena, M. (2001). The peritrophic matrix of hematophagous insects. *Archives of Insect Biochemistry and Physiology*, 47, 119–25.

Short, S.M., Mongodin, E.F., MacLeod, H.J., Talyuli, O.A.C., & Dimopoulos, G. (2017). Amino acid metabolic signaling influences *Aedes aegypti* midgut microbiome variability. *PLoS Neglected Tropical Diseases*, 11, e0005677.

Sommer, F. & Backhed, F. (2013). The gut microbiota—masters of host development and physiology. *Nature Reviews Microbiology*, 11, 227–38.

Song, X., Wang, M., Zhu, H., & Wang, J. (2018). PGRP-LD mediates *A. stephensi* vector competency by regulating homeostasis of microbiota-induced peritrophic matrix synthesis. *PLoS Pathogens*, 14, e1006899.

Stathopoulos, S., Neafsey, D., Lawniczak, M., Muskavitch, M., & Christophides, G. (2014). Genetic dissection of *Anopheles gambiae* gut epithelial responses to *Serratia marcescens*. *PLoS Pathogens*, 10, e1003897.

Strand, M.R. (2017). The gut microbiota of mosquitoes: Diversity and function. In: S.K. Wikel, S. Aksoy, & G. Dimopoulos, eds. *Arthropod Vector: Controller of*

Disease Transmission, London: Academic Press, pp. 185–99.

Su, T. & Mulla, M.S. (1997). Nutritional reserves, body weight, and starvation tolerance of autogenous and anautogenous strains of *Culex tarsalis* (Diptera: Culicidae). *Journal of Medical Entomology*, 34, 68–73.

Tawidian, P., Rhodes, V.L., & Michel, K. (2019). Mosquito-fungus interactions and antifungal immunity. *Insect Biochemistry and Molecular Biology*, 111, 103,182.

Tchioffo, M.T., Boissière, A., Churcher, T.S. et al. (2013). Modulation of malaria infection in *Anopheles gambiae* mosquitoes exposed to natural midgut Bacteria. *PLoS One*, 8, e81663.

Tchioffo, M.T., Boissière, A., Abate, L.,et al. (2016a). Dynamics of bacterial community composition in the malaria mosquito's epithelia. *Frontiers in Microbiology*, 5, 1500.

Tchioffo, M.T., Abate, L., Boissière, A., et al. (2016b). An epidemiologically successful *Escherichia coli* sequence type modulates *Plasmodium falciparum* infection in the mosquito midgut. *Infection, Genetics and Evolution*, 43, 22–30.

Telang, A., Li, Y.P., Noriega, F.G. & Brown, M.R. (2006). Effects of larval nutrition on the endocrinology of mosquito egg development. *Journal of Experimental Biology*, 209, 645–55.

Terenius, O., Lindh, J.M., Eriksson-Gonzales, K. et al. (2012). Midgut bacterial dynamics in *Aedes aegypti*. *FEMS Microbiology Ecology*, 80, 556–65.

Thongsripong, P., Chandler, J.A., Green, A.B. et al. (2017). Mosquito vector-associated microbiota: Metabarcoding bacteria and eukaryotic symbionts across habitat types in Thailand endemic for dengue and other arthropod-borne diseases. *Ecology and Evolution*, 8, 1352–68.

Trexler, J.D., Apperson, C.S., Zurek, L. et al. (2003). Role of bacteria in mediating the oviposition responses of *Aedes albopictus* (Diptera: Culicidae). *Journal of Medical Entomology*, 40, 841–8.

Valzania, L., Coon, K.L., Vogel, K.J., Brown, M.R., & Strand, M.R. (2018a). Hypoxia-induced transcription factor signaling is essential for larval growth of the mosquito *Aedes aegypti*. *Proceedings of the National Academy of Sciences of the United States of America*, 115, 457–65.

Valzania, L., Martinson, V.G., Harrison, R.E. et al. (2018b). Both living bacteria and eukaryotes in the mosquito gut promote growth of larvae. *PLoS Neglected Tropical Diseases*, 12, e0006638.

van Tol, S. & Dimopoulos, G. (2016). Influences of the mosquito microbiota on vector competence. *Advances in Insect Physiology*, 51, 243–91.

Verhulst, N.O., Beijleveld, H., Knols, B.G.J. et al. (2009). Cultured skin microbiota attracts malaria mosquitoes. *Malaria Journal*, 8, 302.

Verhulst, N.O., Andriessen, R., Groenhagen, U, et al. (2010). Differential attraction of malaria mosquitoes to volatile blends produced by human skin bacteria. *PLoS One*, 5, e15829.

Verhulst, N.O., Qiu, Y.T., Beijleveld, H. et al. (2011). Composition of human skin microbiota affects attractiveness to malaria mosquitoes. *PLoS One*, 6, e28991.

Vogel, K.J., Valzania, L., Coon, K.L., Brown, M.R., & Strand, M.R. (2017). Transcriptome sequencing reveals large-scale changes in axenic *Aedes aegypti* larvae. *PLoS Neglected Tropical Diseases*, 11, e0005273.

Walker, E.D., Olds, E.J., & Merritt, R.W. (1988). Gut content analysis of mosquito larvae (Diptera: Culicidae) using DAPI stain and epifluorescence microscopy. *Journal of Medical Entomology*, 25, 551–4.

Walker, E.D., Lawson, D.L., Morgan, W.T., & Klug, M.J. (1991). Nutrient dynamics, bacterial populations, and mosquito productivity in tree hole ecosystems. *Ecology*, 72, 1529–46.

Wang, Y., Gilbreath, T., Kukutla, P., Yan, G., & Xu, J. (2011). Dynamic gut microbiome across life history of the malaria mosquito *Anopheles gambiae* in Kenya. *PLoS One*, 6, e24767.

Wang, S. & Jacobs-Lorena, M. (2013a). Genetic approaches to interfere with malaria transmission by vector mosquitoes. *Trends in Biotechnology*, 31, 185–93.

Wang, Y., Wang, Y., Zhang, J., Xu, W., Zhang, J., & Huang, F. (2013b). Ability of TEP1 in intestinal flora to modulate natural resistance of *Anopheles dirus*. *Experimental Parasitology*, 134, 460–5.

Wang, S., Dos-Santos, A.L.A., Huang, W., Liu, K.C., Oshaghi, M.A., Wei, G., Agre, P., & Jacobs- Lorena, M. (2017). Driving mosquito refractoriness to *Plasmodium falciparum* with engineered symbiotic bacteria. *Science*, 357, 1399–402.

Wang, Z., Liu, T., Wu, Y. et al. (2018). Bacterial microbiota assemblage in *Aedes albopictus* mosquitoes and its impacts on larval development. *Molecular Ecology*, 27, 2972–85.

Wei, G., Lai, Y., Wang, G., Chen, H., Li, F., & Wang, S. (2017). Insect pathogenic fungus interacts with the gut microbiota to accelerate mosquito mortality. *Proceedings of the National Academy of Sciences of the United States of America*, 114, 5994–9.

Wilkerson, R.C., Linton, Y.M., Fonseca, D.M., Schultz, T.R., Price, D.C., & Strickman, D.A. (2015). Making mosquito taxonomy useful: A stable classification of tribe Aedini that balances utility with current knowledge of evolutionary relationships. *PLoS One*, 10, e0133602.

Wu, P., Sun, P., Nie, K. et al. (2019). A gut commensal bacterium promotes permissiveness to arboviruses. *Cell Host Microbe*, 25, 101–12.

Xia, X. & Kung, A.L. (2009). Preferential binding of HIF-1 to transcriptionally active loci determines cell-type specific response to hypoxia. *Genome Biology*, 10, R113.

Xi, Z., Ramirez, J.L., & Dimopoulos, G. (2008). The *Aedes aegypti* toll pathway controls dengue virus infection. *PLoS Pathogens*, 4, e1000098.

Yun, J., Roh, S.W., Whon, T.W. et al. (2014). Insect gut bacterial diversity determined by environmental habitat, diet, developmental stage, and phylogeny of host. *Applied and Environmental Microbiology*, 80, 5254–64.

Zouache, K., Raharimalala, F.N., Raquin, V. et al. (2011). Bacterial diversity of field- caught mosquitoes, *Aedes albopictus* and *Aedes aegypti*, from different geographic regions of Madagascar. *FEMS Microbiology Ecology*, 75, 377–89.

Zouache, K., Michelland, R.J., Failloux, A.B., Grundmann, G.L., & Mavingui, P. (2012). Chikungunya virus impacts diversity of symbiotic bacteria in mosquito vector. *Molecular Ecology*, 21, 2297–309.

SECTION IV

Applications

CHAPTER 14

Direct and Indirect Social Drivers and Impacts of Vector-Borne Diseases

Sadie J. Ryan, Catherine A. Lippi, Kevin L. Bardosh, Erika F. Frydenlund, Holly D. Gaff, Naveed Heydari, Anthony J. Wilson, and Anna M. Stewart-Ibarra

14.1 Introduction

Vector-borne diseases are often thought of by the highly developed nations of the Northern Hemisphere as a 'problem' of the tropics, conjuring images of mosquito-borne diseases, such as malaria, caught on vacation or relegated to the poorest of the poor. In this chapter, we seek to move the conversation beyond the 'neglected tropical disease (NTD)—poverty trap' framing of the interactions between vector-borne disease (VBD) and social contexts. First, we briefly review the poverty trap concept with respect to VBDs and their management to assess the management narratives they may fit. We then present a series of vignettes to illustrate different facets of the broad range of social drivers which interact with VBDs. We explore the repeated emergence and establishment of mosquito-borne arboviruses (e.g. dengue, chikungunya, and Zika) in the urban social-ecological context, and the multiscale challenges to managing these diseases in Latin America and the Caribbean. Next, we explore the lessons learned in developing community-based programs for control of VBDs in Haiti that are centered on trust and community involvement. We then explore the drivers and impact of recent ruminant arbovirus emergence events in Europe, and the extent to which they mirror human VBD events.

These vignettes highlight the importance of reframing VBD narratives about emergence and spread, in terms of their context and system drivers, rather than simply lumping them as 'problems of poverty', in order to identify best practices in moving research from the field onto the desks of decision makers.

14.1.1 The VBD-Poverty Trap Paradigm

The connection of VBDs with the global poor is in many ways an environmental link, with a coincidence of tropical climates and large, exposed populations. In children under five years of age, diseases such as malaria and persistent parasitic infections result in a cycle of fever and stunting, leading to delayed development, low cognitive scoring, and a consistently immunologically compromised workforce. This is thought to contribute to a poverty trap—a cycle of the poor staying poor and sick (Bonds et al. 2010). This logic underpins many global health programs and strategies, and ties development, health goals, and funding together. While this remains a persistent component of many VBD systems, the simplification runs into the hazard of colonial overtones disguised as environmental determinism and overlooks the

Sadie J. Ryan, Catherine A. Lippi, Kevin L. Bardosh, Erika F. Frydenlund, Holly D. Gaff, Naveed Heydari, Anthony J. Wilson, Anna M. Stewart-Ibarra, *Direct and Indirect Social Drivers and Impacts of Vector-Borne Diseases* In: *Population Biology of Vector-Borne Diseases*. Edited by: John M. Drake, Michael B. Bonsall, and Michael R. Strand: Oxford University Press (2021). © Sadie J. Ryan, Catherine A. Lippi, Kevin L. Bardosh, Erika F. Frydenlund, Holly D. Gaff, Naveed Heydari, Anthony J. Wilson, Anna M. Stewart-Ibarra.
DOI: 10.1093/oso/9780198853244.003.0014

Box 14.1 Lessons learned about social drivers and feedbacks of vector-borne diseases

- Vector-borne diseases (VBDs) are often considered diseases of poverty, but this reductive approach does not explore the myriad of social and economic factors that influence transmission systems.
- In addition to economic resources, social and political factors have been shown to have profound influence on the timing and severity of outbreaks.
- The occurrence of disasters, both natural and unnatural, can damage public health infrastructure, deplete resources, and precipitate rapid shifts in housing density and social structures, increasing local VBD transmission.
- Local political and economic histories can influence current social drivers of disease, where public perception and trust determine the success of intervention efforts.
- Social, economic, and cultural influences on zoonotic VBD transmission and management can mirror those observed in human VBD transmission systems.

---

many nuanced ways in which that and additional historical and social contexts shape implementation of intervention.

Many VBD risk factors at the household level result indirectly from the macro-drivers of infectious diseases across the globe: poverty and inequality. This confronts us with the rather large goal of ending poverty to control VBDs. In some instances, forming policies targeted on poverty has been a successful approach to managing VBDs, illustrating the poverty-trap paradigm. For example, the decline in malaria in the Ohio River Valley during the early 20th century was due in large part to reductions in poverty and social inequalities (Farmer 1996). Reduced poverty led to improved housing conditions and access to mosquito control strategies—which reduced transmission and morbidity. While this and other examples underscore a link between broadscale reductions, resigning VBDs to problems arising from poverty may inadvertently lead public health officials to overlook key mechanisms important to controlling diseases on local scales. Conversely, if aid packages aimed at

poverty seek to mitigate VBDs, these may miss countries higher on the development spectrum.

Many of the most intensely managed VBDs are considered Neglected Tropical Diseases (NTDs). Neglected Tropical Diseases are defined as a 'diverse group of communicable diseases' (World Health Organization 2012) found in tropical and subtropical countries, and are described as primarily associated with conditions of poverty. The WHO updated the list of Neglected Tropical Diseases at the 10th meeting of the Strategic and Technical Advisory Group for Neglected Tropical Diseases, in 2017 (see Table 1). As Molyneux (Molyneux 2012) pointed out after the 2012 London Declaration on Neglected Tropical Diseases (http://www.uniting-tocombatntds.org), NTDs had achieved 'brand identity', necessary to attract attention in the Millennium Development Goals (MDGs) independently of HIV and malaria initiatives. Of this updated WHO list of NTDs in 2017, eight are VBDs—six involving insects (mosquitos and flies), one snails, and one, copepods. Slightly more complicated lists of diseases are the Neglected Infections of Poverty (NIPs) and the CDC's Neglected Parasitic Infections of Poverty (NPIPs), of which the CDC has created a list of five priority infectious diseases in the USA (Table 2). Of these, only one, Chagas, is a VBD. In Hotez's 2011 piece, originally describing NIPs in the USA, he makes the policy recommendations to examine VBDs, such as Chagas and dengue, along the Mexico border regions of the USA, and in post-Katrina Louisiana (Hotez 2011). The emphasis on dengue has faded from the CDC's NPIP list and more recent descriptions (Hotez 2014), but recent emergence of both chikungunya and Zika in similar geographic areas, and the associated post-hurricane uptick in concern (e.g. Harvey, Irma, and Maria of 2017), suggest perhaps a revisiting of this connection (Diaz & Stewart-Ibarra, 2018). The poverty-trap paradigm for VBDs can certainly enhance our understanding of these diseases in certain contexts but is not a 'one size fits all' solution. Thus, we aim to present some of the many complex social and cultural interactions that influence VBDs, which may be key components in developing effective public health interventions.

14.2 Social Factors in the Transmission of Vector-Borne Diseases: Reflections from Ecuador and Latin America

Across the Americas, *Aedes aegypti* and *Aedes albopictus* transmitted diseases (dengue fever, chikungunya, and Zika fever) are extending in range and prevalence due to social and environmental conditions. Dengue fever is caused by the dengue viruses (DENV1–4) that can cause illness in people ranging from mild febrile illness to severe hemorrhagic fever and death (WHO 2009). The recent emergence and epidemics of chikungunya (CHIKV) and Zika viruses (ZIKV) in the same populations demonstrate that new viruses can quickly enter the agent-host-environment trio where DENV has flourished for so many decades (Paixão et al. 2018).

Many factors are responsible for the emergence and expansion of dengue fever, with social and demographic changes playing an important role (Gubler 1997). The emergence of DENV infections worldwide in the eighteenth and nineteenth centuries was driven by the expansion of rapid capitalist global commerce, during which humans and mosquitoes were transported long distances by improved nautical technologies, moving along colonial era trade routes (Gubler 1997). The resurgence of the disease in the late twentieth century is correlated with urban population growth, persistent urban inequalities—wherein a mosaic of rich and poor neighborhoods led to, and result from, schisms in access to basic services, the rise of slums, etc.—and global travel, among other factors (Gubler & Meltzer, 1999).

Over the last forty years, social drivers primarily associated with urban areas, such as high population density, poor sanitation, deterioration of public health systems, and lack of effective vector control programs, have contributed to the rise of *Ae. aegypti*-transmitted illnesses in the Americas. *Ae. aegypti* is exceptionally adapted to the urban human environment. Female mosquitoes blood-feed on people in the home during the day, and water-filled containers around the home and patio are the ideal habitat for mosquito larvae. From 1960 to 1980, mid-sized cities in Latin America experienced rapid unplanned growth, resulting in social inequalities to urban infrastructure (e.g. piped water, garbage collection)

and the expansion of slum settlements, often related to deteriorating and/or changing economic and political contexts (Harpham & Molyneux 2001; Kendall et al. 1991; Satterthwaite 2003). Devastating dengue fever outbreaks followed shortly after, and since then the Latin America and Caribbean (LAC) region has reported the highest rise in dengue transmission worldwide (Stanaway et al. 2016). From 2000 to 2006, 68 percent of all cases worldwide were reported from LAC, with regional outbreaks occurring every three to five years (Cafferata et al. 2013).

Ecuador is classified as a highly developed country; the United Nations Development Program (UNDP), using the Human Development Index (HDI), places Ecuador in the 'high human development' category (United Nations Development Program 2018). In Ecuador, *Ae. aegypti*-transmitted illnesses have replaced malaria (transmitted by *Anopheles* mosquitoes) as the most prevalent mosquito-borne diseases ('Dirección Nacional de Vigilancia Epidemiológica', n.d.). Over a five-year period (2012 to 2016), 103,005 cases of DENV were reported in the country, compared to 1,861 cases of malaria. A highly effective malaria control program (Krisher et al. 2016), aimed at controlling *Anopheles* mosquitoes, coupled with growing urban areas, resulted in a decline in malaria transmission and the proliferation of *Ae. aegypti*. This program declared elimination status for malaria in southern Ecuador in 2012, and substantially scaled back surveillance and intervention for malaria, subsequently.

In the southern coast of Ecuador lies El Oro Province, bordered by Peru in the south, the Pacific Ocean in the west, and the Andean foothills in the east. DENV is hyper-endemic in El Oro, with an annual peak in transmission during the hot, rainy season from February to May, when mosquito populations are at their highest (Stewart-Ibarra et al. 2013; Stewart-Ibarra & Lowe 2013). The first outbreaks of CHIKV and ZIKV occurred in 2014–2015 and 2016–2017, respectively. Currently, there are no vaccines for CHIKV or ZIKV (Cohen 2016; Smalley et al. 2016). The DENV vaccine is not available in Ecuador, and is currently the subject of intense international scrutiny following recognition that the vaccine increases risk of severe DENV infections in sero-negative individuals (particularly children)

(Aguiar et al. 2016; Halstead 2017; Larson et al. 2019). Given the lack of vaccines or specific therapeutics, vector control remains the principal means for preventing and controlling *Aedes*-transmitted illnesses and outbreaks. In Ecuador, vector control includes regular household visits by Ministry of Health (MoH) inspectors to eliminate or treat containers with standing water with organophosphate larvicide, and focal control around homes with arbovirus infections using ultra low volume (ULV) fogging and indoor residual spraying (IRS). Over the last few years, the vector control system of the MoH in Ecuador has, from a management and organizational perspective, shifted from a centralized vector control program towards a decentralized program, spread more evenly across the health districts, allowing for greater decision-making power at the local-level. In addition to decentralization, the integrated strategy supported by the Pan American Health Organization (PAHO) is 'expected to produce a qualitative leap forward in prevention and control through stronger partnerships among the State, its various ministries, and governing bodies, private companies; and the range of community and civil groups' (Luis et al. 2007; Morrison et al. 2008a). In comparison to past approaches, this policy emphasizes significant intersectoral coordination and community involvement. However, implementation has been slow and challenging because of the need to reallocate costs to previously unfunded social mobilization initiatives, as well as the need to link knowledge of local social factors that influence disease transmission with operational activities (Leon 2017).

14.2.1 Social Risk in Coastal Ecuador

Studies in Machala, Ecuador, have characterized the complex social and environmental factors driving arbovirus transmission using a social-ecological (SES) systems approach. These studies found that *Ae. aegypti* population dynamics are influenced by an unexpected number of complex social-ecological risk factors at the local level, including risk perceptions, access to municipal services, water storage behaviors, and poor housing conditions (Stewart-Ibarra et al. 2014). Community focus groups conducted in the city identified these local risk factors, which were grouped into bio-physical, political-institutional, and community-household factors

(Stewart Ibarra et al. 2014). In terms of the way that DENV risk was conceptualized by community members, urban development issues were among the most significant. Access to municipal public services and utilities (e.g. garbage collection, sewerage, piped water) were also a major concern, especially in the peripheral areas of the city, where infrastructure inequalities are at their highest. The findings in coastal Ecuador are consistent with our understanding of the behaviors of *Ae. aegypti*, and the ways in which this mosquito is inextricably linked to social conditions. The vector predominates in urban settings where crowded conditions and lack of infrastructure contribute to the transmission of DENV (Gubler 1998; Morrison et al. 2008b). Year round in Machala, household water storage provides the ideal habitat for mosquitoes. It was found that storage of tap water in containers, such as fifty-five-gallon drums, increased the risk of *Ae. aegypti* around the home year round, whereas abandoned containers filled with rainwater increased *Ae. aegypti* abundance during the rainy season. Households with an unreliable piped water supply (frequent water supply interruptions) are more likely to store water in containers, which become potential mosquito larval habitat (Stewart-Ibarra et al. 2013).

Household-level social factors are not the only critical drivers of vectored disease transmission. In fact, political-institutional perceptions were shown to be important social drivers, because attitudes and opinions of citizens determine their willingness to adopt disease control actions and thus have the potential to influence governmental decisions and shifts in policy and management. Dengue epidemics can overwhelm hospitals and clinics, leading to overworked health care providers and sub-standard care for patients. Mesoscale socioeconomic patterns also affect attitudes and perceptions in public health, with communities living in low-income neighborhoods in the periphery of Machala feeling that they were neglected by the government (Stewart Ibarra et al. 2014). Equipped with limited resources to conduct surveillance, as is the case for many dengue endemic countries, local MoH officials are often scrambling to offer a response that has a high impact on community perceptions (Gubler 1998). This has led to the reliance on ultra-low volume spraying (ULV) during epidemics, since it is a highly visible action that demonstrates governments are actively doing something to control the spread of disease (Gubler 1989). Both ULV spraying and larvicide application are highly visible actions, as illustrated in Machala, Ecuador (Figs 14.1, 14.2). Although ULV spraying has the potential to cut transmission during outbreaks, it is an expensive strategy that has limited efficacy in areas with high levels of insecticide resistance (Gubler 1989; Reiter & Newton 1992). In Machala, we

Figure 14.1 MoH staff fumigating a home while the owner watches on (urban periphery, Machala).

Source: Dany Krom https://www.danykrom.com/STORIES/Vector-born-Diseases/

Figure 14.2 MoH staff visiting a home to apply larvicide, while the woman of the household waits.

Source: Dany Krom https://www.danykrom.com/STORIES/Vector-born-Diseases/

found that households reporting fumigating had lower risk of DENV infections (Kenneson et al. 2017), suggesting it is currently a viable strategy, if implemented properly.

The social impact of these events is that communities will demand spraying that may not address the causes of disease risk, or simply may not reduce infections. We recently found evidence of high levels of insecticide resistance (IR) in *Ae. aegypti* among four cities in southern Ecuador, including Machala, indicating that this is an emerging problem (Ryan et al. 2019). The study found that rates of IR were highest in Machala, which has the highest historic dengue burdens among the four cities studied. More nuanced examinations of socioeconomic and sub-city level geographic stratification are needed before we can fully assess the links between demand, implementation of ULV spraying, and IR. If a coincidence of ULV spraying and high levels of IR arise, confidence in vector control may begin to erode, and political will, rather than a reexamination of strategy and efficacy, may dictate the next step.

Whereas environmental conditions are drivers for *Ae. aegypti* population dynamics and dengue incidence at large scales across regional areas, social factors can account for differences at the sub-regional and within-city level. For instance, a study conducted across comparable sites in Arizona, USA, and Sonora, MX, found that *Ae. aegypti* presence was positively associated with highly vegetated areas, and reduced with access to piped water (Hayden et al. 2010). This result appears to contradict our findings in Machala; however, the quality of piped water—in Machala, intermittent service was associated with higher risk—is likely the driver of these differences. In the study comparing Arizona and Sonora, the social factors across the cities varied greatly, with all locations on the U.S. side of the border having access to public health infrastructure (piped water, sewage, sanitation), but many households without access to the same infrastructural amenities on the Mexican side of the border. In Ecuador, the effects of rainfall on dengue risk varied by neighborhood within a city, depending on whether the main larval habitats were water storage containers with tap water (in the urban periphery) or abandoned containers with rain water in the patio (Stewart Ibarra et al. 2014; Stewart-Ibarra et al.

2014, 2013). Studies on malaria prevention have also found similar patterns of VBD intervention access inequalities. Prior to the mass distribution of free insecticide-treated mosquito nets in El Oro, Ecuador, that achieved great equity in coverage, access to bed nets was largely dependent upon wealth (Nuwaha 2002; Rashed et al. 2000; Steketee & Eisele 2009; Wiseman et al. 2006).

Studies in periurban communities in Machala, Ecuador have shown that the economic burden of mosquito prevention is significant for low-income households. Multiple studies conducted in Machala between 2013–2015 have shown that low-income households employ many mosquito control interventions to reduce the burden of *Ae. aegypti* transmitted illness, though the effectiveness of these household-level interventions needs to be evaluated. Members of periurban communities reported economic barriers to DENV prevention in qualitative studies (Stewart Ibarra et al. 2014). This was validated through household surveys, which found that the economic burden of mosquito control was significant in low income households (Heydari et al. 2017). Households spend upwards of 10 percent of discretionary household income, or income that is left for spending after paying for household necessities (e.g. food, housing, gas and water) to prevent mosquito-borne diseases (Heydari et al. 2017). The findings show a healthy and robust market for commercial mosquito products in small neighborhood stores and large, centrally-located supermarkets. Novel interventions that are low cost and effective at reducing the population of *Ae. aegypti* are urgently needed. Innovative control strategies are promising, such as lethal ovitraps, insect growth regulators, and attractive toxic sugar baits (Andrade et al. 2016; Johnson et al. 2017; Paz-soldan et al. 2016; Scott-fiorenzano et al. 2017). However, the success of these strategies will rest on economic latitude and social acceptance and require ongoing trials in the region to determine viability.

The economic situation and levels of poverty play important roles in disease transmission across multiple levels (e.g. macroeconomic level and microeconomic level). At the household level, economic barriers/limitations were the most commonly mentioned barrier to *Ae. aegypti* control across several studies in Machala, Ecuador (Heydari et al. 2017;

Stewart Ibarra et al. 2014). Home ownership and job stability, proxies of poverty, were associated with expenditures on vector control. There can exist striking differences in human housing factors including screened windows and doors, irrigation, and vegetation—all factors that play a key role in the presence or absence of *Ae. aegypti* in and around households (Chang et al. 1997; Kay & Nam 2005; Nagao et al. 2003). To illustrate this juxtaposition of housing quality, a five-star hotel with air-conditioning, screens on all windows, and a garden that is well maintained is located less than five blocks away from several neighborhoods where air-conditioning units and screens are rare, and abandoned lots (Kenneson et al. 2017) are found on every street. Ironically, many of the hotel workers live in the surrounding neighborhoods, meaning that viral transmission is likely also occurring, albeit at less frequency due to reduced biting rates, within wealthier social groups. Family income to purchase *Ae. aegypti* control strategies may lessen the risk of falling ill with arboviruses, although it is not hard to imagine a situation in which a hotel guest sun-bathing poolside is bitten by an *Ae. aegypti* mosquito that has emerged from the cistern of a neighboring household. Improving equity across local sites—that is, bringing whole neighborhoods and cities into an improved standard of living, with improved access to public health and vector control interventions—will be needed when considering economic and public health improvements to fully mitigate VBD transmission, given that the flight radius of the *Ae. aegypti* is around 200m.

14.2.2 Increasing Vulnerability Following a Natural Disaster

There is a well-established connection between human activity and *Aedes* presence (Shragai et al. 2017), but not all *Aedes* risk is caused by humans. Natural disasters, such as earthquakes and hurricanes, damage public health infrastructure and housing, consolidate economic inequalities, and increase the incidence of disease. On 16 April 2016 a 7.8 magnitude earthquake devastated the north-central coastal region of Ecuador just as the ZIKV epidemic was emerging. The earthquake claimed the lives of 663 people, injured over 6,000, displaced

nearly 30,000, and caused severe damage to households and public health infrastructure, leaving families without homes and vulnerable to mosquito-borne infections (Secretaria de Gestion de Risegos 2016). Out of necessity, families stored water in fifty-five-gallon drums, large water jugs, elevated tanks, and other containers that make ideal *Aedes* larval habitat, since water systems were damaged. In areas that already faced problems with inadequate water supply, the earthquake further exacerbated the problem (Ryan et al. 2016). In the earthquake-affected province of Manabi, Ecuador, there was a spike in suspected and confirmed ZIKV cases among the general population and among pregnant women in the aftermath. After the earthquake, the cumulative number of ZIKV cases in Ecuador increased from 103 to 1275 in eleven weeks, with 85 percent of all ZIKV cases in 2016 reported from Manabi Province, near the epicenter (*Zika Gazette SE 38–2017* 2017). This finding is consistent with studies in other disaster zones; for example, with malaria prevalence increasing following an earthquake in Southeast Asia (Feng et al. 2016).

The destruction of property and public health infrastructure is likely to increase *Ae. aegpyti* larval habitat in itself, and the displacement of families and overcrowding is likely to further increase mosquito-borne disease transmission due to an increase in exposure to mosquito bites and infected individuals (Salazar et al. 2016). In fact, most of the increased risk of VBDs following a natural disaster can be attributed to conditions related to population displacement (Watson et al. 2007). Many of the homes that were still standing in Manabi Province were deemed by government officials unsafe to occupy. Families without stable housing migrated into small living quarters, sometimes donated by non-governmental organizations (NGOs), other times a makeshift tent consisting of only simple tarps, fastened into walls. During our last visit to the area, more than one year after the earthquake, some of these 'temporary' living spaces were still in use. Lastly, we observed that during the subsequent efforts to rebuild homes post-disaster, the number of persons living in a household had increased. This rapid shift in housing density could further compound disease transmission risk in the area. Studies in southern Ecuador prior to the earthquake found that higher urban population density, indicated by the number of households per property, was a positive predictor for the presence of *Ae. aegypti* (Stewart-Ibarra et al. 2013).

As a point of comparison, it is worth highlighting the impacts of tropical storms in the Caribbean on arbovirus transmission. In the island country of Dominica, DENV is an ongoing arboviral disease concern, and recorded cases since 1993 show that in epidemic years, there are seasonal signals (Stewart Ibarra et al. 2017); and cases have been reported every year since 2006, according to epidemiological records. In 2014, CHIKV swept through the Caribbean, including Dominica, and in 2016, 1,263 cases of ZIKV were reported during the initial outbreak (Ryan et al. 2017a). Household water storage increased after Tropical Storm Erika in 2015, as reported by local health officials in interviews (Stewart Ibarra et al. 2017; Stewart-Ibarra et al. 2019). The storm damaged the piped water systems and rivers were contaminated, resulting in a scarcity of potable water. As a result, homes began storing water in drums, and this practice continues two years later even though most homes (>90 percent) now had access to reliable piped water, as Dominica is a country with an abundance of freshwater. Dominica and other Caribbean territories were devastated again during the 2017 Atlantic hurricane season, one of the most active seasons on record. It is suspected that the risk of arbovirus transmission increased in the region due to significant damage to housing and other infrastructure, and relocated populations. As observed in Dominica, these amplified social risk factors may remain present years after the disaster.

We hypothesize that psychological distress resulting from natural disasters can also increase the risk of arbovirus transmission by impairing the immune response (Cohen and Williamson 1991; Glaser et al. 1987). Posttraumatic stress disorder (PTSD), suicidal ideation, depression, and anxiety are commonly reported in natural disaster survivors (Chen et al. 2001; Stewart-Ibarra et al. 2017; Zhang et al. 2014). In areas with scarce health system resources, access to mental health care is often limited and socially stigmatized (Borowsky et al. 2000; Harding et al. 1980). Lack of access to mental health care is exacerbated in a post-disaster setting due to competing health priorities. In Ecuador, three months after the earthquake, we found that women, adults, and displaced people who reported more psychological distress symptoms were more likely to report arbovirus symptoms in low-income communities than in higher income communities (Stewart-Ibarra et al. 2017). These findings suggest that there are complex multilevel interactions that affect the mosquito vector and arbovirus transmission following a natural disaster including individual risk factors (e.g. immune response, stress, age, gender) and community-level risk factors (e.g. housing damages, water and healthcare access).

14.2.3 Regional Politics: Unnatural Disasters

It is not easy to pinpoint the moment that Venezuela's economic collapse started, but a combination of falling oil prices and political struggle have meant that the richest economy in the Latin American region has plummeted to a crisis state in under a decade. This has had severe impacts on health, the scope of which are now coming to light—a review of vaccine-preventable disease resurgence in Venezuela pointed out that with an economy suffering 45,000 percent hyperinflation, the health system went into free fall (Paniz-Mondolfi et al. 2019). Measles cases in and from Venezuela represented 68 percent of cases in Latin America in 2018, and Diptheria, previously eradicated, has re-emerged (Paniz-Mondolfi et al. 2019). The implications of this health system collapse for VBDs is two-fold; within Venezuela, the capacity to conduct surveillance, vector control, and treatment has evaporated, and beyond the borders, a massive exodus of the population, migrating through and to surrounding countries, has led to new routes of transmission, and potential spread. In Venezuela, malaria cases rose 359 percent between 2000 and 2015, while neighboring countries in the region were driving toward elimination, and between 2016 and 2017, an increase of 71 percent in malaria cases in Venezuela led to noticeable upticks in imported cases in neighboring countries (Grillet et al. 2019). In 2017, Venezuela contributed 53 percent of the malaria cases in the Americas (World Health Organization 2018).

The current social and political crisis in Venezuela is now influencing vector-borne disease transmission dynamics across the Americas (Grillet et al. 2019). Approximately 3 million people have fled Venezuela, many to other countries in South America. In 2018, the first new locally acquired cases of malaria were reported in El Oro, Ecuador, and Tumbes Region, Peru, following the elimination of malaria in 2011/2012 (Jaramillo-Ochoa et al. 2019). These cases appeared after a series of imported malaria cases were detected in Venezuelan migrants in El Oro and Tumbes. Known malaria mosquito vectors are abundant and present in the region (Ryan et al. 2017b), meaning elimination status is tenuous if regional elimination goals are not maintained. Paralleling malaria importation, prior phylogenetic analyses of DENV from Machala found that they were closely related to DENV from Venezuela (Stewart-Ibarra et al. 2018). This highlights the key role that regional politics and migration can play in disease transmission, and the need for Ecuador and other countries in the region to strengthen integrated vector-borne disease surveillance and control in the border regions, and along migration routes, where human movement poses a major challenge.

14.3 Building Trust in Haiti: Reflections from the *Polisye Kont Moustik* (PKM) Project

Haiti was officially founded in 1804, the first free black republic, after more than a decade of bloody revolt spearheaded by bands of African plantation slaves (Dubois 2012). Founded in 1697, *St. Dominque* was the most profitable slave settlement in the Caribbean, supported by a brutal system of dehumanizing mono-crop agriculture, organized around the supply and demand laws of Europe's burgeoning sugar industry. Newly imported West Africans were segregated based on language and ethnic identity, and made to work until they died from exhaustion, malnutrition and disease, often in combination. A large bronze statue of the *Batay Vètyè* (Battle of Vertières), the last independence battle, led by the legendary Jean Jacques Dessalines, stands in a small park along the main highway connecting Cap-Haitien (Haiti's second largest city) with the capital,

Port-au-Prince. As with other islands in the West Indies, tropical diseases, especially yellow fever and malaria, were a key element of these colonial battles. McNeil, in *Mosquito Empires: Ecology and War in the Greater Caribbean, 1620–1914* (McNeill 2010), argues that while malaria was 'the white man's grave' in West Africa, yellow fever played a more formidable role in hastening in the first Black Republic, acting as 'shields for local populations'. The French forces sent by Napoleon during the *Batay Vètyè* were never able to successfully penetrate the mountainous Haitian landscape. They were decimated by yellow fever, due to a lack of acquired immunity, while suffering from a war-ravaged economy and diminishing supplies (Peterson 1995).

History has a long shadow in places like Haiti. While 'free', the country has seen continued foreign intervention, including an often-forgotten occupation by the United States (1915–34), and an elite 'mulatto' class more interested in maintaining economic and political dominance than in transformative social and democratic change. While the island of Hispaniola is shared with the Dominican Republic (DR), aerial images show a forested DR and a de-forested Haiti, a metaphor for how different histories, cultures and state-citizen relations have produced uneven capabilities (Sheller & León 2016; Stoyan et al. 2016). The 'poorest country in the Western Hemisphere', Haiti has some of the worst health indicators globally; the health delivery landscape is fragmented, dominated by professional and faith-based NGOs and heavily reliant on international aid.

Why is all of this important? How does history and politics play into the current social drivers of VBDs? One of the legacies of this political and economic history is that in modern Haiti, there is little faith and trust in government institutions and foreign aid interventions. For the last two years, *Polisye Kont Moustik* ('Mosquito Police' or PKM), a new community-based vector control program, has been exploring how to build community trust and social relationships as an essential part of an 'effective' public health program, in communes around the old battlefields of *Batay Vètyè* in northern Haiti (Bardosh et al. 2017).

The literature on trust, as a component of vector control, is relatively thin—alluded to but seldom discussed. A study in Peru, for example, pointed to real or imagined burglaries instigated by outreach staff as drivers for community resistance to vector control outreach (Charette et al. 2017). In Ecuador, a community-based randomized *Aedes* control trial intervention, using elementary school-based education and community-wide cleanup and education, found that the strength of local community-based solidarity and political action played a major role in effectiveness (Mitchell-Foster et al. 2015). These, and other studies, conclude that vector control depends on much more than 'knowledge, attitudes and practices' (Launiala 2009), but also on very hard-to-quantify cultures of local municipal governance, community participation and social cohesion. This influences, for example, the willingness of local people to let vector control staff enter their houses, and the level of importance given to interpersonal communication during these visits. Persuasion is a key aspect of social negotiations between vector control staff, families, mosquitoes, and water containers. Encouraging participation and engaging in knowledge dissemination will require different strategies in different contexts. These negotiations and implementations are also political; for example, Nading's ethnography of dengue control in Nicaragua (Nading 2014) showed how vector control outreach focused on 'eliminating breeding sites' can exasperate social divisions between health workers and garbage scavengers—those whose livelihood depends on collecting, storing and selling recycling material.

Our efforts in Haiti, originally funded by Gates Foundation and now USAID, have focused on developing an effective larval source management (LSM) program that engages municipal governments, local community groups and families. Originally, our efforts were focused on integrating LSM with the Haitian Lymphatic Filariasis (LF) Elimination Program. In the face of adversity, the LF program has been hugely successful in Haiti, and is focused on annual rounds of community-wide mass drug administration (MDA) with deworming drugs (Lemoine et al. 2016). But some persistent parasite foci stubbornly remain, mostly in urban areas, raising the prospect that it may be necessary to target the mosquito *Culex quinquefasciatus* which breeds in polluted and stagnant water,

especially in poorly built and maintained cement canals (Burkot et al. 2006). With this mosquito-borne disease now targeted for global elimination by 2020, it is curious that vector control has remained such a peripheral part of the Global LF Elimination Program.

Our original pilot project focused on the small town of Plaine-du-Nord, with roughly 2,500 households and 10,000 people. An important voodoo pilgrimage site surrounded by fertile agricultural land, the town was once the 'bread-basket' of Haiti. During the first nationwide lymphatic filariasis (LF) prevalence survey conducted in the early 2000s, Plaine-du-Nord had a staggering 45 percent filarial antigenemia prevalence, the highest prevalence in Haiti and among the highest globally (de Rochars et al. 2004). But Plaine-du-Nord is also endemic for other VBDs. Malaria has long been endemic in the floodplain. While yellow fever has long vanished, new pathogens like chikungunya and Zika are now widespread, and have integrated themselves into the local disease lexicon. Fig. 14.3 shows an open sewer flowing in the middle of an unpaved road, tangles of electrical wires (that often do not work) and an advertisement from a local 'doctor' for physio-therapy to treat chikungunya, now considered a chronic disease, alongside rheumatism and hypertension, by many in northern Haiti (note that he also provides veterinary care).

When the project was getting off the ground, our first training in the office of the Ministry of Public Health and Population (MSPP) was infested with *Aedes* mosquitoes. Attributed to the chikungunya epidemic in 2014 that swept through the Caribbean, a team member suffered from epileptic-like seizures, and the MSPP medical director suffered chronic joint pain and headaches. In our ethnographic fieldwork, we found that chikungunya was by far the most feared and talked about mosquito-borne disease. Known locally as 'paralysis', nearly 50 percent of people in our household surveys ($n = 293$) reported that they had been infected with the virus in 2014, often sick for weeks.

Building trust and social relationships in Plaine-du-Nord, in the early stages of the PKM project, involved a few interesting and distinct steps. First, the strong community experience with chikungunya meant that, despite our funders being inter-

Figure 14.3 An advertisement from a local 'doctor' for physio-therapy to treat chikungunya, northern Haiti.
Source: Kevin Bardosh

ested in LF, malaria, and then Zika, we found it necessary to emphasize chikungunya in our risk communication and community engagement, to connect to experiential context. This siloed approach to specific intervention targets is a common and unfortunate aspect of vertical disease-specific global health initiatives: for instance, the malaria program can only distribute bednets and target *Anopheles* breeding sites, the LF program can only clean canals, and the Zika program can only do household visits. In order to speak to the priorities and risk perceptions of community members, our focus became more of an integrated vector management approach—an effort to include VBD control strategies with multiple targets, and bundled approaches.

With Haiti's history of colonialism and sense of exploitation from NGOs, it was interesting to see how community members interpreted and responded to foreign American staff visits. In some ways, these were negative. Early on, some community members became angry when their photograph was taken during a food distribution as part of a community-wide larval habitat cleanup campaign. They were concerned that the picture would be shown internationally, and that the American project staff would use it to 'make money'. This was a common theme: the idea that NGOs *used* community members for self-gain, and only did 'little work'. Some rumors began to circulate after this incidence. To address them, we initiated a series of community dialogues, which included inviting religious groups, municipal leaders, health staff, school teachers and informal leaders. From this, the initial 'rumor-spreaders' (a group of politically savvy local artists) became our most vocal and involved advocates. Interestingly, over time, community members asked for more 'foreign oversight' and involvement, citing suspicions of Haitian elites managing health services.

Engagement with the community also meant negotiating a 'partnership' with various local political leaders, each which had their own political agenda. During the early phase of the project, garbage was used as a political weapon, repeatedly dumped on the streets of towns to damage the reputation of the town mayors and to 'sabotage' our first cleanup campaign. Our effort at mosquito control had political value for these leaders. In this way, they collaborated with us. But they did it strategically, often more focused on showing up to large events than contributing anything too substantial (such as staff or funding).

A major area of conflict with local leaders involved our efforts to hire staff. The PKM approach is based on hiring local staff, who conduct daily household visits where they engage in cleanup activities, education, community mobilization and apply Bti (*Bacillus thuringiensis israelensis*) for vector control. It is important that these staff are from the areas that they work in; the project had to consult local leaders for the recruitment process. At first, most recommended workers were not interested or able to work the long hours or did not have the proper inter-personal skills. They had been selected to return political favors during the recent election campaign. It took time to replace these people, without damaging our own relationship with the municipal authorities. Eventually, we developed a recruitment system that was 'democratic'. It included each leader of a *quartier* nominating two to three people, who would undergo training, and then take a practical and written test. Recruitment was based on these test grades, which were posted publicly around the town.

As we scaled-up the project and expanded to neighboring communes, one of our key leadership staff tampered with the test results to promote their friends. People noticed this, and it was the talk of the town, threatening to derail an otherwise carefully planned out process. A quick town-hall meeting was called, with witnesses, accusations, and counter-accusations, leading to dismissal of the staff member. Such occurrences happened continuously over two years, in multiple variations and in many different settings. While we used flyers posters, banners, loudspeakers, radio messages, drama, comedy, murals and competitions to engage community members in mosquito control in northern Haiti, these lessons tell us something different— something relatively simple but hard to put into practice, that has been increasingly discussed in the post-Ebola literature reflecting on the 2014–2016 West African epidemic (Wilkinson et al. 2017). To be effective, these activities need to be implemented with *trust*, something that demands constant atten-

tion, constant tinkering and adapting, alongside local political spheres and the management of everyday life.

14.4 Livestock VBDs in Europe: Similarities to and Differences from Human VBD Emergence Events

14.4.1 Veterinary VBDs: A Primer

Several of the pathogens described already in this chapter are zoonotic: they originate from or circulate within wildlife, companion, and livestock animals, as well as humans. In most cases, however, the zoonotic threat they pose to public health dwarfs the threat to animal welfare, food security and economic security. In contrast with the poverty-trap paradigm, where diseases are a symptom of economic disparities, veterinary VBDs may serve to drive localized poverty through agricultural loss. In this section, we consider some recent outbreaks of VBDs of livestock, focusing primarily on Europe, and we compare and contrast the key policy lessons to those relevant for human VBD outbreaks.

Vector-borne diseases were recognized as a substantial threat to the European livestock and equine industries at a key horizon-scanning event in the 1960s (FAO/OIE 1963). At this event, the 'big three' vector-borne livestock pathogens identified were African horse sickness virus (AHSV), African swine fever virus (ASFV) and bluetongue virus (BTV). African horse sickness virus and bluetongue virus are transmitted by small biting flies in the genus *Culicoides*, commonly called 'midges', but are otherwise epidemiologically 'classical' VBDs rarely transmitted by other routes. In contrast, African swine fever is a highly environmentally-persistent pathogen for which the soft tick vectors (genus *Ornithodoros*) are primarily relevant as a reservoir, contributing a relatively minor force of infection. It spreads readily via environmental contamination including fomites. African horse sickness, as the name suggests, infects equids; bluetongue infects ruminants, and African swine fever infects suids (pigs). All three of these pathogens were first formally described when European settlers introduced 'improved', highly susceptible breeds of livestock into sub-Saharan Africa, where the viruses had each circulated largely sub-clinically in native animals (zebra for AHSV, warthogs for ASFV and various ruminants for BTV) with negligible economic impact.

14.4.2 Social Drivers of Introduction

As with the VBDs discussed elsewhere in this chapter, the frequency with which veterinary VBDs are introduced has increased alongside the increase in international travel and trade over the past decades. These introduction events fall into various classes: the accidental introduction of infected hosts (for example, the introduction of African horse sickness into Spain in 1987 via infected zebra, and the regular introduction of *Leishmania*-infected dogs into Germany (Naucke et al. 2008)), the introduction of infected vectors (either via wind, as with the introduction of bluetongue virus into the UK in 2007, or human transport (Carpenter et al. 2009)) and the introduction of infected materials (the introduction of African swine fever into Portugal as a result of waste from airline flights being fed to pigs near Lisbon airport and the suspected introduction of ASFV into the Black Sea area by food waste discarded from a ship (Stokstad 2017)). The importance of each of these introduction routes is affected by public awareness of their role but also changes in response to changes in public behavior as a result of economic and other social factors. For instance, for other (non-vector-borne) livestock diseases, annual increases in the international movement of live ruminants during religious pilgrimage periods have been recognized as a factor in disease introduction (El-Rahim et al. 2016), and the festival of Eid al-Adha has also been proposed as a possible factor in the emergence of Crimean-Congo haemorrhagic fever in Pakistan (Burki 2012). Recognition of such factors by veterinary authorities may facilitate targeted surveillance or additional support or information services for local veterinarians during high-risk periods.

As well as economic and religious activities, criminal activities may also be a factor in the introduction of important veterinary pathogens, vector-borne or otherwise. Although the initial introduction of African swine fever virus into the Black Sea region is believed to have been accidental, the illegal distribution of contaminated meat from culled animals via wholesale buyers, particularly the military food supply system, have been implicated multiple times in long-distance introduction events (Gogin and Kolbasov 2013).

Socio-political unrest can also be a factor, particularly for pathogens transmitted by ticks which thrive in the high humidity environments afforded by long grass. Pasture in parts of central Anatolia was largely abandoned during the 1990s as a consequence of local terrorist activity; the resumption of farming and hunting was followed by reports of Crimean-Congo hemorrhagic fever virus (CCHFV), and indeed the first ever recognized cases of CCHFV in 1944 followed a similar period of pasture neglect and reoccupation in the Crimea (Hoogstraal, 1979). The emergence of tick-borne encephalitis virus (TBEV) in Eastern Europe also followed the collapse of the Soviet Union and resulting economic and behavioral changes that increased contact rates between hosts and tick vectors (Sumilo et al., 2007).

14.4.3 Social Considerations in the Development and Uptake of Control Measures

The biggest fundamental difference between livestock industries (including, for the purposes of this section, the equine industries) is the economic—and, for horses, social—value of a single animal. This is likely to be a key factor in observed differences in owners' willingness-to-pay for control measures, such as protection from vector attack, vaccination and vector habitat modification. For instance, eliminating contact between vector and host interrupts transmission and is the basis of control strategies such as bed nets and mosquito screening. One equivalent approach for livestock VBDs is the 'Vector-Protected Establishment', which is defined in the OIE Terrestrial Animal Health Code (and derived regional regulations such as EC regulation 1266/2007). European Commission document SANCO/7068/2012 (European Commission 2012) provides guidance on implementing the use of this measure. Although preliminary European studies (Lincoln et al. 2015) have supported the longstanding observation in South Africa (Paton 1863) that stabling can be effective at preventing *Culicoides* attack, the requirements to assure vector protection are quite rigorous, and during the recent emergence of BTV in northern Europe most sites making use of the vector-proof housing regulations have been artificial insemination centers. Engagement with stakeholder groups suggests that the willingness to consider such arrangements would be used much more widely for equids in the event of an outbreak of African horse sickness. This may be partly due to the higher requirement for movement of many such animals, the higher mortality rates seen in most African horse sickness outbreaks, or the higher financial value of the animals, but the social value of companion animals such as horses and dogs is more complex. Another likely example of the effects of perceived social value on control choices was the decision to vaccinate the entire population of California condors (Chang et al. 2007) when West Nile virus appeared likely to arrive in the Western US, despite the cost and unproven safety or efficacy of the vaccine in this endangered species. The social value of this bird is significant, featuring in the culture of several Native American tribes in the region including the Wiyot, Mono, Yokut, and Chumash (Nielsen 2006).

14.4.4 Traditional Farming Practice and Cultural Resistance to Novel Control Strategies

African swine fever was introduced into Europe in Portugal in 1957 and again in 1960. Although it spread to several other countries, it was rapidly eradicated from all locations outside the Iberian peninsula with the exception of Sardinia (Arias & Sánchez-Vizcaíno 2002). However, it was able to persist in parts of Spain and Portugal until the mid-1990s, primarily due to the presence of its soft tick vector/reservoir *Ornithodoros erraticus* (Boinas et al. 2011), which significantly increases the capacity of the virus to persist and is accordingly represented by additional control/quarantine requirements in EU legislation when ticks are suspected to be present (Council Directive 2002/60/EC) (Boinas et al. 2011). The range of these ticks is geographically restricted (Wilson et al. 2013) but even within this range they are predominantly found on farms practicing traditional production methods for the rearing of Alentejano (Iberian black) pigs. Between

1960 and 1990 African swine fever almost destroyed the Alentejano industry (Boinas et al. 2014). However, there is evidence that affordable changes to farming practice such as the adoption of aluminum pig arks for housing (which do not afford resting places for the ticks due to the high temperature they reach during the day) could effectively eliminate tick infestations and significantly reduce the risk from ASFV. Despite this, resistance to modernization among this traditional community remains high. One possible explanation of this is that the market for Alentejano products is relatively high value, low volume, based in part on the social value attributed to traditional farming practices. The Alentejano breed is recognized under the European Union's Protected Designation of Origin (PDO) scheme,[1] which exists to provide consumers with assurance that a product is produced in a specific geographical region using local ingredients and the expertise of local producers. In this case, the resistance of individual farmers to adapting production methods may be a simple economic concern about the risks of moving away from the perception of 'traditional' methods that underlies support for this industry, or it may reflect a cultural preference to this way of life that outweighs the expected benefits of lower-risk farming methods in terms of animal welfare and economic income.

14.4.5 Conclusions and Directions for Future Research

The epidemiology of vector-borne livestock diseases is not simply a function of the distribution of vector or host or of simple environmental factors such as temperature, but arises through differences in population structure, livestock movements, land use patterns and farming practice that may in turn be determined by economic, religious, political or traditional beliefs or behaviors. Attempting to understand recent VBD emergence events in or close to Europe and affecting or involving livestock without appreciating the role of these factors reduces our capacity to understand and mitigate them, to predict future emergence risks, and to develop and effect-

---

[1] https://tradicional.dgadr.gov.pt/en/categories/meat/pig-meat/366-carne-de-porco-alentejano-dop-en

ively deploy appropriate control strategies. The political, economic, cultural, and psychological factors that must be considered when developing management strategies for VBDs in animal systems mirror those involved in transmission cycles that include humans. While veterinary VBDs are an important focus of research in their own right, they may also serve as model systems for better understanding the complexity of human VBDs, providing insights into leverage points for intervention.

14.5 Summary

In this chapter, we explored how social processes in systems may shape drivers or contexts for vector-borne diseases. We used three contrasting systems to illustrate that a poverty-trap framing may generalize the issues complicating VBD control, in the face of history, politics, intervention successes and failures, natural and unnatural disasters, behaviours, perceptions, practices, and financial interests in control, independent of those receiving it. Vector-borne diseases have far larger and more intricate social lives than can be captured by current health management frameworks. While it can be beneficial to use existing frameworks to leverage attention and assistance (e.g. labels such at NTDs, or framing as a poverty-related issue for aid purposes), as we have described here, our frameworks do not capture the range of underlying socioeconomic scope, and political context. It is therefore essential that the instruments of social science research, the recognition of political history, understanding of local context, and the framing of the system, be integral parts of VBD research and intervention practice.

References

Aguiar, M., Stollenwerk, N., & Halstead, S.B. (2016). The risks behind Dengvaxia recommendation. *Lancet Infection and Disease*, 16, 882–3.

Andrade, P.P. De, Lima, J., Colli, W. et al. (2016). Use of Transgenic *Aedes aegypti* in Brazil: Risk Perception and Assessment. *Bulletin of the World Health Organization*, 94(10), 766–71.

Arias, M. & Sánchez-Vizcaíno, J.M. (2002). African swine fever eradication: The Spanish model. In: *Trends in Emerging Viral Infections of Swine*, A. Morilla, K.J. Yoon, and J.J. Zimmerman eds. Iowa: Iowa State University Press, pp. 133–9.

Bardosh, K.L., Jean, L., Beau De Rochars, V.M. et al. (2017). Polisye kont moustik: A culturally competent approach to larval source reduction in the context of lymphatic filariasis and malaria elimination in Haiti. *Tropical Medicine Infection and Disease*, 2, 39.

Boinas, F., Ribeiro, R., Madeira, S. et al. (2014). The medical and veterinary role of *Ornithodorous erraticus* complex ticks (Acari: Ixodida) on the Iberian Peninsula. *Journal of Vector Biology*, 39, 238–48.

Boinas, F.S., Wilson, A.J., Hutchings, G.H., Martins, C., & Dixon, L.J. (2011). The persistence of African swine fever virus in field-infected ornithodoros erraticus during the ASF Endemic Period in Portugal. *PLoS One*, 6, e20383.

Bonds, M.H., Keenan, D.C., Rohani, P., & Sachs, J.D. (2010). Poverty trap formed by the ecology of infectious diseases. *Proceedings of the Royal Society B: Biological Sciences*, 277, 1185–92.

Borowsky, S.J., Rubenstein, L.V., Meredith, L.S., Camp, P., Jackson-Triche, M., & Wells, K.B. (2000). Who is at risk of nondetection of mental health problems in primary care? *Journal of General Internal Medicine*, 15, 381–8.

Burki, T.K. (2012). Ticks and Turkey. *The Lancet*, 380, 1897–8.

Burkot, T., Durrheim, D., Melrose, W., Speare, R., & Ichimori, K. (2006). The argument for integrating vector control with multiple drug administration campaigns to ensure elimination of lymphatic filariasis. *Filaria Journal*, 5, 10.

Cafferata, M.L., Bardach, A., Rey-Ares, L. et al. (2013). Dengue Epidemiology and Burden of Disease in Latin America and the Caribbean: A Systematic Review of the Literature and Meta-Analysis. *Value in Health Regional Issues*, 2, 347–56.

Carpenter, S., Wilson, A., & Mellor, P. (2009). Bluetongue virus and Culicoides in the UK: The impact of research on policy. *Outlooks in Pest Management*, 20, 161–4.

Chang, G.-J.J., Davis, B.S., Stringfield, C., & Lutz, C. (2007). Prospective immunization of the endangered California condors *Gymnogyps californianus*) protects this species from lethal West Nile virus infection. *Vaccine*, 25, 2325–30.

Chang, M.S., Hii, J., Buttner, P., & Mansoor, F. (1997). Changes in abundance and behaviour of vector mosquitoes induced by land use during the development of an oil palm plantation in Sarawak. *Transactions of the Royal Society of Tropical Medicine and Hygiene*, 91, 382–6.

Charette, M., Berrang-Ford, L., Llanos-Cuentas, E.A., Cárcamo, C., & Kulkarni, M. (2017). What caused the 2012 dengue outbreak in Pucallpa, Peru? A socio-ecological autopsy. *Social Science and Medicine*, 174, 122–32.

Chen, C.-C., Yeh, T.-L., Yang, Y.K. et al. (2001). Psychiatric morbidity and post-traumatic symptoms among survivors in the early stage following the 1999 earthquake in Taiwan. *Psychiatry Research*, 105, 13–22.

Cohen, J. (2016). The Race for a Zika Vaccine is On. *Science*, 351(6273), 543–44.

Cohen, S. & Williamson, G.M. (1991). Stress and infectious disease in humans. *Psychological Bulletin*, 109, 5.

de Rochars, M.V.B., Milord, M.D., Jean, Y.S. et al. (2004). Geographic distribution of lymphatic filariasis in Haiti. *American Journal of Tropical Medicine and Hygiene*, 71, 598–601.

Diaz, A. & Stewart-Ibarra, A.M. (2018). Zika Virus Infections and Psychological Distress Following Natural Disasters. *Future Virology*, 13(6), 379–83.

Dirección Nacional de Vigilancia Epidemiológica, n.d. http://www.salud.gob.ec/direccion- nacional-de-vigilancia-epidemiologica/ (accessed 0.22.17).

Dubois, L. (2012). *Haiti: The Aftershocks of History*. Metropolitan Books, New York.

El-Rahim, I.A., Asghar, A., Mohamed, A., & Fat'hi, S. (2016). The impact of importation of live ruminants on the epizootiology of foot and mouth disease in Saudi Arabia. *Revue Scientifique et Technique Epizootics*, 35, 2.

European Commission. (2012). GUIDANCE DOCUMENT to assist Member States or the implementation of the criteria for 'Vector Protected Establishments' for blue-tongue (No. SANCO/7068/2012 Rev 3), laid down in Annex II of Commission Regulation (EC) No 1266/2007 as amended by Commission Regulation (EC) No 456/2012 of 30 May 2012. Euopean Commission HEALTH AND CONSUMERS DIRECTORATE-GENERAL Directorate G- Veterinary and International affairs Unit G2 – Animal Health.

FAO/OIE, 1963. II - Seconde Conference de la Commission Permanente de l'OIE pour l'Europe. Lisbon, Portugal.

Farmer, P. (1996). Social inequalities and emerging infectious diseases. *Emerging and Infectious Diseases*, 2, 259–69.

Feng, J., Xia, Z., Zhang, L., Cheng, S., & Wang, R. (2016). Risk assessment of malaria prevalence in Ludian, Yongshan, and Jinggu Counties, Yunnan Province, after 2014 earthquake disaster. *American Journal of Tropical Medicine and Hygiene*, 94, 674–8.

Glaser, R., Rice, J., Sheridan, J., Fertel, R. et al. (1987). Stress-related immune suppression: Health implications. *Brain Behavior and Immunity*, 1, 7–20.

Gogin, A. & Kolbasov, D. (2013). African swine fever in the Russian Federation: Risk factors for Europe and beyond. *Empres Watch*, 28, 1–14.

Grillet, M.E., Hernández-Villena, J.V., Llewellyn, M.S. et al. (2019). Venezuela's humanitarian crisis, resurgence of vector-borne diseases, and implications for spillover in the region. *Lancet Infectious Diseases*, 19(5), 149–161.

Gubler, D.J. (1998). Dengue and Dengue Hemorrhagic Fever. *Clinical Microbiology Reviews*, 11(3), 480–96.

Gubler, D.J. (1997). Dengue and dengue hemorrhagic fever; its history and resurgence as a global public health problem. In *Dengue and Dengue Hemorrhagic Fever*, D.J. Gubler and G. Kuno eds., London, UK: CAB International, pp. 1–22.

Gubler, D.J. (1989). *Aedes aegypti* and *Aedes aegypti*-Borne Disease Control in the 1990s: Top Down or Bottom Up. *American Journal of Tropical Medicine and Hygiene*, 40, 571–8.

Gubler, D.J. & Meltzer, M. (1999). The impact of dengue/dengue hemorrhagic fever on the developing world. *Advances in Virus Research*, 53, 35–70.

Halstead, S.B. (2017). Dengvaxia sensitizes seronegatives to vaccine enhanced disease regardless of age. *Vaccine*, 35, 6355–8.

Harding, T.W., De Arango, V., Baltazar, J. et al. (1980). Mental disorders in primary health care: A study of their frequency and diagnosis in four developing countries. *Psychological Medicine*, 10, 231–41.

Harpham, T. & Molyneux, C. (2001). Urban health in developing countries: A review. *Progress in Development Studies*, 1, 113–37.

Hayden, M.H., Uejio, C.K., Walker, K. et al. (2010). Microclimate and human factors in the divergent ecology of *Aedes aegypti* along the Arizona, US/Sonora, MX border. *EcoHealth*, 7, 64–77.

Heydari, N., Larsen, D., Neira, M., Beltrán Ayala, E. et al. (2017). Household dengue prevention interventions, expenditures, and barriers to *Aedes aegypti* control in Machala, Ecuador. *International Journal of Environmental Research and Public Health*, 14, 196. https://doi.org/10.3390/ijerph14020196

Hoogstraal, H. (1979). Review Article1: The epidemiology of tick-borne crimean-congo hemorrhagic fever in Asia, Europe, and Africa 23. *Journal of Medical Entomology*, 15, 307–417.

Hotez, P.J. (2014). Neglected parasitic infections and poverty in the United States. *PLoS Neglected Tropical Diseases*, 8, e3012.

Hotez, P.J. (2011). Neglected infections of poverty in the United States of America. In: Institute of Medicine (US) Forum on Microbial Threats. The Causes and Impacts of Neglected Tropical and Zoonotic Diseases: Opportunities for Integrated Intervention Strategies. National Academies Press, Washington D.C., p. A8.

Jaramillo-Ochoa, R., Sippy, R., Farrell, D. et al. (2019). Effects of Political Instability in Venezuela on Malaria Resurgence at Ecuador-Peru Border, 2018. *Emerging and Infectious Diseases*, 25: 834–36.

Johnson, B.J., Ritchie, S.A., & Fonseca, D.M. (2017). The State of the Art of Lethal Oviposition Trap-Based Mass Interventions for Arboviral Control. *Insects*, 8, 1–16.

Kay, B. & Nam, V.S. (2005). New strategy against *Aedes aegypti* in Vietnam. *The Lancet*, 365, 613–17.

Kendall, C., Hudelson, P., Leontsini, E., Winch, P., Lloyd, L., & Cruz, F. (1991). Urbanization, dengue, and the health transition: Anthropological contributions to international health. *Medical Anthropology Quarterly*, 5, 257–68.

Kenneson, A., Beltrán-Ayala, E., Borbor-Cordova, M.J. et al. (2017). Social-ecological factors and preventive actions decrease the risk of dengue infection at the household-level: Results from a prospective dengue surveillance study in Machala, Ecuador. *PLoS Neglected Tropical Diseases*, 11, e0006150.

Krisher, L.K., Krisher, J., Ambuludi, M., Arichabala, A. et al. (2016). Successful malaria elimination in the Ecuador–Peru border region: Epidemiology and lessons learned. *Malaria Journal*, 15, 573.

Larson, H.J., Hartigan-Go, K., & de Figueiredo, A. (2019). Vaccine confidence plummets in the Philippines following dengue vaccine scare: Why it matters to pandemic preparedness. *Human Vaccines and Immunotherapy*, 15, 625–7.

Launiala, A. (2009). How much can a KAP survey tell us about people's knowledge, attitudes and practices? Some observations from medical anthropology research on malaria in pregnancy in Malawi. *Anthropology Matters*, 11, 1–13.

Lemoine, J.F., Desormeaux, A.M., Monestime, F. et al. (2016). Controlling Neglected Tropical Diseases (NTDs) in Haiti: Implementation strategies and evidence of their success. *PLoS Neglected Tropical Diseases*, 10, e0004954.

Leon, T.O. (2017). Personal Communication.

Lincoln, V.J., Page, P.C., Kopp, C. et al. (2015). Protection of horses against Culicoides biting midges in different housing systems in Switzerland. *Vet. Parasitol.*, 210, 206–14.

Luis, J., Martín, S., & Brathwaite-dick. O., (2007). La Estrategia de Gestión Integrada para la Prevención y el Control del Dengue en la Región de las Américas. *Reviews Panamericana de Salud Pública*, 21(1), 55–63.

McNeill, J.R. (2010). *Mosquito Empires: Ecology and War in the Greater Caribbean, 1620–1914*. Cambridge University Press, Cambridge, UK.

Mitchell-Foster, K., Ayala, E.B., Breilh, J., Spiegel, J., Wilches, A.A., Leon, T.O., & Delgado, J.A. (2015). Integrating participatory community mobilization processes to improve dengue prevention: An eco-bio-social scaling up of local success in Machala, Ecuador. *Transactions of the Royal Society of Tropical Medicine and Hygiene*, 109, 126–33.

Molyneux, D.H. (2012). The 'Neglected Tropical Diseases': Now a Brand Identity; Responsibilities, Context and Promise. *Parasites & Vectors*, 5, 23.

Morrison, A.C., Zielinski-gutierrez, E., Scott, T.W., & Rosenberg, R. (2008a). Defining challenges and proposing solutions for control of the virus vector *Aedes aegypti*. *PLoS Medicine*, 5, e68.

Nading, A.M. (2014). *Mosquito Trails: Ecology, Health, and the Politics of Entanglement*. University of California Press, Berkeley, California.

Nagao, Y., Thavara, U., Chitnumsup, P., Tawatsin, A., & Chansang, C. (2003). Climatic and social risk factors for *Aedes* infestation in rural Thailand. *Tropical Medicine and International Health*, 8(7), 650–9.

Naucke, T.J., Menn, B., Massberg, D., & Lorentz, S. (2008). Sandflies and leishmaniasis in Germany. *Parasitology Research*, 103, 65–8.

Nielsen, J. (2006). *Condor: To the Brink and Back—the Life and Times of One Giant Bird*. Harper Collins, New York.

Nuwaha, F. (2002). People's perception of malaria in Mbarara, Uganda. *Trop. Med. Int. Health*, 7, 462–70.

Paixão, E.S., Teixeira, M.G., & Rodrigues, L.C. (2018). Zika, chikungunya and dengue: The causes and threats of new and re-emerging arboviral diseases. *BMJ Global Health*, 3, e000530.

Paniz-Mondolfi, A.E., Tami, A., Grillet, M.E., Márquez, M., Hernández-Villena, J., Escalona-Rodríguez, M.A., Blohm, G.M., Mejías, I., Urbina-Medina, H., & Rísquez, A. (2019). Resurgence of vaccine-preventable diseases in Venezuela as a regional public health threat in the Americas. *Emerging and Infectious Diseases*, 25, 625.

Paton, T. (1863). The 'horse sickness' of the Cape of Good Hope. *Veterinarian*, 36, 489–94.

Paz-soldan, V.A., Yukich, J., Soonthorndhada, A. et al. (2016). Design and Testing of Novel Lethal Ovitrap to Reduce Populations of *Aedes* Mosquitoes: Community-Based Participatory Research between Industry, Academia and Communities in Peru and Thailand. *PLoS One*, 11(8), e0160386.

Peterson, R.K. (1995). Insects, disease, and military history. *American Entomologist*, 41, 147–61.

Rashed, S., Johnson, H., Dongier, P. et al. (2000). Economic impact of febrile morbidity and use of permethrin-impregnated bed nets in a malarious area II. Determinants of febrile episodes and the cost of their treatment and malaria prevention. *American Journal of Tropical Medicine and Hygiene*, 62, 181–6.

Reiter, P. & Newton, E.A.C. (1992). A Model of the transmission of dengue fever with an evaluation of the impact of ultra-low volume (ULV) insecticide applications on dengue epidemics. *American Journal of Tropical Medicine and Hygiene*, 47, 709–20.

Ryan, S.J., Carlson, C.J., Stewart-Ibarra, A.M. et al. (2017a). Outbreak of Zika Virus Infections, Dominica, 2016. *Emerging and Infectious Diseases*, 23, 1926–7.

Ryan, S.J., Lippi, C.A., Boersch-Supan, P.H. et al. (2017b). Quantifying seasonal and diel variation in Anopheline and *Culex* human biting rates in Southern Ecuador. *Malaria Journal*, 16, 479.

Ryan, S.J., Mundis, S.J., Aguirre, A. et al. (2019). Seasonal and geographic variation in insecticide resistance in *Aedes aegypti* in southern Ecuador. *PLoS Neglected and Tropical Diseases*, 13, e0007448.

Ryan, S.J., Rheingans, R., Amratia, P. et al. (2016). WASH Poverty Diagnostic; Poverty Risk Model Assessment: Ecuador. (Report commissioned by the 'Poverty Risk Models (PRM) for water, sanitation and health (WASH) project' for World Bank,). World Bank, Washington, D.C.

Salazar, M.A., Pesigan, A., Law, R., & Winkler, V. (2016). Post-disaster health impact of natural hazards in the Philippines in 2013. *Global Health Action*, 9, 31320.

Satterthwaite, D. (2003). The links between poverty and the environment in urban areas of Africa, Asia, and Latin America. *Annals of the American Academy of Political and Social Science*, 590, 73–92.

Scott-fiorenzano, J.M., Fulcher, A.P., Seeger, K.E., Allan, S.A., Kline, D.L., Koehler, P.G., Müller, G.C., & Xue, R. (2017). Evaluations of dual attractant toxic sugar baits for surveillance and control of *Aedes aegypti* and *Aedes albopictus* in Florida. *Parasites & Vectors*, 1–9.

Secretaria de Gestion de Risegos, (2016). Informe de situacion, No. 65: Terremoto 7.8 - Pedernales (No. 65). Secretaria de Gestion de Risegos, Quito, Ecuador.

Sheller, M. & León, Y.M. (2016). Uneven socio-ecologies of Hispaniola: Asymmetric capabilities for climate adaptation in Haiti and the Dominican Republic. *Geoforum*, 73, 32–46.

Shragai, T., Tesla, B., Murdock, C., & Harrington, L.C. (2017). Zika and chikungunya: Mosquito-borne viruses in a changing world. *Annals of the New York Academy of Sciences*, 1399(1): 61–77.

Smalley, C., Erasmus, J.H., Chesson, C.B., Beasley, & D.W. (2016). Status of research and development of vaccines for chikungunya. *Vaccine*, 34, 2976–81.

Stanaway, J.D., Shepard, D.S., Undurraga, E.A., Halasa, Y.A., Coffeng, L.E., Brady, O.J., Hay, S.I., Bedi, N., Bensenor, I.M., Castañeda-Orjuela, C.A., Chuang, T.-W., Gibney, K.B., Memish, Z.A., Rafay, A., Ukwaja, K.N., Yonemoto, N., & Murray, C.J.L. (2016). The global burden of dengue: An analysis from the Global Burden of Disease Study 2013. *Lancet Infectious Diseases*, 16, 712–23.

Steketee, R.W. & Eisele, T.P. (2009). Is the scale up of malaria intervention coverage also achieving equity? *PLoS One*, 4(12), e8409.

Stewart Ibarra, A.M., Luzadis, V.A. et al. (2014). A social-ecological analysis of community perceptions of dengue fever and *Aedes aegypti* in Machala, Ecuador. *BMC Public Health* 14, 1135.

Stewart Ibarra, A.M., Ryan, S.J. et al. (2017). A spatio-temporal modeling framework for *Aedes aegypti* transmitted diseases in the Caribbean. United States Agency for International Development's (USAID) Programmeme for Building Regional Climate Capacity in the Caribbean (BRCCC Programmeme).

Stewart-Ibarra, A., Hargrave, A., Diaz, A. et al. (2017). Psychological distress and Zika, dengue and chikungunya symptoms following the 2016 earthquake in Bahía de Caráquez, Ecuador. *International Journal of Environmental Research and Public Health*, 14, 1516.

Stewart-Ibarra, A.M. & Lowe, R. (2013). Climate and non-climate drivers of dengue epidemics in southern coastal Ecuador. *American Journal of Tropical Medicine and Hygiene*, 88, 971–81.

Stewart-Ibarra, A.M., Muñoz, Á.G., Ryan, S.J. et al. (2014). Spatiotemporal clustering, climate periodicity, and social-ecological risk factors for dengue during an outbreak in Machala, Ecuador, in 2010. *BMC Infectious Diseases*, 14, 610.

Stewart-Ibarra, A.M., Romero, M., Hinds, A.Q. et al. (2019). Co-developing climate services for public health: Stakeholder needs and perceptions for the prevention and control of Aedes-transmitted diseases in the Caribbean. *PLoS Neglected Tropical Diseases*, 13(10), e0007772.

Stewart-Ibarra, A.M., Ryan, S.J., Beltran, E., Mejia, R., Silva, M., & Munoz, A. (2013). Dengue vector dynamics (*Aedes aegypti*) influenced by climate and social factors in Ecuador: Implications for targeted control. *PLoS One*, 8.

Stewart-Ibarra, A.M., Ryan, S.J., Kenneson, A. et al. (2018). The burden of dengue fever and chikungunya in southern coastal Ecuador: Epidemiology, clinical presentation, and phylogenetics from the first two years of a prospective study. *American Journal of Tropical Medicine and Hygiene*, 98, 1444–59.

Stokstad, E. (2017). Deadly virus threatens European pigs and boar. *Science*, 358, 1516.

Stoyan, A.T., Niedzwiecki, S., Morgan, J., Hartlyn, J., & Espinal, R. (2016). Trust in government institutions: The effects of performance and participation in the Dominican Republic and Haiti. *International Political Science Reviews*, 37, 18–35.

Sumilo, D., Asokliene, L., Bormane, A., Vasilenko, V., Golovljova, I., & Randolph, S.E. (2007). Climate change cannot explain the upsurge of tick-borne encephalitis in the Baltics. *PLoS One*, 2, e500.

United Nations Development Program. (2018). Human Development Indices and Indicators; 2018 Statistical Update. Geneva: United Nations Development Program.

Watson, J.T., Gayer, M., & Connolly, M.A. (2007). Epidemics after natural disasters. *Emerging and Infectious Diseases*, 13, 1–5.

WHO (2009). *Dengue: Guidelines for diagnosis, treatment, prevention and control.* Geneva: World Health Organization.

Wilkinson, A., Parker, M., Martineau, F., & Leach, M. (2017). Engaging 'communities': Anthropological insights from the West African Ebola epidemic. *Philosophical Transactions of the Royal Society B*, 372, 20160305.

Wilson, A.J., Ribeiro, R., & Boinas, F. (2013). Use of a Bayesian network model to identify factors associated with the presence of the tick *Ornithodoros erraticus* on pig farms in southern Portugal. *Preventative Veterinary Medicine*, 110, 45–53.

Wiseman, V., McElroy, B., Conteh, L., & Stevens, W. (2006). Malaria prevention in The Gambia: Patterns of expenditure and determinants of demand at the household level. *Tropical Medicine and International Health*, 11, 419–31.

World Health Organization (2018). World Malaria Report 2018. Geneva: WHO.

World Health Organization (2012). Accelerating work to overcome the global impact of neglected tropical diseases: A roadmap for implementation. Geneva: WHO.

Zhang, W., Liu, H., Jiang, X., Wu, D., & Tian, Y. (2014). A longitudinal study of posttraumatic stress disorder symptoms and its relationship with coping skill and locus of control in adolescents after an earthquake in China. *PLoS One*, 9, e88263.

Zika Gazette SE 38–2017, *Enfermedades Transmitidas por Vectores. National Subsecretary for Public Health Surveillance*. Ecuador: Ministry of Health of Ecuador.

CHAPTER 15

Vector Control, Optimal Control, and Vector-Borne Disease Dynamics

Michael B. Bonsall

15.1 Introduction

Vector-borne diseases have huge social, economic, and geopolitical effects.

Billions of people live with vector borne diseases (VBDs). Approximately half the world's population are at risk of malaria and dengue infections. The global burden of these diseases, in terms of mortality and morbidity, is huge. For malaria, in the 1990s it was projected that, along with other vector borne diseases, incidence would fall (ranked eleventh in 1996 projected to be twenty-fifth by 2020—Murray & Lopez 1996). This has partly been realised; to date, malaria deaths (of, mainly, children under the age of five) are around 500,000 per year and the burden of malaria has declined by more than 30 percent over the last decade (GBD 2017). Targeted control, better healthcare, and appropriate governance are testament to the success of this decline in malaria (Feachem et al. 2019). But there remains more to do to control this preventable VBD. In contrast, dengue infections have increased by 65 percent over the last decade (GBD 2017). The lack of substantial integrated vector management programmes, complex epidemiology, and the spread of invasive mosquitoes that carry flaviviruses have all contributed to the increased incidence of dengue. Dengue is the fastest growing VBD and new, innovative control methods are needed.

Economic development has also been linked to the spread of VBDs (Alsan 2015). The limited economic expansion and availability of agricultural surplus in precolonial Africa has been linked to the spread and prevalence of trypanosomiasis by tsetse (*Glossina*). Sleeping sickness (and its domestic animal equivalent—nagana) limited livestock farming expansion, delayed the formation of political infrastructures, and restricted human population growth across sub-Saharan Africa. Still, today, in the twenty-first century, human african trypanosomiasis (HAT) continues to influence the socio-economics and development of communities and individuals across Africa (Bukachi et al. 2017).

Targeting invertebrate vectors (such as mosquitoes, sandflies, tsetse flies, ticks, and triatomine bugs) that transmit pathogens has long been a focus of managing VBDs. Approaches to vector control that emerged in the early twentieth century focused on the application of mathematical models to understanding disease spread (see Reiner & Smith—Chapter 2 of this volume). Ronald Ross (1911, 1915, 1916a,b; Ross & Hudson 1916) showed how mathematical models could be used to understand the spread of malaria and developments by Lotka, Sharpe, and MacDonald all illustrated that reducing vector bites would be critical to reducing disease spread. Contemporary assessments of vector-borne disease dynamics have repeatedly

Michael B. Bonsall, *Vector Control, Optimal Control, and Vector-Borne Disease Dynamics* In: *Population Biology of Vector-Borne Diseases.*
Edited by: John M. Drake, Michael B. Bonsall, and Michael R. Strand: Oxford University Press (2021). © Michael B. Bonsall.
DOI: 10.1093/oso/9780198853244.003.0015

confirmed this, through expressions such as the reproductive number or invasion potential of a VBD (Smith et al. 2012):

$$-\frac{(\mu_h + \mu_v)}{2} + \sqrt{(\mu_h + \mu_v)^2 + 4\frac{V}{H}a^2 b\mu_h\mu_v} > 0.$$

These expressions highlight the importance of bite rate, a, as a critical (nonlinear) parameter in disease spread and control; but also emphasize that vector densities (V), vector mortality (μ_v), host recovery (μ_h), host density (H), and the interaction of the parasite within the vector and host (b) are all factors in VBD control and spread (e.g. Aron & May 1982; Smith et al. 2012).

Successful vector control, therefore, is likely to be effective if based on a detailed understanding of the ecological context of VDBs. Vector ecology is central to vector control. In this chapter, I review contemporary approaches to genetics-based methods for vector control. The chapter begins with a brief overview of the novel developments in genetics-based approaches for mosquito vector control and then focuses on the critical aspects of vector ecology (dispersal, intraspecific competition, interspecific competition) necessary for effective vector control. The importance of economic (optimal control) decisions on ecological and epidemiological outcomes is highlighted as critical for achieving integrated vector control together with the policy implications of genetics-based vector control approaches.

15.2 Genetics-Based Approaches to Vector Control

Chemical control interventions (e.g. betnets, indoor residual spraying, larvicides) provide a mainstay for vector control. While these are well utilized, classic and more contemporary genetics-based methods also provide a set of tools for vector control. Building on and extending on from the sterile insect technique (SIT) (Knipling, 1955), these methods use knowledge of population genetics and modern molecular methods to engineer species to achieve sustained (or limited) population control (Alphey & Bonsall, 2014a).

Classic SIT uses the mass rearing of vectors, inducing genetic mutations using radiation (into germlines to induce effective sterility), and releasing these insects such that any matings between these modified vectors and their wildtype counterparts do not produce viable offspring. This leads to a reduction or suppression in the population size of the vector or pest insect. While proof of principle for SIT was achieved for agricultural pests (Knipling 1955), there has been and continues to be interest in using this approach for vector control (e.g. see Dame et al. 2009 for a review of SIT).

Using principles from the population dynamics outcomes of SIT, modern genetics-based methods are building on this approach and can be broadly divided into two classes of population-level control: those that are self-limiting or those that are self-sustaining. Self-limiting methods are expected to provide restricted limited temporal and/or spatial control (Table 15.1). Whereas self-sustaining methods will achieved long-lasting approaches to vector control (Table 15.1).

Self-limiting technologies act over limited ranges of time and/or space and without continual re-introduction, the modified insects fade out. SIT is a self-limiting technology. Recent advances in molecular biology have allowed refinements of this approach for vector control. Dominant lethal synthetic genes have been developed and inserted into a range of insect pests and vectors (Thomas et al. 2000; Scheletig et al. 2007). These genetic constructs function by activating and inducing lethality at a particular stage in the life-cycle of the insect. Developed for *Aedes aegypti* (Phuc et al. 2007), dominant lethal genetic constructs have been tested in field trials (e.g. Carvalho et al. 2015) as a tool for population suppression and dengue disease mitigation.

Mathematical models provide tools to evaluate the population-level outcomes of different genetics-based control interventions. While these tools are reviewed throughout this chapter, from an ecological perspective a simple model that captures the dynamics of a self-limiting, population-suppression technology is:

$$\frac{dN}{dt} = r\frac{N}{(N+A)}f(N)N$$

where r is the population growth rate, N is the population size of wildtype insects, A is the density of released modified (SIT) insects, and $f(N)$ is a function describing the density feedback effects operat-

Table 15.1 Genetics-based methods of modification for vector control, the mechanisms of action and the expected outcome in terms of vector control.

Modification method	Mechanism	Expected population-level outcome
Radiation-based SIT	Self-limiting inudative approach. Offspring production is disrupted through sterilisation of sperm in *Aedes* mosquitoes.[1]	No viable offspring production.
Genetics-based SIT	Self-limiting, inundative method. Larval survival is disrupted by a dominant lethal gene in *Aedes* mosquitoes.[2]	Increased larval mortality.
Gene Drive	Self-limiting method. X-shredder gene in male *Anopheles* leads to only Y-bearing gametes. Leads to male only offspring. Self-limiting construct as modified gene engineered to be only carried on an autosome (not an allosome/sex chromosome).[3]	Biased sex ratio.
Underdominance	Self-sustaining, high threshold drive. Frequency dependent gene inheritance leading to heterozygote disadvantage achieved through translocating chromosomes in *Aedes* mosquitoes.[4]	No viable offspring production.
MEDEA	Self-sustaining, high threshold drive. MEDEA (Maternal Effect Dominant Embryonic Arrest). Synthetic genes that prevent embryonic development. Requires gene expression of a toxin-antidote system to be effective for vector control.[5]	No viable offspring production.
Wolbachia	Self-sustaining, high threshold drive. Not a gene modification *per se* but introduction of a specific intracellular bacteria (WMelPop strain) in *Aedes* mosquitoes reduces longevity, fecundity and egg survival.[6]	Affects adult and larval densities.
Gene Drive	Self-sustaining, low threshold drive. Nuclease-based gene that cuts *doublesex* that controls sex development in *Anopheles*. Disrupts female phenotype and fertility. Use CRISPR-Cas system to spread in non-Mendelian ways.[7]	No viable offspring production.

Sources: [1] Bellini et al. (2007), [2] Phuc et al. (2007), [3] Galizi et al. (2014), [4] Lorimer et al. (1972), [5] Chen et al. (2007), [6] McMeniman et al. (2009), [7] Kyrou et al. (2018).

ing, implicitly, through juvenile stages. Critical in formulating this population model is $(N/(N+A))$, a term describing mating disruption (at the population rather than behavioral level) by the modified insects (A). The *per capita* change in wildtype population size $((1/Nd)N/dt = r(N/(N+A)f(N))$ illustrates that as the density of modified insects increases (A), there is a critical growth threshold below which the population declines to extinction (an Allee threshold) (Fig. 15.1). It is also important to note that the density of modified insects (A) may not be constant through time and may vary due to species attributes (e.g. longevity) ecological effects (e.g. competition, predation) and/or persistence of the programme of releases. Temporally varying densities of modified insects alter the shape and magnitude of the Allee effect (Fig. 15.1).

Self-sustaining technologies act over more extensive periods of time and/or space. Originally developed in the 1940s (Serebrovsky 1940), these approaches build on ideas that non-Mendelian patterns of inheritance, that bias the frequency of particular genes in the gametes, can be used to spread costly traits through populations. In natural systems, these selfish genetic elements bias their own inheritance and the associated patterns of spread have been well studied (Burt & Trivers 2008). For instance, a transposable element, the so-called P-element in *Drosophila melanogaster* is a well-known example of a selfish genetic element that rapidly spread through all natural *D. melanogaster* populations during the latter part of the twentieth century (Burt & Trivers 2008). The theory on these selfish genetic elements is also well advanced (Hamilton 1967; Hastings 1994; Burt & Trivers 2008; Charlesworth & Charlesworth 2010) and this provides the conceptual basis for the application of these biased inheritance, self-sustaining approaches to the genetic control of mosquitoes.

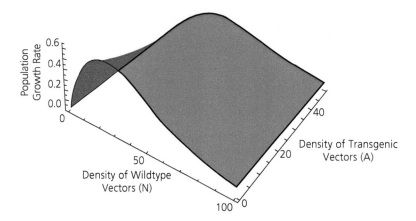

Figure 15.1 Changes in population growth ($(1/N)dN/dt$) of wildtype vectors under SIT-type control. Changes in wildtype population sizes are affected by the strength of intraspecific competition, here for illustration determined by $\dfrac{1}{(1+(0.02*N)^3)}$. As population density of wildtype vectors increases (N increases), population growth rate is expected to decrease towards zero as the population reaches a carrying capacity. Changes in the density of released insects (A) determines the magnitude strength of the Allee effect and the Allee threshold (where population growth rate turns downwards at low densities of wildtype vectors). More transgenic vectors (increases in A) restrict overall population growth of the wildtype vector.

Original work by Curtis demonstrated how a genetic pattern of inheritance where heterozygotes are less fit than the homozygotes (underdominance) could be used for vector control (Curtis 1968; Curtis & Hill 1971). Mathematically, underdominant systems can be described by a simple diploid genetic recursion model (Curtis 1968; Gillespie 2010):

$$p_{t+1} = \frac{p_t^2 + (1-t)p_t(1-p_t)}{p_t^2 + (1-t)2p_t(1-p_t) + (1-p_t)^2}$$

where p is the frequency of the dominant (wildtype) allele and t is the strength of selection against the heterozygote. This genetic system is characterized by an unstable internal equilibrium when $p = 0.5$ and only the two extremes (when $p = 1$ (all dominant) or $p = 0$ (all recessive)) are stable states. The trick for adapting this idea to a vector control system was to tip the system towards the heterozygote state by engineering insects such that progeny (heterzygotes) would not be viable (Curtis 1968). Based on chromosome translocation techniques, Curtis argued that while translocation homozygotes (say *TT*) (and homozygote wildtypes (say ++)) might be fully fertile, offspring from matings with homozygote wildtypes (*T*+) would not be viable. As a proof of principle, this approach was developed by Lorimer et al. (1972) for *Aedes aegypti* vector control. Using chromosome translocations,

Lorimer et al. (1972) generated viable homozygotes based on a sex-linked translocation involving chromosomes 1 and 3 and an autosomal translocation involving chromosomes 2 and 3. The sex-linked translocation homozygotes were trialed for genetic stability (Rai et al. 1973) and rolled out in larger field experiments (Lorimer et al., 1976; McDonald et al. 1977; Petersen et al. 1977; Lorimer 1981) through the 1970s as a method for *Aedes* control in Kenya.

Modern methods of self-sustaining genetic technologies have built on this approach (Table 15.1) to investigate a wide range of different forms of non-Mendelian patterns of inheritance for vector control (Sinkins & Gould 2006); and molecular methods provide a wide range of options for both sustaining and limiting spread of engineered genetic constructs to achieve necessary levels of control and/or to meet necessary regulatory requirements.

Broadly, these self-sustaining technologies are known as 'gene drives' and work by converting wildtype homozygotes into modified (recessive) homozygotes carrying (lethal) traits. This non-Mendelian pattern of spreading costly traits through a population (that would otherwise be lost due to being recessive) can achieve population suppression depending on the action of the introduced trait. Gene drive technologies can be grouped into high threshold or low threshold drives depending on the critical frequency threshold of released modified insects needed to ensure the drive is able to spread (Table 15.1).

Examples of high threshold drives include underdominance, maternally inherited MEDEA and *Wolbachia*. Mating incompatibilities and extreme fitness differences determine whether these high threshold gene drives will spread successfully (e.g. Sinkins & Gould 2006; Chen et al. 2007; Akbari et al. 2014; Oberhofer et al. 2019). Low threshold drives, such as homing endonuclease genes (HEGs), or CRISPR-Cas based engineered synthetic genes are expected to spread under very low frequency releases. Each of these genetics-based technologies is currently under development as tools for vector control (Table 15.1).

The principles of how gene drives systems work can be studied using mathematical models. The dynamics of vector control using self-sustaining technologies can be described by:

$$\frac{dN_i}{dt} = \frac{rv_i(t-\tau)\phi_i}{1+(\sum_{j=1}^{3}a_jv_j(t-\tau)\phi_j)^b} - \mu N_i(t)$$

where N_i represents the population density of different genotypes v_i, r is the population intrinsic rate of increase, τ is a developmental lag and μ is the density-independent population death rate. Density dependence operates on genotype births (modulated by fitness effects associated with the different genotypes, ϕ_i) and is determined by threshold (a_i) and magnitude b parameters (Alphey & Bonsall 2014b). Population genetic analysis (Deredec et al. 2008; Alphey & Bonsall 2014b) reveals that (i) the frequency of the gene drive depends on the relative cost of the genetic disruption (0 → no effect; 1 → fully lethal), (ii) the effects of the fitness costs on the heterozygotes, and (iii) the efficacy of the gene drive at biasing its own inheritance.

If the gene drive is high enough to outweigh fitness cost effects, then the engineered construct will spread to fixation and wildtype insects will be driven extinct. Phenotypic properties of the gene drive and the efficacy of the drive determines alternative population genetic outcomes where a stable polymorphism occurs between the drive and the wildtype insects, the drive goes extinct or the outcome is unstable (Fig. 15.2). While developing richer understanding of the genetic potential of these technologies, appreciating the details of the

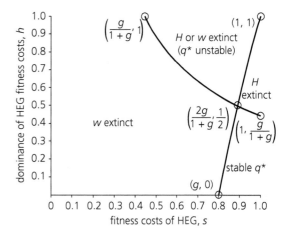

Figure 15.2 Population genetic outcomes for gene-drive constructs for vector control. The population genetic outcome depends on the phenotypic fitness properties of the drive (s, h) and the homing rate (g). The graph shows the regions of parameter space defined by s (the gene drive fitness penalty) and h (the dominance of that fitness cost), plotted for $g = 0.8$.

Source: from Alphey & Bonsall 2014b.

ecology of vectors and their environments is critical to determining whether the tools will achieve success or not.

15.3 Ecology of Vector Control

Whilst we might be able to achieve and develop highly efficacious genetic constructs for vector control, the success of these biotechnological innovations is only as good as their performance once released in an integrated vector management control programme. The success of these technologies relies on a rich and detailed understanding of vector ecology. Ecology is focused on understanding the distribution and abundance of organisms, and their relationship with the environment.

At a basic level, the population ecology of vector control is determined by the net effects of births, deaths and dispersal and the ecological attributes of behaviors (such as dispersal, assortative mating) on within-species competition and between-species. All of these factors may impact on the success of vector control programmes.

Dispersal is a critical behavior affecting the spatial and temporal distribution of vectors. Dispersal affects vector dynamics at different spatial scales and underpins all aspects of vector control and spread. Within patches, dispersal is critical to the distribution of vector abundance and the effectiveness of vector control. For instance, Manoranjan & van den Driessche (1986) modelled the efficacy of SIT using a reaction (population dynamics)—diffusion (spread) model for mosquitoes. They concluded that the number of modified insects required for population elimination was dependent on the demographic parameters of the mosquito population, the dimension of the spatial area and the initial population distribution. Building on this, Ferreria et al. (2008) investigated discrete-spatial diffusion between patches supporting sterile and wild type vectors. Under this scenario, control for spatially heterogeneous vector distributions is difficult to achieve as patches of high vector densities can escape control. Yakob & Bonsall (2009) developed a more spatially explicit model to investigate the optimal timing and the effects of sex-specificity of lethal transgenes on control. Optimal release strategies depended on aspects of population growth, mosquito population structure, the intensity of intraspecific competition (see below) and dispersal. High rates of mosquito dispersal reduce the effectiveness of genetics-based methods. In a more comprehensive study, Legros et al. (2012) applied the 'skeeter buster' model (Magori et al. 2009)—a stochastic, spatially explicit model of *Aedes aegypti*—to understand the spatial effects of daily dispersal on vector suppression. Population elimination is only feasible in small geographical areas unless the habitat is uniform and releases of sterile insects is also spatially uniform across the habitat.

With all of this it is critical to understand how mosquitoes respond to control interventions. Spatial spread is the key element in the application of conventional and genetics-based methods for vector control.

To understand the small scale consequences of control under different dispersal conditions, integro-difference equations (IDEs), which are discrete in time and continuous in space and incorporate a dispersal kernel for the spatial distribution of vectors, provide an appropriate framework (Kot & Schaffer 1986; Neubert et al. 1995; Zhou & Kot 2011; Reimer et al. 2016; Kura et al. 2019) in which vector

population growth dynamics and vector dispersal dynamics can be separated in time.

IDEs allow a range of dispersal conditions to be explored and generally take the form:

$$N_{t+1}(x) = \int_{\omega} k(x,y) f(N_t(y); y) dy$$

where $N(x)$ is the number of vectors at spatial location x, $k(x, y)$ is a probability density function and represents the probability that an individual starting at point y and settles at point x in space in the next time step representing the dispersal kernel, and $f(N(y))$ represents the population dynamics processes at spatial location y. Under SIT control, an IDE framework (Kura et al. 2019) is:

$$N_{t+1}(x) = \int_{\omega} \frac{1}{1 + r_t(y)N^* / N_t} k(x,y) f(N_t(y); y) dy.$$

where $\dfrac{1}{1 + r_t(y)N^* / N_t}$ is the proportion of matings disrupted by sterile insect releases (in proportion to wild type insects (N^*/N_t) at equilibrium, N^*). Using this framework, Kura et al. (2019) showed that the (optimal) control of vectors depends on the spatial distribution and that, due to the redistribution of wildtype mosquitoes, local eradication is often difficult to achieve.

Beyond the small scale, spatial heterogeneity at a regional (say village) scale influences vector control outcomes. Within villages (differences between dwellings), dispersal heterogeneity in relation to key environmental features (such as proximity to breeding sites) affects the spatial variation in vectorial capacity (Cano et al. 2006). Vector dispersal behavior is a critically important factor in determining the outcome of a genetics-based approach to control. For instance, *Aedes aegypti* is a relatively short-dispersing species with a large percentage of individuals living within a single house or moving between neighboring houses very infrequently (Harrington et al. 2005; Hemme et al. 2010). In contrast, *Anopheles gambiae* disperses much more widely (Taylor et al. 2001; Thomson et al. 1995; Dao et al. 2014). Across simple networks, differences in average dispersal distance and the levels of dispersal determine the spatial extent of genetics-based vector control programmes. For slowly dispersing species (such as *Aedes*), local elimination and/or eradication may be possible as repopulation by wildtype vectors is expected to be too weak. Spatial containment of genetics-based control methods thus depends critically on the combination of fitness costs imposed by the control technology and the vector's dispersal range (Khamis et al., unpublished); it is expected that spatial containment using genetic control technologies may be easier in *Aedes* than *Anopheles*.

At a broader spatial scale, non-random distribution of vectors across landscapes can hinder the success of vector control programmes (Yakob et al. 2008a,b) as more densely populated patches may not receive sufficient densities of genetically modified insects required for control to be successful (Barclay 1982). However, patch connectivity and landscape structures are significant to this between-patch variability in vector control. Network models of vector metapopulations suggest that understanding the coverage proportion (propensity of released modified insects to inhabit patches occupied by wild types) is central to predicting vector control outcomes (Yakob et al. 2008b). Equal establishment of modified vectors in highly clustered sets of populations might be less likely to occur. The spread of modified vectors across clustered networks might take longer than in evenly distributed metapopulations, as adjacent populations are much more likely to have already been colonised. Increasing dispersal rates can mitigate this establishment as coverage proportions tend to match the spatial configuration of the vector populations (Yakob et al. 2008b). Targeting highly connected populations therefore might provide limited improvement in vector control unless the modified released vectors are able to reach a critical coverage proportion. Without adequate attention to the spatial structure, the coverage proportion and the dispersal biology of the vector, control through the use of transgenics will not be expected to be successful.

Equally important as vector dispersal is the movement of humans (Martens & Hall 2000). As mosquitoes have reasonably limited dispersal (limited to kilometers), human movement can be an important factor in the spread (e.g. Tatem et al. 2006) and control (e.g. Hendron & Bonsall 2016) of vector-borne diseases. Movement of individual hosts alters the relative density of susceptible and infected people and this can have important consequences for vector

and disease control. Changes in population sizes can affect the spread of disease. For instance, R_0 calculations are based on a ratio of vectors to hosts and if host population size fluctuates, R_0 expressions for vector-borne disease spread are more complex (Smith et al. 2012; Hendron & Bonsall 2016). Across spatial networks where towns and cities are of varying sizes, the probability of mosquito bites may be reduced and vector borne disease transmission may be lower through herd immunity generated by human movement (Hendron & Bonsall 2016). This suggests that control strategies that focus on towns may be potentially more efficient than city-only control interventions.

Density dependence is the fundamental ecological phenomenon that prevents populations from increasing in an unbounded way. We lack even a basic understanding of density-dependence in the majority of insect vectors, particularly those that spread diseases with high public health burdens. Those studies that have attempted to unravel density dependence in, for instance, *Aedes* and *Anopheles*, remain rather deficient; limited by poor design, inadequate sample sizes and misunderstandings of the concepts of density dependence. Given the fundamental importance of intraspecific competition in determining population-dynamic outcomes, it is imperative that greater empirical attention is given to this set of ecological processes and mechanisms.

Detecting density dependence (in any ecological population) is often difficult. For instance, the rank importance of density dependence with respect to other ecological processes such as density independent biotic and abiotic processes varies across time and space. The tacit assumption that density dependent processes are the same across the whole spatial and temporal range of a vector is a convenient fiction. Appreciating that density-mediated effects might be vague (Strong 1986) and/or variable is essential in determining the potential success of vector control.

The population dynamic consequences of genetics-based vector control are critically determined by the mechanism, type and strength of density dependence. Analysis and simulation of mathematical models highlight that the interplay between the timing of ecological processes (such as those related to/resulting in density dependence) and the genetic-induced lethality or mortality can affect the efficacy of control.

Alphey & Bonsall (2014b) investigated the role of genetics and competition on successful *Anopheles* control. Both the type of intraspecific competition and its magnitude could be strain specific (such that transgenic insects have different resource requirements or competitive abilities). Using a general form of density dependence (Maynard Smith & Slatkin 1973): $f(N) = 1 / (1 + (aN)^b)$ where b determines the type of intraspecific competition (when $b = 1$, intraspecific competition is compensating; $b > 1$ competition is overcompensating) and $1/a$ is a threshold density above which the density dependent processes start to operate ($0 < a <= 1$), different dynamical outcomes might be anticipated to occur.

If the genetic effects operate before the ecological process of competition, particularly if the competition is highly overcompensating, this can lead to vector populations increasing in the presence of control (Fig. 15.3a). This idea that increasing mortality can increase population sizes is not new (e.g. Rogers & Randolph 1984; Yakob et al. 2008a; Abrams 2009) and operates by mortality alleviating (or reducing) the effects of density dependence and allowing the population to achieve growth close to or at its maximum rate. By appreciating the natural history and ecology of the vector, timing the effects of genetic interventions to operate after the effects of the intraspecific competition is mostly likely to improve control success (Fig. 15.3b) even if the dynamics are highly overcompensatory or cyclic. As vector control is successful and a species declines towards extinction, it is the consequences of the Allee effect that are mostly likely to predominate, not factors associated with the lack of wildtype intraspecific competition. As noted, simple population models (e.g. Fig. 15.1) illustrate that it is the relative densities of wildtype and modified strains that determine the success of vector eradication and elimination programmes.

Interspecific competition is a major force structuring ecological communities (Hardin 1960) and is predicted to lead to the exclusion of species that share similar niches. Coexistence is expected only when the effects of intraspecific competition are stronger (act sooner) than the effects of interspecific competition (e.g. Begon et al. 2006). Underpinning

these broader processes are a range of specific mechanisms including niche partitioning, behavioral segregation and aggregation, and life-history differences (e.g. Begon et al. 2006).

The population dynamics of interspecific competition under genetics-based vector control have been investigated (Bonsall et al. 2010; Paton & Bonsall 2019). A framework exploring competition and vector control is a straightforward extension of the Lotka–Volterra competition model. If N_i is the vector species being controlled with genetically modified insect releases then general vector interspecific competition dynamics (extending Dye 1984; Bonsall et al. 2010) can be described by:

$$\frac{dN_i}{dt} = \rho_i N_i(t - \tau_i)\left(1 - \alpha_i N_i(t - \tau_i)E_i - \beta_{ij}N_j(t - \tau_j)E_j\right)$$
$$- \mu_i N_i(t)$$

where N_i and E_i are the adult and egg densities, respectively. α_i and β_{ij} are the strengths of intraspecific and interspecific competition, respectively, τ_i is the developmental time lag, and ρ_i and μ_i are the birth and death rates of the ith species (Bonsall et al. 2010).

The effects of vector control are described by reductions in net birth rate:

$$\frac{N_i(t - \tau_i)}{N_i(t - \tau_i) + IN_i^*}$$

where I is the overflooding rate of genetically modified vectors (based on the pre-control equilibrium levels of the wildtype vector—N_i^*).

Isoclines (boundaries between positive and negative population growth) can be used to determine the outcome of interspecific competition. Under Lotka–Volterra competition, these isoclines are linear and only if the intraspecific effects of competition outweigh the interspecific effects do species coexist. Vector control introduces additional nonlinear effects on net population growth and this affects both the shape of the isoclines and the relative balance of intraspecific versus interspecific competitive effects (Fig. 15.4). Vector control leading to non-linear patterns in net growth (and the shape of the isoclines) introduce the potential for multiple coexistence points (both stable and unstable) and can generate the potential for novel interspecific competitive interactions (Fig. 15.4).

As vector control programmes could lead to a range of stable and unstable equilibria, this also has

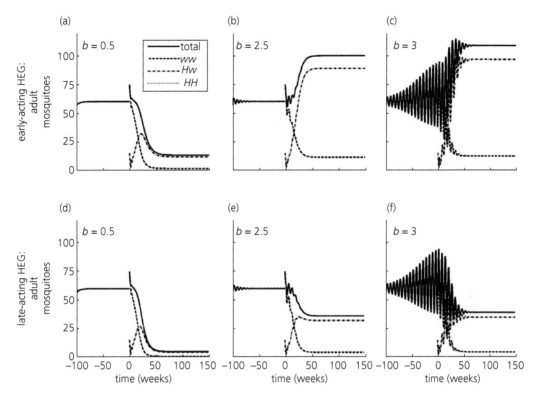

Figure 15.3 Dynamical behavior of gene-drive constructs for vector control. The gene-drive knockout is recessive lethal ($s = 1$, $h = 0$, homing rate $g = 0.8$). Heterozygous insects are released at time $t = 0$, fifteen per host (i.e. + 25 percent of the natural population equilibrium sixty per host). The number of adults (w—wildtype, H—HEG) over time is shown for each genotype (ww: dashed line, Hw: dashed-dotted line, HH: dotted line, here always zero) and in total (solid line). Strength of density dependence b is 0.5 (weak, undercompensating, a,b), 2.5 (underdamped before release, c,d), or 3 (overcompensating, oscillating before release, e,f), the HEG is early acting (a,c,e) or late-acting (b,d,f).

Source: from Alphey & Bonsall 2014b.

the potential to generate a wide range of transient dynamics. Transient dynamics have important implications for community ecology (Hastings 2001) particularly over ecologically relevant temporal and spatial scales. Transient dynamics are generated by saddle points (unstable equilibria), difference in life histories between competing species and/or time-delayed population-level effects. These time-delays and non-linear competitive interactions lead to transient and/or aperiodic dynamics in vector species interactions (Fig. 15.5).

Interaction between vectors can mediate interspecific effects in more diffusive ways. *Aedes aegypti* and *Aedes albopictus* are closely related vectors where reproductive interference can be an important determinant of interspecific effects. Competition at the larval stage for food and/or space is an obvious point in the life cycle for intra- and interspecific

interactions to operate in a density-dependent way on mortality and survival (Juliano 2009, 2010). However, interspecific interactions between vectors might also affect fecundity through reproduction interference; this occurs when heterospecifics engage in mating activities that reduce the fitness of one (or both) species (Gröning & Hochkirch 2008). In *Aedes* mosquitoes, reproductive interference can lead to non-viable interspecific matings between *Aedes aegypti* and *Aedes albopictus* (Tripet et al. 2011); *Aedes albopictus* females which mismate with *Aedes aegypti* males often have the potential to re-mate. However, mismatings between *Aedes albopictus* males and *Adedes aegypti* females leads to those females being refractory to further mating attempts (Tripet et al. 2011).

This mechanism of interspecific interactions has implications for vector coexistence (Paton &

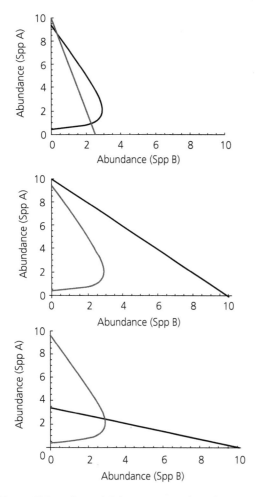

Figure 15.4 Isocline analysis for vector competition under late-acting, self-limiting (RIDL) control for different competitive interactions. Self-limiting vector control introduces non-linearities into the shape of the isoclines that affect the potential for coexistence. Different competitor interactions can lead to (a) stable and unstable equilibria, (b) no coexistence—exclusion of focal control vector, or (c) unstable equilibrium (saddle point). [Parameter values: $P_a = 5$, $l = 2$, $a_a = 0.1$, $\beta_{ab} = 0.2$ $\mu_a = 0.1$, $E_a = 1.0$, $A^* = 10$]. Legend: grey line: target species isocline; black line non-target species isocline

Source: from Bonsall et al. 2010.

Bonsall 2019). A simple model of reproductive interference is:

$$\frac{dN_1}{dt} = r_1 N_1 \left(\frac{N_1}{N_1 + \delta_1 N_2} \right) - \alpha_1 N_1(t)[N_1(t) + \beta_1 N_2(t)] - d_1 N_1$$

$$\frac{dN_2}{dt} = r_2 N_2 \left(\frac{N_2}{N_2 + \delta_2 N_1} \right) - \alpha_2 N_2(t)[N_2(t) + \beta_2 N_1(t)] - d_2 N_2$$

where N_1 and N_2 are the population densities of the competitive species, r_i and d_i are birth and deaths rates of the ith species, and δ_i is the strength of reproductive interference on the jth species. α_i and β_i are the strengths of intraspecific and interspecific competition, respectively.

In regions of phase space, unstable (saddle) points occur when the interspecific competitive effects outweigh the intraspecific competitive effects. At low densities, reproductive interference leads to an Allee effect and the frequency-dependent costs of this reproductive interference preclude species invading when rare. Extending this to be more representative of *Aedes* vector dynamics by including (i) delayed adult recruitment, (ii) larval competition, and (iii) behaviorally mediated reproductive interference driven by the way in which hosts are used in proportion to their availabilities (Real 1977; Yakob 2016) affects the likelihood of vector coexistence. Alterations in the behavioral response shifting from linear preference for human hosts (anthropophilic) through to preferring non-human hosts (zoophilic), and the availability of hosts can affect the likelihood of vector coexistence. As reproductive interference strengthens (increases in δ_i) the potential for coexistence declines. Reducing host availabilities can expand the region for vector coexistence as the level of reproductive interference and larval competition interact—however, this effect varies if the behavioral response to different hosts is asymmetric (Fig. 15.6).

Even though the long-term outcome of vector control is expected to be elimination or eradication, altering species interactions over ecological, epidemiological and economically relevant time scales requires critical evaluation of the importance of the population-level consequences and dynamics.

15.4 Optimal Control

Related to the calculus of variations (Kamien & Schwartz 2012), optimal control is a set of techniques for determining the solution to a dynamical system subject to some constraints. In a general sense, suppose the time dynamics are determined by:

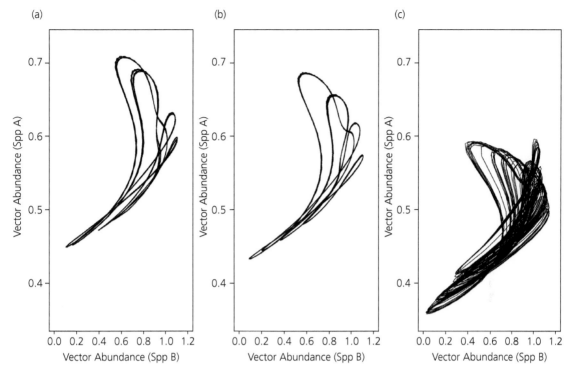

Figure 15.5 Phase space plots for vector abundances under standard self-limiting control for increasing levels of overflooding ((a) $I = 0.0$, (b) $I = 2.0$, (c) $I = 8.0$). Increasing levels of overflooding can affect the absolute abundance of both the focal vector species and its competitors, and the structure of the dynamical attractor. [Parameter values $P_a = 3.0$, $P_b = 5.0$, $\alpha_a = \alpha_b = 0.9$, $\beta_{ab} = \beta_{ba} = 0.5$, $E_a = E_b = 1.0$, $\mu_a = \mu_b = 0.95$, $\tau_a = \tau_b = 2.0$].

Source: from Bonsall et al. 2010.

$$\frac{dy}{dt} = g(z(t), y(t), t)$$

where $z(t)$ is time-dependent but externally controlled. The problem then is to chose $z(t)$ so as to maximise (or minimise) some pre-determined constraint:

$$\int_0^T f(z(t), y(t), t)dt$$

subject to the time dynamics. The mathematics for optimal control have been well developed (e.g. Kamien & Schwartz 2012) and in a vector control scenario, if we presume that the dynamics of the vector—disease interaction follow a Ross-MacDonald framework, for example:

$$\frac{dh}{dt} = ab\frac{V}{H}(1 - h)v - vh$$

$$\frac{dv}{dt} = a(1 - v)h - \mu v$$

where h and v are the proportion of infected hosts and vectors, respectively, H is the density of hosts and V is the density of vectors, a is the bite rate and b is the propensity of a bite to cause an infection in a host, then if the host recovery rate v is under (external) control (say through the availability of some disease curing drug), the optimal control approach asks what is the *best* way to control this parameter (host recovery rate) to achieve some pre-determined outcome (say minimizing proportion of people getting infected).

This simple vector-disease scenario sets the scene for optimal control (or dynamical optimisation) in which control variables constrain the optimal solution to a dynamical (time, space to time-space) problem. Widely used in economics, optimal control has been applied to ecological (e.g. fisheries and conservation) (e.g. Clark 2010) and epidemiological (e.g. Rowthorn et al. 2009; Khamis et al. 2018) problems.

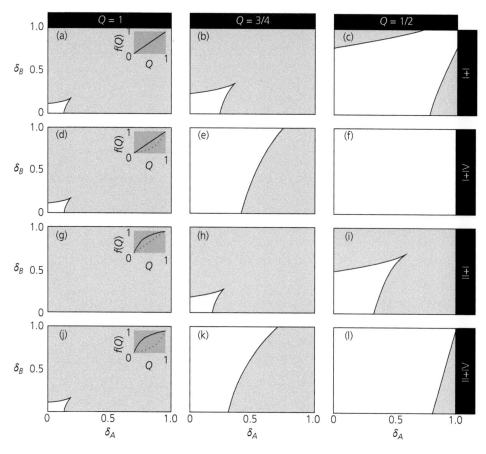

Figure 15.6 Parameter-space plots showing the boundaries of coexistence for competing mosquitoes as a function of the strength of reproductive interference suffered by species A (δ_A) and B (δ_B). Regions where coexistence is possible are denoted in white. Panel rows show different combinations of behavioral responses for A and B (A is given first, then B), which can be linear, anthropophilic or zoophilic. Panel columns show different levels of host availability (Q = 1, (3/4), (1/2)). Small interior plots show how the behavioral responses differ in each of the rows, with the response for A shown as a solid line and that for B as a dotted line. In all cases, the possibility of coexistence increases as host availability is reduced. In panels e, f, h, i, k and l, the behavioral responses differ between species and $Q < 1$. Here species B is able to suffer higher costs of reproductive interference while still coexisting with A.

Source: from Paton & Bonsall 2019.

While the best outcome is that found from an unconstrained optimisation problem, in (bio)economics, a 'good-enough' solution is often that associated with a constraint linked to a pre-determined outcome, which is then often solved over a *finite* time horizon. Furthermore, time-discounting (where future benefits are discounted to reflect the idea that future returns are less valuable than returns received in the present) is essential in determining optimal (economic) outcomes for disease

control. For instance, a critical issue surrounding time-discounting is the *value* of people currently infected in comparison to the *value* of infections still yet to come.

For self-limiting vector control programmes (such as SIT), one of the most critical parameters is the release ratio (or overflooding rate)—how many modified mosquitoes must be released (for each wildtype mosquito) to disrupt matings (or spread dominant lethal genes) and cause vector population decline? Ecologically, this necessitates knowing (i) how mating disruption occurs and (ii) a knowledge of the size of the population (to be suppressed). These ecological parameters become more important when economics are incorporated, as it is then essential to find the most cost efficient releases strategy.

Dynamic optimization of bioeconomic models for the genetics-based approaches of mosquito control has focused on the the release ratio as a key control parameter (Kura et al. 2019; Khamis et al. 2018). In heterogeneous environments, optimal control approaches seek to identify the most efficient releases at different points in space and time to be determined (Kura et al. 2019). Appropriate constraints for vector control focus on the economic costs and include the costs associated with the wildtype mosquitoes (in causing disease burden, health care, loss of tourism etc.), the production costs of rearing modified mosquitoes and the release costs. For discrete-time mosquito dynamics where dispersal and population growth are separated (the integro-difference approach), a cost-function based constraint can be defined as (Kura et al. 2019)

$$J(r) = \sum_{t=0}^{T-1} \int_{\Omega} \left[m_t N_t(x) + n_t r(x)_t N^*(x) + s_t r(x)_t^2 N^*(x)^2 \right] dx$$

where, over the spatial domian, Ω, and finite-time horizon T, with the linear costs of wildtype vectors $(N_t(x))$ on health (m_t) and non-linear costs of wildtype mosquito control $(n_t r(x)_t N^*(x) + s_t r(x)_t^2 N^*(x)^2)$, the aim is to minimize overall mosquito management costs by finding $r = r^*$ such that $J(r^*) = min_{r \geq 0} J(r)$. The optimal release strategies depend on (i) the spatial distribution of wildtype mosquitoes; (ii) the dispersal-redistribution processes and (iii) the costs of control (Kura et al. 2019). Dispersal and redistribution of wildtype mosquitoes together with high costs of control can limit local eradication and suppression of mosquitoes, particularly in heterogeneous habitats. Furthermore, uniform releases are not optimal either in time or space. Across spatially heterogeneous habitats, releasing more modified mosquitoes in high population densities is more efficacious than uniform releases for vector control. As populations decline through control, complete wildtype

eradication (through modified mosquito releases) may not be a cost-effective, optimal solution. Building more integrated vector control programmes is essential to achieving cost-efficient control strategies (Hackett & Bonsall 2018; Kura et al. 2019).

Optimal control is a powerful approach for understanding the cost effectiveness of combining control interventions. For instance, dengue is a debilitating and devastating viral infection spread by *Aedes* mosquitoes with current estimates suggesting (approximately) 390 million people are infected each year (Bhatt et al. 2013). Although vaccine development is advancing, because of dengue's atypical epidemiology (causing increased susceptibility to disease by second strains), immune-based therapies must be efficacious against all dengue serotypes. As such, vector control remains the primary method of mitigating this heavy public health burden.

While insecticides, particularly larvicides, and the management of stagnant water sources have been the focus of control, there is increasing focus on the use of genetics-based methods for vector control (Alphey 2014). Combining these vector intervention methods with the availability and distribution of dengue vaccinations could provide an optimally economic approach to managing dengue disease burden.

Hendon & Bonsall (2016) developed a mathematical framework for investigating dengue vaccination and SIT-based vector control approaches for managing dengue on small networks. Investigating how people move between central city hubs and smaller towns, they showed how (i) disease spread could be mitigated by combining vector control and vaccination, (ii) how vector control could be used to offset imperfect levels of herd immunity from vaccination, and (iii) understanding how people move around landscapes is critical to the management of dengue infections. Building on this work, Rawson et al. (2020) consider how cost-effective combined control approaches can be achieved for dengue management. Using an optimal control approach, they consider the following cost constraint function:

$$J(x) = min \int_0^T \sum_{I=1}^n C_1(I_A^2 + I_{A2}^2 + I_B^2 + I_{B2}^2)C_2u_1^2 + C_3u_2^2 dt$$

where C_1, C_2 and C_3 are the costs associated with treating those infected with dengue, vaccinating a population and releasing SIT mosquitoes, respectively. I_k is the number infected with particular strains of dengue and whether is a first or second infection. Here, both the proportion of individuals vaccinated (u_1^2) and the proportion of total possible SIT mosquitoes being released (u_2^2) in a settlement (across the network) are considered as control variables. The aim is to minimise the total number of individuals infected with dengue across the network of n settlements. Nonlinear control constraints provide more realistic and versatile cost structure (Rawson et al. 2020) and the choice of cost function on optimal control outcomes for vector-borne disease control allow more relevant economic scenarios to be investigated (Khamis et al. 2018).

As might be anticipated to control on-going dengue, delivering fully efficacious vaccination and SIT vector control as early as possible provides the most optimal approaches to minimising infections. Reducing the level of these controls as (i) the susceptible numbers are converted to recovered class individuals through vaccination, and (ii) mosquito numbers fall below the entomological threshold to sustain disease spread, provides cost-efficient strategies to managing dengue infections. Further, reducing insect release ratios as disease management is achieved before reducing levels of vaccinations is considered more optimal.

Across a small network where people move temporarily between outer settlements and a central city hub, optimal control approaches suggest that targeted vaccination and vector control strategies achieved cost-effect disease control compared to uniform vaccination and vector control programmes. Maximum vaccination coverage across all settlements achieves effective disease control, but only if the coverage is reduced at a slower rate in outer settlements compared to the central city hub. Vector control is best targeted in the high density

city hub but with minimal release levels in the outer settlements. In general, combined interventions are most effective but, for cost-effective strategies, how best to combine vector control with other management strategies requires more careful investigations.

Combining vector interventions with drug or vaccine based strategies may be extremely advantageous is both suppressing disease burden while also managing resistance to conventional intervention methods. Theoretical work (Alphey at al. 2007, 2009, 2011a) has shown that combining genetics-based suppression technologies can be used as an effective resistance management tool (at least for agricultural pests; e.g. Zhou et al. 2018, 2019). Optimal control approaches can be used to evaluate the economics of combining vector control approaches and drug-based interventions (Khamis et al. 2018).

Combining vector control and drug therapies is the most effective and efficient use of resources, and optimizing implementation strategies can substantially reduce costs. Evaluating the optimal control strategy under stage-structured vector and disease dynamics (Khamis et al. 2018) requires a cost constraint function which accounts for multiple nonlinear controls (health care costs h, drug administration costs w and transgenic vector costs u):

$$J(\mathbf{X}(u,w)) = \int_0^T exp(-\phi t)\left[\theta_1 h^{m_1} + \theta_2 hw^{m_2} + \theta_3 hu^{m_3}\right]dt$$

where T is the finite-time horizon for control, ϕ is the discount rate and, the price factors θ_i and the cost function exponents m_i govern the scale and behavior of cost for each contributing factor, illustrates that combining multiple control inventions achieves more efficacious control (Fig. 15.7).

Evaluating the constrained dynamical outcomes highlights that the optimal use of vaccination and vector control strategies is substantially better than either the single use of vaccination or vector control (Fig. 15.8). Furthermore, alleviating the substantial economic costs for optimal strategy for vector control and disease intervention requires focus on healthcare and drug administration costs. Vector control strategies can be implemented at minimal economic costs (Alphey

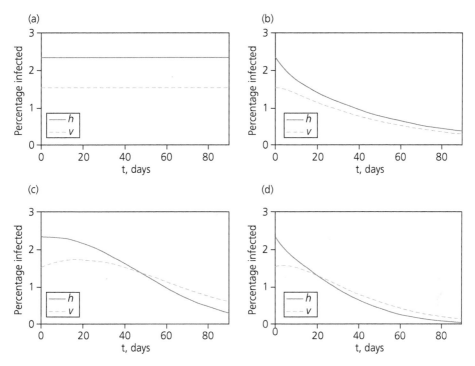

Figure 15.7 Optimal vector control. Vector control and drug therapies are most effective when used in tandem. Proportions of humans (*h*) and vectors (*v*) infected are plotted under four control scenarios. a no control, $w = 0$, $u = 0$; b only drug treatment, $w = 0.05$ (5 percent drug coverage), $u = 0$; c only vector control, $w = 0$, $u = 0.2$ (releasing modified males at a rate of 20 percent of the wild male populations per day); d both drug treatment and vector control, $w = 0.05$, $u = 0.2$. The total number of infected mosquitoes at time t=90 days for each scenario is a 6125 b 1145 c 238 d 55; the vector control suppresses the vector population significantly. Early-acting SIT is assumed
Source: from Khamis et al. 2018.

et al. 2011b) once economics of scale associated with the capital costs of mass rearing facilities are achieved.

While combined vector management programmes are always better than single interventions, the precise underlying cost structures affect the economics of these intervention programme. In the cost constraint function (given previously), different values on the exponents (*m*) can give rise to linear, accelerating or decelerating (at the origin) functions. Reducing the exponent in the cost constraint (from accelerating to linear to decelerating) increases the *per capita* cost of health care when optimal drug and vector control programmes are implemented. Small exponents (*m* < 1) describe conditions where treating a small proportion of the population is disproportionally expensive. In contrast, large

exponents (*m* > 1) describe cost structures where treating large proportion of the population is relatively cheap. Differences in cost structures for vector control have little effect on the *per capita* costs due to low costs associated with releasing modified mosquitoes (Khamis et al. 2018).

As noted previously, (section: density dependence) poor appreciation of the timing of the self-regulatory mechanisms and the genetics can lead to increase in population sizes due to over-compensatory density dependence mechanisms. Mistiming genetics-based control with respect to the ecology also has economic implications leading to increasing costs when the genetics operate before the self-limiting processes. Linking ecology, genetics, and economics for vector control programmes needs more attention.

(a)

(b)

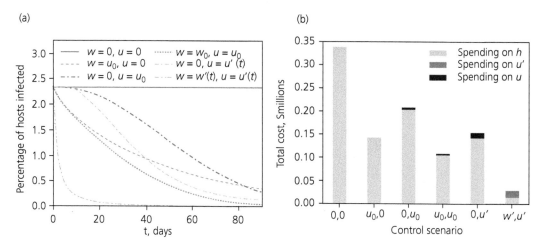

Figure 15.8 Optimal strategies for releases and drug treatments can substantially reduce the cost of managing malaria. a The proportion of infected humans under six control scenarios (where w is the host treatment proportion and u is the insect release ratio with respect to the wild vector population): no control (x-axis label 0,0), w = w0 , u = 0 (x-axis label w0,0"), w = 0 , u = u0 (x-axis label 0,u0"), w = w0 , u = u0 (x-axis label w0,u0') w = 0 , u = u*(t) , w = w*(t) , u = u*(t) (x-axis label "w*," "u*") where w0 = 0.05 , u0 = 0.2 (5 percent drug coverage and releasing 20 percent of the wild male population per day) and w* and u* are the optimal control strategies. b The total cost of the scenario, including spending on traditional healthcare (h), spending on drug treatment (w) and spending on insect releases (u). Early-acting SIT is assumed. Quadratic cost functions of h, w and u are assumed.

Source: from Khamis et al. 2018.

15.5 Policy Perspectives

The development and application of these genetics-based methods for vector control has stimulated considerable discussion around appropriate regulatory frameworks. Emerging from the Cartagena Protocol on Biosafety (Secretariat of the Convention on Biological Diversity 2000), an international agreement of the use of modified organisms resulting from biotechnology, many jurisdictions now mandate risk assessments for the release of genetically modified organisms. These risk assessments are predicated on the precautionary principle and the legislative frameworks tend to focus on evaluating risks associated with the releases of genetically modified organisms on biodiversity and/or human health.

Legal frameworks have been developed for environmental risk assessments of GM plants, however only in the last few years, has policy guidance been developed for GM insects (EFSA 2013; House of Lords 2016) and specifically the release of GM mosquitoes for the control of VBDs (WHO 2015).

The WHO has led the development of guidance for GM mosquitoes. Specifically, this is focused on

(i) efficacy, (ii) biosafety, and (iii) stakeholder engagement. Emerging guidance recommends a tiered framework for the testing and risk assessment of GM mosquitoes (Fig. 15.9). At the early stage of development of a genetically modified mosquito, where the focus is on molecular genetic research and ecological laboratory cage experiments, risk analyses are focused on *efficacy*. That is, GM mosquitoes must be effective in achieving the desired aim (e.g. reducing vector numbers). Once these risks have been highlighted and managed or mitigated, the tiered approach focuses more on *biosafety* (risks to the environment where releases will be made; effects of releases on human health) as the approaches move out from the laboratory to field cage trials, small-scale field trials and onwards towards deployment/commercialization.

In all stages, risk evaluation requires a clear risk hypothesis (a plausible pathway to environmental harm and/or a detrimental effect on human health), clear endpoints and appropriate methods to ensure endpoints have been achieved. For example, for GM mosquitoes, endpoints might be measured through entomological and epidemiological assess-

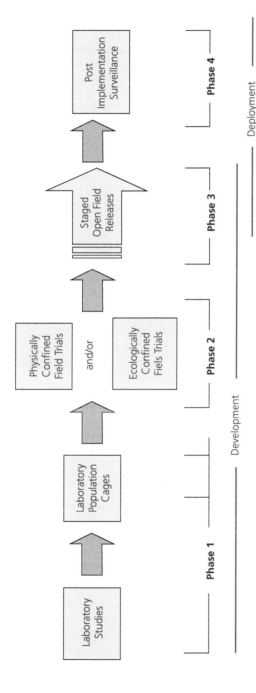

Figure 15.9 Phased testing pathway for genetically-modified mosquitoes (adapted from WHO 2015). A four-tiered phase approach allows risks (hazards by exposures) to be divided into those aspects of transgenic vector control development associated with efficacy (Phases 1,2, & 3) and those aspects of vector control products associated with biosafety deployment (Phases 3 & 4). Phased testing allows risk associated with the transgenic vector to be identified and mitigated through laboratory molecular characterization and laboratory ecological studies. Phases 2 & 3 allows risks associated with the ecological aspects of transgenic vector control to be determined. Phase 4 allows unintended risks to be identified and monitored through appropriate case-specific monitoring, following deployment of a transgenic vector for mosquito control.

ment. Entomological endpoint assessments focus on the reduction of disease transmission measured through mosquito traits (e.g. entomological inoculation rates; vectorial capacity) whereas epidemiological endpoints focus on assessing reductions in the public health burden of infections or disease incidences.

In defining risk hypotheses (and pathways to harm) for GM mosquitoes, it is clear that appreciation for definitions of risk are essential. Risk is often defined as a product of hazard (the event) and exposure (the *likelihood* of the event). However, evaluating risk might be better thought through as a Bayesian problem: updating probabilistic expectations given prior information and likelihood of (risk-based) harms. Developing more robust probabilistic, Bayesian based approaches to risk evaluation are clearly warranted. This should be the focus for pragmatic risk assessments of genetics-based methods for vector control. While regulatory frameworks are necessary, legislation and policy should be proportionate, logically consistent and flexible to promote innovation for achieving the goal of reducing the burden of devastating vector-borne diseases.

References

Abrams, P.A. (2009). When does greater mortality increase population size? The long history and diverse mechanisms underlying the hydra effect. *Ecology Letters*, 12, 462–74.

Akbari, O.S., Chen C.-H., Marshall, J.M., Huang, H., Antoshechkin, I., & Hay, B.A. (2014). Novel synthetic MEDEA selfish genetic elements drive population replacement in *Drosophila*; a theoretical exploration of MEDEA-dependent population suppression. *American Chemical Society Synthetic Biology*, 3, 915–28.

Alsan, M. (2015). The effect of tse-tse fly on African development. *American Economic Review*, 105, 382–410.

Alphey, L. (2014). Genetic control of mosquitoes. *Annual Review of Entomology*, 59, 205–24.

Alphey, N., Coleman, P.G., Donnelly, C.A., & Alphey, L. (2007). Managing insecticide resistance by mass release of engineered insects *Journal of Economic Entomology*. 100, 1642–9.

Alphey, N., Bonsall, M.B., & Alphey, L. (2009). Combining pest control and resistance management: Synergy of engineered insects with Bt crops. *Journal of Economic Entomology*, 102, 717–32.

Alphey. N., Bonsall, M.B., & Alphey, L. (2011a). Modeling resistance to genetic control of insects. *Journal of Theoretical Biology*, 270, 42–55.

Alphey, N., Alphey, L., & Bonsall, M.B. (2011b). A model framework to estimate impact and cost of genetics-based sterile insect methods for dengue vector control. *PLoS One*, 6, e25384.

Alphey, N. & Bonsall, M.B. (2014a). *RIDL: Modelling Release of Insects Carrying a Dominant Lethal*. CABI Books, London.

Alphey, N. & Bonsall, M.B. (2014b). Interplay of population genetics and dynamics in the genetic control of mosquitoes. *Journal of the Royal Society Interface* 11, 20131071.

Aron, J.L. & May, R.M. (1982). The population dynamics of malaria. In: Anderson, R.M., eds. *Population Dynamics and Infectious Disease*, pp. 139–79. Chapman and Hall, London.

Barclay, H.J. (1982). The sterile release method with unequal male competitive ability. *Ecological Modelling*, 15, 251–63.

Begon, M., Townsed, C.R., & Harper, J.L, (2006). *Ecology: From Individuals to Ecosystems*. John Wiley & Sons , Oxford, New York.

Bellini, R., Calvitti, M., Medici, A., Carrieri, M., Celli, G., & Maini, S. (2007). Use of the sterile insect technique against *Aedes albopictus*. In: M.J.B. Vreysen, A.S. Robinson, & J. Hendrichs eds., *Area-Wide Control of Insect Pests*. Springer, Dordrecht.

Bhatt, S., Gething, P.W., Brady, O.J. et al. (2013). The global distribution and burden of dengue. *Nature*, 496, 504–7.

Bonsall, M.B., Yakob, L., Alphey, N., & Alphey, L. (2010). Transgenic control of vectors: The effects of interspecific interactions. *Israel Journal of Ecology and Evolution*, 56, 353–70.

Bukachi, S.A., Wandibba, S., & Nyamongo, I.K. (2017). The socio-economic burden of human African trypanosomiasis and the coping strategies of households in south western Kenya foci. *PLoS Neglected Tropical Diseases*, 11, e0006002.

Burt, A. & Trivers, R. (2008). *Genes in Conflict: The Biology of Selfish Genetic Elements*. Belknap Press, Cambridge, Massachusetts.

Cano, J., Descalzo, M.A., Moreno, M., et al. (2006). Spatial variability in the density, distribution and vectorial capacity of Anopheline species in a high transmission village (Equatorial Guinea). *Malaria Journal*, 5, 21.

Carvalho, D.O., McKemey, A.R., Garziera, L., et al. (2015). Suppression of a field population of *Aedes aegypti* in Brazil by sustained release of transgenic male mosquitoes. *PLoS Neglected Tropical Diseases*, 9, e0003864.

Charlesworth, B. & Charlesworth, D. (2010). *Elements of Evolutionary Genetics*. Geneva: WHO.

Chen, C.-H., Haixia, H., Ward, C.M., et al. (2007). A synthetic maternal-effect selfish genetic element drives population replacement in *Drosophila*, *Science*, 316, 597–600.

Clark, C.W. (2010). *Mathematical Bioeconomics: The Mathematics of Conservation* (3rd edn). Wiley, New York.

Curtis, C.F. (1968). Possible use of translocations to fix desirable genes in insect pest populations. *Nature*, 218, 368–69.

Curtis, C.F., Hill, W.G. (1971). Theoretical studies on the use of translocations for the control of Tsetse flies and other disease vectors. *Theoretical Population Biology*, 2, 71–90.

Dame, D.A., Curtis, C.F., Benedict, M.Q., Robinson, A.S., & Knols B.G.J. (2009). Historical applications of induced sterilisation in field populations of mosquitoes. *Malaria Journal*, 8, S2.

Dao, A., Yaro, A.S., Diallo, M., et al. (2014). Signatures of aestivation and migration in Sahelian malaria mosquito populations. *Nature*, 516, 387–90.

Deredec. A., Burt, A., & Godfray, H.C.J. (2008). The population genetics of using homing endonuclease genes in vector and pest management. *Genetics*, 179, 2013–26.

Dye, C. (1984). Models for the population dynamics of the yellow fever mosquito, *Aedes aegypti*. *Journal of Animal Ecology*, 53, 247–68.

EFSA (2013). Guidance on the environmental risk assessment of genetically modified animals. *European Food Safety Authority Journal*, 11, 3200.

Feachem, R.G.A, Chen, I., & Akbari, O., et al. (2019). Malaria eradication within a generation: Ambitious, achievable, and necessary. *The Lancet 394, 1056–1112*.

Ferreira, C.P., Yang, N.M., & Esteva, L. (2008). Assessing the suitability of sterile insect technique applied to *Aedes aegypti*. *Journal of Biological Systems*, 16, 565–77.

Galizi, R., Doyle, L.A., Menichelli, M., et al. (2014). A synthetic sex ratio distortion system for the control of the human malaria mosquito. *Nature Communications*, 5, 3977.

GBD (2017). Global, regional, and national age-sex-specific mortality for 282 causes of death in 195 countries and territories, 19802017: A systematic analysis for the Global Burden of Disease Study 2017, *The Lancet*, 392, 10159.

Gillespie, J. (2010). *Population Genetics: A Concise Guide*. John Hopkins University Press, Baltimore, Maryland.

Gröning, J. & Hochkirch, A. (2008). Reproductive interference between animal species. *Quarterly Review of Biology*, 83, 257–82.

Hackett, S.C. & Bonsall, M.B. (2018), Management of a stage-structured insect pest: An application of approximate optimization. *Journal of Ecological Applications*, 28, 938–52.

Hamilton, W.D. (1967). Extraordinary sex ratios. A sex-ratio theory for sex linkage and inbreeding has new implications in cytogenetics and entomology. *Science*, 156, 477–88.

Hardin, G. (1960). The competitive exclusion principle. *Science*, 131, 1292–7.

Harrington, L.C., Scott, T.W., Lerdthusnee, K., et al. (2005). Dispersal of the dengue vector *Aedes aegypti* within and between rural communities. *American Journal of Hygiene and Tropical Medicine*, 72, 209–220.

Hastings, A. (2001). Transient dynamics and persistence of ecological systems. *Ecology Letters* 4, 215–20.

Hastings, I.M. (1994). Selfish DNA as a method of pest control. *Philosophical Transactions of the Royal Society B*, 344.

Hemme, R.R., Thomas, C.L., Chadee, D.D., & Severson, D.W. (2010). Influence of urban landscapes on population dynamics in a short-distance migrant mosquito: Evidence for the dengue vector *Aedes aegypti*. *PLoS Neglected Tropical Diseases*, 4, e634.

Hendron, R.S. & Bonsall, M.B. (2016). The interplay of vaccination and vector control on small dengue networks. *Journal of Theoretical Biology*, 407, 349–61.

House of Lords (2016). *Genetically Modified Insects*. Science and Technology Select Committee Report. London: The Stationery Office Limited, p. 56.

Juliano, S.A. (2009.) Species interactions among larval mosquitoes: Context dependence across habitat gradients. *Annual Review of Entomology*, 54, 37–56.

Juliano, S.A. (2010). Coexistence, exclusion, or neutrality? A meta-analysis of competition between *Aedes albopictus* and resident mosquitoes. *Israel Journal of Ecology and Evolution*, 56, 325–51.

Kamien, M.I. & Schwartz, N.L. (2012). Dynamic Optimization: The Calculus of Variations and Optimal Control in Economics and Management. Second Edition. Dover Books on Mathematics. Dover Publications, Dover.

Khamis, D., El Mouden, C., Kura, K., & Bonsall, M.B. (2018). Optimal control of malaria: Combining vector interventions and drug therapies. *Malaria Journal*, 17, 174.

Knipling, E.F. (1955). Possibilities of insect control or eradication through use of sexually sterile males. *Journal of Economic Entomology*, 48, 459–62.

Kot, M. & Schaffer, W.M. (1986). Discrete-time growth-dispersal models. *Mathematical Biosciences*, 80, 109–36.

Kura, K., Khamis, D., El Mouden, C., & Bonsall, M.B. (2019). Optimal control for disease vector management in SIT models: An integrodifference equation approach. *Journal of Mathematical Biology*, 78, 1821–39.

Kyrou, K., Hammond, A., Galizi, R., et al. (2018). A CRISPR-Cas9 gene drive targeting *doublesex* causes complete population suppression in caged *Anopheles gambiae* mosquitoes. *Nature Biotechnology*, 36, 1062–66.

Legros, M., Xu, C., Okamoto, K., et al. (2012). Assessing the feasibility of controlling *Aedes aegypti* with transgenic methods: A model-based evaluation. *PLoS One*, 7, e52235.

Lorimer, N. (1981). Long-term survival of introduced genes in a natural population of *Aedes aegypti* (Diptera: Culicidae). *Bulletin of Entomological Research*, 71, 129–32.

Lorimer, N., Hallinan, E., & Rai, K.S. (1972). Translocation homozygotes in the yellow fever mosquito. *Aedes aegypti. Journal of Heredity*, 63, 158–66.

Lorimer, N., Lounibos, P., & Petersen, J.L. (1976). Field trials with a translocation homozygote in *Aedes aegypti* for population replacement. *Journal of Economic Entomology*, 69, 405–9.

McDonald, P.T., Hausermann, W., & Lorimer, N. (1977). Sterility introduced by release of genetically altered males to a domestic population of *Aedes aegypti* at the Kenya coast. *The American Journal of Tropical Medicine and Hygiene*, 16, 553–61.

McMeniman, C.J., Lane, R.V., Cass, B.N., et al. (2009) Stable introduction of a life-shortening Wolbachia infection into the mosquito Aedes aegypti. *Science*, 323, 141–4.

Magori, K., Legros, M., Puente, M.E. et al. (2009). Skeeter Buster: A stochastic, spatially explicit modeling tool for studying *Aedes aegypti* population replacement and population suppression strategies. *PLoS Neglected Tropical Diseases*, 3, e508.

Manoranjan, V.S. & van den Driessche, P. (1986). On a diffusion model for sterile insect release. *Mathematical Biosciences*, 79, 199–208.

Martens, P. & Hall, L. (2000). Malaria on the move: Human population movement and malaria transmission. *Emerging Infectious Diseases*, 6, 103–9.

Maynard Smith, J. & Slatkin, M. (1973). The stability of predator prey systems. *Ecology*, 54, 384–91.

Murray, J.L. & Lopez, A.D. (1996). *The Global Burden Of Disease. A Comprehensive Assessment of Mortality and Disability from Diseases, Injuries, and Risk Factors in 1990 and Projected to 2020*. Geneva: WHO Publications.

Neubert, M.G., Kot, M., & Lewis, M.A. (1995). Dispersal and pattern formation in a discrete-time predatorprey model. *Theoretical Population Biology*, 8, 7–43.

Oberhofer, G., Ivy, T., & Hay, B.A. (2019). Cleave and rescue, a novel selfish genetic element and general strategy for gene drive. *Proceedings of the National Academy of Sciences of the United States of America*, 116, 6250–9.

Paton, R.S. & Bonsall, M.B. (2019). The ecological and epidemiological consequences of reproductive interference between the vectors *Aedes aegypti* and *Aedes albopictus*. *Journal of the Royal Society Interface*, 16, 20190270.

Petersen, J.L., Lounibos, P., & Lorimer, N. (1977). Field trails of double translocation heterozygotes males for genetic control of Aedes aegypti (L.) (Diptera: Culicidae). *Bulletin of Entomological Research*, 67, 313–24.

Phuc, H.K., Andreasen, M.H., Burton, R.S. et al. (2007). Late-acting dominant lethal genetic systems and mosquito control. *BMC Biology*, 5, 11.

Rai, K.S., Grover, K.K., & Suguna, S.G. (1973). Genetic manipulation of Aedes aegypti: Incorporation and maintainance of a genetic marker and a chromosomal translocation in natural populations. *Bulletin of the World Health Organization*, 48, 49–56.

Rawson, T., Wilkins, K.E., & Bonsall, M.B. (2020) Optimal control approaches for combining medicines and mosquito control in tackling dengue. *Royal Society Open Science*, 7: 181843.

Real, L. (1977). The kinetics of functional response. *The American Naturalist*, 111, 289–300.

Reimer, J.R., Bonsall, M.B., & Maini, P.K. (2016). Approximating the critical domain size of integrodifference equations. *Bulletin of Mathematical Biology*, 78, 72–109.

Rogers, D.J. & Randolph, S.E. (1984). From a case study to a theoretical basis for tsetse control. *Insect Science and Its Application*, 5, 419–23.

Ross, R. (1911). *The Prevention Of Malaria. 1911*. John Murray, London.

Ross, R. (1915) Some a priori pathometric equations. *British Medical Journal*, 1, 546–447.

Ross, R. (1916a). An application of the theory of probabilities to the study of a priori pathometry—I. *Proceedings of the Royal Society*, A92, 204–30.

Ross, R. (1916b). An application of the theory of probabilities to the study of a priori pathometry—II. *Proceedings of the Royal Society*, A93, 212–25.

Ross, R. & Hudson, H.P. (1916). An application of the theory of probabilities to the study of a priori pathometry—III. *Proceedings of the Royal Society*, A93, 225–40.

Rowthorn, R.E, Laxminarayan, R., & Gilligan, C.A. (2009). Optimal control of epidemics in metapopulations. *Journal of The Royal Society Interface*, 6, 1135–44.

Schetelig, M.F., Horn, C., Handler, A.M., & Wimmer, E.A. (2007). Development of an embryonic lethality system for transgenic SIT in the fruit pest *Ceratitis capitata*. In, Vreysen, M.J.B., Robinson, A.S., & Hendricks, J. eds., *Are Wide Control of Insect Pests: From Research to Field Implementation*, pp. 85–93. Springer, Dordrecht, The Netherlands.

Secretariat of the Convention on Biological Diversity (2000). *Cartagena protocol on biosafety to the convention on biological diversity*: Text and annexes. Montreal: Secretariat of the Convention on Biological Diversity.

Serebrovskii, A.S. (1940). On the possibility of a new method of insect control. *Zoologicheskii Zhurnal*, 19, 618–31 (in Russian).

Sinkins S.P. & Gould, F. (2006). Gene drive systems for insect vectors. *Nature Reviews Genetics*, 7, 427–35.

Smith, D.L., Battle, K.E., Hay, S.I., Barker, C.M., Scott, T.W., & McKenzie, F.E. (2012). Ross, Macdonald, and a theory for the dynamics and control of mosquito-transmitted pathogens. *PLoS Pathogens,* 8, e1002588.

Strong, D.R. (1986). Density-vague population change. *Trends in Ecology and Evolution,* 1, 39–42.

Tatem, A.J., Rogers, D.J., & Hay, S.I. (2006). Global transport networks and infectious disease spread. *Advances in Parasitology,* 62, 293–343.

Taylor, C., Touré, Y.T., Carnahan, J., et al. (2001). Gene flow among populations of the malaria vector, *Anopheles gambiae,* in Mali, West Africa. *Genetics,* 157, 743–50.

Thomas, D.D., Donnelly, C.A., Wood, R.J, & Alphey, L.S. (2000). Insect population control using a dominant, repressible, lethal genetic system. *Science,* 287, 2474–6.

Thomson, M.C., Connor, S.J., Quinones, M.L., Jawara M., Todd, J, & Greenwood, B.M. (1995). Movement of *Anopheles gambiae* s.l. malaria vectors between villages in The Gambia. *Medical and Veterinary Entomology,* 9, 413–19.

Tripet, F., Lounibos, L.P., Robbins, D., Moran, J., Nishimura, N., & Blosser, E.M. (2011). Competitive reduction by satyrization? Evidence for interspecific mating in nature and asymmetric reproductive competition between invasive mosquito vectors. *The American Journal of Tropical Medicine and Hygiene,* 85, 265–70.

WHO (2015). *Biosafety for Human Health and the Environment in the Context of the Potential Use of Genetically Modified Mosquitoes (GMMs).* Geneva: World Health Organization. p. 242.

Yakob, L. (2016). How do biting disease vectors behaviourally respond to host availability? *Parasites & Vectors,* 9, 1–9.

Yakob, L. & Bonsall, M.B. (2009). Importance of space and competition in optimizing genetic control strategies. *Journal of Economic Entomology,* 102, 50–57.

Yakob, L., Alphey, L., & Bonsall, M.B. (2008a). *Aedes aegypti* control: The concomitant role of competition, space and transgenic technologies. *Journal of Applied Ecology,* 45, 1258–65.

Yakob, L., Kiss, I.Z., & Bonsall, M.B. (2008b). A network approach to modeling population aggregation and genetic control of pest insects. *Theoretical Population Biology,* 74, 324–31.

Zhou, L., Alphey, N., Walker, A.S. et al. (2018). Combining the high-dose/refuge strategy and self-limiting transgenic insects in resistance management—a test in experimental mesocosms. *Evolutionary Applications,* 11, 727–38.

Zhou, L., Alphey, N., Walker, A.S., et al. (2019). The application of self-limiting transgenic insects in managing resistance in experimental metapopulations. *Journal of Applied Ecology,* 56, 688–98.

Zhou, Y. & Kot, M. (2011). Discrete-time growth-dispersal models with shifting species ranges. *Theoretical Ecology,* 4, 13–25.

Index

Aedes aegypti
  carry-over effects 158, 159, 160, 161, 162
  datasets 103
  gut microbiota 233, 235, 236, 237, 238
  insecticide resistance 254
  mosquito–virus interactions 198, 199, 200, 201, 202
  rainfall sensitivity 92
  social drivers 251, 252, 254
  social-ecological systems 252
  sterile male release 103–4
  temperature sensitivity 90
  urban areas 95
  vector control 272, 274, 277
  Wolbachia-infected 103, 204
Aedes albopictus
  carry-over effects 158, 159, 160, 161
  datasets 103
  genetic mutations 199
  land use 96
  mosquito–virus interactions 199, 200, 201
  social drivers 251
  vector control 277
African horse sickness virus 260–1, 262
African swine fever virus 260–1, 262
age
  confounding effects 30–1
  survival 179, 180, 183–4
allometrics 167
amplification effects 97, 137
Anaplasma phagocytophilum 136, 141
animal African trypanosomiasis 177–8
Anopheles mosquitoes
  carry-over effects 157, 158, 159, 160, 161, 162, 182
  deforestation 95
  gut microbiota 233
  mosquito–virus interactions 199, 205

vector control 275, 276
aphids
  life cycle 88
  rainfall sensitivity 94–5
  temperature sensitivity 90, 91
  wind sensitivity 95
apoptosis 198, 199
auto-regressive integrated moving average (ARIMA) models 50–1
axenic mosquitoes 235

Babesia microti 136, 138, 141
bacteriome 237
basal lamina 196
basic reproduction number 15, 17, 30, 121–4, 137, 140–1
'baton effect' 91
Bayesian frameworks 37
bed bugs, *Trypanosoma cruzi* 224
biodiversity-buffers-disease hypothesis 142–3
biting rate
  carry-over effects 158–9, 160, 162, 163–4
  human 18
  mammal 73, 74
  tsetse 179–82
bluetongue virus 91, 94, 119, 122–3, 124–8, 260–1; *see also Culicoides*-borne viruses
Borrelia burgdorferi 135, 137, 138, 140, 141, 142, 146
Borrelia mayonii 136, 140
Borrelia miyamotoi 136, 138
breeding sites 87, 95

carbon dioxide level 95
carry-over effects 155–69
  allometrics 167
  competition 160–2
  mechanisms 168
  microbiome 162–4
  modelling 164–7
  multiple 168

nutrition 159–60, 168
predation 164
Ross–Macdonald model 165–6
silver spoon effect 155–6
stoichiometrically-driven 168
temperature 155, 157–9, 164–5, 167–8, 182
toxins 164
catch per unit effort 73
causation inference 101
Chagas disease
  control 218–19
  drug treatment 218, 223
  habitat and land use 96
  historical background 217
  phases of disease 218
  see also Trypanosoma cruzi
chickens, *Trypanosoma cruzi* 223–4
Chikungunya virus
  gene–environment interaction 204
  genetic variation 199, 201
  Haiti 258
  rainfall sensitivity 92
  social drivers 251
  spatial analyses 32
climate 47, 54–5, 86–7, 89–95, 140, 182; *see also* temperature sensitivity
cofeeding 88, 138
co-infection 203, 205
communication, forecasting results 62
compartment models 49–51, 164
competition
  carry-over effects 160–2
  vector control 276–8
condors, vaccination program 262
continuous ranked probability score 58
control
  locally-appropriate strategies 38
  mass drug administration 38
  optimal control 278–83
  spatial analyses 35–7
  see also vector control

coupled natural-human systems
144–5, 148–9
coverage rate 57
Crimean-Congo haemorrhagic fever
virus 261
cross-validation methods 59–60
Culex mosquito
apoptosis 198
carry-over effects 163
co-infection 203
evasion of RNAi 200
temperature sensitivity 90
see also West Nile virus
Culicoides-borne viruses 119–30
basic reproduction number 121–4
life cycles 87
overwintering 120, 128–9, 130
rainfall sensitivity 92, 94
temperature sensitivity 90, 123,
125–7
transmission cycle 120
transmission routes 120–1
transplacental transmission 129,
130
wind sensitivity 95
cytochrome *bd* oxidase 236

daily reproductive rate 15, 17
databases 103
data-mining 51
DDT-based indoor residual
spraying 15
deforestation 95, 96, 98
dengue virus
acidification 200
ENSO metrics 94
evasion of RNAi 200
force of infection 32
forecasting 51, 53, 59, 61, 62
genetic variation 199, 201, 202
host mobility 36
humidity sensitivity 92
immune suppression 199, 200
microbiome 203–4
optimal control 281–3
political factors 257
rainfall sensitivity 92, 94, 254
social drivers 251, 252
spatial analyses 32
temperature sensitivity 90
transgenic animals 206
density forecasts 56
Diebold–Mariano test 58
digital epidemiology 54
dilution-by-deer effect 138
dilution effect 96, 137–8, 142, 146
dispersal 273–5

dissemination 74
dogs, *Trypanosoma cruzi* 220
dominant lethal synthetic genes 270
drought 92, 94

earth observations data 102–3
earthquakes 255–6
ecology
social-ecological systems 252
tick-borne pathogens 141–2
tsetse 182–6
vector control 23, 270, 273–8
economic burden, mosquito
control 254–5
ecosystem services 104, 147
edge effects 146–7
Ehrlichia muris eauclairensis 136, 138,
140
Elizabethkingia meningoseptica 162
El Niño Southern Oscillation
(ENSO) 94
emergence
forecasting 52
T. cruzi, Arequipa, Peru 219
tick-borne pathogens 136–42
two-host system 137–8
ensemble models 55
entomological inoculation rate
(EIR) 16, 18, 19
environmental factors 85–104, 182
biotechnology input 103–4
carbon dioxide level 95
climate 47, 55, 86–7, 89–95, 140, 182
database use 103
earth observations data 102–3
ecosystem services 104
habitat 20, 85, 95–8, 145–6, 147–8
humidity 85, 91–2
interacting or correlated drivers 99
interdisciplinary collaboration 101
land use 85, 87, 95–8, 142–8
mechanistic models 99–100
natural disasters 255–6
nonlinear transmission 99
rainfall 85, 92–5, 254
sea level 95
semi-mechanistic models 101
statistical models 99–100
synthetic analysis 103
temperature 85, 89–91, 98, 123,
125–7, 155, 157–9, 164–5, 167–8,
182, 184–6, 202–3
temporal and spatial variations 99
uncertainty 100–1
vector–virus interaction 200–1,
202–4
wind 21–2, 95

epidemiology
digital 54
spatial analyses 30–5
expert opinion 54
exponential smoothing techniques 51
extrinsic incubation period (EIP) 14,
15, 17, 89, 180

fecundity 157–8, 159, 160–1, 162–3
feeding synchrony 138, 140
flooding 92
food resources 96, 98
force of infection (FOI) 16, 18–19, 32,
70–1, 72–3, 75–9, 98
forecasting 45–63
accuracy 57
ARIMA models 51
climate data 47, 54–5
communication of results 62
comparison tests 58–9
compartment models 50–1
components 55–61
consistency 62
cross-validation 59–60
data 48
data-mining 51
data sparsity 47
definition 46
density forecasts 56
digital epidemiology 54
emerging outbreaks 53
ensemble models 55
evaluation 57–8
expert opinion 54
exponential smoothing
techniques 51
external data sources 53–5
feedback loop 47
forecast horizon 49
frequency 48
generalized linear regression
models 51
historical background 45–6
interval forecasts 56–8
kernel conditional density
estimation 51
k-step-ahead 48
laboratory data 53–4
machine-learning approaches
51
mechanistic models 47,
49–52
nowcasts 46, 61–2
operationalizing 61–2
out-of-sample data 57
point forecasts 56, 57
prediction and 46

probabilistic forecasts 46, 48, 56, 58
proper scoring rules 57
quality 62
quasi-mechanistic model 53
regression-style methods 51
reporting delays 61–2
SARIMA models 51
scale-independent metrics 56–7
scoring 56–7
SEIR model 53
SIR models 49–50
snapshots of data 62
statistical models 49, 51–53
system complexity 46–7
targets 48, 56
time 49
time scale 49
training and testing 59–61
usefulness 62
value of 62
vector data 53
fragmentation of habitat 145–6, 147–8

Gamma distribution 22
Garki Project 20
gene drives 272–3
generalized linear regression
    models 51
genetics
    vector control 270–3, 284–6
    vector–virus interaction 199, 200,
        201–2, 204
Global Malaria Eradication
    Programme 15, 19
Glossina spp., *see* tsetse
gnotobiotic mosquitoes 235
Google Flu 54
guinea pigs, *Trypanosoma cruzi*
    219–22
gut microbiome, *see* microbiome

habitat 20, 85, 95–8, 145–6, 147–8
Haiti, *Polisye Kont Moustik*
    project 257–60
'happenings' rate 18–19
hazard rate 18–19
heat islands 98
herd immunity 33–4
heterogenous biting 22
heterogenous transmission 19–23
Holt–Winters models 51
hotspots for transmission 38–9
household biting density 19
household infective density 19
human African trypanosomiasis 178
human biting rate 19
human blood index 15

humidity 85, 91–2
hydrology 20–1
hypoxia-inducible transcription
    factors 236

Imd pathway 198
immune function 158, 159, 161,
    197–8, 199–200, 234, 256
indoor residual spraying 15, 218
infection thresholds 37
infective biting density 19
infectivity 74
influenza
    forecasting 54, 55
    Google Flu 54
inoculation rate 19
insecticide resistance 202, 254
integro-difference equations 274, 281
intermediate disturbance
    hypothesis 146
interval forecasts 56–8
Ixodes scapularis 135, 136–7, 138, 141,
    146
ixodid ticks
    habitat and land use 96
    humidity sensitivity 92
    life cycles 88, 136–7
    temperature sensitivity 90
    see also tick-borne pathogens

joint threshold abundance curves 137

kernel conditional density
    estimation 51
k-step-ahead forecasts 48

land use 85, 87, 95–8, 142–8
leafhoppers
    temperature sensitivity 90
    wind sensitivity 95
leishmaniasis
    habitat and land use 96
    rainfall sensitivity 94
    socio-economic factors 101
    surveillance sensitivity 35
life cycles 87–8, 136–7, 155, 182
livestock vector-borne disease 260–2
log score 58
Lotka, Alfred 14
Lotka–Volterra competition
    model 276
Lyme disease 96–7, 98, 135, 142–3,
    147–8
Lyme disease risk 148
lymphatic filariasis
    elimination programs 258
    infection thresholds 35

Macdonald, George 14–15
machine-learning, forecasting 51
malaria
    deforestation 95
    Global Malaria Eradication
        Programme 15, 19
    hotspots 38
    Malaria Atlas Project 24, 30–1
    malaria rate 18
    political factors 256–7
    rainfall sensitivity 92, 94
    source-sink dynamics 35–6
    temperature sensitivity 89, 90
    zooprophylaxis 98
mammal biting rate 73, 74
mechanistic models
    environmental drivers of
        transmission 99–100
    forecasting 47, 49–52
    spatial analyses 30, 39
microbiome 229–39
    carry-over effects 162–4
    larval growth and moulting 235–6
    mosquito–virus interactions 203–4
    nutrition source 236–7
    vector competence and pathogen
        transmission 233–5
microclimates 98
micro-filaraemia prevalence 38
Microsoft, Project Premonition 104
midgut escape barrier 195, 196–7
midgut infection barrier 195–6
minimum infection rate 73–4
miR-281 200
miRNA 199
mobile phone call data 35, 37
mobility, human 35–6, 275
mosquito-borne pathogens
    acidification 200
    axenic mosquitoes 235
    biotechnology input 103–4
    carry-over effects 155–69
    co-infection 203, 205
    competition 160–2
    deforestation 96, 98
    dispersal and control 274
    economic burden of control 254–5
    ENSO metrics 94
    environmental noise 21
    evasion of RNAi 200
    food resources 98
    genetically modified
        mosquitoes 284–6
    genetic variations 199, 200, 201–2,
        204
    gnotobiotic mosquitoes 235
    heterogenous biting 22

mosquito-borne pathogens (*Cont.*)
heterogenous transmission 19–23
humidity sensitivity 92
immune system 158, 159, 161,
197–8, 199–200, 234
indoor residual spraying
control 15
insecticide resistance 202, 254
life cycle 87, 155
microbiome 162–4, 203–4, 229–39
midgut escape barrier 195, 196–7
midgut infection barrier 195–6
modelling 164–7, 204–6
mosquito–virus interactions 193–206
nutrition 159–60, 168, 236–7
optimal control 281–3
physical barriers of infection 194–7
predation 164
rainfall sensitivity 92, 94
Ross–Macdonald model 15–17, 178
salivary gland escape barrier 195,
197
salivary gland infection
barrier 195, 197
seasonality 20
skin microbiota 238
social drivers 251
spatial dynamics 21–2
temperature sensitivity 89, 90, 155,
157–9, 164–5, 167–8, 202–3
theories 14–19
toxins 164
viral mechanisms enhancing
infection 198–200
wind sensitivity 21–2, 95
myxoma virus 88

natural disasters 255–6
neglected infections of poverty 250
neglected parasitic infections of
poverty 250
neglected tropical diseases 250
noise 21
nonlinear transmission 98–9
nonpersistent pathogens 88
nowcasts 46, 61–2
nutrition 159–60, 168, 236–7

onchocerciasis 96
optimal control 278–83
overwintering 120, 128–9, 130

palm trees 95, 96, 218
parasite rate 18
patch connectivity 147
peritrophic matrix 196
PIWI-interacting RNA 198

Plasmodium falciparum malaria
climate change 90
global distribution 30, 31
parasite rate (PfPR) 31
Plasmodium vivax malaria,
deforestation 95
plume models 21
point forecasts 56–8
policy 187, 284–6
Polisye Kont Moustik project 257–60
political unrest 256–7, 261
population immunity 31–3
poverty trap paradigm 249–50
Powassan virus 136, 138, 141
prediction 46
probabilistic forecasts 46, 48, 56, 58
Project Premonition, Microsoft 104
psychological distress 256

quasi-mechanistic forecasting
model 53

rainfall 85, 92–5, 254
regression models 51
relative mean absolute error 57–8
reservoir hosts 96
Rhodnius prolixius 218, 219
Rift Valley fever 92, 94, 182
RNA interference 197–8, 200
RNA viruses 198–9
rolling-origin-recalibration window 60
Ross, Ronald 14
Ross–Macdonald model 15–17, 50,
72, 165–6, 178–9, 220
Ross River fever
humidity sensitivity 92
rainfall sensitivity 92, 94
sea level 95
temperature sensitivity 90

St Louis encephalitis 92
salivary gland escape barrier 195, 197
salivary gland infection barrier
195, 197
sand flies 87, 94, 96, 98
Schmallenberg virus 119–20, 122–3,
124–8; *see also Culicoides*-borne
viruses
sea level 95
seasonal ARIMA (SARIMA)
models 51
seasonality 20
SEIR model 53
selfish genetic elements 271
self-limiting methods 270–1
self-sustaining methods 270, 271–3
semi-mechanistic models 101

silver nanoparticles 164
silver spoon effect 155–6
Sindbis virus 196, 199, 200
SIR models 49–50
skin microbiome 238
small interfering RNA 197–8
social-ecological systems 252
social factors 101, 249–63
control measures 252, 254–5,
257–60, 261–2
insecticide resistance 254
livestock vector-borne disease 260–2
natural disasters 255–6
political unrest 256–7, 261
poverty trap paradigm 249–50
social-ecological systems
approach 252
transmission of diseases 251–7
source-sink dynamics 35–6
spatial analyses 29–40
age-related confounding 30–1
control 35–9
epidemiological patterns 30–5
herd immunity 33–4
hotspots for transmission 38–9
mapping 29, 30
mechanistic approaches 30, 39
population immunity 31–3
spatial dynamics 21–2
spatial heterogeneity 34–5, 37–8, 79
spatial regressions 29
statistical approaches 29–30, 39
surveillance sensitivity 34–5, 39
tick-borne pathogens 141–2
spectral radius method 72
spillover infections 88
spleen rate 18
sporozoite rate 14, 17, 19, 20
statistical models
environmental drivers of
transmission 99–100
forecasting 49, 51–3
spatial analyses 29–30, 39
sterile insect technique 103–4, 270–1
stochastic transmission 100–1
stoichiometry, carry-over effects 168
superensembles 100
superinfection 15–16
surveillance sensitivity 34–5, 39
susceptible-exposed-infected-
recovered (SEIR) model 53
susceptible-infectious-recovered (SIR)
model 49–50

temperature sensitivity 85, 89–91, 98,
123, 125–7, 155, 157–9, 164–5,
167–8, 182, 184–6, 202–3

tick-borne encephalitis virus 138, 140, 261
tick-borne pathogens 135–49
  basic reproduction number 137, 140–1
  canonical seasonal feeding inversion 138
  climate 140
  cofeeding 88, 138
  coupled natural-human systems 144–5, 148–9
  ecological determinants 141–2
  ecosystem services 147
  emergence 136–42
  habitat 96, 98, 145–6, 147–8
  horizontal transmission 138
  host community impacts 142–8
  host types 140
  human-tick interactions 147–8
  humidity sensitivity 92
  immature feeding synchrony 138, 140
  land use 96, 98, 142–8
  life cycles 88, 136–7
  patch connectivity 147
  phenology and transmission pathways 138–42
  socio-economic factors 101
  spatial distribution 141–2
  temperature sensitivity 90–1
  vertical transmission 138
Toll pathway 198
transient dynamics 277
transmission
  competence 88
  heterogenous 19–23
  measurement 18–19
  nonlinearity 98–9
  potential 31–3
  stochastic 100–1
transplacental transmission 129, 130
Triatoma infestans 218, 219
tropical storms 256
Trypanosoma cruzi 217–24
  bed bug vectors 224
  burden 218
  chicken hosts 223–4
  control 218–19, 222

discovery 217–18
dog hosts 222
dormancy 222–3
drug treatment 218, 223
guinea pig hosts 219–22
host ranges 218
Ross–Macdonald equation 220
sylvatic vectors 223
vertical transmission 222
zooprophylaxis 224
trypanosomiasis
  animal African 177–8
  human African 178
  rainfall sensitivity 92
  temperature sensitivity 90
tsetse 177–88
  age effects 183–4
  biology 182–3
  biting rate 179–82
  breeding sites 95
  deforestation 98
  environmental factors 182
  extrinsic incubation period 180
  life expectancies 179–82
  life history 182–6
  model-based policy decisions 187
  rainfall sensitivity 94
  reproduction 182–3, 184
  temperature sensitivity 182, 184–6
  transmission dynamics 178–82
  vector ecology 182–6

ultra-low volume spraying 252–4
uncertainty 100–1

vector competence 73, 74
vector control 269–86
  coverage proportion 275
  decentralization 252
  density dependence 275–6
  dispersal 273–5
  ecology 23, 270, 273–8
  genetics-based approaches 270–3, 284–6
  heterogeneity 23, 37–8
  household economic burden 254–5
  human movement 35–7, 275
  indoor residual spraying 15, 218

interspecific competition 276–8
optimal control 278–83
policy issues 284–6
Polisye Kont Moustik project, Haiti 257–60
self-limiting methods 270–1
self-sustaining methods 270, 271–3
social factors 252, 254–5, 257–60, 261–2
sterile insect technique 103–4, 270–1
Trypanosoma cruzi 218–19, 222
vectorial capacity 15, 17–19, 156–7, 235
Vector-Protected Establishment 261
veterinary vector-borne disease 260–2
viral suppressors of RNAi 200

West Nile virus 69–79
  condor vaccination 262
  evasion of RNAi 200
  force of infection 70–1, 72–3, 75–9
  forecasting 53
  genetic variation 199, 201
  habitat and land use 96
  immune suppression 199–200
  rainfall sensitivity 92, 94
  spatial heterogeneity 79
  temperature sensitivity 90
  variation in outbreak size 69–70, 71–3, 75, 79
whiteflies 88, 90
white-footed mice 96, 137, 142, 146
white-tailed deer 137
wind 21–2, 95
Wolbachia 103, 162, 163, 204

yellow fever
  Haiti 257
  immune suppression 199–200

Zika virus 193
  genetic variation 201
  natural disasters 255
  social drivers 251
  spatial analyses 32–4
zooprophylaxis 98, 224